Methods in Enzymology

Volume 397
ENVIRONMENTAL MICROBIOLOGY

METHODS IN ENZYMOLOGY

EDITORS-IN-CHIEF

John N. Abelson   Melvin I. Simon

DIVISION OF BIOLOGY
CALIFORNIA INSTITUTE OF TECHNOLOGY
PASADENA, CALIFORNIA

FOUNDING EDITORS

Sidney P. Colowick and Nathan O. Kaplan

Methods in Enzymology

Volume 397

Environmental Microbiology

EDITED BY

Jared R. Leadbetter

DEPARTMENT OF ENVIRONMENTAL SCIENCE & ENGINEERING
CALIFORNIA INSTITUTE OF TECHNOLOGY
PASADENA, CALIFORNIA

AMSTERDAM • BOSTON • HEIDELBERG • LONDON
NEW YORK • OXFORD • PARIS • SAN DIEGO
SAN FRANCISCO • SINGAPORE • SYDNEY • TOKYO
Academic Press is an imprint of Elsevier

ELSEVIER

Elsevier Academic Press
525 B Street, Suite 1900, San Diego, California 92101-4495, USA
84 Theobald's Road, London WC1X 8RR, UK

This book is printed on acid-free paper. ∞

For all information on all Elsevier Academic Press publications visit our Web site at www.books.elsevier.com

ISBN-13: 978-0-12-182802-8
ISBN-10: 0-12-182802-6

PRINTED IN THE UNITED STATES OF AMERICA
05 06 07 08 09 9 8 7 6 5 4 3 2 1

Table of Contents

Section I. Cultivation

Section II. Physiological Ecology

Section III. Nucleic Acid Techniques

Contributors to Volume 397

Article numbers are in parantheses and following the names of Contributors.
Affiliations listed are current.

ODED BÉJÀ (22), *Department of Biology, Technion – Israel Institute of Technology, Haifa 32000, Israel*

ANDREAS BRUNE (11), *Max Planck Institute for Terrestrial Microbiology, D-35043 Marburg, Germany*

PHYLLIS H. CHAN (15), *Department of Chemistry, Occidental College, Los Angeles, California 90041*

LUDMILA CHISTOSERDOVA (27), *Chemical Engineering, University of Washington, Seattle, Washington 98195*

GREG CLARK (7), *Diversa Corp., San Diego, California 92121*

IRENE A. DAVIDOVA (2), *Department of Microbiology, University of Oklahoma, Norman, Oklahoma 73019*

JAAP SINNINGHE DAMSTÉ (3), *Royal Netherlands Institute for Sea Research (NIOZ), Department of Marine Biogeochemistry and Toxicology, 1790 AB Den Burg, The Netherlands*

MARC G. DUMONT (25), *Department of Biological Sciences, University of Warwick, Coventry CV4 7AL, United Kingdom*

LIN K. DUONG (15), *Department of Chemistry, Occidental College, Los Angeles, California 90041*

DAVID EMERSON (6), *American Type Culture Collection, Manassas, Virginia 20110*

KATRIN FICHTL (20), *Lehrstuhl für Mikrobiologie, Technical University Munich, D-85354 Freising, Germany*

MELISSA MERRILL FLOYD (6), *American Type Culture Collection, and George Mason University, Manassas, Virginia 20110*

MICHAEL W. FRIEDRICH (26, 29), *Max Planck Institute for Terrestrial, Microbiology, D-35043 Marburg, Germany*

GEORG FUCHS (12), *Mikrobiologie, Faculty of Biology, University of Freiburg, 79104 Freiburg Germany*

MICHAEL HÜGLER (12), *Mikrobiologie, Faculty of Biology, University of Freiburg, 79104 Freiburg, Germany*

IMKE HALLER (7), *Diversa Corp., San Diego, California 92121*

JONG-IN HAN (19), *Department of Civil and Environmental Engineering, The University of Michigan, Ann Arbor, Michigan 48109-2125**

TREVIN HOLLAND (7), *Diversa Corp., San Diego, California 92121*

JENNIFER B. HUGES (17), *Department of Ecology and Evolutionary Biology, Brown University, Providence, Rhode Island 02912*

MIKE JETTEN (3), *Department of Microbiology, Institute for Water & Wetland Research, Radboud University Nijmegen, 6525 ED Nijmegen, The Netherlands*

*Current affiliation: Rensselaer Polytechnic Institute, Jonsson Engineering Center, Troy, New York 12180.

MICHAEL KÜHL (10), *Marine Biological Laboratory, Institute of Biology, University of Copenhagen, DK-3000 Helsingør, Denmark*

MARINA G. KALYUZHNAYA (27), *Chemical Engineering, University of Washington, Seattle, Washington 98195*

ANDREAS KAPPLER (4), *Department of Geomicrobiology, Center for Applied Geosciences, University of Tübingen, 72074 Tübingen, Germany*

MARTIN KELLER (7), *Diversa Corp., San Diego, California 92121*

MICHAEL KLEIN (29), *University of Vienna, Department of Microbial Ecology, A-1090 Wien, Austria*

FRANK E. LÖFFLER (5, 13), *School of Civil and Environmental Engineering and School of Biology, Georgia Institute of Technology, Atlanta, Georgia 30332-0512*

ADAM B. LEAPHART (28), *Division of Diagnostic Microbiology, Bureau of Laboratories, South Carolina Department of Health and Environmental Control, Columbia, South Carolina 29223*

NATUSCHKA LEE (29), *Department of Microbiology, Technical University Munich, D-85354 Freising, Germany*

CHARLES R. LOVELL (28), *Department of Biological Sciences, University of South Carolina, Columbia, South Carolina 29208*

ALEXANDER LOY (29), *University of Vienna, Department of Microbial Ecology, A-1090 Wien, Austria*

WOLFGANG LUDWIG (20), *Lehrstuhl für Mikrobiologie, Technical University Munich, D-85354 Freising, Germany*

TERENCE L. MARSH (18), *Department of Microbiology & Molecular Genetics, Michigan State University, East Lansing, Michigan 48824*

MARK O. MARTIN (15), *Department of Chemistry, Occidental College, Los Angeles, California 90041*

ERIC J. MATHUR (7), *Diversa Corp., San Diego, California 92121*

J. COLIN MURRELL (25), *Department of Biological Sciences, University of Warwick, Coventry CV4 7AL, United Kingdom*

MEGAN E. NÚÑEZ (15), *Department of Chemistry, Mount Holyoke College, South Hadley, Massachusetts 01075*

PER HALKJÆR NIELSEN (14), *Section of Environmental Engineering, Department of Life Sciences, Aalborg University, DK-9000 Aalborg, Denmark*

JEPPE LUND NIELSEN (14), *Section of Environmental Engineering, Department of Life Sciences, Aalborg University, DK-9000 Aalborg, Denmark*

GREGORY D. O'MULLAN (24), *Department of Ecology and Evolutionary Biology, Princeton University, Princeton, New Jersey 08540*

JÖRG OVERMANN (8), *Bereich Mikrobiologie, Fakultät für Biologie, Ludwig Maximilians-Universität, München, D-80638 München, Germany*

ANNELIE PERNTHALER (21), *Department of Molecular Ecology, Max Planck Institute for Marine Microbiology, 28359 Bremen, Germany**

JAKOB PERNTHALER (21), *Department of Molecular Ecology, Max Planck Institute for Marine Microbiology, 28359 Bremen, Germany*

*Current affiliation: Division of Geological and Planetary Sciences, California Institute of Technology, Pasadena, California 91125.

MICHAEL PESTER (11), *Max Planck Institute for Terrestrial Microbiology, Karl-von-Frisch-Straße, D-35043 Marburg, Germany*

CAROLINE PLUGGE (1), *Laboratory for Microbiology, Microbial Physiology Group, Wageningen University, 6703 CT Wageningen, The Netherlands*

KEVIN J. PURDY (16), *School of Biological Sciences, University of Reading, Reading RG6 6AJ, United Kingdom*

NIELS B. RAMSING (29), *Department of Microbial Ecology, University of Aarhus, DK-8000 Aarhus, Denmark*

NIELS PETER REVSBECH (9), *Department of Microbial Ecology, Institute of Biology, University of Aarhus, DK-8000 Aarhus, Denmark*

KIRSTI M. RITALAHTI (5), *School of Civil and Environmental Engineering, Georgia Institute of Technology, Atlanta, Georgia 30332-0512*

ROBERT A. SANFORD (5, 13), *Department of Geology, University of Illinois at Urbana/Champaign, Urbana, Illinois 61801*

BERNHARD SCHINK (4), *Department of Biology, University of Konstanz, 78457 Konstanz, Germany*

MARKUS SCHMID (3), *Department of Microbiology, Institute for Water & Wetland Research, Radboud University Nijmegen, 6525 ED Nijmegen, The Netherlands*

JEREMY D. SEMRAU (19), *Department of Civil and Environmental Engineering, The University of Michigan, Ann Arbor, Michigan 48109*

STEVEN M. SHORT (23), *Department of Biological Sciences, University of Denver, Denver, Colorado 80208*

JAY M. SHORT (7), *Diversa Corp., San Diego, California 92121*

ANIL R. SINDHURAKAR (15), *Department of Chemistry, Norris Hall of Chemistry, Occidental College, Los Angeles, California 90041*

EILEEN M. SPAIN (15), *Department of Chemistry, Norris Hall of Chemistry, Occidental College, Los Angeles, California 90041*

DAVID A. STAHL (29), *Department of Civil and Environmental Engineering, University of Washington, Seattle, Washington 98195-2700*

KRISTINA L. STRAUB (4), *Department of Geomicrobiology, Center for Applied Geosciences, University of Tübingen, 72074 Tübingen, Germany*

MARC STROUS (3), *Department of Microbiology, Institute for Water & Wetland Research, Radboud University Nijmegen, 6525 ED Nijmegen, The Netherlands*

JOSEPH M. SUFLITA (2), *Department of Microbiology, University of Oklahoma, Norman, Oklahoma 73019*

GERARDO TOLEDO (7), *Diversa Corp., San Diego, California 92121*

KATINKA VAN DE PAS-SCHOONEN (3), *Department of Microbiology, Institute for Water & Wetland Research, Radboud University Nijmegen, 6525 ED Nijmegen, The Netherlands*

MICHAEL WAGNER (29), *University of Vienna, Department of Microbial Ecology, A-1090 Wien, Austria*

MARION WALCHER (7), *Diversa Corp., San Diego, California 92121*

BESS B. WARD (24), *Department of Geosciences, Princeton University, Princeton, New Jersey 08544*

GARY WOODNUTT (7), *Diversa Corp., San Diego, California 92121*

JONATHAN P. ZEHR (23), *Department of Ocean Sciences, University of California-Santa Cruz, Santa-Cruz, California 95064*

GILL ZEIDNER (22), *Department of Biology, Technion – Israel Institute of Technology, Haifa 32000, Israel*

KARSTEN ZENGLER (7), *Diversa Corp., San Diego, California 92121*

KATRIN ZWIRGLMAIER (20), *Lehrstuhl für Mikrobiologie, Technical University Munich, D-85354 Freising, Germany*

Preface

Environmental Microbiology per se has not previously been the dedicated subject of its own volume in this long running series. The desire to understand the activities of microbes in natural environmental settings and samples has longstanding roots. Take, for example, van Leeuwenhoek's first observations of the microbial communities of the human mouth. Or take the "crude analytical technique" employed by Alessandro Volta during his investigations of the biogenesis of methane in sediments. Environmental Microbiology has a rich history that not only has continued up through the present, but that has markedly accelerated in pace in recent years.

The last 30 year period has witnessed incredible advances in the study of microbes in the environment. From the brilliant, first isolation of many diverse microbes having novel forms, functions, and phylogenies. To the continued development of techniques used to image cells and the physical, chemical, and biological gradients within which they reside. To the large-scale sequencing efforts that are now beginning to characterize the gene contents of literally hundreds of microbes in one fell swoop. Environmental Microbiology remains in a golden age of discovery and of methods development.

Because of the current maturity of, and methodological richness in the study of microbes in the environment, no single volume on the subject could be complete. In choosing the topics covered in this volume, I have attempted to cover a broad swath of methods that I feel represent, and thus serve to expose field- and bench-enthusiasts to a reasonable diversity of approaches that are currently being used to tackle a great variety of issues of interest. The overall aim of this volume is to continue to encourage the development of methods and investigations encompassing a great variety of experimental flavors and colors. I wish to thank all of the authors as well as Cindy Minor, whose experience in and guidance of such endeavors proved so invaluable.

I dedicate this volume to my parents, Gloria Ann and Edward Renton Leadbetter. Their 50 year partnership has not only served to mentor the development their own four children, but also (during the course of their travels in life from Austin to Amherst to Storrs to Woods Hole and to the many other hot spots of microbiology they have visited in between) served to foster the development of countless other early-career environmental microbiologists of this nation and from abroad. Thank You.

JARED R. LEADBETTER

xiii

METHODS IN ENZYMOLOGY

Section I

Cultivation

[1] Anoxic Media Design, Preparation, and Considerations

By CAROLINE M. PLUGGE

Abstract

Exclusion of oxygen from growth media is essential for the growth of anoxic prokaryotes. In general, anaerobic techniques focus on the use of deaerated boiled growth media. Successful enrichment, isolation, and cultivation of anoxic prokaryotes critically depend on the choice of appropriate growth media and incubation conditions. This chapter discusses the requirements of anoxic prokaryotes for growth in the laboratory and different existing methods for their cultivation.

Introduction

> *"Study of obligate anaerobes continually provides that extra challenge for all procedures, but for some of us, challenge is what it is all about."*
> Ralph S. Wolfe (1999)

Outside the laboratory, bacterial communities are nearly always consortia composed of a wide variety of species. There are many instances where interactions between different species may alter the outcome in cultivability or activity assays. For example, the presence or absence of a symbiotic relation between species could be involved. Metabolic cross-protection or cross talk between signal molecules or molecules participating in natural competition plays a role. It is appropriate to consider the relevance of these factors to the outcome of culturability in the laboratory.

The Need for Cultivation?!

Cultivability in the laboratory is dependent on many issues. Two reviews deal with the relationship between viability and culturability (Barer and Harwood, 1999; Mukamolova *et al.*, 2003). Barer and Harwood concluded that culturability is the only secure and operational definition of viability, although an overwhelming number of "culture-independent" methods are available to demonstrate the presence of microorganisms (Hugenholtz, 2002; Kaksonen *et al.*, 2004; Spring *et al.*, 2000).

METHODS IN ENZYMOLOGY, VOL. 397
0076-6879/05 $35.00
DOI: 10.1016/S0076-6879(05)97001-8

In fluctuating natural environments, bacteria detect external changes and respond to them rapidly (such as negative chemotaxis or production of protective compounds) and after that with long-term changes (lowering metabolic activity, dormancy, or sporulation). Due to a variety of stressful conditions such as nutrient depletion, extreme temperature or pH, changes in pO_2, light, or the presence of toxic compounds, bacteria may change from the viable-and-culturable state to a temporal or permanent viable-but-nonculturable (VBNC) state. The majority of the microorganisms in natural environments have never been knowingly cultivated in the laboratory. In many instances, key reactions in bio(geo)chemical cycles appear to be mediated by microorganisms that are not among those that are frequently isolated—they are quite often phylogenetically distinct from isolated strains and also often among the not yet cultured. Currently, approximately half of the almost 100,000 rRNA sequences in the public databases represent cultured organisms (Ribosomal Database project RDP-II, http://rdp.cme.msu.edu/html, July 2004) (Cole et al., 2003).

Some ecosystems harbor whole groups of entirely unstudied bacteria and archaea, which most likely are important for a number of key biochemical transformations at those sites (Hugenholz et al., 1998; McGarvey et al., 2004). Reasons for this discrepancy have been discussed extensively (Amann et al., 1995; Kell and Young, 2000). If we assume that, at least in principle, all microbes can be cultivated (Leadbetter, 2003), we must conclude that the appropriate cultivation conditions have not yet been found for many microorganisms. Thus, scientists need to put continuous effort in developing novel, innovative strategies to enrich and isolate novel "kleine diertjes" (little animals, as Anthonie van Leeuwenhoek described his microscopic observation of microorganisms in his pepper tube experiment in 1680) to make them available for detailed physiological and molecular studies and to understand their significant role in Nature. Those who persevere may discover novel biochemical reactions on Earth.

Anoxic biodegradation processes are usually considered to be more difficult to study than aerobic ones. Anaerobic enrichment cultures not infrequently show very long apparent lag phases of 6 months or 1 year (or even longer) before significant biodegradation is initiated. After repeated transfer, lag phases often shorten. However, many years may be required to optimize conditions in order to achieve significantly rapid rates of biodegradation and those may or may not ever approach the rates of comparable aerobic biodegradation processes. In any event, the basis for successful growth and enrichment is the first step in such endeavors and involves the strategic design of the growth medium for the anoxic microorganism of physiotype of interest.

Anoxic Environments

Anoxic niches are found in soil, marshes, organic sludges, riverbeds, and ocean floors and in the intestines of many living animals. Anoxic environments are either already present or formed by aerobes that reduce all the available oxygen. This is essential for the survival of obligate anaerobes because of the toxicity to them of traces of oxygen. Other anaerobes are facultative, using oxygen preferentially, but change to alternative oxidants when anoxic conditions prevail. Growth of microorganisms in anoxic environments may proceed by fermentation or by anaerobic respiration. The thermodynamic principle that governs anaerobic respiration is a suitable large redox potential difference between donor substrate and oxidant to support cell maintenance and growth.

Constituents of Anoxic Media

Major Constituents of Anoxic Media

A large variety of media for the cultivation of anoxic microorganisms are available from a number of commercial suppliers. Descriptions of an overwhelming amount of media can be found in general microbiology journals. Furthermore, at the Web sites of the German Culture Collection (DSMZ, Braunschweig, Germany, www.dsmz.de) and the American Type Culture Collection (ATCC, Manassas, VA, www.atcc.org) and other strain collections worldwide, large databases are available with media descriptions for microorganisms deposited at those institutions. However, to work successfully with those media and grow anoxic microorganisms, certain aspects need to be taken into consideration.

The design of a medium for the successful growth of a specific anoxic microorganism is crucial. The conventional approach to cultivate anoxic microorganisms is to prepare a liquid or solid growth medium that contains a carbon source, energy source, and an electron donor and acceptor. The medium is inoculated with a source of microorganisms and incubated until visible growth of microorganisms is observed. This approach leads to the growth and enrichment of the fastest growing microorganisms, often referred to as microbial weeds. A similar, more quantitative approach leading to the growth and enrichment of the most predominant organisms is to dilute the source of microorganism to extinction prior to inoculation in their growth medium (the most probable number technique, MPN). These approaches are still commonly used worldwide and continue to yield novel genera and species of anoxic prokaryotes.

Media for the growth of anoxic microbes often are formulated using a defined mixture of components to mimic as closely as possible the natural environment. The essential problem is to find the conditions that are suitable to enable growth, especially if the desired microorganisms originate from environments with relatively little cultured representatives. The challenge of culturing is to make successful adaptations to conventional growth media and enrichment strategies. Factors that can be taken into consideration are pH, salinity, temperature, light, and shaking.

As a basis for the formulation of appropriate growth media, knowledge of the composition of the general microbial biomass is necessary. The dry weight of microorganisms consists of over 95% of the macro elements C, H, O, N, P, and S and the metals Ca, Cl, Fe, K, Mg, and Na (Table I). Very often the simplified formula $<C_4H_8O_2N>$ is used to define biomass composition, although more detailed formulae have been reported, e.g., $C_4H_{6.4}O_{1.5}NP_{0.09}S_{0.024}$ (Ingraham *et al.*, 1983). As a consequence, these major constituents must be present in growth media in sufficient amounts.

The Internet is an inexhaustible source of information. A selection of worldwide Web sites relevant for the formulation of appropriate growth media based on element ratios in general has been presented by Wackett (2005).

TABLE I

TYPICAL COMPOSITION OF ELEMENTS THAT MAY BE PRESENT IN PROKARYOTIC CELLS[a]

Element	% of dry mass	
	Range	Typical
C	45–55	50
O	16–22	20
N	12–16	14
H	7–10	8
P	2–5	3
S	0.8–1.5	1
K	0.8–1.5	1
Na	0.5–2.0	1
Ca	0.4–0.7	0.5
Mg	0.4–0.7	0.5
Cl	0.4–0.7	0.5
Fe	0.1–0.4	0.2
All others	0.2–0.5	0.3

[a] From Tchobanoglous and Burton (1991), with permission.

Minor and Other Constituents of Anoxic Media

In addition to macroelements, a number of microelements are also required for the growth of anoxic microorganisms. The concentration and composition of the microelements used in growth media of anoxic microorganisms differ significantly (www.dsmz.de/media, www.atcc.org). Furthermore, it is very difficult to demonstrate the micronutrient requirement of a microorganism. Often, contaminants from water, glassware, and regular media components are already adequate to sustain growth. However, changing the composition and concentration of micronutrients in growth media may lead to the growth of microbes with special characteristics. It has also been reported that trace elements inhibit certain biochemical processes; e.g., nitrogen fixation by termite spirochetes was inhibited by the presence of tungsten in the growth medium (Leadbetter, 2003).

Numerous microorganisms grow with only one carbon source added to the medium. However, additional (organic) compounds may stimulate growth of the microbes. These growth factors are the building blocks of cell constituents and are classified as amino acids, vitamins, or pyrimidines and purines. The presence of these growth factors can also be an obligatory factor for growth. It is very difficult to find out which growth factor is essential for each microorganism. Therefore, the addition of growth factors of unknown or undefined composition is often preferred. Yeast extract, peptone, casitone, casamino acids, clarified rumen fluid, sludge supernatant, or fermented yeast extract are examples of these growth factors with undefined composition. Environmental factors such as temperature, redox conditions, pH, and osmolarity need to be taken into account as well when a growth medium is designed.

Solidifying agents that are generally used for solid media are agar, agarose, and gelrite (phytagel). For oligotrophic microbes, agar is usually purified to remove inhibiting substances. The agar is washed repeatedly in large volumes of MilliQ water and is subsequently extracted with acetone and ethanol. An agar that is very pure and suitable for the cultivation of anaerobes, especially methanogens, is agar noble, available commercially from Difco (www.difco.com). To grow thermophilic and hyperthermophilic microorganisms on solid media in general gelrite is used (Lin and Casida, 1984). The concentration of agar that has to be used to keep these media solid at such high temperatures is often toxic for microorganisms.

The gel strength of gelrite (Sigma; http://www.sigmaaldrich.com) is determined by the choice and concentration of divalent cations (Mg^{2+}, Ca^{2+}). The gels are stable even at high temperature (up to $100°$) and can be autoclaved without a significant loss of gel strength. The gels are

TABLE II

REDOX HALF-POTENTIALS FOR REDUCTANTS USED FREQUENTLY IN ANOXIC
CULTURE MEDIA[a]

Compound	$E_0'^b$ (mV)
Dihydroascorbate/ascorbate	58
Dithioglycollate disulfide/thioglycollate	−140
S^0/H_2S	−250
S^0/HS^-	−270
Cystine/2 cysteine	−325
Dithiothreitol (DTT) disulfide/DTT	−330
Titanium(IV) citrate/titanium(III) citrate	−480
$2 SO_3^{2-}/S_2O_4^{2-}$	−574
Resofurin (pink)/dihydroresofurin (colorless)	− 80[c]

[a] From Veldkamp (1970).
[b] Defined at pH 7 and 298 °K.
[c] Redox potential at which 1 mg/liter resofurin (resazurin) turns
colorless (concentration generally used in anoxic media).

transparent, which facilitates visual screening of the plates. The preparation of gelrite-containing media is similar to those that contain agar (see earlier discussion). The final concentration of gelrite in the media may vary from 0.8 to 2%.

Creating and Maintaining Anoxic Conditions

For the cultivation of anoxic microorganisms, the addition of reducing agents to the media is necessary. Reductants are used to create the appropriate anoxic or reducing conditions in the media. The removal of O_2 only by physical means (flushing) is not sufficient. The redox potential of the growth media is related to the overall availability of electrons in the media, specifically the ratio of positive and negative ions in the solution. Table II lists the redox half-potentials of frequently used reductants. Final concentrations of the reductants are relatively low (<0.05%, w/v) due to the inhibitory effect of them at higher concentrations.

Preparation of Liquid and Solid Media

General Aspects

Procedures for the growth of anaerobes, even nowadays, are still often based on the pioneering work of Hungate (1950, 1969). In 1976, Balch made modifications to the Hungate technique, developing a pressurized atmosphere in vials or tubes. This technique was generally accepted more user-friendly one than previous techniques. Pfennig (1981) and Widdel (1999) reviewed procedures for cultivation techniques of anoxygenic phototrophs and sulfate-reducing bacteria that can be applied easily to other anoxic prokaryotes. Procedures and manifolds described here are modifications of the original techniques used by Hungate and Balch.

Procedures

The composition and preparation of a general, multipurpose, bicarbonate-buffered mineral salts freshwater medium for the cultivation of anaerobes in pressurized vials are given.

Composition per Liter Medium

Add before autoclaving (use a clean Erlenmeyer): 500 ml dH$_2$O, 0.5 mg resazurin (redox indicator), 0.4 g KH$_2$PO$_4$, 0.53 g Na$_2$HPO$_4$, 0.3 g NH$_4$Cl, 0.3 g NaCl, 0.1 g MgCl$_2$·6H$_2$O, and 1 ml of trace elements stock solutions I and II (each)

Acid stock solution (I) contains per liter: 50 mM HCl, 1mM H$_3$BO$_3$, 0.5 mM MnCl$_2$, 7.5 mM FeCl$_2$, 0.5 mM CoCl$_2$, 0.1 mM NiCl$_2$, and 0.5 mM ZnCl$_2$

Alkaline stock solution (II) contains per liter: 10 mM NaOH, 0.1 mM Na$_2$SeO$_3$, 0.1 mM Na$_2$WO$_4$, and 0.1 mM Na$_2$MoO$_4$ (or any other trace metal mixture)

1. Make up to 950 ml with dH$_2$O.
2. Bring to a full boil to remove O$_2$ (do not boil longer than 20 s to prevent evaporation).
3. Cool down under an oxygen-free N$_2$ flow to prevent oxygen diffusion into the medium.
4. Disperse into serum bottles; serum bottles and penicillium flasks should be of the highest quality (pressure and temperature resistant).
5. Stopper and seal the bottles using butyl rubber stoppers (Rubber B. V.; Hilversum, The Netherlands) and aluminum crimp seals.
6. Change to the desired gas phase (N$_2$/CO$_2$ or H$_2$/CO$_2$; 80/20 mixture) and pressurize the bottles to 1.5–1.8 atm. When other mixtures of gases are used, recalculations should be made with the new CO$_2$ concentrations to obtain the desired pH.
7. Autoclave for 20 min at 121°.
8. After autoclaving from *sterile* stock solutions, add 1.1% (v/v) of a filter-sterilized calcium/vitamin solution; make up per liter medium: 10 ml CaCl$_2$ (11 g/liter) + 1 ml vitamin solution. The vitamin stock solution contains per liter: 20 mg biotin, 200 mg nicotinamide, 100 mg

p-aminobenzoic acid, 200 mg thiamin (vitamin B_1), 100 mg panthothenic acid, 500 mg pyridoxamine, 100 mg cyanocobalamin (vitamin B_{12}), and 100 mg riboflavin(or any other vitamin mixture).

9. Add the carbon source (electron donor) and electron acceptor.

10. Finally, to reduce the medium completely, add 5% (v/v) of a sterilized reducing solution; make up per liter medium: 50 ml $NaHCO_3$ solution (80 g/liter) + 1 ml $Na_2S.9 H_2O$ solution (240 g/liter) + 0.5 g cystein·HCl. Under certain circumstances, better growth is obtained if the "reducing solution" is filter sterilized because H_2S can react with the butyl rubber stoppers at higher temperatures. The final pH of the medium is 7.0–7.2. The medium is now ready to be inoculated.

This medium can be adapted for the cultivation of marine organisms by increasing the salt strength to the level of seawater or brackish water. It is important to take into consideration that the high salt strength in these media may lead to the formation of insoluble precipitates after autoclaving the media. Therefore it is better to add the minerals $CaCl_2$, $MgCl_2$, and KCl after autoclaving from sterile stock solutions.

This medium can also be used as a basis for agar-containing solid media. Because the minerals NH_4Cl, NaCl, $CaCl_2$, $MgCl_2$, and KCl interact with agar, it is advisable to add them after autoclaving to the media from sterile stock solutions.

For the cultivation of thermophilic and extreme thermophilic anoxic prokaryotes on solid gelrite media, the following growth medium can be applied.

1. Add before autoclaving:
   a. Solution I: 500 ml dH_2O, 0.5 mg resazurin (redox indicator), MOPS 12 g, Na_2HPO_4 0.3 g, NH_4Cl 0.3 g, NaCl 0.3 g and 1 ml of trace elements stock solutions I and II (composition: see multipurpose liquid medium)
2. Bring to a full boil to remove O_2 (do not boil longer than 20 s to prevent evaporation).
3. Cool down under an oxygen-free N_2 flow to prevent oxygen diffusion into the medium.
4. Stopper and seal the bottle using butyl rubber stoppers and aluminum crimp seals or screw caps. Flush the headspace with N_2.
5. Autoclave solution I for 20 min at 121°.
6. After autoclaving, keep solution I at >60°.
7. Add from *sterile* stock solutions to solution I: 20 ml filter sterilized calcium/magnesium/vitamin solution (20 ml contains 0.2 g $CaCl_2$,

2 g $MgSO_4$ plus 1 ml vitamin solution (composition vitamins: see multipurpose liquid medium).
8. Add the carbon source (electron donor) and electron acceptor.
9. Finally, to reduce the medium completely, add 2 ml $Na_2S \cdot 9 \, H_2O$ solution (240 g/liter).
    b. Solution II: 20 g gelrite (phytagel) in 500 ml dH_2O that has been made oxygen free by boiling prior to use
10. Stopper and seal the bottle using butyl rubber stoppers.
11. Autoclave solution II for 20 min at 121° and keep it at >60° afterward to prevent the gelrite from solidifying.

To finish the medium, place both hot solutions in an anaerobic glove box. Mix both solutions gently but quickly and pour the completed medium into sterile *glass* petri dishes. At high temperatures plastic petri dishes release toxic compounds to the medium. When the gelrite has become solid the plates are ready to be inoculated.

The plates are inoculated either by streaking them with the desired microorganism or by embedding the microorganisms in a 0.5% (w/v) phytagel solution that is poured as an overlay on top of the base layer.

Plates are incubated up side down in the anaerobic glove box at the desired temperature or in heat-resistant anaerobic jars outside the glove box.

Media containing water-insoluble carbon sources can be applied with a two liquid phase system. The water-insoluble component is dissolved in a suitable inert liquid phase and, through diffusion, the carbon source is slowly released in the water phase, where the microbes can use it. Hexadecane is used frequently as the second liquid phase (Holliger, 1992).

Butylrubber stoppers are not resistant to all chemicals. In this case, viton stoppers are a good alternative to use (Rubber B.V., Hilversum, The Netherlands).

All manipulations to anoxic flasks, tubes, and vials are made via the stoppers using disposable syringes and needles (Becton Dickenson; www. bd.com). The choice of the length and diameter of the needles is dependent on the viscosity or size of the particles in the substratum to be added. In general, the smaller the diameter of the needle, the less risk there is of leakages in the stoppers. To minimize the introduction of oxygen into the anoxic bottles, the syringes can be flushed with sterile N_2 before use.

Useful Equipment for Anoxic Media Preparation

The cultivation of microorganisms in an anoxic atmosphere often takes place using anaerobic jars, an anaerobic glove cabinet, or a CO_2 incubator. The massive space-occupying chambers (anaerobic cabinets or

CO_2 incubators) with fixed environments consume huge quantities of gas. These incubators are incapable of cultivating microorganisms under different environmental conditions. To have more flexibility in the laboratory, two types of gas exchange devices have been developed and are available commercially: the Anaerobic Lap System and the Anoxomat.

The Anaerobic Lap System (Fig. 1) has been designed at the Laboratory of Microbiology of Wageningen University by Mr. Frits Lap in close collaboration with microbiologists whose wish it was to improve efficiency in working with anaerobes. The system is available commercially at GR Instruments B.V., Wijk bij Duurstede, The Netherlands (http://www.grin-struments.com/).

The system has been designed to produce large quantities of bottles with anoxic media with a defined atmosphere. It automatically evacuates the gas phase of stoppered flasks, vials, or anaerobic jars and replaces it with the desired gas phase at the desired pressure. The manifold can be

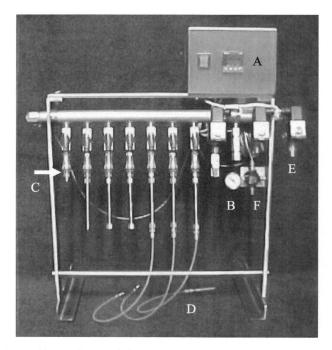

FIG. 1. Anaerobic Lap System: (A) electronic control box, (B) digital pressure regulator, including overpressure security valve, (C) stainless-steel gasing probes, with quick-snap connection; different stages, (D) complete quick connector, (E) line to vacuum pump, and (F) line to gas supply.

connected easily to any gas tank. Because the system operates fast and is fully automatic, it is possible to obtain the same conditions in each vial. It is possible to autoclave the quick connectors (Fig. 1D) and perform the gas exchange procedure under aseptic conditions.

The Anoxomat (Fig. 2) is designed for similar purposes. It is available commercially at Mart Microbiology B.V., Lichtenvoorde, The Netherlands (www.anoxomat.com). However, in contrast to the Anaerobic Lap System, it cannot be connected to stoppered flasks and vials. It is used most often in clinical laboratories to create anoxic atmospheres in anaerobic jars that contain agar plates (Summanen *et al.*, 1999). Instead of flushing out the oxygen, the air is replaced from a confined environment, thus limiting the consumption of gas. As the system operates fully automatically, it is possible to exactly repeat the process so that reproducibility can be achieved. The Anoxomat system is enable to document, control, and retrace the sample-processing history in the laboratory. Both systems need very little maintenance.

A semiautomatic gas exchange manifold according to Balch (1976) can be constructed relatively easily. The body of the manifold consists of gas-tight and pressure-resistant material such as Teflon tubing that is assembled

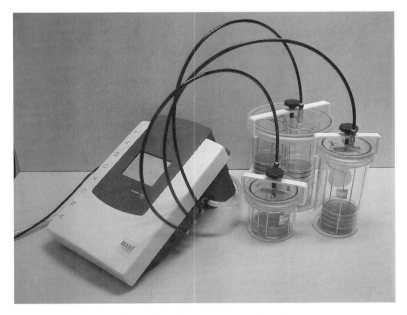

Fig. 2. The Anoxomat, type Mark II.

together with stainless-steel connectors. This results in an octopus-like set-up where needles at the end of the tentacles are used to connect flasks and vials to the system. A three-way stopcock is inserted at the other end of the device to connect the device to a vacuum pump and a gas tank. By manually turning back and forth the valve from the line connected to the vacuum pump and the gas supply, the connected vials and flasks can be made anoxic.

The Widdel flask (Widdel, 1999) is used frequently to make up larger quantities of the same medium, which is divided into smaller portions in vials without headspace. The medium is prepared in the flask and auto-claved. After autoclaving the flask is cooled, either in an anaerobic cham-ber or by flushing the medium with oxygen-free nitrogen, carbon dioxide, or a mixture of gases. Finally, vitamins, electron donor and acceptor, minerals, and reducing agent are added aseptically to the flask to complete the medium. After this the medium is transferred into sterile, anoxic vials. The vials are closed with screw caps and are now ready to be inoculated.

The use of gases for the cultivation of anaerobes is essential. It is important that these gases are of the highest purity, as impurities in the gas may lead to unsuccessful cultivation of the microorganisms. Traces of oxygen can be removed from the gas prior to use by leading the gas through hot copper wire (Hungate, 1969).

Novel Developments

Conventional cultivation of anoxic microorganisms is laborious, time-consuming, and selective and is biased for the growth of specific micro-organisms due to the artificial conditions inherent in most culture media (e.g., extremely high substrate concentrations or the lack of specific nutri-ents required for growth). Successful novel developments to culture the uncultured organisms focused on the application of relatively low concen-trations of nutrients, nontraditional sources of nutrients, signaling mole-cules or inhibitors, and relatively lengthy periods of incubation. Zengler *et al.* (2002) described a high-throughput cultivation method based on the combination of a single cell encapsulation procedure with flow cytometry that enables cells to grow with nutrients that are present at environmental concentrations. This gel microdroplet method was originally developed for aerobic marine organisms. However, with modifications it can be adapted easily for the cultivation of anoxic bacteria. The gel microdroplet method is performed in an open, continuously fed system instead of a closed batch system that also simulates the open condition of most natural environ-ments. When each cell is encapsulated, all of the encapsulated cells from the environment are cultured together in a single vessel. This proximity simulates to some extent the natural environment; because the pore size of the gel micromatrix is large, it allows the exchange of metabolites and other molecules such as signaling molecules. In the case of interactions between different species of microorganisms, it is possible to add the partner organisms to the microdroplets. As an example, the obligate syntrophic interaction between the anaerobic propionate-converting acetogen *Syntrophobacter fumaroxidans* and the hydrogen-consuming methanogen *Methanospirillum hungatei* can be mentioned (Schink and Stams, 2002). The hydrogen produced by the acetogen must be transferred to the methanogen via interspecies hydrogen transfer for thermodynamic reasons. These interactions require that the physical distances between the organisms are very short, which can be achieved by encapsulating bacteria in the gel microdroplets.

Acknowledgment

The author was supported by the Research Council for Earth and Life Sciences (ALW) with financial aid from the Netherlands Organization for Scientific Research (NWO).

References

Amann, R., Ludwig, W., and Schleifer, K. (1995). Phylogenetic identification and *in situ* detection of individual microbial cells without cultivation. *Microbiol. Rev.* **59,** 143–169.

Balch, W. E., and Wolfe, R. S. (1976). New approach to the cultivation of methanogenic bacteria: 2-Mercaptoethanesulfonic acid (HS-CoM)-dependent growth of *Methanobacterium ruminantium* in a pressurized atmosphere. *Appl. Environ. Microbiol.* **32,** 781–791.

Barer, M. R., and Harwood, C. R. (1999). Bacterial viability and culturability. "Advances in Microbial Physiology," Vol. 41, pp. 93–137.

Cole, J. R., Chai, B., Marsh, T. L., Farris, R. J., Wang, Q., Kulam, S. A., Chandra, S., McGarrell, D. M., Schmidt, T. M., Garrity, G. M., and Tiedje, J. M. (2003). The Ribosomal Database Project (RDP-II): Previewing a new autoaligner that allows regular updates and the new prokaryotic taxonomy. *Nucleic Acids. Res.* **31,** 442–443.

Hugenholtz, P. (2002). Exploring prokaryotic diversity in the genomic era. *Genome Biol.* **3**.

Hugenholtz, P., Goebel, B. M., and Pace, N. R. (1998). Impact of culture-independent studies on the emerging phylogenetic view of bacterial diversity. *J. Bacteriol.* **180,** 4765–4774.

Hungate, R. E. (1950). The anaerobic mesophilic cellulolytic bacteria. *Bacteriol. Rev.* **14,** 1–49.

Hungate, R. E. (1969). A roll tube method for cultivation of strict anaerobes. *In* "Methods in Microbiology" (J. R. Norris and W. D. Ribbons, eds.), Vol. 3B, pp. 117–132. Academic Press, London.

Ingraham, J. L., Maaløe, O., and Neidhardt, F. C. (1983). "Growth of the Bacterial Cell." Sinauer, Sunderland, MA.

Kaksonen, A. H., Plumb, J. J., Franzmann, P. D., and Puhakka, J. A. (2004). Simple organic electron donors support diverse sulfate-reducing communities in fluidized-bed reactors treating acidic metal- and sulfate-containing wastewater. *FEMS Microbiol. Ecol.* **47,** 279–289.

Kell, D. B., and Young, M. (2000). Bacterial dormancy and culturability: The role of autocrine growth factors: Commentary. *Curr. Opin. Microbiol.* **3,** 238–243.

Leadbetter, J. R. (2003). Cultivation of recalcitrant microbes: Cells are alive, well and revealing their secrets in the 21st century laboratory. *Curr. Opin. Microbiol.* **6,** 274–281.

Lin, C. C., and Casida, L. E., Jr. (1984). Gelrite as a gelling agent in media for the growth of thermophilic bacteria. *Appl. Environ. Microbiol.* **47,** 427–429.

McGarvey, J. A., Miller, W. G., Sanchez, S., and Stanker, L. (2004). Identification of bacterial populations in dairy wastewaters by use of 16S rRNA gene sequences and other genetic markers. *Appl. Environ. Microbiol.* **70,** 4267–4275.

Mukamolova, G. V., Kaprelyants, A. S., Kell, D. B., and Young, M. (2003). Adoption of the transiently non-culturable state: A bacterial survival strategy? *In* "Advances in Microbial Physiology," Vol. 47, pp. 65–129.

Pfennig, N., and Trüper, H. G. (1981). Isolation of members of the families Chromatiaceae and Chlorobiaceae. *In* "The Prokaryotes" (M. P. Starr, H. Stolp, H. G. Trüper, A. Balows, and H. G. Schlegel, eds.), Vol. I, pp. 279–289. Springer-Verlag, Berlin.

Schink, B., and Stams, A. J. M. (2002). Syntrophism among prokaryotes. *In* "The Prokaryotes: An Evolving Electronic Resource for the Microbiological Community" (M. Dworkin *et al.*, eds.), 3rd Ed. Springer-Verlag, New York. http://link.springer-ny.com/link/service/books/10125/.

Spring, S., Schulze, R., Overmann, J., and Schleifer, K.-H. (2000). Identification and characterization of ecologically significant prokaryotes in the sediment of freshwater lakes: Molecular and cultivation studies. *FEMS Microbiol. Rev.* **24,** 573–590.

Summanen, P. H., McTeague, M., Vaisanen, M.-L., Strong, C. A., and Finegold, S. M. (1999). Comparison of recovery of anaerobic bacteria using the Anoxomat(®), anaerobic chamber, and GasPak(®)jar systems. *Anaerobe* **5,** 5–9.

Wackett, L. P. (2005). Element ratios and bacteria. *Environ. Microbiol.* **7,** 147–148.

Wolfe, R. S. (1999). Anaerobic life: A centennial view. *J. Bacteriol.* **131,** 3317–3320.

Zengler, K., Toledo, G., Rappe, M., Elkins, J., Mathur, E. J., Short, J. M., and Keller, M. (2002). Cultivating the uncultured. *Proc. Natl. Acad. Sci. USA* **99,** 15681–15686.

Further Readings

Tchobanoglous, and Burton (1991). "Wastewater Engineering; Treatment, Disposal and Reuse." McGraw-Hill, New York.

Veldkamp, H. (1970). Enrichment cultures of prokaryotic organisms. *In* "Methods in Microbiology" (J. R. Norris and D. W. Ribbons, eds.), Vol. 3A, pp. 305–361. Academic Press, London.

Widdel, F., and Bak, F. (1999). Gram-negative mesophilic sulfate-reducing bacteria. *In* "The Prokaryotes: An Evolving Electronic Resource for the Microbiological Community" (M. Dworkin *et al.*, eds.), 3rd Ed. Springer-Verlag, New York. http://link.springer-ny.com/link/service/books/10125/.

[2] Enrichment and Isolation of Anaerobic Hydrocarbon-Degrading Bacteria

By IRENE A. DAVIDOVA and JOSEPH M. SUFLITA

Abstract

Recent progress in microbiology resulted in the enrichment and isolation of anaerobic bacteria capable of the biodegradation of various hydrocarbons under a variety of electron-accepting conditions. Problems challenging the enrichment and isolation of anaerobic hydrocarbonclastic organisms required new approaches and modifications of conventional microbiological techniques. This chapter summarizes the collective experience accumulated in this area starting from anaerobic sampling precautions and includes all stages of cultivation from the construction of initial incubations to final isolation steps and the evaluation of culture purity.

Introduction

Research over the last decade has demonstrated convincingly that many classes of hydrocarbons are amenable to microbial attack under a variety of anaerobic conditions. Such studies have major implications for understanding the transport and fate of these substances in the environment and how microbial metabolism can impact the economic value of petroliferous deposits and help ameliorate the risks associated with accidental releases (Gieg and Suflita, 2002). Enrichment cultures capable of coupling the biodegradation of straight or branch-chained, alicyclic, aromatic, and polynucleararomatic hydrocarbons with sulfate, nitrate, or iron reduction or methanogenesis have been described (Burland and Edwards, 1999; Edwards *et al.*, 1992; Grbic-Galic and Vogel, 1987; Kropp *et al.*, 2000; Rabus *et al.*, 1999; Rueter *et al.*, 1994; Townsend *et al.*, 2003, 2004; Ulrich and Edwards, 2003; Weiner and Lovley, 1998; Zengler *et al.*, 1999; Zhang and Young, 1997). Despite the exigent problems of low aqueous solubility and fairly high toxicity of the substrates, as well as relatively slow culture growth rates, attempts at the enrichment and isolation of pure cultures of hydrocarbon-degrading anaerobes have paid important dividends. In particular, insights on the pathways employed by novel bacterial isolates capable of metabolizing various classes of hydrocarbons are now known. The vast majority of these bacteria couple the biodegradation of a hydrocarbon to the reduction of either sulfate or nitrate.

METHODS IN ENZYMOLOGY, VOL. 397
0076-6879/05 $35.00
DOI: 10.1016/S0076-6879(05)97002-X

Thus, studies of toluene degradation under nitrate-reducing conditions led to the discovery of several members of the genera *Thauera* and *Azoarcus* (Anders *et al.*, 1995; Chee-Sanford *et al.*, 1996; Dolfing *et al.*, 1990; Shinoda *et al.*, 2004; Song *et al.*, 1999; Zhou *et al.*, 1995). Several pure sulfate-reducing cultures capable of toluene mineralization were also isolated, including *Desulfobacula toluolica* (Rabus *et al.*, 1993), strain PRTOL1 (Beller *et al.*, 1996), and strains oXyS1 and mXyS1 (Harms *et al.*, 1999). Only a single bacterium, *Geobacter metallireducens* GS 15, is known to couple the metabolism of toluene to the utilization of ferric iron as an electron acceptor (Lovley and Lonergan, 1990), while several denitrifying and sulfate-reducing bacteria capable of *m*-xylene and *o*-xylene biodegradation have been described (Dolfing *et al.*, 1990; Fries *et al.*, 1994; Harms *et al.*, 1999; Hess *et al.*, 1997; Rabus and Widdel, 1995). Three denitrifying cultures and one sulfate-reducing bacterium are known for the ability to completely mineralize ethylbenzene (Ball *et al.*, 1996; Kniemeyer *et al.*, 2003; Rabus and Widdel, 1995).

Unlike the other so-called BTEX hydrocarbons, benzene is the least reactive substrate, but there is no doubt that it can be anaerobically metabolized by microorganisms. The isolation of a pure benzene-degrading culture has been most enigmatic. Until very recently, only two benzene-degrading bacterial strains in the genus *Dechloromonas* have been described (Coates *et al.*, 2001). These organisms oxidize benzene with nitrate as an electron acceptor, but the metabolic pathway must be different from that suggested previously (Suflita *et al.*, 2004).

Other anaerobic hydrocarbon-degrading pure cultures are also known. Studies have shown that naphthalene can be completely mineralized by pure cultures of sulfate-reducing and -dentrifying bacteria (Galushko *et al.*, 1999; Rockne *et al.*, 2000). Several sulfate-reducing and -denitrifying bacterial strains (Aeckersberg *et al.*, 1991, 1998; Cravo-Laureau *et al.*, 2004; Ehrenreich *et al.*, 2000; Rueter *et al.*, 1994; So and Young, 1999) are capable of the complete conversion of *n*-alkanes to carbon dioxide. We have had recent success in the isolation of alkane-degrading sulfate reducing bacteria, strain ALDC obtained from previously described enrichments (Davidova *et al.*, 2002 and Kropp *et al.*, 2000) and strain Lake, from an Oklahoma oil field sample. The metabolic pathways for anaerobic oxidation of hydrocarbons have been reviewed repeatedly in recent years (Spormann and Widdel, 2000; Suflita *et al.*, 2004; Widdel and Rabus, 2001).

A few novel hydrocarbon-degrading bacterial strains have been fully characterized, and available data are reviewed elsewhere (Heider *et al.*, 1999; Spormann and Widdel, 2000). Because *n*-alkanes are quantitatively the most important constituents of crude oil and refined petroleum products, this chapter focuses on isolates capable of the biodegradation of these substrates. The majority of known alkane-degrading, nitrate-reducing bacteria belong to the β-*Proteobacteria Azoarcus/Thauera* group. They are oval shaped, have more slender rods, and most have been partially characterized. All of the sulfate-reducing isolates are short oval-shaped rods belonging to the δ subclass of *Proteobacteria*. None of the known alkane-degrading strains have been fully characterized, and published characteristics are summarized in Table I. Morphologically and physiologically, the sulfate-reducing isolates (with the exception of the thermophilic strain TD3) are generally quite comparable. However, a phylogenetic comparison reveals substantive differences among the organisms (Fig. 1). Based on 16S rRNA gene sequences, the sulfate reducers capable of anaerobic *n*-alkane oxidation can be divided into three groups. The first group clusters close to strain AK-01 and includes the mesophilic Pnd3, Hxd3, and the recently isolated CV 2803 strains. It has been suggested that these strains be combined into a new genus, *Desulfatibacilum* gen. nov. (Cravo-Laureau *et al.*, 2004). The second group includes strains ALDC and Lake, which are phylogenetically close to each other, but not to members of the suggested *Desulfatibacilum* genus. The closest known relatives include *Desulfacinum* sp. and *Syntrophobacter* sp. However, the alkane-degrading strains share only 93–95% and 92–95% sequence similarity with the aforementioned genera, respectively. Finally, the thermophilic strain TD3 constitutes a relatively deep branch in the δ subdivision of Proteobacteria.

Enrichment Cultures

Anaerobic hydrocarbon-degrading organisms, more specifically alkane-degrading bacteria, have been isolated from a variety of environments, including freshwater and marine sediments, deep marine hydrothermal sediments, aquifers, oil field production waters, and oily sludges amassed during the course of petroleum waste disposal activities. Hydrocarbon-degrading activity can be most easily demonstrated and cultures obtained from environments chronically exposed to hydrocarbons. In fact, our search for evidence of anaerobic benzene decomposition in anaerobic aquifers not chronically exposed to hydrocarbons proved unproductive (unpublished result). The importance of hydrocarbon exposure should not be underestimated. In fact, the physical association of bacteria with an immiscible hydrocarbon layer has been observed with many hydrocarbon-degrading cultures. In our hands, when a mineral medium was inoculated with oil field production water and either sterile or nonsterile oil from the same formation, hydrocarbon biodegradation occurred only in incubations containing the nonsterile oil (unpublished observation). Such observations on freshly sampled oil production fluids suggest that the association

TABLE I
PURE CULTURES THAT CAN DEGRADE *n*-ALKANES UNDER ANAEROBIC CONDITIONS

Organism	Source	Alkanes metabolized	Doubling time	Temperature (°)	pH	Salinity (g/liter)	Reference
Denitrifying strains							
HxN1	Ditch sediments	C_6-C_8	11 h	28	7.1	No data	Ehenreich et al. (2000)
OcN1	Ditch sediments	C_8-C_{12}	No data	28	7.1	No data	Ehenreich et al. (2000)
HdN1	Activated sludge	C_{14}-C_{20}	No data	28	7.1	No data	Ehenreich et al. (2000)
Sulfate-reducing strains							
Hxd3	Oily sludge	C_{12}-C_{20}	5–7 weeks on C_{16}	28–30 (opt)	No data	No growth in fresh water	Aeckersberg et al. (1991)
Pnd3	Marine sediments	C_{14}-C_{17}	No data	28–30 (opt)	No data	≥20	Aeckersberg et al. (1998)
AK-01	Estuarine sediments	C_{13}-C_{18}	3 days on C_{16}	26–28 (opt)	6.9–7.0 (opt)	1–60, 10 (opt)	So and Young (1999)
Desulfatibacillum aliphaticivorans (CV2803)	Marine sediments	C_{13}-C_{18}	No data	15–40, 28–35 (opt)	6.6–7.8, 7.5 (opt)	6–45, 24 (opt)	Cravo-Laureau et al. (2004)
TD3	Guaymas Basin sediments	C_6-C_{16}	No data	55–65 (opt)	6.8 (opt)	≥20	Rueter et al. (1994)
ALDC	Oily sludge	C_6-C_{12}	5.3 days on C_{10}	20–45, 37 (opt)	4.5–8.2, 6.5 (opt)	1–50	
Lake	Produced water	C_6-C_{10}	No data	20–55	6.8–8.2	14–24 (opt), 1–25	

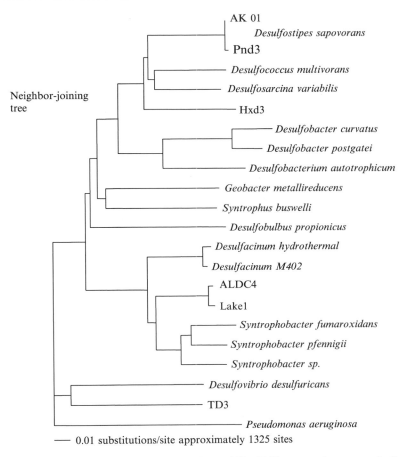

FIG. 1. Phylogenetic relationships based on 16S rRNA gene sequences of alkane-degrading sulfate-reducing bacteria within the δ subclass of Proteobacteria. The dendrogram was constructed on the basis of evolutionary distances from approximately 1300 aligned nucleotides used for analysis.

between oil-degrading bacteria and a hydrocarbon phase may be particularly common.

It was also noted that the metabolic capacity of a complex hydrocarbon-degrading microbial community was surprisingly diverse in that it could utilize a range of *n*-alkanes to which it was never previously exposed (Townsend *et al.*, 2003). The same study demonstrated that the range and degree of *n*-alkane utilization was comparable under both sulfate-reducing and methanogenic conditions. These were the predominant electron-accepting conditions in the environment to which the organisms were regularly exposed. However, when nitrate was provided in a parallel experiment, very little hydrocarbon utilization could be demonstrated (unpublished observation). Such findings point out the desirability of conducting enrichment and bacterial isolation attempts under the predominant electron-accepting conditions in the environment being explored. Thus, hydrocarbon-degrading cultures originating from marine sediments were all enriched and purified under sulfate-reducing conditions (Aeckersberg et al., 1998; Cravo-Laureau et al., 2004; Galushko et al., 1999; Rueter et al., 1994). For guidance in determining in situ predominant redox conditions, the reader is referred to Smith (1997) and references therein.

Nevertheless, the expression of anaerobic hydrocarbon-degrading activity under a particular electron-accepting regime does not preclude the prospect of obtaining an active culture coupled with the consumption of an alternate terminal electron acceptor. Ulrich and Edwards (2003) obtained a methanogenic benzene-degrading culture from a formerly sulfate-reducing enrichment and discovered both nitrate- and sulfate-reducing hydrocarbon metabolizing enrichments from the same environmental matrix.

Sampling

The most obvious precaution in sampling anaerobic hydrocarbon-degrading cultures from environmental samples is to minimize the exposure of the requisite organisms to oxygen. This is usually accomplished without elaborate procedures, but prudent measures are recommended. For instance, liquid and solid samples should be taken in nonplastic vessels (usually glass) that can be filled to capacity to avoid a gaseous headspace and tightly sealed. Consolidated or unconsolidated cores can be kept in the barrels used during recovery operations, and the ends of the core barrels are typically capped. We have had good success in obtaining hydrocarbon-degrading enrichments without the use of oxygen-free gases during field-sampling efforts. However, as immediate as practical, we generally place all samples in metal ammunition containers. These containers are readily available in sporting supply or military surplus stores, come in a variety of sizes, have excellent sealing properties, and can be fitted with a septum to allow for the field flushing and partial pressurization of the containers with an inert gas (N_2, Ar, etc). Normally, the sample containers are then chilled, shipped to the laboratory, and stored at 4° until use. When required, samples can be opened and processed inside an anaerobic glove box under an oxygen-free atmosphere, usually N_2:H_2 (90:10).

Initial Incubations

Enrichment cultures are initiated using the environmental sample as an inoculum source. Relatively large incubations are recommended because they tend to be more tolerant of trace exposures to oxygen. Typically, at least 50 g of sediment or 100 ml of aqueous phase is employed in initial incubations. The volume can be scaled to accommodate individual needs, but the solids/liquid ratio is generally maintained from 1:1 to 1:2 (g/v). Filter-sterilized anoxic water from the environmental site of interest makes an excellent aqueous phase for initial incubations. When desired, this aqueous phase can be replaced with a mineral medium chosen to mimic the environment originally sampled. Thus, a marine medium of comparable salinity and sulfate composition might be used in lieu of filter-sterilized seawater. For detection of the desired microbial activity and to select for the requisite microorganisms, incubations should be supplemented with the substrates and/or electron acceptor of interest. All culture manipulations are routinely conducted under an anaerobic atmosphere provided by either a gasing station (Balch, *et al.*, 1979) or an anaerobic glove box. If natural liquids used for aqueous phase are not anoxic, they can be flushed with N_2 and reduced with reducing agents (Caldwell *et al.*, 1998; Rios-Hernandez *et al.*, 2003; Townsend *et al.*, 2003, 2004). Resazurin is routinely used as a sensitive indicator (Hungate, 1969; Tanner, 1997) of anaerobiosis, as the dye is colorless when reduced and pink when oxidized. All amendments should be added from anoxic sterile stock solutions prepared according to strict anaerobic techniques (see later). Because hydrocarbons readily partition to soft rubber closures, incubations can be sealed with Teflon-lined stoppers, secured in place with aluminum crimps, and incubated in the inverted position. After removing the sealed incubations from the glove box, the culture headspace can be exchanged to the desired composition (e.g., N_2:CO_2/80:20) with a gasing station. Incubations are then monitored for electron acceptor consumption or methane production as a function of time. This information is then coupled with a substrate depletion assay. Of course, the inoculum source itself may contain enough endogenous substrate to confound the interpretation of such assays. Therefore, it is imperative to incorporate both substrate-unamended and sterile controls in the experimental design. Substrate-unamended controls, without the exogenous addition of the hydrocarbon of interest, allow for the distinction of the degree of electron-accepting consumption occurring at the expense of endogenous substrates. Sterile controls account for potential abiotic losses of the substrate or electron acceptors. Significant losses of an electron acceptor (or methane accumulation) relative to the substrate-unamended and sterile controls indicate the occurrence of the desired activity. Positive incubations are used in further enrichment procedures.

Sediment-Free Enrichments

The repeated transfer of positive incubations in the cultivation medium that supported the initial metabolism of interest can be done to obtain sediment-free enrichments. The transfers to fresh media should not be less than 10% and are more typically 20% (v/v). Eventually, the sediment is reduced to the point where it is an insignificant portion of the incubations, as it is essentially diluted by the repeated transfer of the culture. Before transfers are made, the hydrocarbon biodegradation activity in the incubation should be confirmed.

Media. The bicarbonate-buffered mineral medium described by Widdel and Bak (1992), modified for the appropriate salinity conditions, has been widely used for freshwater, brackish, and saltwater enrichments. This medium is supplemented with a trace metals and vitamins solution as in Widdel and Bak (1992) or as described by Tanner (1997). Techniques for the preparation of anaerobic media and amendments are well established (Balch *et al.*, 1979; Bryant, 1972; Widdel and Bak, 1992). In our cultural work we never use organic amendments such as yeast extracts or peptone. Even in low concentrations, these supplements promote the growth of heterotrophic bacteria that are generally incapable of hydrocarbon decay. For best results we use the individual hydrocarbon compound or mixture of interest as the exogenous source of carbon and energy. Thus, the culture conditions are highly selective for hydrocarbonclastic bacteria. We are also careful to remove any carryover of hydrogen that may have been introduced to the cultures during glove box manipulation steps. Hydrogen stimulates the growth of hydrogenotrophic bacteria that may tend to outcompete hydrocarbon degraders for the available electron acceptor in the medium.

When an enrichment culture on a particular class of hydrocarbons is desired, a few precautions should be kept in mind. For instance, it is known that ethylbenzene can interfere with the enrichment of a benzene-degrading culture because the former substrate is utilized preferentially (Krumholz *et al.*, 1996). However, when enriching for alkane-degrading organisms, the provision of a hydrocarbon mixture containing a range of alkane carbon chain lengths may prove desirable. Thus far, all known anaerobic alkane-degrading bacteria exhibit a rather narrow range of substrates they can metabolize. That is, bacteria that grow at the expense of hexadecane cannot utilize hexane and vice versa. When the bacterial metabolic capacities are unknown, a broad range of alkanes often allows the investigator to avoid false-negative enrichment attempts.

Hydrophobic hydrocarbon substrates have generally low aqueous solubilities and often fail to form true solutions. Regardless of the way they are applied (see later), the substrates often represent another phase, typically an overlay or as droplets in the aqueous phase. To maximize bacterial contact with the substrate, we avoid tubes or small vials for obtaining sediment-free enrichments. Rather, we typically employ 160-ml serum bottles containing 25–50 ml of the cultivation medium. Further, we incubate the cultures at an angle and, as indicated earlier, in the inverted position to avoid contact with the bottle closures. In our experience, we have never found the need for continuous agitation of the cultures. Cultures are only shaken periodically during routine inspection and sampling events.

Substrates. Hydrocarbons are hydrophobic, poorly soluble, and sometimes inhibitory substrates. In order to cope with these characteristics, hydrocarbons are often added dissolved in the carrier as an overlay on the medium. The most common carrier used in this regard is 2,2,4,4,6,8,8-heptamethylnonane (HMN). Because the metabolism of this branched alkane has not yet been reported, it is considered "nonbiodegradable" and essentially inert. Flushing the compound with N_2 or by repeated vacuuming/flushing with an oxygen-free gas and autoclaving should precede the use of HMN in an experiment. When sparingly soluble compounds are investigated, such as phenanthrene, pyrene, or dibenzothiophene, they can be applied as solid crystals (Annweiler *et al.*, 2000) or in a solution in HMN (Annweiler *et al.*, 2001; Galushko *et al.*, 1999). To prepare the latter, solid crystals of the polyaromatic substrate are added to anoxic HMN under a stream of nitrogen and autoclaved gently (5 min). The resulting suspension is then added to the anaerobic incubation as an overlay using a N_2-flushed syringe and needle through the closure or with a glass pipette under a stream of oxygen-free gas.

The hydrocarbon in HMN forms the overlay that serves as a continuous source of the substrate. The amount of hydrocarbon that partitions from the overlay to the aqueous phase of the incubation mixture never exceeds its water solubility. The hydrocarbon degraded in the aqueous phase is replenished from the HMN overlay. Thus, the precise concentration of poorly soluble hydrocarbon in the HMN overlay is relatively unimportant. Typically, we apply hydrocarbons in the overlay in amounts that facilitate their gas chromatographic analysis. In this way, the use of an overlay has the advantage of allowing for the monitoring of hydrocarbon loss from the HMN layer without the need to destructively sacrifice the incubations.

A small portion of an overlay (from 1 to 2 μl) can be sampled via a microliter syringe and injected directly onto the gas chromatograph. This technique requires a certain thickness of the HMN layer in that it should be easily reached with the syringe needle without penetration into the underlying aqueous phase. For poorly soluble substrates, we typically use 3 mg hydrocarbon per milliliter of HMN and 3 to 4 ml of overlay per 25 ml of media. However, with more water-soluble hydrocarbons, caution should be exercised, as the aqueous phase concentration can sometimes reach inhibitory levels. For instance, with naphthalene-growing enrichments, we tested 5–25 mg/ml HMN and found that incubations supplemented with 10 mg/ml HMN degraded naphthalene and reduced sulfate most reliably.

Experience shows that in most cases, the system of supplying a PAH compound dissolved in HMN provides more reliable biodegradation results than merely the placement of crystals in the incubation. This may be a function of better mass transfer of the substrate to the requisite organisms. The monoaromatic hydrocarbons collectively known as BTEX (benzene, toluene, xylene, and ethylbenzene) hydrocarbons are reasonably water soluble and can be supplied as neat compounds or dissolved in an inert carrier phase such as HMN. In our hands, neat benzene (up to 150 μM) and toluene (up to 500 μM) elicited no obvious inhibitory effect on background electron-accepting processes or biodegradation.

Alkanes are most often added neat without a carrier phase from anoxic filter-sterilized or autoclaved preparations (Aeckersberg et al., 1991; Kropp et al., 2000; So and Young, 1999). Sometimes alkanes shorter than dodecane are added using an inert carrier (Ehrenreich et al., 2000) to avoid even the prospect of microbial inhibition by these substances. However, we have found that this was generally unnecessary, at least with sulfate-reducing bacteria. Even alkanes as small as hexane in concentrations up to 25 mM did not produce any obvious inhibitory effect and could be used without a carrier. One should recognize that this amount of hydrocarbon is well above the aqueous solubility limit of n-alkanes and represents an absolute value rather than true solution concentration. Similarly, when crude oil is used as a starting material, repeated vacuuming and flushing with N_2 gas should be done to remove any traces of oxygen prior to sterilization of the hydrocarbon mixture.

Purification and Isolation

Enrichment cultures demonstrating a consistent stoichiometric relationship between the amounts of substrate utilized and electron-acceptor reduced can be chosen for further purification and isolation attempts. It is not uncommon for cultures with well-documented hydrocarbon biodegradation patterns to have relatively few bacterial cells and exhibit only scant optical density. Transfer of such cultures to fresh media results in the slowing of biodegradation and often such transfers fail. It is therefore helpful to improve the reliability of bacterial growth before isolation attempts. To this end, the increase in actual cell numbers transferred and optimization of growth conditions can be equally important. Variations in environmental parameters such as temperature, pH, and salinity not only improve growth, but also aid in purification attempts by increasing the selective pressure on the requisite organism. Inocula can also be concentrated in different ways. If cells are not strongly attached to a hydrocarbon layer they can be concentrated before transfer by centrifugation using gastight sterile centrifuge tubes flushed with an anoxic atmosphere. Biomass collected from a culture can be resuspended in a small amount of fresh medium and used as an inoculum. If cells are strongly attached, a hydrocarbon layer with a small portion of aqueous phase should be taken for reinoculation. Transfers that received concentrated inoculum will grow faster, yield more biomass, and lead to a greater likelihood that isolation attempts will prove successful. Centrifugation in Percoll gradients can be useful for highly enriched cultures where different morphotypes can be readily distinguished under the microscope (Beaty et al., 1987). Such gradients have been useful for the separation of Syntrophomonas wolfei cells from Methanospirillum hungatii cells (Beaty et al., 1987). This method will hardly yield a pure culture but can result in considerable enrichment of one type of cell relative to the other. We used this procedure as an intermediate step in eventually obtaining the pure alkane-degrading sulfate-reducing bacterium strain ALDC. In brief, cells from a 120-ml culture are collected anaerobically by centrifugation, washed, and resuspended in 2 ml of a fresh reduced medium. A Percoll gradient (70–55%, v/v) is constructed in disposable centrifuge tubes (2 ml), amended with a 0.5-ml concentrated cell suspension, and centrifuged at 3000g for 1 h. Following centrifugation, there is a visible ring of cells on top of the tube and some precipitate at the bottom. About 0.5 ml of the upper ring is resuspended into 5 ml of reduced mineral medium. The remainder of the supernatant is discarded, while the precipitate is resuspended in 10 ml of a mineral medium. The suspensions are amended with dodecane, and only the subculture obtained from the upper ring of cells proved capable of metabolizing this hydrocarbon. Microscopic examination confirmed that cells were highly concentrated, although still not completely pure. However, further isolation steps were focused on this preparation.

There are at least two approaches for the isolation of hydrocarbon-degrading bacteria that have proven to be successful, although each has their particular advantages and drawbacks. The first approach employs the use of water-soluble substrates, often presumed metabolic intermediates of hydrocarbons, instead of the hydrocarbon molecule per se. The assumption is that the water-soluble substrate is likely to be metabolized by the same organism in a manner consistent with the hydrocarbon of interest. Without prior knowledge of the metabolic capability of an isolate, such an approach can be risky and result in the isolation of bacteria that lack hydrocarbon-degrading ability. However, the use of water-soluble substrates that can be incorporated easily into solid media allows for the implementation of more conventional techniques, such as the agar dilution series (serial dilutions in anoxic agar), roll-tube isolation, or the use agar plates/bottles that are used frequently for the isolation of anaerobic bacteria (Hermann *et al.*, 1986; Hungate, 1969; Widdel and Bak, 1992). Another approach to isolation uses the hydrocarbon itself as the sole source of carbon and energy. While it directly targets hydrocarbon-degrading bacteria, the incorporation of hydrophobic substrates into solid media can often be problematic. It is not unusual for a combination of the aforementioned approaches to be utilized.

Most promising is isolation of the denitrifying bacterium strain HdN1 by repeated streaking on a gelrite-solidified medium containing emulsified hexadecane (Ehrenreich *et al.*, 2000). We also used this approach for obtaining colonies of strain ALDC. Either hexane or decane is used independently as emulsified growth substrates and is dispersed by sonication in a prewarmed (40°) mineral medium that is inoculated with an alkane-degrading enrichment culture. The entire volume containing the emulsified alkane and cells is then added to a liquid gelrite medium (0.8% and kept at 70°), mixed thoroughly, and placed immediately on ice. After solidification, the bottles are incubated at 31° in an inverted position. We obtained several colonies and all required repeated purification by streaking and/or the combination with other purification techniques (serial dilution in liquid media, dilutions in anoxic agar). Our experience would suggest that there are no approaches that are 100% failsafe and that every culture requires the development of its own tactic. Thus, an alkane-degrading denitrifying bacterium, strain HdN1, was obtained by repetitive restreaking on a gelrite-solidified medium with C_{16} as the hydrocarbon substrate and then with valerate as a carbon source (Ehrenreich *et al.*, 2000). However, other alkane-degrading denitrifying strains were isolated in different ways. Strain OcN1 was isolated by the agar dilution series with caproate as a carbon source, and strain HxN1 was obtained by anoxic liquid dilution series incorporating hexane (Ehrenreich *et al.*, 2000). We opted to conduct further purification of an alkane-degrading enrichment obtained after centrifugation in a Percoll gradient or by passing through solid medium in anoxic liquid medium dilutions with alkanes as a growth substrate. A search for optimal environmental parameters did not result in conditions that selectively favored the growth of the bacterium of interest. Rather, it became obvious that conditions promoting accumulation of the requisite cells included a basal saltwater mineral medium (Widdel and Bak, 1992) with decane or hexane as a growth substrate. All further purification steps for strain ALDC were conducted with liquid dilution series with C_{10} or C_6 as growth substrates. Dilution series were carried out in 25-ml vials containing 5 ml of basal saltwater mineral medium and 0.01 ml of hydrocarbon. A 10% culture volume was serially transferred up to the 10^{-9} dilution. Strict anaerobic techniques were employed throughout this exercise, although we were less concerned with hydrocarbon absorbance on stoppers, vessel walls, and so on, as the stoichiometry of substrate utilization was already established. The vials were incubated inverted at 31° and positive growth was determined by microscopy. The highest dilution with positive growth was used as the starting culture for the next round of serial dilutions. Typically, we observed strain ALDC cells growing up to the 10^{-8} dilution. However, culture development depended on inoculum size, and growth in the higher dilutions was considerably slower. Even with these procedures, we found it difficult to separate strain ALDC cells from minor contaminant organisms observed by microscopy. We also noted that ALDC cells failed to proliferate at the highest dilutions, where they were inoculated as a monoculture. We then started to amend the higher dilutions ($\geq 10^{-6}$) with 0.01% yeast extract, and positive growth was detected more readily up to the 10^{-8} dilution. We never added yeast extract in the lower dilutions because it promoted growth of the contaminant organism, complicating isolation of the cell of interest. After several more serial dilution attempts using the 10^{-7} dilution as a starting inoculum, we obtained a pure culture that contained cells of just one morphotype.

Purity Control

Like most microbiological investigations, there are several levels of purity control. Checking the culture morphology regularly by phase-contrast microscopy is highly recommended. This allows for the relatively rapid evaluation of culture purification steps and offers particularly good results when the bacteria of interest are easily distinguished from others. Most known hydrocarbonclastic organisms are oval rods, which often have a refractile bright core. In most cases, these organisms can be distinguished from contaminant cells. However, the lack of contaminants by microscopy is not a guarantee of culture purity. The bacterium of particular interest can be in numbers that easily overwhelm the contaminant (often orders of magnitude) and culture impurities can simply escape visual detection. Another level of purity assessment involves inoculating the culture into a medium containing readily degradable substrates, such as glucose, lactate, and growth supplements such as yeast extract. The goal is to reveal the presence of contaminant organotrophic anaerobes. We check the purity of strain ALDC and other hydrocarbon-degrading anaerobes routinely by inoculating the cultures into their usual cultivation medium containing glucose (10 mM), lactate (10 mM), and 0.1% yeast extract instead of the hydrocarbon of interest. Lack of growth indicates the absence of spurious contaminants. In early studies of anaerobic alkane degradation, particularly with the first described anaerobic alkane-degrading bacterium (strain Hxd3), another routine check was for the presence of aerobic hydrocarbon-degrading contaminants (Aeckersberg et al., 1991). However, anaerobic hydrocarbon decay is now well established, and enrichments and pure cultures obtained under strict anaerobic procedures are unlikely to harbor the presence of aerobes. Finally, cultures can be monitored routinely and purity assessed by denaturing gradient gel electrophoresis (DGGE) (Duncan et al., 2003; Muyzer et al., 1993). Cultures that exhibit a single morphology, do not harbor organotrophic contaminants, and show only a single DGGE band pattern can be assumed to be pure.

Concluding Remarks

Isolation of anaerobes has proved to be a difficult task, but one that has yielded to the procedures outlined earlier. The isolates provide precious material for probing the metabolic pathway associated with anaerobic hydrocarbon decay and for understanding the ecological boundaries associated with this type of metabolism. Nevertheless, the authors believe that the study of this general topic is still in its infancy. Many other hydrocarbon-degrading cells have yet to be uncovered, and information on the predominant metabolic pathways utilized has yet to fully emerge. Creative approaches to the isolation and identification of organisms responsible for the biotransformation of highly hydrophobic substrates will certainly unfold in the years ahead. The authors are convinced that the approaches will only be limited by the imagination of the investigators.

Acknowledgments

We are grateful to Dr. Kathleen Duncan for her help with phylogenetic analysis and to Dr. Kevin Kropp for his fruitful cooperation.

References

Aeckersberg, F., Bak, F., and Widdel, F. (1991). Anaerobic oxidation of saturated hydrocarbons to CO_2 by a new type of sulfate-reducing bacterium. *Arch. Microbiol.* **156,** 5–14.

Aeckersberg, F., Rainey, F., and Widdel, F. (1998). Growth, natural relationships, cellular fatty acids and metabolic adaptation of sulfate-reducing bacteria that utilize long-chain alkanes under anoxic conditions. *Arch. Microbiol.* **170,** 361–369.

Anders, A., Kaetzke, A., Kampfer, P., Ludwig, W., and Fuchs, G. (1995). Taxonomic position of aromatic degrading denitrifying pseudomonad strains K172 and KB740 and their description as new members of the genera *Thauera*, *T. aromatica* sp. nov., and *Azoarcus*, *A. evansii* sp. nov., respectively, members of the beta subclass of Proteobacteria. *Int. J. Syst. Bacteriol.* **45,** 327–333.

Annweiler, E., Materna, A., Safinowski, M., Kappler, A., Richnow, H. H., Michaelis, W., and Meckenstock, R. U. (2000). Anaerobic degradation of 2-methylnaphthalene by a sulfate-reducing enrichment culture. *Appl. Environ. Microbiol.* **66,** 5329–5333.

Annweiler, E., Michaelis, W., and Meckenstock, R. U. (2001). Anaerobic cometabolic conversion of benzothiophene by a sulfate-reducing enrichment culture and in a tar-oil-contaminated aquifer. *Appl. Environ. Microbiol.* **67,** 5077–5083.

Balch, W. E., Fox, G. E., Magrum, L. J., Woese, C. R., and Wolfe, R. S. (1979). Methanogens: Reevaluation of a unique biological group. *Microbiol. Rev.* **43,** 260–296.

Ball, H. A., Johnson, H. A., Reinhard, M., and Spormann, A. M. (1996). Initial reactions in anaerobic ethylbenzene oxidation by a denitrifying bacterium, strain EB1. *J. Bacteriol.* **178,** 5755–5761.

Beaty, P. S., Wofford, N. Q., and McInerney, M. J. (1987). Separation of *Syntrophomonas wolfei* from *Methanospirillum hungatii* in syntrophic cocultures by using Percoll gradients. *Appl. Environ. Microbiol.* **53,** 1183–1185.

Beller, H. R., Spormann, A. M., Sharma, P. K., Cole, J. R., and Reinhard, M. (1996). Isolation and characterization of a novel toluene-degrading sulfate-reducing bacterium. *Appl. Environ. Microbiol.* **62,** 1188–1196.

Bryant, M. P. (1972). Commentary on the Hungate technique for culture of anaerobic bacteria. *Am. J. Clin. Nutr.* **25,** 1324–1328.

Burland, S. M., and Edwards, E. A. (1999). Anaerobic benzene biodegradation linked to nitrate reduction. *Appl. Environ. Microbiol.* **65,** 529–533.

Caldwell, M. E., Garrett, R. M., Prince, R. C., and Suflita, J. M. (1998). Anaerobic biodegradation of long-chain *n*-alkanes under sulfate-reducing conditions. *Environ. Sci. Technol.* **32,** 2191–2195.

Chee-Sanford, J. C., Frost, J. W., Fries, M. R., Zhou, J., and Tiedje, J. (1996). Evidence for acetyl-coenzyme A and cinnamoyl coenzyme A in the anaerobic toluene mineralization pathway in *Azoarcus tolulyticus* Tol-4. *Appl. Environ. Microbiol.* **62,** 964–973.

Coates, J. D., Chakraborty, R., Lack, J. G., O' Connor, S. M., Cole, K. A., Bender, K. S., and Achenbach, L. A. (2001). Anaerobic benzene oxidation coupled to nitrate reduction in pure culture by two strains of *Dechloromonas*. *Nature* **411,** 1039–1043.

Cravo-Laureau, C., Matheron, R., Cayol, J.-L., Joulian, C., and Hirschler-Réa, A. (2004). *Desulfatibacillum aliphaticivorans* gen. nov., sp. nov., an n-alkane- and n-alkene-degrading, sulfate-reducing bacterium. *Int. J. Syst. Evol. Microbiol.* **54,** 77–83.

Davidova, I. A., Kropp, K. G., Duncan, K. E., and Suflita, J. M. (2002). Anaerobic biodegradation of *n*-alkanes by sulfate-reducing bacterial cultures. *In* "International Symposium on Subsurface Microbiology." Copenhagen, Denmark.

Dolfing, J., Zeyer, J., Binder-Eicher, P., and Schwarzenbach, R. P. (1990). Isolation and characterization of a bacterium that mineralizes toluene in the absence of molecular oxygen. *Arch. Microbiol.* **154,** 336–341.

Duncan, K., Jennings, E., Buck, P., Wells, H., Kolhatkar, R., Sublette, K., Potter, W., and Todd, T. (2003). Multi-species ecotoxicity assessment of petroleum-contaminated soil. *Soil Sediment Contam. Int. J.* **12,** 181–206.

Edwards, E. A., Wills, L. E., Reinhard, M., and Grbic-Galic, D. (1992). Anaerobic degradation of toluene and xylene by aquifer microorganisms under sulfate-reducing conditions. *Appl. Environ. Microbiol.* **58,** 794–800.

Ehrenreich, P., Behrends, A., Harder, J., and Widdel, F. (2000). Anaerobic oxidation of alkanes by newly isolated denitrifying bacteria. *Arch. Microbiol.* **173,** 58–64.

Fries, M. R., Zhou, J., Chee-Sanford, J., and Tiedje, J. M. (1994). Isolation, characterization, and distribution of denitrifying toluene degraders from a variety of habitats. *Appl. Environ. Microbiol.* **60,** 2802–2810.

Galushko, A., Minz, D., Schink, B., and Widdel, F. (1999). Anaerobic degradation of naphthalene by a pure culture of a novel type of marine sulphate-reducing bacterium. *Environ. Microbiol.* **1,** 415–420.

Gieg, L. M., and Suflita, J. M. (2002). Detection of anaerobic metabolites of saturated and aromatic hydrocarbons in petroleum-contaminated aquifers. *Environ. Sci. Technol.* **36,** 3755–3762.

Grbic-Galic, D., and Vogel, T. (1987). Transformation of toluene and benzene by mixed methanogenic cultures. *Appl. Environ. Microbiol.* **53,** 254–260.

Harms, G., Zengler, K., Rabus, R., Aeckersberg, F., Minz, D., Rossello-Mora, R., and Widdel, F. (1999). Anaerobic oxidation of *o*-xylene, *m*-xylene, and homologous alkylbenzenes by new types of sulfate-reducing bacteria. *Appl. Environ. Microbiol.* **65,** 999–1004.

Heider, J., Spormann, A. M., Beller, H. R., and Widdel, F. (1999). Anaerobic bacterial metabolism of hydrocarbons. *FEMS Microbiol. Rev.* **22,** 459–473.

Hermann, M., Noll, K. M., and Wolfe, R. S. (1986). Improved agar bottle plate for isolation of methanogens or other anaerobes in a defined gas atmosphere. *Appl. Environ. Microbiol.* **51,** 1124–1126.

Hess, A., Zarda, B., Hahn, D., Häner, A., Stax, D., Höhener, P., and Zeyer, J. (1997). *In situ* analysis of denitrifying toluene- and *m*-xylene degrading bacteria in a diesel fuel-contaminated laboratory aquifer column. *Appl. Environ. Microbiol.* **65,** 2136–2141.

Hungate, R. E. (1969). A roll tube method for cultivation of strict anaerobes. *In* "Methods in Microbiology" (J. R. Norris and D. W. Robbins, eds.), Vol. 3B, pp. 117–132. Academic Press, New York.

Kniemeyer, O., Fischer, T., Wilkes, H., Glöckner, F. O., and Widdel, F. (2003). Anaerobic degradation of ethylbenzene by a new type of marine sulfate-reducing bacterium. *Appl. Environ. Microbiol.* **69,** 760–768.

Kropp, K. G., Davidova, I. A., and Suflita, J. M. (2000). Anaerobic oxidation of *n*-dodecane by an addition reaction in a sulfate-reducing bacterial enrichment culture. *Appl. Environ. Microbiol.* **66,** 5393–5398.

Krumholz, L. R., Caldwell, M. E., and Suflita, J. M. (1996). Biodegradation of "BTEX" hydrocarbons under anaerobic conditions. *In* "Bioremediation: Principles and Applications" (R. L. Crawford and D. L. Crawford, eds.), pp. 61–99. Cambridge University Press, Cambridge.

Lovley, D. R., and Lonergan, D. J. (1990). Anaerobic oxidation of toluene, phenol and p-cresol by the dissimilatory iron reducing organism GS-15. *Appl. Environ. Microbiol.* **56,** 1858–1864.

Muyzer, G., De Waal, E. C., and Uitterlinden, A. G. (1993). Profiling of microbial populations by denaturing gradient gel electrophoresis analysis of polymerase chain reaction-amplified genes coding for 16S rRNA. *Appl. Environ. Microbiol.* **59,** 695–700.

Rabus, R., Nordhaus, R., Ludwig, W., and Widdel, F. (1993). Complete oxidation of toluene under strictly anaerobic conditions by a new sulfate-reducing bacterium. *Appl. Environ. Microbiol.* **59,** 1444–1451.

Rabus, R., and Widdel, F. (1995). Anaerobic degradation of ethylbenzene and other aromatic hydrocarbons by new denitrifying bacteria. *Arch. Microbiol.* **163,** 96–103.

Rabus, R., Wilkes, H., Schramm, A., Harms, G., Behrends, A., Amann, R., and Widdel, F. (1999). Anaerobic utilization of alkylbenzenes and *n*-alkanes from crude oil in an enrichment culture of denitrifying bacteria affiliating with the ß-subclass of *Proteobacteria.* *Environ. Microbiol.* **1,** 145–157.

Rios-Hernandez, L. A., Gieg, L. M., and Suflita, J. M. (2003). Biodegradation of an alicyclic hydrocarbon by a sulfate-reducing enrichment from a gas condensate-contaminated aquifer. *Appl. Environ. Microbiol.* **69,** 434–443.

Rockne, K. J., Chee-Sanford, J. C., Sanford, R. A., Hedlund, B. P., Staley, J. T., and Strand, S. E. (2000). Anaerobic naphthalene degradation by microbial pure cultures under nitrate-reducing conditions. *Appl. Environ. Microbiol.* **66,** 1595–1601.

Rueter, P., Rabus, R., Wilkes, H., Aeckersberg, F., Rainey, F. A., Jannasch, H. W., and Widdel, F. (1994). Anaerobic oxidation of hydrocarbons in crude oil by new types of sulphate-reducing bacteria. *Nature* **372,** 455–458.

Shinoda, Y., Sakai, Y., Uenishi, H., Uchihashi, Y., Hiraishi, A., Yukawa, H., Yurimoto, H., and Kato, N. (2004). Aerobic and anaerobic toluene degradation by a newly isolated denitrifying bacterium *Thauera* sp. strain DNT-1. *Appl Environ. Microbiol.* **70,** 1385–1392.

Smith, R. L. (1997). Determining the terminal electron-accepting reactions in the saturated subsurface. *In* "Manual of Environmental Microbiology" (C. J. Hurst, G. R. Knudsen, M. J. McInerney, L. D. Stetzenbach, and M. V. Walter, eds.), pp. 577–585. ASM Press, Washington, DC.

So, C. M., and Young, L. Y. (1999). Isolation and characterization of a sulfate-reducing bacterium that anaerobically degrades alkanes. *Appl. Environ. Microbiol.* **65,** 2969–2976.

Song, B., Haggblom, M. M., Zhou, J., Tiedje, J. M., and Palleroni, N. J. (1999). Taxonomic characterization of denitrifying bacteria that degrade aromatic compounds and description of *Azoarcus toluvorans* sp. nov. and *Azoarcus toluclasticus* sp. nov. *Int. J. Syst. Bacteriol.* **49,** 1129–1140.

Spormann, A. M., and Widdel, F. (2000). Metabolism of alkylbenzenes, alkanes, and other hydrocarbons in anaerobic bacteria. *Biodegradation* **11,** 85–105.

Suflita, J. M., Davidova, I. A., Gieg, L. M., Nanny, M., and Prince, R. C. (2004). Anaerobic hydrocarbon biodegradation and the prospects for microbial enhanced energy production. *In* "Petroleum Biotechnology: Developments and Perspectives" (R. Vazquez-Duhalt and R. Quintero-Ramirez, eds.), pp. 283–306. Elsevier Science.

Tanner, R. S. (1997). Cultivation of bacteria and fungi. *In* "Manual of Environmental Microbiology" (C. J. Hurst, G. R. Knudsen, M. J. McInerney, L. D. Stetzenbach, and M. V. Walter, eds.), pp. 52–60. ASM Press, Washington, DC.

Townsend, G. T., Prince, R. C., and Suflita, J. M. (2003). Anaerobic oxidation of crude oil hydrocarbons by the resident microorganisms of a contaminated anoxic aquifer. *Environ. Sci. Technol.* **37,** 5213–5218.

Townsend, G. T., Prince, R. C., and Suflita, J. M. (2004). Anaerobic biodegradation of alicyclic constituents of gasoline and natural gas condensate by bacteria from an anoxic aquifer. *FEMS Microbiol. Ecol.* **49,** 129–135.

Ulrich, AC., and Edwards, E. A. (2003). Physiological and molecular characterization of anaerobic benzene-degrading mixed cultures. *Environ. Microbiol.* **5,** 92–102.

Weiner, J., and Lovley, D. R. (1998). Anaerobic benzene degradation in petroleum contaminated aquifer sediments after inoculation with a benzene-oxidizing enrichment. *Appl. Environ. Microbiol.* **64,** 775–778.

Widdel, F., and Bak, F. (1992). Gram-negative mesophilic sulfate-reducing bacteria. *In* "The Prokaryotes" (A. Balows, H. G. Trüper, M. Dworkin, W. Harder, and K. H. Schleifer, eds.), 2nd Ed., pp. 3352–3378. Springer-Verlag, New York.

Widdel, F., and Rabus, R. (2001). Anaerobic biodegradation of saturated and aromatic hydrocarbons. *Curr. Opin. Biotechnol.* **12,** 259–276.

Zengler, K., Richnow, H. H., Roselló-Mora, R., Michaelis, W., and Widdel, F. (1999). Methane formation from long-chain alkanes by anaerobic microorganisms. *Nature* **401,** 266–269.

Zhang, X., and Young, L. Y. (1997). Carboxylation as an initial reaction in the anaerobic metabolism of naphthalene and phenanthrene by sulfidogenic consortia. *Appl. Environ. Microbiol.* **63,** 4759–4764.

Zhou, J.-Z., Fries, M. R., Chee-Sanford, J. C., and Tiedje, J. M. (1995). Phylogenetic analyses of a new group of denitrifiers capable of anaerobic growth of toluene and description of *Azoarcus tolulyticus* sp.nov. *Int. J. Syst. Bacteriol.* **45,** 500–506.

[3] Anammox Organisms: Enrichment, Cultivation, and Environmental Analysis

By Mike Jetten, Markus Schmid, Katinka van de Pas-Schoonen, Jaap Sinninghe Damsté, and Marc Strous

Abstract

Anaerobic ammonium oxidation (anammox) is the microbial oxidation of ammonium with nitrite to dinitrogen gas under strict anoxic conditions mediated by planctomycete-like bacteria. Anammox is not only important in the oceanic nitrogen cycle, but can also contribute substantially to nitrogen removal in municipal and industrial wastewater treatment. This chapter addresses the enrichment and cultivation of anammox bacteria in a sequencing batch reactor (SBR) and a gas lift reactor. The reactors can be operated anoxically as an anammox reactor or as an oxygen-limited "completely autotrophic nitrogen removal over nitrite" (CANON) system. Pure cultures of anammox organisms have not yet been obtained, but anammox cells can be purified to more than 99.5% using a Percoll density gradient centrifugation protocol. Furthermore, we show how anammox communities in natural and man-made ecosystems can be identified and characterized using molecular methods such as fluorescence *in situ* hybridization (FISH) or *Planctomycetes*/anammox-specific polymerase chain

METHODS IN ENZYMOLOGY, VOL. 397
0076-6879/05 $35.00
DOI: 10.1016/S0076-6879(05)97003-1

reaction (PCR). Both techniques are based on retrieved 16S rRNA gene sequences. In addition to 16S rRNA, unique anammox ladderane lipids can also serve as biomarkers to determine the abundance of anammox organisms in environmental samples.

Introduction

The notion that ammonium could be oxidized under anoxic conditions came from calculations based on the Redfield ratio in marine ecosystems (Richards, 1965) and from theoretical considerations based on thermodynamic calculations (Broda, 1977). The anammox reaction is the oxidation of ammonium under anoxic conditions with nitrite as the electron acceptor and hydrazine as the intermediate (Jetten *et al.*, 1998). The responsible organism for this reaction was identified as a member of *Planctomycetes*, named *Candidatus* "Brocadia anammoxidans" (Strous *et al.*, 1999). Enrichment cultures containing about 80% of these anammox organisms were cultivated for physiological studies in sequencing batch reactors (Strous *et al.*, 1998). Due to the long generation time of 11 days and the cell density of anammox bacteria, it was so far not possible to isolate the anammox organisms in pure culture. However, anammox cells can be enriched to about 99.5% or more by Percoll density centrifugation (Strous *et al.*, 1999).

At the time of discovery, little was known about the habitat and abundance of planctomycetes. However, initial 16S rRNA gene libraries using planctomycete-specific PCR primers of European treatment plants with high nitrogen input showed not only the high abundance (up to 50% of the biomass) of anammox cells, but also indicated a high genus-level diversity. The dominant species in these plants, *Candidatus* "Kuenenia stuttgartiensis," had less than 90% similarity to *Candidatus* "Brocadia anammoxidans" on the 16S rRNA gene level (Schmid *et al.*, 2000). Additionally, this observation showed that the anammox reaction was not the idiosyncrasy of a single bacterial strain. Members of the new anammox genus *Candidatus* "Scalindua" have been discovered in the Black Sea (Kuypers *et al.*, 2003), in a Danish fjord (Risgaard-Petersen *et al.*, 2004), and in a wastewater treatment plant in England (Schmid *et al.*, 2003). Remarkably, the marine species *Candidatus* "Scalindua sorokinii" could be responsible for as much as 50% of the nitrogen conversion in the world's oceans (Jetten *et al.*, 2003; Kuypers *et al.*, 2003; Ward, 2003). Furthermore, nutrient profiles and ^{15}N tracer studies in suboxic marine and estuarine environments underline the important role of the anammox reaction in the marine nitrogen cycle (Dalsgaard *et al.*, 2003; Thamdrup and Dalsgaard, 2002; Trimmer *et al.*, 2003). Hence, anammox organisms are, at the moment, separated into three branches. While the two strict freshwater species, Candidatus *"Brocadia anammoxidans"* and Candidatus *"Kuenenia stuttgartiensis,"* play an important role in the sustainable removal of nitrogen from wastewater, *Candidatus* "Scalindua species" seem to appear both in marine environments and in freshwater systems (Jetten *et al.*, 2003; Strous and Jetten, 2004). Because the genus *Candidatus* "Scalindua" is also the deepest branch within anammox organisms, there is reason to speculate that the original anammox habitats are the anoxic and microaerophilic zones of the oceans.

Another interesting aspect of anammox organisms is the presence of a prokaryotic organelle, the anammoxosome, in their cytoplasma (Lindsay *et al.*, 2001). This compartment is surrounded by a membrane, that contains the unique anammox ladderane lipids. The cyclobutane/cyclohexane ring system of these lipids makes the anammoxosome membrane much more impermeable (Sinninghe Damsté *et al.*, 2002; van Niftrik *et al.*, 2004). Analysis of ladderane lipids in environmental samples can serve as a good biomarker for the presence of anammox organisms (Kuypers *et al.*, 2003).

This chapter describes how anammox organisms can be cultivated and enriched in anoxic or oxygen-limited reactor systems. Furthermore, we provide a purification protocol for anammox cells using Percoll density gradient centrifugation. We also discuss how the anammox reaction and organisms can be detected and identified in environmental samples by rRNA-based detection methods and analysis of ladderane lipids.

Enrichment and Cultivation of Anammox Organisms

Reactor Systems for Enrichment and Cultivation of Anammox Organisms

The Sequencing Batch Reactor as Standard for the Cultivation of Anammox Organisms. The standard method for the cultivation and study of anammox organisms is the sequencing batch reactor. This reactor system has very efficient biomass retention and thus provides favorable conditions to slowly growing organisms. Furthermore, a homogeneous distribution of substrates and biomass over the reactor is possible under stable substrate (nitrite) limited conditions (Strous *et al.*, 1998). An anammox SBR is usually set up in a 2- to 15-liter vessel equipped with a water jacket, which is connected to a water bath for keeping the temperature at 30–37°. The vessel is stirred at 80 ± 10 rpm (six-bladed turbine stirrer, diameter one-third of the vessel diameter). Anoxic conditions are maintained by flushing the reactor with about 6 ml/min Ar/CO_2 (95/5%). The oxidation/reduction potential can be monitored by using a combined redox electrode. The CO_2 present in the gas is sufficient to buffer the solution to a pH between 7.0

and 8.0. Fresh anoxic mineral medium is applied to the SBR continuously over a period of 11.5 h. After the filling period, the stirrer and the influent pump are stopped and the aggregates are allowed to settle for 15–20 min. In the remaining 10–15 min of the total cycle, about half the reactor volume is removed by an effluent pump. To prevent entry of air during this period, a 10-liter gas buffer filled with Ar/CO_2 should be provided. The mineral medium consists of (in g liter^{-1}) KHCO3, 1.25 g; NaH$_2$PO4, 0.05 g; CaCl$_2$.2H$_2$O, 0.3 g; MgSO$_4$.7H$_2$O, 0.2 g; FeSO4, 0.00625 g; EDTA, 0.00625 g; and trace element solution, 1.25 ml liter^{-1} (Van de Graaf et al., 1996). The trace element solution contains (g liter^{-1}) EDTA, 15 g; ZnSO$_4$.7H$_2$O, 0.43 g; CoCl$_2$.6H$_2$O, 0.24 g; MnCl$_2$. 4H$_2$O, 0.99, CuSO$_4$.H$_2$O, 0.25 g; Na-MoO$_4$.2H$_2$O, 0.22 g; NiCl$_2$.6H$_2$O, 0.19 g; NaSeO$_4$. 10H$_2$O, 0.21 g; H$_3$BO$_4$, 0.014 g; and NaWO$_4$.2H$_2$O, 0.050 g. The components of the medium are autoclaved separately. The medium concentrations of NaNO$_2$, (NH$_4$)$_2$SO$_4$, and NaNO$_3$ are initially set to 5, 5, and 10 mM each. Suitable start material for the SBR is activated sludge from a treatment plant with a long sludge age (Dapena et al., 2004; Pynaert et al., 2003; van Dongen et al., 2001). Typically, the biomass concentration should stay above 1 g dry weight per liter to prevent loss of activity during enrichment (Dapena-Mora et al., 2004). The ammonia and nitrite concentrations are increased gradually (to a maximum of 30 mM each) as long as the nitrite is completely consumed. Figure 1 depicts such an enrichment of anammox bacteria (Van Dongen et al., 2001). The nitrogen load to the reactor can be increased further by applying more cycles per day. The upper limits of nitrogen loading of anammox reactors are about 9 kg N m^{-3} reactor day^{-1} (Sliekers et al., 2003). Prolonged exposure to nitrite concentrations of 10 mM or higher will lead to severe inhibition of the biomass and failure to enrich the responsible bacteria (Strous et al., 1999). Due to the long generation time of

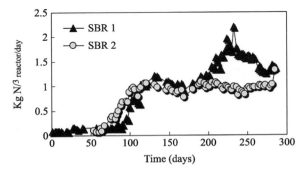

FIG. 1. Enrichment of anammox biomass from activated sludge using 2-liter SBR systems inocculated with activated sludge. The ammonium and nitrite loads are increased gradually. The percentage of anammox cell was monitored with FISH.

anammox bacteria, enrichments usually take 200 days, although longer times up to 600 days have been reported (Pynaert *et al.,* 2003; Toh and Ashbolt, 2003). Analysis of the biomass with molecular probes specific for anammox bacteria will help document the progressive enrichment of the responsible bacteria (see later).

Cocultivation of Aerobic Ammonia Oxidizers and Anammox Organisms in a CANON System. In order to remove ammonia from the supplied medium, anammox bacteria must be provided with sufficient nitrite. The nitrite can be added from a chemical stock, but it may also be produced from ammonia by aerobic autotrophic ammonia-oxidizing bacteria (AOB) in one oxygen-limited reactor system. Thus, anammox organisms and AOB can coexist in one reactor, provided that the system is kept oxygen limited. The process is called CANON (Schmidt *et al.,* 2002a,b; Sliekers *et al.,* 2002, 2003; Third *et al.,* 2001). The CANON process can best be operated in a gas lift reactor with high oxygen mass transfer rates. In the gas lift reactor, gas is supplied from the bottom of the reactor at a maximum gas flow of 200 ml min^{-1} for fluidization of the biomass. The same mineral medium but without nitrite as for the anammox SBR is also used in the CANON. An anammox enrichment culture can serve as anammox biomass source to reduce start-up time. In the first period, the reactor is kept anoxic in order to adapt the anammox bacteria. In this initial phase, Ar/CO_2 (95/5%) and $(NH_4)_2SO_4$ (6.4 g $liter^{-1}$), as well as $NaNO_2$ (6.75 g $liter^{-1}$) containing medium, are supplied to the reactor. Only in this period is the effluent of the reactor returned to the system. After the initial period, the medium is switched to contain $(NH_4)_2SO_4$ (7.3 g $liter^{-1}$) only. Nitrifying activated sludge can serve as a source for AOB to shorten start-up of the oxygen-limited mode. The Ar/CO_2 is mixed carefully and gradually with air to support the activity and growth of AOB. The oxygen concentration can be measured using a Clark-type electrode and is controlled manually. When only Ar/CO_2 (95/5%) is used, the pH does not need to be controlled. However, when air is supplied, the pH needs to be controlled and adjusted with 0.5 *M* Na_2CO_3 and 1 *M* HCl to about pH 7.5. The nitrogen loading of CANON in gas lift reactors is about 1.5 kg N removal m^{-3} reactor per day (Sliekers *et al.,* 2003).

Physical Separation of Anammox Organisms from Other Members of an Anammox Enrichment Culture by Density Gradient Centrifugation

Attempts to isolate anammox bacteria in pure culture have not yet been successful. Due to the long generation time of 11 days and the cell density-dependent activity, anammox organisms are most likely incompatible with any currently existing isolation protocols. Therefore, anammox cells were purified and separated from other organisms present in the community using a Percoll gradient centifugation (Strous *et al.*, 1999, 2002). About 100 ml of anammox aggregates from an SBR is washed in a 75 m*M* 4-(2-hydroxyethyl)-1-piperazineethanesulfonic acid (HEPES) buffer containing 5 m*M* bicarbonate with a final pH of 7.8. Subsequently, the anammox cells are concentrated in 40 ml of this buffer and applied in 5-ml portions to mild sonication (30 s, 150 W, tip diameter 9.5 nm, 20–33°). This results in a mixture of some remaining large aggregates and single cells. After centrifugation (5°, 5 min, 10,000*g*), the larger aggregates form a light red pellet on top of the dark red pellet of the single cells. The aggregates are removed, the single cells are resuspended in 3.1 ml HEPES/bicarbonate buffer, and 6.9 ml of Percoll is added. This mixture is centrifuged for 60 min at 10,000*g* at 5°. Anammox cells form a red band at the lower part of the gradient (density 1.107–1.138 g liter^{-1}). After extraction of the band, the Percoll is removed by washing with HEPES/bicarbonate buffer. The resulting cell suspension contains usually only one contaminating organism for every 200–800 target organisms (Fig. 2). For *Candidatus* "Kuenenia stuttgartiensis," a different method is required because this bacterium is very sensitive to sonication and single, intact cells are not released from biofilm aggregates in significant quantities. An alternative protocol is currently developed.

Analysis of Environments Harboring Anammox Organisms

Characterization of Anammox Communities by rRNA-Based Methods

Polymerase Chain Reaction, Cloning, and Phylogenetic Analysis of Anammox Organisms. The initial difficulty of the detection of anammox bacteria with PCR in highly enriched samples was that anammox 16S rRNA genes did not amplify very well with general bacterial primers (Schmid *et al.*, 2000). One explanation could be a reduced recovery of anammox DNA by the applied DNA extraction method as it was observed earlier for aerobic ammonia oxidizing bacteria (Juretschko *et al.*, 1998). Additionally, PCR amplification might also introduce biases in the anammox template to product ratio. Thus, a planctomycete-specific PCR approach involving the phylum-specific probe Pla46 (*Escherichia coli* positions 46–63; Table II; Neef *et al.*, 1998) was applied. Pla46 was used as an unlabeled forward primer (Table I) in combination with the universal reverse primers 1390R (*E. coli* positions 1390–1407; Table I; Zheng *et al.*, 1996, Schmid *et al.*, 2000) or 630R (*E. coli* positions 1529–1545; Juretschko *et al.*, 1998; Schmid *et al.*, 2003). For an even more specific amplification

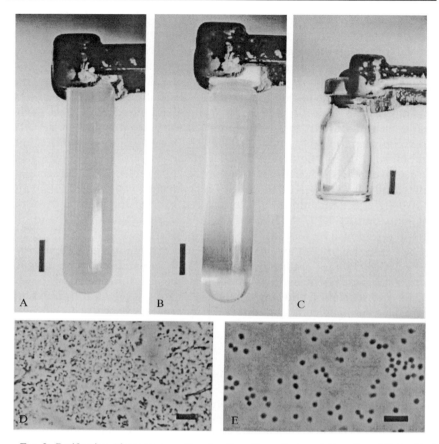

Fɪɢ. 2. Purification of anammox cells using Percoll gradient centrifugation. (A) Sonified cells (3.1 ml) are mixed with 6.9 ml Percoll. After 1 h centrifugation, a red band of anammox cells appears at the bottom of the centrifuge tube (B). For the activity test, this band is excised and washed with HEPES buffer, and cells are concentrated to the desired volume and incubated anoxically (C). A phase-contrast picture of starting material is shown in (D) and the resulting purified preparation is depicted in (E).

of anammox organisms, some probes originally constructed for a specific detection of anammox organisms by FISH (Table II) can be used as forward or reverse primers, respectively (see Table I for specific combinations).

Planctomycete and anammox-specific PCRs are usually performed in thin-walled PCR tubes in a total volume of 50 μl of the reaction mixture containing 2 mM MgCl$_2$, 10 nmol of each deoxynucleoside triphosphate, 15

TABLE I

PRIMERS USED FOR PCR AMPLIFICATION AND SEQUENCING OF 16S rRNA AND 23S rRNA GENES OF ANAMMOX ORGANISMS WITH PRIMER SEQUENCES, SPECIFICITY, AND ANNEALING TEMPERATURE

Primer name[a]	Specificity	Sequence 5'-3'	Target site[b]	Optimal annealing temperature
Pla46F (Neef et al., 1998; Schmid et al., 2000)	Planctomycetes	GGATTAGGCA-TGCAAGTC	16S rRNA gene, 46–63	58°
1274F (Ludwig et al., 1992; Schmid et al., 2001)	All Bacteria	GCGTRCCTTTT-GTAKAATG	23S rRNA gene, 559–577	45°c
AMX 820R (Egli et al., 2001; Schmid et al., 2000)	Genera Candidatus "Brocadia" and Candidatus "Kuenenia"	AAAACCCCTC-TACTTAGTGCCC	16S rRNA gene, 820–841	56°
1390R (Schmid et al., 2000; Zheng et al., 1996)	Bacteria, Archaea, and Eukarya	GACGGGCGG-TGTGTACAA	16S rRNA gene, 1390–1407	58°
630R (Juretschko et al., 1998; Schmid et al., 2001, 2003)	All Bacteria	CAKAAAGG-AGGTGATCC	16S rRNA gene, 1529–1545	58°

(continued)

TABLE I (*continued*)

Primer name[a]	Specificity	Sequence 5′-3′	Target site[b]	Optimal annealing temperature
1035R (Ludwig et al., 1992; Schmid et al., 2001)	All *Bacteria*	TTCGCTCG-CCRCTAC	23S rRNA gene, 242–256	45°[c]
1020R (Ludwig et al., 1992; Schmid et al., 2001)	All *Bacteria*	TCTGGGY-TGTTYCCCT	23S rRNA gene, 975–990	45°[c]
1037R (Ludwig et al., 1992; Schmid et al., 2001)	All *Bacteria*	CGACAAGG-AATTTCGCTAC	23S rRNA gene, 1930–1948	58°

[a] *E. coli* numbering.
[b] These primers were used exclusively as sequencing primers.
[c] In addition to the primer name, references are given. The first reference in each row is the original publication of the primer. All other references are publications in which these primers were used for the amplification of anammox organisms.

TABLE II

OLIGONUCLEOTIDE PROBES USED FOR THE DETECTION OF ANAMMOX ORGANISMS WITH OPD DESIGNATION, PROBE SEQUENCES, SPECIFICITY, AND FORMAMIDE CONCENTRATIONS IN THE HYBRIDIZATION BUFFER AS WELL AS NaCl CONCENTRATIONS IN WASHING BUFFER REQUIRED FOR SPECIFIC *IN SITU* HYBRIDIZATION

OPD[a] designation	Specificity	Sequence 5'-3'	% Formamide/ mM [NaCl][b]
S-P-Planc-0046-a-A-18 (Pla46; Neef et al., 1998)	*Planctomycetales*	GACTTGCATGCCTAATCC	25/159
S-P-Planc-0886-a-A-19 (Pla 886 ; Neef et al., 1998)	*Isosphaera, Gemmata, Pirellula, Plantomyces*	GCCTTGCGACCATACTCCC	30/112[b]
S-D-Bact-0338-b-A-18 (Eub338II; Daims et al. 1999)	Bacterial lineages not covered by probe EUB338 and EUB338II	GCAGCCACCCGTAGGTGT	0/900
S-*-Amx-0368-a-A-18 (Schmid et al., 2003)	All anammox organisms	CCTTTCGGGCATTGCGAA	15/338
L-*-Amx-1900-a-A-21 (Schmid et al., 2001)	Genera Candidatus "Brocadia" and Candidatus "Kuenenia"	CATCTCCGGCTTGAACAA	30/112
S-*-Amx-0820-a-A-22 (AMX 820, Schmid et al., 2000)	Genera *Candidatus* "Brocadia" and *Candidatus* "Kuenenia"	AAAACCCCTCTACTTAGTGCCC	40/56
S-G-Sca-1309-a-A-21 (Schmid et al., 2003)	Genus *Candidatus* "Scalindua"	TGGAGGCGAATTTCAGCCTCC	5/675
S-*-Scabr-1114-a-A-22 (Schmid et al., 2003)	*Candidatus* "Scalindua brodae"	CCCGCTGGTAACTAAAAACAAG	20/225
S-*-BS-820-a-A-22 (Kuypers et al., 2003)	*Candidatus* "Scalindua wagneri" *Candidatus* "Scalindua sorokinii"	TAATTCCCTCTACTTAGTGCCC	40/56[c]
S-S-Kst-0157-a-A-18 (Schmid et al., 2001)	*Candidatus* "Kuenenia stuttgartiensis"	GTTCCGATTGCTCGAAAC	25/159

(continued)

TABLE II (continued)

OPD[a] designation	Specificity	Sequence 5'-3'	% Formamide/ mM [NaCl][b]
S-*-Kst-1275-a-A-20 (Schmid et al., 2000)	Candidatus "Kuenenia stuttgartiensis"	TCGGCTTTATAGGTTTCGCA	25/159
S-S-Ban-0162(B.anam.)-a-A-18 (Schmid et al., 2000)	Candidatus "Brocadia anammoxidans"	CGGTAGCCCCAATTGCTT	40/56
S-*-Amx-0156-a-A-18 (Schmid et al., 2000)	Candidatus "Brocadia anammoxidans"	CGGTAGCCCCAATTGCTT	40/56
S-*-Amx-0223-a-A-18 (Schmid et al., 2000)	Candidatus. "Brocadia anammoxidans"	GACATTGACCCCTCTG	40/56
S-*-Amx-0432-a-A-18 (Schmid et al., 2000)	Candidatus. "Brocadia anammoxidans"	CTTAACTCCGACAGTGG	40/56
S-*-Amx-0613-a-A-22 (Schmid et al., 2000)	Candidatus. "Brocadia anammoxidans"	CCGCCATTCTTCCGTTAAGCGG	40/56
S-*-Amx-0997-a-A-21 (Schmid et al., 2000)	Candidatus "Brocadia anammoxidans"	TTTCAGGTTTCTACTTCTACC	20/225
S-*-Amx-1015-a-A-18 (Schmid et al., 2000)	Candidatus. "Brocadia anammoxidans"	GATACCGTTCGTCGCCCT	60/14
S-*-Amx-1154-a-A-18 (Schmid et al., 2000)	Candidatus. "Brocadia anammoxidans"	TCTTGACGACAGCAGTCT	20/225
S-*-Amx-1240-a-A-23 (Schmid et al., 2000)	Candidatus. "Brocadia anammoxidans"	TTTAGCATCCCTTTGTACCAACC	60/14

[a] Oligonucleotide probe database designations according to Alm et al. (1996); trivial names of probes and original reference are given in parentheses.
[b] Percentage formamide in the hybridization buffer and mM NaCl in the washing buffer, respectively, required for specific in situ hybridization.
[c] Originally published with 20% formamide, but can be used with up to 40% formamide and should be used as a competitor for S-*-Amx-0820-a-A-22.

to 20 pmol of each primer, 100 ng of template DNA, and 1.5 U of *Taq* DNA polymerase. An optimal performance of the PCR can be achieved by using *Taq* DNA polymerase provided by Promega (Madison) in combination with $10\times$ PCR buffer containing $MgCl_2$ and $10\times$ bovine serum albumin provided by Idaho Technologies (Idaho Falls).

The usually used PCR program for the amplification of planctomycete and anammox 16S rRNA is started with an initial denaturation (10 min at 94°), followed by 30 cycles of denaturation at 94° for 45 s, annealing at different temperatures (Table I) for 50 s, and elongation at 72° for 3 min. Cycling is completed by a final elongation step at 72° for 10 min.

Amplification of the nearly complete 16S–23S rRNA operon does not require a change of the reaction mixture or the program. However, for a higher yield of amplificate, it may be favorable to increase the cycle denaturation time and the annealing time to 90 s. Primers and their annealing temperatures, which were used in amplification of 16S–23S rRNA operon, are listed in Table I.

Negative controls (no DNA added) and positive controls should be included in all sets of amplifications. The presence and size of amplification products are determined by agarose (0.8% for large and 1.7% for small DNA fragments) gel electrophoresis of 5- to 10-μl aliquots of the PCR products.

The planctomycete or anammox-specific primer pairs can be used in various PCR- based applications, such as denaturing gradient gel electrophoresis (DGGE), terminal restriction fragment length polymorphism (TRFLP), or quantitative PCR. However, it is strongly recommended to test the obtained PCR products by either direct sequencing or dot blot analysis with 16S rRNA-specific probes (see Table II) to confirm the phylogenetic affiliation of the obtained sequences.

The PCR products are usually cloned with the TOPO TA cloning kit provided by Invitrogen (Carlsbad, CA). This proves to yield the highest amount of transformed *E. coli* cells.

For phylogenetic analyses, many computer programs and internet services are available, but the ARB program package developed at Technical University Munich, Germany, is our preferred choice (www.arb-home.de; Ludwig *et al.*, 2004). Its sequence database currently encompasses about 28,000 16S rRNA and about 2500 23S rRNA entries. Sequences in these data sets are aligned according to the primary and secondary structures of rRNA. In the latest version of the 16S rRNA data set, the extensions of helix 9, which are unique structural features for *Candidatus* "Kuenenia stuttgartiensis" and *Candidatus* "Brocadia anammoxidans," are considered in the secondary structure information (Schmid *et al.*, 2001). This facilitates a correct alignment of anammox sequences with the alignment tools.

New anammox rRNA gene sequences are added to the respective 16S or 23S rRNA sequence data set of the ARB program package. Then 16S and 23S rRNA gene sequences are aligned automatically. Subsequently, the alignments are corrected by visual inspection. Phylogenetic analyses of rRNA sequences are performed by applying neighbor joining, ARB parsimony, and maximum likelihood analyses using the respective tools of the ARB and PHYLIP (Phylogeny Inference Package, version 3.57c; J. Felsenstein, Department of Genetics, University of Washington, Seattle) program packages and the fastDNAml program (Maidak *et al.*, 1996). It is imperative to infer the phylogeny of organisms by more than one treeing method. Furthermore, a variety of conservation filters (using different degrees of conservation) should be applied, which allow removing highly variable sequence positions to different degrees. Highly variable sequence positions can influence positions of nodes and branch lengths (Brochier and Philippe, 2002; Ludwig *et al.*, 1998). Usually, for phylogenetic inference of anammox organisms, the three aforementioned treeing methods are used. They are applied with no filter and a 50% conservation filter for *Bacteria* and *Planctomycetes*. In some cases, 25 and 75% conservation filters were also used in these calculations to check the robustness of the tree calculations. All tree information is put together in one consensus tree, which reflects the closest interpretation of the phylogeny. An example of such a 16S rRNA-based tree is depicted in Fig. 3. Bootstrapping values for the nodes can be obtained using the PHYLIP parsimony or the

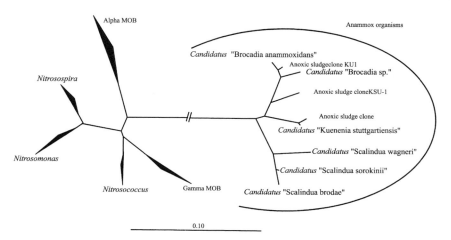

FIG. 3. Phylogenetic relationship of anammox planctomycetes based on retrieved 16S rRNA gene sequences.

neighbor-joining tool. New sequences affiliated to anammox organisms should also be checked for possible chimeric sequences. In ARB this can be done by adding sequences to an existing tree with filters for the beginning, middle, and end parts of the sequences. The phylogenetic position of the newly added sequence parts should be the same.

FISH for the Detection of Anammox Organisms in Environmental Samples. Fluorescence *in situ* hybridization is the standard technique used to detect anammox organisms in environmental samples (Kuypers *et al.*, 2003; Risgaard-Petersen *et al.*, 2004; Schmid *et al.*, 2000, 2003). However, it is often hampered by its detection limit of about 1000 cells per milliliter (Kuypers *et al.*, 2003). Furthermore, the high diversity of anammox organisms on the 16S rRNA level often masks the discovery of new anammox species (Schmid *et al.*, 2003). Thus, it is essential to reevaluate the anammox-specific probe set continuously and to construct new probes for new anammox species.

The ARB program package provides an easy-to-use tool to design probes targeting rRNA. The output of this so-called probe design tool is a list of possible probes for one organism or a group of organisms with their calculated GC content and their melting temperature. To evaluate the strength and position of mismatches, such a suggested probe can be compared with other sequences of the rRNA data set with the probe match tool of the ARB program package. The result is a differential alignment of target and nontarget organisms with information about the quality of the mismatches to nontarget organisms. Probes can also be checked manually in the ARB sequence editor. In some cases, probes have to be adjusted in length or their position has to be shifted on the alignment to increase their specificity or stringency.

Probes are usually purchased as Cy3, Cy5, and 5(6)-carboxyfluorescein-*N*-hydroxysuccinimide ester (FLUOS)-labeled derivatives. Sample fixations and hybridizations of anammox organisms are performed as described previously (Amann *et al.*, 1995). Simultaneous hybridizations with probes requiring different stringency are realized by a successive hybridization procedure (Wagner *et al.*, 1994). Optimal hybridization conditions for newly designed probes are determined by using the hybridization and wash buffers described by Manz *et al.* (1992). The *in situ* probe dissociation temperatures are evaluated by measuring the relative fluorescence intensity of bacteria after hybridization at different stringencies as described by Daims *et al.* (1999). For information about anammox-specific probes, refer to Table II or ProbeBase.net (Loy *et al.*, 2003).

A commonly occurring problem with FISH of anammox samples is the low amount of adherence of anammox cells on the microscopic slides. Poly-lysine or gelatine coating of the slides may be of limited use. An alternative to these procedures is the embedding of cells in agarose. After a fixed sample is dropped onto the slides and air dried as is usual in the FISH procedure, 0.1% agarose in water at 50° is applied to each well. After cooling of the agarose, the slide is dehydrated in ethanol (50, 80, and 100% for 3 min each). Subsequently, the slide is air dried until the agarose forms a thin layer over the sample. Then FISH is performed as described earlier. The results of such a procedure are depicted in Fig. 4.

DNA staining of cells is usually performed with 4,6-diamidino-2-phenylindole (DAPI) as described previously (Juretschko *et al.*, 1998).

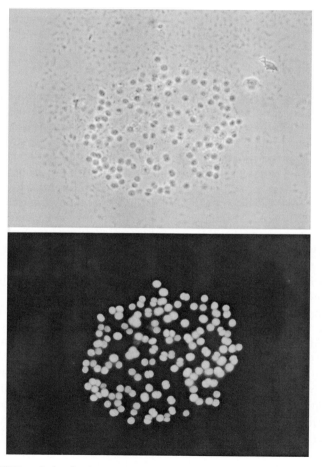

FIG. 4. FISH analysis of a freshwater sample from Ninja Jinya (Uganda). (Top) Phase contrast of anammox clusters. (Bottom) The same cluster stained with Fluos label S-*-Amx-0820-a-A-22 (AMX820).

However, anammox biofilm material sometimes showed a very weak DAPI-conferred fluorescence. Thus, SYBR Green I (FMC Bioproducts, Rockland) can be used as an alternative stain for DAPI. For preparation of a working solution, SYBR Green I is 10,000-fold diluted with double-distilled H_2O. Twenty microliters of the SYBR Green I working solution is applied to each well of the microscopic slide and incubated in the dark for 10 min at room temperature. Slides are washed briefly with ddH_2O, air dried, and embedded in embedding medium for epifluorescence microscopy. The disadvantage of SYBR Green I is that its excitation and emission spectrum are close to those of FLUOS. Therefore, it can only be used as a DNA counterstain in combination with Cy3- and Cy5-labeled oligonucleotide probes.

Image acquisitions of anammox organisms are performed with a Zeiss LSM 510 laser- scanning confocal microscope (Zeiss, Jena, Germany) or a Zeiss Axioplan II epifluorescence microscope together with the standard software packages delivered with the instruments. The CLSM should be equipped with a UV laser (351 and 364 nm), an Ar ion laser (458 and 488 nm), and two HeNe lasers (543 and 633 nm). Both CLSM and epifluorescence microscopes possess various excitation filters, as well as long-pass and band-pass filters for the emission wavelengths of the dyes used for FISH and DNA staining.

Special FISH Approaches to Link Abundance with the Physiology of Anammox Organisms.

FISH TARGETING THE INTERGENIC SPACER REGION BETWEEN 16S AND 23S RRNA OF ANAMMOX ORGANISMS. The classical FISH approach is a tool for qualitative and quantitative determination of a microbial population. Additionally, for many fast-growing organisms, the FISH intensity can be used to assess their activity. For these organisms, the FISH signal intensity is directly proportional to the concentration of ribosomes and precursor-rRNA molecules in the cells (Poulsen *et al.*, 1993). However, anammox organisms do not decrease their ribosome content significantly during periods of inhibition with oxygen (Schmid *et al.*, 2001). Hence, there is no correlation between cellular rRNA content and physiological activity of anammox organisms. For these rather slow-growing bacteria, the precursor rRNA concentrations might be linked more directly to the ribosome turnover rate in the cells (Cangelosi and Brabant, 1997). To assess the precursor rRNA concentration in a cell, the intergenic spacer region (ISR) between 16S rRNA and 23S rRNA is targeted by fluorescently labeled oligonucleotide probes (Schmid *et al.*, 2001).

The procedure for ISR FISH with anammox organisms does not differ significantly from the classical FISH approach (Amann *et al.*, 1995). However, some slight adjustments had to be made. Because the abundance of

ISR compared to ribosomes is significantly lower in active anammox cells, four probes targeting the ISR have to be applied together to get a sufficient signal. Tests with different target sites also showed that no probe binding occurred when the two tRNA-coding regions of the ISR of anammox organisms were chosen as probe targets. The formamide concentration in the hybridization buffer for all probe sets was set to 10% for hybridization with the ISR probes. The ISR probe sets should be always used together with a 16S rRNA targeting probe to verify that the ISR probe signal is originating from anammox cells. It is also necessary to include a negative control with inhibited anammox cells in the experiments. Signal intensities of the negative control and the original sample must be compared to assess if the cells are really active or if they just exhibit autofluorescence. Two sets of ISR probes are available so far for *Candidatus* "Brocadia anammoxidans" and *Candidatus* "Kuenenia stuttgartiensis," respectively (Table III).

TABLE III

OLIGONUCLEOTIDE PROBE SETS USED FOR THE DETECTION OF THE INTERGENIC SPACER OF *CANDIDATUS* "BROCADIA ANAMMOXIDANS" AND *CANDIDATUS* "KUENEIA STUTTGARTIENSIS," RESPECTIVELY, WITH SPECIFICITY, OPD DESIGNATION, AND PROBE SEQUENCES

Specificity	OPD designation	Sequence 5'-3'
Candidatus "Brocadia anammoxidans"	I-*-Ban-0071(B.anam.)-a-A-18 (Schmid *et al.*, 2001)	CCCTACCACAAACCTCGT
	I-*-Ban-0108(B.anam.)-a-A-18 (Schmid *et al.*, 2001)	TTTGGGCCCGCAATCTCA
	I-*-Ban-0222(B.anam.)-a-A-19 (Schmid *et al.*, 2001)	GCTTAGAATCTTCTGAGGG
	I-*-Ban-0389(B.anam.)-a-A-18 (Schmid *et al.*, 2001)	GGATCAAATTGCTACCCG
Candidatus "Kuenenia stuttgartiensis"	I-*-Kst-0031(K.stutt.)-a-A-18 (Schmid *et al.*, 2001)	ATAGAAGCCTTTTGCGCG
	I-*-Kst-0077(K.stutt.)-a-A-18 (Schmid *et al.*, 2001)	TTTGGGCCACACTCTGTT
	I-*-Kst-0193(K.stutt.)-a-A-19 (Schmid *et al.*, 2001)	CAGACCGGACGTATAAAAG
	I-*-Kst-0288(K.stutt.)-a-A-20 (Schmid *et al.*, 2001)	GCGCAAAGAAATCAAACTGG

However, the ISR-FISH approach has some limitations. ISR sequences lack evolutionary pressure, which causes a very high diversity of ISR sequences, maybe even within one species of anammox organisms. Consequently, for every newly discovered anammox organism or new environmental sample, the ISR sequences have to be determined and, if necessary, the ISR probe sets have to be updated.

COMBINATION OF FISH AND MICROAUTORADIOGRAPHY (FISH-MAR). The combination of FISH and microautoradiography is a FISH-based approach, which directly links the uptake of a radioactively labeled substrate with uncultured organisms in a complex environmental sample (Lee *et al.*, 1999). Because a wide variety of radioactively labeled substrates can be applied, a substrate uptake pattern of certain organisms can be determined. Because anammox organisms have not been grown in pure culture, the involvement of other nonanammox bacteria in CO_2 fixation could not be fully excluded. Consequently, FISH-MAR experiments were performed first to prove the chemolithoautotrophic life style of anammox bacteria (see later).

To ensure sufficient uptake of radioactively labeled substrate in all experiments, anammox cells have to be incubated in a small reactor system under nitrite limitation.

For batch incubations, the biomass from a sequencing batch reactor (containing 30–40% *Candidatus* "Kuenenia stuttgartiensis," 20–30% aerobic nitrifiers) (Third *et al.*, 2001) is diluted to a final concentration of 0.2 mg protein/ml in 2 ml mineral medium per incubation and incubated oxically with 9 mM ammonium and anoxically with 9 mM ammonium and 4.5 mM nitrite. Per incubation, 0.2 mM NaH^{14}CO$_3$ is added (2072 MBq/mmol). Consumption of ammonium and nitrite and production of nitrate are measured as described previously (Strous *et al.*, 1998). After 24 h, the biomass is harvested and fixed with a 4% paraformaldehyde solution.

For continuous incubations, the biomass (0.5 g protein/liter final concentration) from an anammox sequencing batch reactor (75–80% *Candidatus* "Brocadia anammoxidans") is incubated in a 50-ml continuous, down-flow fixed bed reactor (30°) fed with mineral medium (Strous *et al.*, 1998) containing 5 mM NH$_4^+$, 3 mM NO$_2^-$, 5 mM NO$_3^-$, and 0.75 mM KHCO$_3$ supplemented with 20 mM (final concentration) HEPES buffer, pH 7.8. Consumption of ammonium and nitrite and production of nitrate are measured as described previously (Strous *et al.*, 1998). After adaptation of the biomass to the reactor for 8 days, the medium is exchanged for the same medium with 0.75 mM NaH^{14}CO$_3$ (107 MBq/mmol). Additionally, NaH^{14}CO$_3$ is added directly to the reactor (final concentration 3.8 mM bicarbonate, 108 MBq/mmol). Labeled medium is supplied continuously for 24 h (190 ml), after which the biomass is harvested and fixed with a 4% paraformaldehyde solution.

After incubation, the biomass containing 30–40% *Candidatus* "Kuenenia stuttgartiensis" and 20–30% aerobic nitrifiers is cryosectioned. Samples from the down-flow fixed bed reactor are processed directly.

The autoradiographic procedure was performed according to the protocol published previously (Lee *et al.*, 1999). For developing the samples from the down-flow fixed bed reactor, Ilford Phenisol (Ilford Imaging, New Jersey) rather than Kodak D19 was used. No differences in the quality of the final results were observed. Hybridizations and image acquisitions are performed as described earlier.

Lipid Analysis

The presence of a membrane-surrounded organelle in a prokaryote, the anammoxosome, underlines the unique status of anammox organisms. Furthermore, this membrane contains unique ladderane lipids with cyclobutane/cyclohexane ring systems, which make the anammoxosome membrane very impermeable (Sinninghe Damsté *et al.*, 2002; Van Niftrik *et al.*, 2004). This ladderane lipid can also be used as an indicator for the presence of anammox organisms in environmental samples (Kuypers *et al.*, 2003; Van Niftrik *et al.*, 2004).

For the detection of ladderane lipids, anammox organisms containing biomass are ultrasonically extracted with methanol (MeOH), MeOH/dichloromethane (DCM), and DCM (Schmid *et al.*, 2003). The extracts are combined and the solvent is removed by a rotary evaporator. Subsequently, an aliquot of the extract is methylated with BF_3/methanol. This fraction is separated over a small silica column. Ethyl acetate is used as the eluent to remove very polar material. This fraction is silylated with BSTFA in pyridine at 60° for 30 min to convert alcohols in TMS ethers and is analyzed with gas chromatography and gas chromatography/mass spectrometry (GC/MS) (Sinninghe Damsté *et al.*, 2002). Compound-specific stable carbon isotope ratios are determined using a Thermofinnigan Delta plus XL isotope ratio- monitoring GC-IR-MS system. The gas chromatograph is equipped with a fused silica capillary column (25 m × 0.32 mm) coated with CP Sil-5 (0.12 mm film thickness) and uses helium as the carrier gas. Samples dissolved in ethyl acetate are injected at 70°, and the oven is programmed to 130° at 20°/min and then to 320° at 4°/min, followed by an isothermal hold for 10 min. All values are usually determined by averaged duplicate analyses. The $\Delta\ ^{13}C$ values of fatty acids and alcohols are obtained by correcting their measured $\Delta\ ^{13}C$ values for the isotopic composition of carbon added during the derivatization step as described by Schouten *et al.* (2004).

Measurement of Anammox Activity

The activity of anammox bacteria can be assessed in several ways. Anammox bacteria are unique in their ability to convert excess hydroxylamine to hydrazine under anoxic conditions (Jetten *et al.*, 1998; Schmid *et al.*, 2003). An anammox activity test can be carried out in anoxic batch incubations in 100-ml serum bottles containing 50 ml medium [2.5 mM $(NH_4)_2SO_4$ and 5 mM $NH_2OH.H_2SO4$] and about 150 mg protein of an anammox biomass sample. Ammonium, hydrazine, and hydroxylamine should be measured colorimetrically at appropriate time intervals (Schmid *et al.*, 2003). The results of such an activity test are depicted in Fig. 5.

When the contribution of anammox to the total nitrogen conversion in an environmental sample is estimated, the isotope pairing method should

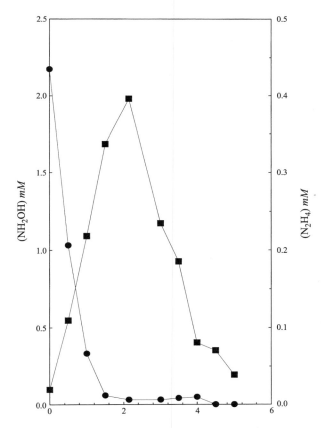

FIG. 5. Anammox activity test with hydroxylamine (●), resulting in transient accumulation of hydrazine (■).

be applied (Dalsgaard *et al.*, 2003; Kuypers *et al.*, 2003; Nielsen, 1992; Risgaard *et al.*, 2003, 2004; Thamdrup and Dalsgaard, 2002; Trimmer *et al.*, 2003). The method makes use of the unique anammox combination of ^{15}N-NH_4^+ with unlabeled nitrite or nitrate into $^{29}N_2$. Samples are prepared by transferring approximately 1 ml of homogenized material (e.g., sediment) to 12-ml gas-tight glass vials. The vials are placed in a He or N_2 atmosphere. Furthermore, the head space of the vials containing the biomass samples is purged with He or N_2. The samples are left untreated for 4 h to eliminate the background concentration of nitrate or nitrite. Then the vials are filled completely with anoxic medium and are supplied with 100 μM $^{15}NH_4^+$ and ^{14}N-nitrate or ^{14}N-nitrite. The appropriate controls do not contain labeled ammonia or contain labeled nitrite or nitrate to estimate the amount of denitrification in the sample. The vials are transferred to a gas-tight bag filled with He or N_2 and placed on a shaker. The samples are incubated for up to 48 h at *in situ* temperatures. Samples can be taken after 0, 12, 24, and 48 h by removing 1 ml water while replacing it with He or N_2 and adding 200 μl of a 7 M $ZnCl_2$ solution (alternatively $HgCl_2$) to the samples to stop biological activity. The ratio of $^{28}N_2$, $^{29}N_2$, and $^{30}N_2$ in the gas samples is measured by combined gas chromatography and isotope ratio mass spectrometry (Risgaard-Petersen and Rysgaard, 1995). Production of $^{29}N_2$ gas is evidence for the presence of the anammox reaction. The contribution of anammox and denitrification to total nitrogen gas production are calculated from the ratio in the production of $^{29}N_2$ and $^{30}N_2$ as described elsewhere (Risgaard-Petersen *et al.*, 2003, 2004).

Acknowledgments

The research on anaerobic ammonium oxidation (for an overview, see www.anammox. com) was financially supported by the European Union EESD EVK1-CT-2000–00054, the Foundation for Applied Sciences (STW), the Foundation of Applied Water Research (STOWA), the Netherlands Foundation for Earth and Life Sciences (NWO-ALW), the Royal Netherlands Academy of Arts and Sciences (KNAW), DSM-Gist, and Paques BV. Finally, we thank the many co-workers and students who contributed to the anammox research.

References

Alm, E. W., Oerther, D. B., Larsen, N., Stahl, D. A., and Raskin, L. (1996). The oligonucleotide probe database. *Appl. Environ. Microbiol.* **62**, 3557–3559.

Amann, R., Ludwig, W., and Schleifer, K-H. (1995). Phylogenetic identification and *in situ* detection of individual microbial cells without cultivation. *Microbiol. Rev.* **59**, 143–169.

Brochier, C., and Philippe, H. (2002). Phylogeny: A nonthermophilic ancestor for bacteria. *Nature* **417**, 244.

Broda, E.. (1977). Two kinds of lithotrophs missing in nature. *Z. Allg. Mikrobiol.* **17**, 491–493.

Cangelosi, G. A., and Brabant, W. H. (1997). Depletion of pre-16S rRNA in starved *Escherichia coli* cells. *J. Bacteriol.* **179**, 4457–4463.

Daims, H., Brühl, A., Amann, R., Schleifer, K-H., and Wagner, M. (1999). The domain-specific probe EUB338 is insufficient for the detection of all bacteria: Development and evaluation of a more comprehensive probe set. *Syst. Appl. Microbiol.* **22,** 434–444.

Dalsgaard, T., Canfield, D. E., Petersen, J., Thamdrup, B., and Acuña-González, J. (2003). Anammox is a significant pathway of N_2 production in the anoxic water column of Golfo Dulce, Costa Rica. *Nature* **422,** 606–608.

Dapena-Mora, A., Campos, J. L., Mosquera-Corral, A., Jetten, M. S. M., and Méndez, R. (2004). Stability of the anammox process in a gas-lift reactor and a SBR. *J. Biotechnol.* **110,** 159–170.

Egli, K., Franger, U., Alvarez, P. J. J., Siegrist, H. R., Vander Meer, J. R., and Zehnder, A. J. B. (2001). Enrichment and characterization of an anmmox bacterium from a rotating biological contractor treating ammonium-rich leachate. *Arch. Microbiol.* **175,** 198–207.

Jetten, M. S. M., Sliekers, A. O., Kuypers, M. M. M., *et al.* (2003). Anaerobic ammonium oxidation by marine and fresh water planctomycete-like bacteria. *Appl. Microbiol. Biotechnol.* **63**(2), 107–114.

Jetten, M. S. M., Strous, M., Van de Pas-Schoonen, K. T., Schalk, J., Van Dongen, U., Van de Graaf, A. A., Logemann, S., Muyzer, G., Van Loosdrecht, M. C. M., and Kuenen, J. G. (1998). The anaerobic oxidation of ammonium. *FEMS Microbiol. Rev.* **22,** 421–437.

Juretschko, S., Timmermann, G., Schmid, M. C., Schleifer, K-H., Pommerening-Röser, A., Koops, H-P., and Wagner, M. (1998). Combined molecular and conventional analyses of nitrifying bacterium diversity in activated sludge: *Nitrosococcus mobilis* and *Nitrospira*-like bacteria as dominant populations. *Appl. Environ. Microbiol.* **64,** 3042–3051.

Kuypers, M. M. M., Sliekers, A. O., Lavik, G., Schmid, M., Jørgensen, B. B., Kuenen, J. G., Sinninghe Damste, J. S., Strous, M., and Jetten, M. S. M. (2003). Anaerobic ammonium oxidation by anammox bacteria in the Black Sea. *Nature* **422,** 608–611.

Lee, N., Nielsen, P. H., Andreasen, K. H., Juretschko, S., Nielsen, J. L., Schleifer, K-H., and Wagner, M. (1999). Combination of fluorescent *in situ* hybridization and microautoradiography: A new tool for structure-function analyses in microbial ecology. *Appl. Environ. Microbiol.* **65,** 1289–1297.

Lindsay, M. R., Webb, R.I, Strous, M., Jetten, M. S. M., Butler, M. K., Forde, R. J., and Fuerst, J. A. (2001). Cell compartmentalisation in planctomycetes: Novel types of structural organisation for the bacterial cell. *Arch. Microbiol.* **175,** 413–429.

Loy, L. A., Horn, M., and Wagner, M. (2003). probeBase: An online resource for rRNA-targeted oligonucleotide probes. *Nucleic Acid Res.* **31,** 514–516.

Ludwig, W., Kirchhof, G., Klugbauer, N., Weizenegger, M., Betzl, D., Ehrmann, M., Hertel, C., Jilg, S., Tatzel, R., Zitzelsberger, H., Liebl, S., Hochberger, M., Shah, J., Lane, D., Wallnoefer, P. R., and Schleifer, K.-H. (1992). Complete 23S ribosomal RNA sequences of Gram-positive bacteria with a low DNA G+C content. *Syst. Appl. Microbiol.* **15,** 487–501.

Ludwig, W., Strunk, O., Klugbauer, S., Klugbauer, N., Weizenegger, M., Neumaier, J., Bachleitner, M., and Schleifer, K.-H. (1998). Bacterial phylogeny based on comparative sequence analysis. *Electrophoresis* **19,** 554–568.

Ludwig, W., Strunk, O., Westram, R., Richter, L., Meier, H., Yadhukumar, Buchner, A., Lai, T., Steppi, S., Jobb, G., Forster, W., Brettske, I., Gerber, S., Ginhart, A. W., Gross, O., Grumann, S., Hermann, S., Jost, R., Konig, A., Liss, T., Lussmann, R., May, M., Nonhoff, B., Reichel, B., Strehlow, R., Stamatakis, A., Stuckmann, N., Vilbig, A., Lenke, M., Ludwig, T., Bode, A., and Schleifer, K. H. (2004). ARB: A software environment for sequence data. *Nucleic Acid Res.* **32,** 1363–1371.

Maidak, B. L., Olsen, G. J., Larsen, N., Overbeck, R., Mc Caughey, M. J., and Woese, C. R. (1996). The ribosomal database project (RDP). *Nucleic Acid Res.* **24,** 82–85.

Manz, W., Amann, R., Ludwig, W., Wagner, M., and Schleifer, K.-H. (1992). Phylogenetic oligodeoxynucleotide probes for the major subclasses of *proteobacteria*: Problems and solutions. *Syst. Appl. Microbiol.* **15,** 593–600.

Neef, A., Amann, R. I., Schlesner, H., and Schleifer, K-H. (1998). Monitoring a widespread bacterial group: *In situ* detection of planctomycetes with 16S rRNA-targeted probes. *Microbiology* **144**, 3257–3266.

Nielsen, L. P. (1992). Denitrification in sediments determined from nitrogen isotope pairing. *FEMS Microbiol. Eco.* **86**, 357–362.

Poulsen, L. K., Ballard, G., and Stahl, D. A. (1993). Use of rRNA fluorescence *in situ* hybridization for measuring the activity of single cells in young and established biofilms. *Appl. Environ. Microbiol.* **59**, 1354–1360.

Pynaert, K., Smets, B. F., Wyffels, S., Beheydt, D., Siciliano, S. D., and Verstraete, W. (2003). Characterization of an autotrophic nitrogen-removing biofilm from a highly loaded lab-scale rotating biological contactor. *Appl. Environ. Microbiol.* **69**, 3626–35.

Richards, F. A. (1965). Anoxic basins and fjords. *In* "Chemical Oceanography" (J. P. Ripley and G. Skirrow, eds.), pp. 611–645. Academic Press, London.

Risgaard-Petersen, N., Meyer, R. L., Schmid, M. C., Jetten, M. S. M., Enrich-Prast, A., Rysgaard, S., and Revsbech, N. P. (2004). Anaerobic ammonia oxidation in an estuarine sediment. *Aq. Microbiol. Ecol.* **36**, 293–304.

Risgaard-Petersen, N., Nielsen, L. P., Rysgaard, S., Dalsgaard, T., and Meyer, R. L. (2003). Application of the isotope pairing technique in sediments where anammox and denitrification coexist. *Limnol. Ocean. Method* **1**, 63–73.

Risgaard-Petersen, N., and Rysgaard, S. (1995). Nitrate reduction in sediments and waterlogged soils measured by ^{15}N techniques 287–296. *In* "Methods in Applied Soil Microbiology" (K. Alef and P. Nannipieri, eds.). Academic Press, San Diego.

Schmid, M. C., Schmitz-Esser, S., Jetten, M. S. M., and Wagner, M. (2001). 16S–23S rDNA intergenic spacer and 23S rDNA of anaerobic ammonium oxidising bacteria: Implications for phylogeny and in situ detection. *Environ. Microbiol.* **7**, 450–459.

Schmid, M. C., Twachtmann, U., Klein, M., Strous, M., Juretschko, S., Jetten, M. S. M., Metzger, J., Schleifer, K-H., and Wagner, M. (2000). Molecular evidence for genus level diversity of bacteria capable of catalyzing anaerobic ammonium oxidation. *Syst. Appl. Microbiol.* **23**, 93–106.

Schmid, M., Walsh, K., Webb, R., Rijpstra, W. I. C., van de Pas-Schoonen, K. T., Verbruggen, M. J., Hill, T., Moffett, B., Fuerst, J., Schouten, S., Sinninge Damste, J. S., Harris, J., Shaw, P., Jetten, M. S. M., and Strous, M. (2003). *Candidatus* "Scalindua brodae", sp. nov., *Candidatus* "Scalindua wagneri", sp. nov., two new species of anaerobic ammonium oxidizing bacteria. *Syst. Appl. Microbiol.* **26**, 529–538.

Schmidt, I., Hermelink, C., Van de Pas-Schoonen, K. T., Strous, M., Op den Camp, H. J. M., Kuenen, J. G., and Jetten, M. S. M. (2002a). Anaerobic ammonia oxidation in the presence of nitrogen oxides (NO_x) by two different lithotrophs. *Appl. Environ. Microbiol.* **68**, 5351–5357.

Schmidt, I., Sliekers, A. O., Schmid, M. C., Cirpus, I. Y., Strous, M., Bock, E., Kuenen, J. G., and Jetten, M. S. M. (2002b). Aerobic and anaerobic ammonia oxidizing bacteria: Competitors or natural partners? *FEMS Microbiol. Ecol.* **39**, 175–181.

Schouten, S., Strous, M., Kuypers, M. M. M., Rijpstra, W. J. C., Baas, M., Schubert, C. J., Jetten, M. S. M., and Sinninghe Damsté, J. S. (2004). Stable carbon isotopic fractionations associated with inorganic carbon fixation by anaerobic ammonium oxidizing bacteria. *Appl. Environ. Microbiol.* **40**, 3785–3788.

Sinninghe Damsté, J. S., Strous, M., Rijpstra, W. I. C., Hopmans, E. C., Geenevasen, J. A. J., Van Duin, A. C. T., Van Niftrik, L. A., and Jetten, M. S. M. (2002). Linearly concatenated cyclobutane lipids form a dense bacterial membrane. *Nature* **419**, 708–712.

Sliekers, A. O., Derwort, N., Campos, L., Kuenen, J. G., Strous, M., and Jetten, M. S. M. (2002). Completely autotrophic ammonia removal over nitrite in a single reactor system. *Water Res.* **36**, 2475–2482.

Sliekers, A. O., Third, K., Abma, W., Kuenen, J. G., and Jetten., M. S. M. (2003). CANON and Anammox in a gas-lift reactor. *FEMS Microbiol. Lett.* **218,** 339–344.

Strous, M., Heijnen, J. J., Kuenen, J. G., and Jetten, M. S. M. (1998). The sequencing batch reactor as a powerful tool to study very slowly growing microorganisms. *Appl. Microbiol. Biotechnol.* **50,** 589–596.

Strous, M., and Jetten, M. S. M. (2004). Anaerobic oxidation of ammonium and methane. *Annu. Rev. Microbiol.* **58,** 99–117.

Strous, M., Kuenen, J. G., Fuerst, J. A., Wagner, M., and Jetten, M. S. M. (2002). The anammox case: A new experimental manifesto for microbiological eco-physiology. *Antonie van Leeuwenhoek* **81,** 693–702.

Strous, M., Kuenen, J. G., and Jetten, M. S. M. (1999). The key physiological parameters of the anaerobic ammonium oxidation process. *Appl. Environ. Microbiol.* **65,** 3248–3250.

Thamdrup., B., and Dalsgaard, T. (2002). Production of N_2 through anaerobic ammonium oxidation coupled to nitrate reduction in marine sediments. *Appl. Environ. Microbiol.* **68,** 1312–1318.

Third, K., Sliekers, A. O., Kuenen, J. G., and Jetten, M. S. M. (2001). The CANON system (completely autotrophic nitrogen-removal over nitrite) under ammonium limitation: Interaction and competition between three groups of bacteria. *Syst. Appl. Microbiol.* **24,** 588–596.

Trimmer, M., Nicholls, J. C., and Deflandre, B. (2003). Anaerobic ammonium oxidation measured in sediments along the Thames estuary, United Kingdom. *Appl. Environ. Microbiol.* **69,** 6447–6454.

Van de Graaf, A. A., De Bruijn, P., Robertson, L. A., Jetten, M. S. M., and Kuenen, J. G. (1996). Autotrophic growth of anaerobic ammonium-oxidizing micro-organisms in a fluidized bed reactor. *Microbiology* **142,** 2187–2196.

Van Dongen, U., Jetten, M. S. M., and van Loosdrecht, M. C. M. (2001). The Sharon-anammox process for the treatment of ammonium rich wastewater. *Wat. Sci. Technol.* **44,** 153–160.

Van Niftrik, L. A., Fuerst, J. A., Sinninghe Damste, J. S., Kuenen, J. G., Jetten, M. S., and Strous, M. (2004). The anammoxosome: An intracytoplasmic compartment in anammox bacteria. *FEMS Microbiol. Lett.* **233,** 7–13.

Wagner, M., Erhart, R., Manz, W., Amann, R., Lemmer, H., Wedi, D., and Schleifer, K.-H. (1994). Development of an rRNA-targeted oligonucleotide probe for the genus *Acinetobacter* and its application for *in situ* monitoring in activated sludge. *Appl. Environ. Microbiol.* **60,** 792–800.

Ward, B. B. (2003). Significance of anaerobic ammonium oxidation in the ocean. *Trend. Microbiol.* **11,** 408–410.

Zheng, D., Alm, E. W., Stahl, D. A., and Raskin, L. (1996). Characterization of universal small-subunit rRNA hybridization probes for quantitative molecular microbial ecology studies. *Appl. Environ. Microbiol.* **62,** 4504–4513.

Further Readings

Polz, M. F., and Cavanaugh, C. M. (1998). Bias in template to product ratios in multitemplate PCR. *Appl. Environ. Microbiol.* **64,** 3724–3730.

Strous, M., Fuerst, J., Kramer, E., Logemann, S., Muyzer, G., van de Pas-Schoonen, K. T., Webb, R., Kuenen, J. G., and Jetten, M. S. M. (1999a). Missing lithotroph identified as new planctomycete. *Nature* **400,** 446–449.

Toh, S. K., Webb, R. I., and Ashbolt, N. J. (2002). Enrichment of autotrophic anaerobic ammonium-oxidizing consortia from various wastewaters. *Microbial Ecol.* **43**(1), 154–167.

[4]   Enrichment and Isolation of Ferric-Iron- and Humic-Acid-Reducing Bacteria

By Kristina L. Straub, Andreas Kappler, and
Bernhard Schink

Abstract

In anoxic habitats, ferric iron oxides and humic acids are widespread, and ferric-iron- and humic-acid-reducing microorganisms presumably play an important role in the oxidation of organic matter. Representative strains of ferric-iron- or humic-acid-reducing bacteria were isolated from a wide range of freshwater or marine environments. Most of them are strict anaerobes, and facultatively anaerobic microorganisms reduce ferric iron oxides or humic acids only after oxygen has been consumed. Hence, anaerobic techniques have to be used for the preparation of media as well as for the cultivation of microorganisms. Furthermore, special caution is needed in the preparation of ferric iron oxides and humic acids.

Introduction

Iron is the most prevalent element on Earth and the fourth most abundant one in the Earth's crust. In soils and sediments, it can make up a few percent per weight of the total dry mass. Different ferric iron oxides provide the ochre to the brown color of soils in temperate zones or the red color of tropical soils (Cornell and Schwertmann, 1996). Because iron is present in soils and sediments in the range of several 10 mM, ferric iron Fe(III) is the most important electron acceptor for microorganisms under anoxic conditions, especially in water-logged soils, natural wetlands, rice paddies, and freshwater lake sediments. However, ferric iron oxides are only poorly soluble at neutral pH with concentrations of soluble Fe^{3+} $\leq 10^{-9}$ M (Kraemer, 2004). Hence, microbial reduction of ferric iron oxides under these conditions has to cope with a practically insoluble electron acceptor. With increasing acidity, the solubility of ferric iron oxides increases, and Fe^{3+} is well soluble at pH < 2.5. Coincidentally with this solubility change, the actual redox potential of the transition between Fe(II) and Fe(III) changes dramatically, from about 100 mV at pH 7 to 770 mV below pH 2.5 (Fig. 1). Therefore, microorganisms involved in the reduction of ferric iron at neutral or at acidic pH are essentially dealing with entirely different substrates (Straub *et al.*, 2001).

METHODS IN ENZYMOLOGY, VOL. 397   0076-6879/05 $35.00
DOI: 10.1016/S0076-6879(05)97004-3

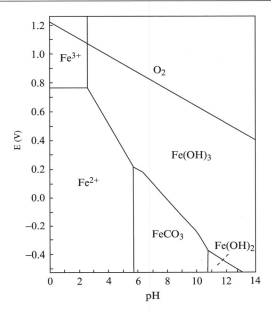

FIG. 1. Eh–pH diagram considering iron and carbonate at concentrations of 10 mM each, and oxygen at atmospheric pressure (0.21 atm). The thin line in the upper right refers to reduction of oxygen to water.

To overcome experimental problems of low ferric iron solubility at neutral pH, many laboratories use complexed iron species, i.e., Fe(III)-citrate, Fe(III)-NTA, or Fe(III)-EDTA as substitutes of ferric iron minerals. This may be justified for certain purposes, e.g., to grow major amounts of cell mass. However, one should be aware of pitfalls associated with the fact that a dissolved electron carrier might enter the periplasm and perhaps also the cytoplasm and may therefore be reduced in a different way than insoluble ferric iron oxides are (Straub and Schink, 2004a).

Humic compounds represent the majority of all organic matter, not only in soils but also in aquatic environments; e.g., more than 95% of all dissolved organic carbon in lake water is present as a yellow to brownish polymeric material, which is formed by the polymerization of aromatic and aliphatic precursors and is slowly decomposed by microbial activity or by photochemical reactions (Wetzel, 2001). Humic substances can be reduced by many microorganisms, such as fermenting, ferric-iron- and sulfate-reducing bacteria, and methanogenic, or hyperthermophilic archaea (Benz *et al.*, 1998; Cervantes *et al.*, 2002; Coates *et al.*, 1998; Kappler *et al.*, 2004; Lovley *et al.*, 1996, 1998, 2000). They contain quinoid structures, which

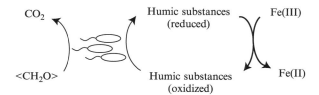

FIG. 2. Electron shuttling by humic substances: microbial oxidation of organic matter (CH_2O) with humic substances as electron acceptor and subsequent chemical reduction of Fe(III) by the reduced humic substances.

were suggested to represent their main electron-accepting moieties (Nurmi and Tratnyek, 2002; Scott *et al.*, 1998). However, complexed metal ions may also be involved in accepting electrons (Benz *et al.*, 1998; Lovley and Blunt-Harris, 1999; Struyk and Sposito, 2001), as well as conjugated aromatic constituents, which could take up electrons in delocalized π electron systems (Chen *et al.*, 2003).

Reduced humic compounds can transfer electrons directly to dissolved Fe^{3+} and various ferric iron minerals such as ferrihydrite, hematite, goethite, and ferruginous smectite (Chen *et al.*, 2003; Lovley *et al.*, 1998; Royer *et al.*, 2002; Szilágyi, 1971) (Fig. 2). Because humic substances are reoxidized during this chemical reaction, humic substances can serve again as electron acceptors for microorganisms, thus acting as electron shuttles. This allows an indirect reduction of solid-phase ferric iron without direct contact to the bacterial cells in the presence of low concentrations of humic substances, thus enhancing the rates of microbial metal reduction (Kappler *et al.*, 2004; Lovley *et al.*, 1996, 1998).

This chapter describes methods for the enrichment and cultivation of bacteria involved in the reduction of ferric iron oxides, either directly or indirectly via humic compounds. We focus primarily on the reduction of iron oxide minerals; growth media applying complexed iron, e.g., Fe(III)-citrate, have been described elsewhere (e.g., Lovley, 2000).

Preparation and Application of Ferric Iron Oxides

General Aspects

Sixteen different ferric iron oxides, hydroxides, or oxide hydroxides are known today, and they are often collectively referred to as iron oxides (Cornell and Schwertmann, 1996). Goethite, lepidocrocite, hematite, and ferrihydrite represent major iron oxides that are widespread in soils and sediments. Accordingly, these iron oxides are used frequently in laboratory studies. In particular, ferrihydrite, the least crystallized ferric iron oxide, is used in most studies on microbial ferric iron reduction. The term ferrihydrite embraces a variety of structurally ill-defined, poorly crystallized ferric iron species, which are often also termed amorphous iron oxide, hydrous ferric oxide, or ferric (hydr)oxide. Goethite, hematite, and lepidocrocite can be either purchased or prepared in the laboratory (Schwertmann and Cornell, 1991). In contrast, ferrihydrite is not available commercially and always needs to be synthesized in the laboratory. We recommend verifying the identity of the prepared ferric iron oxides (e.g., by X-ray or electron diffraction analysis), particularly if one is not yet experienced in synthesizing ferric iron oxides.

Ferrihydrite is frequently also a by-product in the synthesis of goethite, hematite, or lepidocrocite. Ferrihydrite impurities can therefore hamper studies with the better crystallized ferric iron oxides. Hence, it is recommended to check for ferrihydrite impurities by selective dissolution (e.g., with hydroxylamine hydrochloride) and extraction from better crystallized ferric iron oxides. Extraction methods using oxalate or hydroxylamine hydrochloride as extracting agents have been described by various authors (e.g., Chao and Zhou, 1983; McKeague and Day, 1966; Schwertmann *et al.*, 1982). In control experiments with artificial mixtures of ferrihydrite and better crystallized ferric iron oxides (e.g., goethite), we found that both extracting agents leached most but not all ferrihydrite. It is therefore important to check the extraction efficiency and to consider quantitative aspects of experiments, i.e., to verify that ferrous iron was produced from the bulk ferric iron oxide and not only from small ferrihydrite impurities.

For storage and usage, synthesized ferric iron oxides can either be (freeze-) dried and added dry to experiments or be kept in aqueous suspension. We prefer to keep ferric iron oxides in suspension because suspensions can be deoxygenated thoroughly by stirring under vacuum and repeated flushing of the headspace with N_2. Ferric iron oxides are added from stock suspensions to cultures by means of syringes.

Except for short-term experiments (e.g., cell suspension experiments), it is usually necessary to sterilize ferric iron oxides. Unfortunately, no sterilization method is ideally suited because heat, pressure, or radiation might induce changes of the ferric iron crystal structure, particle size, and surface area. To limit possible changes, we sterilize ferric iron oxide suspensions only after thorough deoxygenation. Ferrihydrite after autoclaving once for 15 min was reduced by *Geobacter* species (*G. bremensis* and *G. pelophilus*) at the same rate as ferrihydrite that was not autoclaved. However, when we autoclaved ferrihydrite a second or third time the reduction rates declined notably. Alternatively, we boil ferric iron oxide suspensions for 30 min, incubate the suspensions for 2 days at 28° (to allow spores to germinate), boil the suspension again for 30 min, and repeat the incubation and boiling procedure once more.

Preparation of Ferrihydrite

As ferrihydrite transforms with time to the better crystallized iron oxides goethite and/or hematite, we prefer to synthesize small amounts of ferrihydrite frequently and not to store them too long (≤ 6 months). For synthesis of ferrihydrite, we modified protocols of Ryden et al. (1977) and Lovley and Phillips (1986).

Prepare a 0.4 M solution of $FeCl_3$ in distilled water. This solution will be acidic, orange in color, and transparent. Stir the solution vigorously and adjust the pH slowly to 7.0 by adding 1 M NaOH drop by drop. We prefer to add NaOH at this low concentration and low rate because pH values above 7 (even only locally or momentarily) will induce the formation of other ferric iron oxides. Upon addition of NaOH, precipitates form, which dissolve again at the beginning of the neutralization process. Later the precipitates will persist and the color of the suspension changes to dark brown. When the pH has stabilized at pH 7.0, the synthesis of ferrihydrite is completed. After washing the produced ferrihydrite five times with a 10-fold volume of distilled water to remove sodium and chloride, the precipitate is spun down by centrifugation (5000g, 10 min). After the last centrifugation step, the ferrihydrite is resuspended in an adequate volume of distilled water to obtain a suspension of appropriate concentration, e.g., 0.5 M. The resulting suspension is transferred into glass test tubes together with a stirring bar. Each tube is sealed with a thick butyl rubber septum, and the ferrihydrite suspension is deoxygenated carefully by stirring under vacuum and repeated flushing of the headspace with N_2. Finally, the suspension is sterilized by autoclaving at 121° for 15 min.

Considerations on Ferrihydrite Concentrations

In batch culture experiments with iron-reducing bacteria, ferrihydrite is often used at concentrations in the range of 50 to 100 mM. At these concentrations, ferrihydrite gives a dark brown color to the medium, and the typical end product of ferrihydrite reduction under these conditions is magnetite, a black mixed-valence iron oxide (Fe_3O_4). Accordingly, the color of the medium changes from dark brown to black, and therefore other iron mineral phases that form transiently or as end product(s) may be masked and overlooked. The reason why magnetite forms under these conditions is still being discussed (Bell et al., 1987; Brookins, 1988; Glasauer et al., 2003; Jeon et al., 2003). However, magnetite is not necessarily the only end product of microbial ferrihydrite reduction: In growth experiments supplied with ferrihydrite at lower concentrations (5 to 10 mM), ferrihydrite was reduced completely to the ferrous state by various *Geobacter* species (*G. bremensis*, *G. metallireducens*, *G. pelophilus*, and *G. sulfurreducens*; Straub *et al.* 1998; Straub and Schink, 2004a). Medium containing only 5 to 10 mM ferrihydrite is orange-brown in color. Concomitant with complete reduction of ferrihydrite, the color of the iron precipitates changes from orange-brown to white because of the formation of ferrous iron precipitates with carbonate (siderite, $FeCO_3$) and phosphate [vivianite, $Fe_3(PO_4)_2$].

Enrichment of Ferric-Iron-Reducing Bacteria

Ferric iron oxides are reduced predominantly in anoxic habitats, and most ferric-iron-reducing bacteria are strict anaerobes. Facultatively anaerobic bacteria use ferric iron as an electron acceptor only after oxygen has been consumed. Hence, techniques for the preparation of media and cultivation of bacteria under strictly anoxic conditions should be used in studies on microbial ferric iron reduction; these techniques have been described in detail elsewhere (e.g., Widdel and Bak, 1992). In anoxic habitats, ferric-iron-reducing bacteria coexist with anaerobic bacteria that grow by fermentation or by anaerobic respiration, e.g., by reduction of nitrate or sulfate. To limit growth of fermentative bacteria in enrichment cultures, we use defined mineral media with electron donors/carbon sources that cannot be fermented easily, e.g., alcohols or fatty acids, and avoid complex nutrients such as yeast extract or peptone.

Medium Composition and Growth Conditions

Most media (freshwater, brackish, or seawater) contain substantial amounts of sulfate according to the composition of natural waters and/or sulfur source. In order to limit the growth of sulfate-reducing bacteria in enrichments for ferric-iron-reducing bacteria, the sulfate concentration is lowered to 0.1 mM, which is a sufficient sulfur supply for cell mass formation. Replacement of sulfate by organic sulfur sources (e.g., cysteine) is problematic in enrichment cultures, as they may be used as an additional source of organic substrate. Even if the sulfate concentration of the medium is as low as 0.1 mM, an indirect reduction of ferric iron oxides via sulfur cycling cannot be ruled out; it is therefore necessary to check isolated strains for direct or indirect ferric iron reduction (Straub and Schink, 2004b). Table I describes the mineral composition of freshwater and artificial seawater medium used in the enrichment and isolation of

TABLE I
COMPOSITIONS OF MINERAL MEDIA USED FOR ENRICHMENT AND ISOLATION OF
FERRIC-IRON-REDUCING BACTERIA

Compound	Freshwater medium (g per liter)	Artificial seawater medium (g per liter)
KH_2PO_4	0.6	0.4[a]
NH_4Cl	0.3	0.25[a]
$MgSO_4 \times 7H_2O$	0.025	0.025
$MgCl_2 \times 6H_2O$	0.4	11
$CaCl_2 \times 2H_2O$	0.1	1.5
KBr	—	0.09
KCl	—	0.7
NaCl	—	26.4

[a] Autoclaved separately in 500-fold concentrated stock solutions and added to the medium after cooling.

ferric-iron-reducing bacteria. These media resemble other media that have been used successfully by other laboratories. Routinely, our media are buffered with 30 mM bicarbonate/CO_2 and are supplemented further with a mixture of seven vitamins, selenite, tungstate, and a solution of eight trace elements (Widdel and Bak, 1992). We prefer to use trace element solutions without chelators such as ethylene diamine tetraacetate (EDTA) or nitrilotriacetate (NTA), as they might unintentionally facilitate microbial ferric iron reduction via the formation of iron complexes.

Addition of a reducing agent will allow growth of oxygen-sensitive ferric-iron-reducing bacteria. Usage of an organic reducing agent such as cysteine or ascorbate is not recommended in enrichment cultures as it would represent an additional source of organic substrate. As an alternative, ferrous iron (2 mM) can be used as a mild reducing agent. To avoid an increase in sulfate concentration, we use $FeCl_2$ instead of $FeSO_4$, although the $FeCl_2$ salt is less stable. $FeCl_2$ is highly hygroscopic, and moist $FeCl_2$ oxidizes rapidly. Oxidation of ferrous iron salts is accompanied by a color change from green to orange or brown. To avoid oxidation, we store $FeCl_2$ salts or solutions under anoxic conditions (e.g., in an anoxic glove box).

Ferric-iron-reducing cultures are preferentially incubated in flat bottles or tubes, which are placed horizontally to provide a large surface area of the ferric iron oxides. Because it is still unclear how ferric-iron-reducing bacteria deliver electrons to ferric iron oxides, we avoid exposing the cells to excessive shearing on shakers, but rather shake them gently by hand every other day to allow an even distribution of iron minerals and bacteria and to avoid diffusion limitation.

Considerations on Redox Potentials

Ferric iron oxides differ in their midpoint potentials of reduction (e.g., Thamdrup 2000). In general, chelated ferric iron forms, e.g., Fe(III)-citrate or Fe(III)-NTA, have considerably higher redox potentials than ferric iron oxide minerals (Table II). With ferric iron minerals, the situation is even more complex because their particle size also influences their actual redox potential; e.g., the redox potentials for the reduction of goethite particles of 10 or 100 nm are -200 or -270 mV, respectively (Thamdrup, 2000). The redox potentials of ferric iron oxides are influenced further by the pH of the medium as they increase by 59 mV per unit decrease in pH (Thamdrup, 2000). Figure 1 gives an overview of stability fields and changes in redox potential in dependence of the prevailing pH for a system that includes iron (with ferrihydrite as the most generally used iron oxide), carbonate, and oxygen. More Eh-pH diagrams that involve further ferric iron species (e.g., magnetite, hematite) and/or additional elements (e.g., silicon, sulfur) are available in the literature (e.g., Bell *et al.*, 1987; Brookins, 1988). The redox potentials of the ferric iron oxides have to be considered for the choice of an appropriate electron donor or vice versa; e.g., at pH 7.0, molecular

TABLE II

REDOX POTENTIALS FOR REDUCTION OF VARIOUS IRON OXIDES AT pH 7.0 AND 25°[a]

System	E'_0 (mV)	Reference
Fe(III)-NTA/Fe(II) NTA[b]	385	Thamdrup (2000)
Fe(III)-citrate/Fe(II) citrate	372	Thamdrup (2000)
Fe(III)-EDTA/Fe(II) EDTA	96	Wilson (1978)
Ferrihydrite/Fe^{2+}	-100 to 100	Brookins (1988); Thamdrup (2000); Widdel *et al.* (1993)
γ-FeOOH (lepidocrocite)/Fe^{2+}	-88	Thamdrup (2000)
α-FeOOH (goethite)/Fe^{2+}	-274	Thamdrup (2000)
α-Fe$_2$O$_3$ (hematite)/Fe^{2+}	-287	Thamdrup (2000)
Fe$_3$O$_4$ (magnetite)/Fe^{2+}	-314	Thamdrup (2000)

[a] Redox potentials depend strongly on pH value, temperature, concentration of reactants, and thermodynamic data chosen for calculations. For details, see related references.
[b] Nitrilotriacetate.

hydrogen ($2H^+/H_2$, $E_0' = -414$ mV) or formate (CO_2/formate, $E_0' = -432$ mV) can serve as an electron donor even for the reduction of better crystallized ferric iron oxides such as goethite, hematite, or magnetite, whereas the oxidation of acetate ($2CO_2$/acetate, $E_0' = -290$ mV) is energetically not favorable with these ferric iron oxides.

Monitoring Microbial Ferric Iron Reduction

Enrichment of ferric-iron-reducing bacteria can be monitored by measuring the production of ferrous iron. Before samples are taken with syringes under oxygen exclusion, cultures should be shaken vigourously to ensure a homogeneous distribution of ferrous and ferric iron species. Samples are acidified immediately with HCl (final concentration 1 M) to stabilize ferrous iron and to detach adsorbed ferrous iron from mineral phases. Ferrous iron can be quantified by a colorimetric assay with ferrozine (Stookey, 1970) or o-phenanthroline (Fachgruppe Wasserchemie, 1991; Straub $et\ al.$, 1999). Both assays allow quantification of ferrous iron in the micromolar range. o-Phenanthroline is classified as cancerogenic and hence is not so popular in ferrous iron determination as is ferrozine. Ferrozine stock solutions are prepared in distilled water with ammonium acetate (50%, w/v) containing 0.1% (w/v) ferrozine. As the ferrozine stock solution is light sensitive, it should be stored at 4° in the dark and can then be kept for approximately 2 weeks. As the sensitivity of the photometric assay with ferrozine is high, it is often necessary to dilute the samples with 1 M HCl before measurement. A sophisticated method to quantify ferrous and ferric iron by ion chromatography was developed by Schnell $et\ al.$ (1998).

Isolation of Ferric-Iron-Reducing Bacteria

In general, methods that separate single cells on or inside solidified media are superior for the isolation of pure cultures over a simple dilution to extinction in liquid medium. Ferric-iron-reducing bacteria may be purified by streaking on ferric iron-containing agar plates that are incubated in anoxic jars. All preparations should be carried out inside an anoxic chamber, as highly oxygen-sensitive strains may die upon exposure to air. However, not all bacteria grow on surfaces, and a dilution inside semisolid agar might be necessary for the separation of single cells. The technique of anoxic agar dilutions has been described in detail by Widdel and Bak (1992). Unfortunately, mineral ferric iron oxides are opaque and therefore cannot be used in agar dilution series. Although one may see cleared zones in the ferric iron-containing agar (indicating ferric-iron-reducing bacteria), one will not recognize faint neighboring colonies formed by noniron-reducing contaminants. It is also difficult to recognize such faint colonies in agar dilutions prepared with soluble ferric iron [e.g., Fe(III) citrate], as such medium is transparent but colored. To circumvent this problem, alternative electron acceptors that are soluble and colorless (e.g., fumarate, nitrate, dimethyl sulfoxide) are used for purification in agar dilution series. For example, *G. bremensis* and *G. pelophilus* were isolated with fumarate as an alternative electron acceptor (Straub *et al.*, 1998). After isolation of ferric-iron-reducing bacteria with an alternative electron acceptor, cultures are transferred back to the ferric iron oxide that was chosen initially for enrichment. Stock cultures should always be grown with ferric iron minerals as electron acceptor to maintain a selective pressure to use these insoluble electron acceptors.

Properties of Humic Substances

Humic substances, colloquially also called humus or humics, are redox-active polymeric organic compounds formed during the degradation of biopolymers such as lignin, proteins, and carbohydrates and are present in virtually all aquatic and terrestrial environments (Aiken *et al.*, 1985; Frimmel and Christman, 1988). A representative humic substance mole-cule is shown in Fig. 3 illustrating the heterogeneous structure of these compounds. Based on their solubility, humic substances are grouped into different fractions: humins (insoluble at alkaline pH), humic acids (soluble at alkaline and insoluble at acidic pH), and fulvic acids (soluble at alkaline and acidic pH) (Stevenson, 1994). The borderlines between these opera-tionally defined fractions are not sharp and depend strongly on the exact pH and on the concentration and ionic strength of the solution/suspension. Chemical analyses showed that humins are more hydrophobic, have less polar functional groups, and a higher molecular weight than humic acids, which in turn are less soluble, higher in molecular weight, and less polar than fulvic acids (Stevenson, 1994).

Because of their heterogeneous structure and composition, which both depend on the chemical characteristics of the precursor compounds (Hayes *et al.*, 1989; Stevenson, 1994), humic substances contain various redox-active functional groups of different redox potentials. Thus, no defined redox potential can be assigned to humic substances but rather a range of redox potentials over which humic substances can accept or release elec-trons. Using different methods and different humic substances, ranges of redox potentials at pH 7 from -200 to 300 mV (Straub *et al.*, 2001) but also exact values of 320 to 380 mV (Visser, 1964), 700 mV (Matthiessen, 1995; Szilagyi, 1971), and 778 mV (at pH 5; Struyk and Sposito, 2001) have been

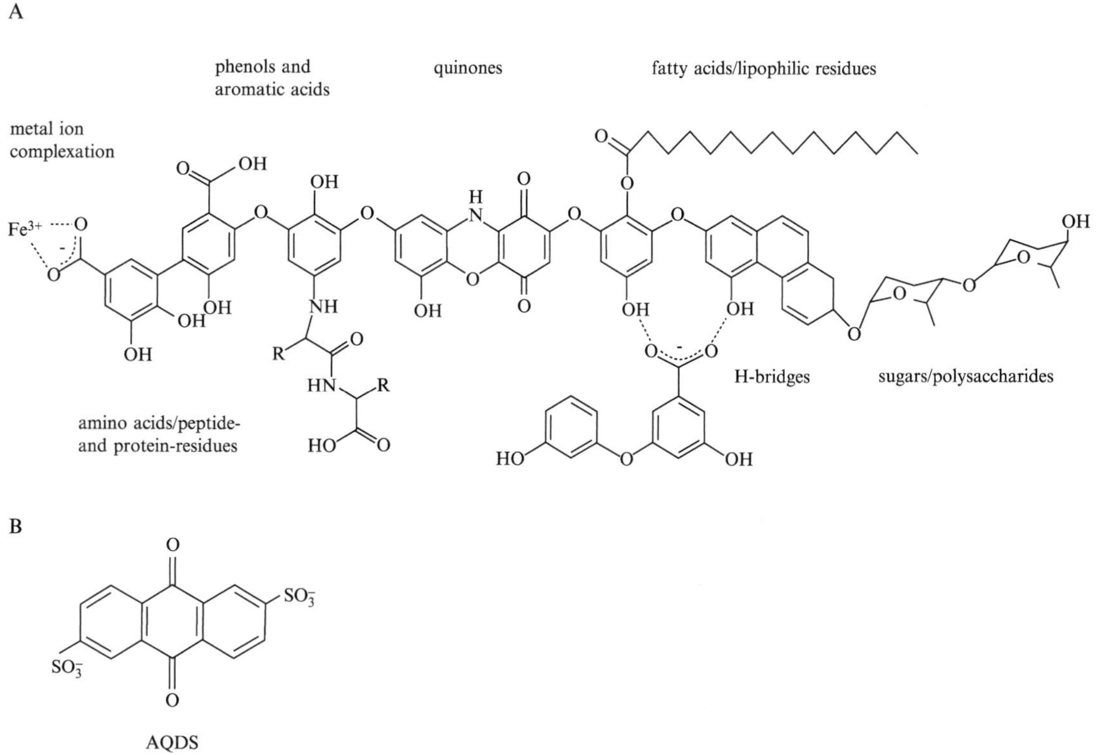

Fig. 3. Molecular structures of a representative humic substance molecule (A) and anthraquinone-2,6-disulfonate (AQDS), a model compound for quinone moieties in humic substances (B).

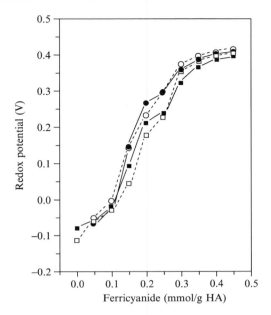

FIG. 4. Redox potentials recorded with a Pt redox electrode during titration of Pd/H$_2$-reduced humic acid preparations (Aldrich, 1 g/liter, phosphate buffer, pH 7) with potassium ferricyanide {K$_3$[Fe(CN)$_6$]}. Four replicates are shown.

reported. This wide range of values illustrates the difficulties in assigning accurate redox potentials to an insufficiently defined class of organic compounds. Because of their structural and compositional heterogeneity, a range of redox potentials between -200 and 300 mV is most likely, also with respect to a possible involvement in electron transfer to ferric iron oxides. Titration of reduced humic acid preparations with K$_3$[Fe(CN)$_6$] and concomitant recording of the apparent redox potential showed that indeed most of the reducing equivalents present in reduced humic acids are transferred to Fe(III) at redox potentials between -100 and 400 mV (Fig. 4).

Isolation, Purification, and Application of Humic Substances

Isolation of Humic Substances

Humic substances are isolated from soils and sediments after the removal of small rocks and roots by alkaline extraction with 0.1 M NaOH at a solvent-to-soil ratio of 4:1 to 10:1 (v/w) for 24 h at 30° under shaking, if possible, under anoxic conditions to avoid oxidative polymerization and degradation reactions. Inorganic constituents and the insoluble humin fraction are removed by centrifugation (30 min at 12,000g), and the supernatant is acidified with 1 M HCl to pH $\sim$1 and kept at 4° for 24 h to precipitate humic acids. The fulvic acids remain in solution. The precipitated humic acids are separated by centrifugation and freeze-dried to be stored as a dry powder. To isolate the fulvic acids, the supernatant is passed through a column filled with XAD-8. This nonionic, macroporous methyl methacrylate ester resin should be washed with 1 M NaOH and rinsed with distilled water several times before use. The effluent is discarded, and the adsorbed fulvic acids are washed with 0.5 column volumes of distilled water. The fulvic acids are removed from the column by elution with 0.1 M NaOH. The eluate is freeze-dried either directly to obtain the Na-salt or after a passage through a cation-exchange resin where the Na^+ is replaced by H^+.

To isolate fulvic and humic acids from natural waters, the water is filtered (0.45 μm membrane filter) and the pH is adjusted to $\sim$2 with HCl. The sample is passed over an XAD-8 column, and the sorbed fulvic and humic acids are eluted by 0.1 M NaOH. After acidification of the eluate to pH 1 with 1 M HCl, the precipitated humic acids can be separated by centrifugation. The fulvic acid fraction is passed through an XAD-8 column, washed with one column volume of distilled water to remove the salt, and eluted with 0.1 M NaOH. After passing the eluate through a cation-exchange resin column, the proton-saturated fulvic acids are obtained as powder by freeze drying.

In addition to separation of humic substances according to their solubility, humic substances can also be separated into polyphenolic and carbohydrate-rich fractions (Chen *et al.*, 2003; Gu and Chen, 2003).

Purification of Humic Substances

Because humic compounds after this isolation still contain significant amounts of inorganic constituents (quantifiable by the ash content), further purification may be necessary. The International Humic Substance Society (IHSS) provides a protocol that includes the removal of inorganic constituents by hydrofluoric acid treatment and dialysis to obtain highly pure humic substances (Swift, 1996). However, because these inorganic constituents may be important for the activities of humic substances in the environment, extensive purification is usually not required for experiments studying the effect of humic compounds on microbial redox processes. Furthermore, the IHSS offers standard humic compounds purified after the IHSS protocol (http://www.ihss.gatech.edu/) that have been well characterized and were already used in many studies.

Inexpensive humic compounds are also available commercially in large quantities (Aldrich). However, because this material is prepared from lignite, it represents highly humified and altered organic material and contains high amounts of inorganic compounds; thus, it is hardly representative of humic compounds from recent soils and waters (Malcolm and MacCarthy, 1986).

Determination of the Redox State of Humic Substances

Depending on how many redox-active groups are reduced/oxidized, humic substances possess a certain capacity to take up electrons (by the oxidized functional groups) or to release electrons from reduced functional groups to an electron acceptor of a more positive redox potential. The reducing and electron-accepting capacity of a certain humic substance sample can be determined by quantifying the reducing capacity (with ferric iron as an electron acceptor) of a native (untreated) vs a chemically reduced humic substance sample. For that, humic substances are dissolved (1 mg/ml) in 50 mM anoxic phosphate buffer (pH 7.0). Two separate 6-ml aliquots are incubated under H_2 and N_2 atmospheres, both in air-tight vessels in the presence of a Pd catalyst (5% Pd on activated charcoal; 1 mg/ml) on a rotary shaker at 30° for 24 h. After reduction, the H_2 headspace is replaced by N_2 by several cycles of vacuum degasing and N_2 flushing. Of both aliquots, 1 ml is mixed with 0.5 ml of a 3 mM potassium hexacyanoferrate ($K_3[Fe(CN)_6]$) solution (three replicates) and incubated on a shaker in the dark at 30°. After 24 h, 1 ml is taken per sample, mixed with 1 ml of HCl (1 M), and incubated for 30 min to precipitate the humic acids. After centrifugation (15,000g, 5 min), the absorbance at a wavelength of 416 nm is measured. Two 1-ml samples of humic acids to which water is added instead of hexacyanoferrate (followed also by the HCl precipitation step) give the background absorption of the humic acids that stay in solution after precipitation. A sample set with phosphate buffer and 0–3 mM hexacyanoferrate serves as calibration standard. The electron-accepting capacity of the humic acids can be calculated as the difference between the amounts of electrons transferred to hexacyanoferrate by the reduced and the nonreduced humic acid preparation.

A wide range of electron-accepting capacities of 0.1–0.9 mequiv electrons per gram sedimentary or soil humic acid were described (e.g., Chen et al., 2003; Kappler and Haderlein, 2003; Royer et al., 2002; Scott et al., 1998; Struyk and Sposito, 2001). By electrochemical reduction of humic compounds and subsequent transfer of electrons to ferric iron or by chemical reduction and subsequent titration with I_2 as oxidant, electron-accepting capacities of 5.6 (Kappler and Haderlein, 2003) or even 11.5 mequiv/g humic acids (Struyk and Spositio, 2001) were determined for natural humic acids, indicating that the obtained values depend on the reduction method (Pd/H$_2$ vs electrochemical) and the oxidant [Fe(III) vs I$_2$] used.

The redox state of microbially reduced dissolved humic compounds can be quantified after filtration (to remove the cells) by incubation of the filtrate with Fe(III)-citrate for 15 min and quantification of ferrous iron formed by the ferrozine method (see earlier discussion).

Application of Humic Substances

For microbiological experiments with humic substances, mostly the humic acid fraction has been used because humic acids can be obtained easily at large amounts and are well soluble at neutral pH at concentrations of up to ~1 mg/ml (remember that they are isolated by alkaline extraction and precipitation at acidic pH). In contrast, fulvic acid concentrations in soils/sediments are usually much lower, and the humin fraction is not soluble and difficult to handle.

For microbiological experiments with defined cultures of bacteria or under nongrowth conditions (e.g., cell suspension experiments), humic compounds are usually used without sterilization. Nonetheless, for enrichments or isolation, humic substances can be sterilized either by autoclaving or by filtration (membrane filter, 0.2 μm). Because of the limited solubility of humic acids at neutral pH, they can form precipitates and/or colloids at high concentrations and thus cause material loss during filtration. For autoclaving, humic acids are suspended in water or in a buffer solution similar to the buffer system used in the culture medium (e.g., 30 mM anoxic bicarbonate or phosphate buffer, pH 7.0) to 50 mg/ml under a N$_2$ headspace, and autoclaved at 121° for 25 min (Kappler et al., 2004).

Enrichment, Isolation, and Quantification of Humic-Acid-Reducing Bacteria

In principle, media and cultivation techniques described earlier for ferric-iron-reducing bacteria can also be applied for the enrichment and cultivation of humic acid-reducing bacteria. However, some specific aspects of humic substances and humic acids need consideration. After purification of humic substances, all low molecular weight compounds have been removed, and the remaining polymeric humic compounds are not degradable to any significant extent by microorganisms (as shown, e.g., for *Shewanella alga* in the presence/absence of lactate; Lovley et al., 1996). It is not advisable to use humic substances directly as sole electron acceptors for enrichment or isolation because of (i) the relative low solubility of humic substances at neutral pH, leading to a relative low electron accepting capacity by the humic substances; (ii) the problems of measuring protein content or counting cells in the presence of significant concentrations of humic substances; and (iii) the high cost of large amounts of highly purified humic substances. Therefore, anthraquinone-2,6-disulfonic acid (AQDS, 5 mM) in combination with 2 mM acetate was used to obtain humic acid-reducing bacteria from diverse sediments in enrichment cultures (Coates et al., 1998). AQDS (Fig. 3) is a model compound for quinone moieties in humic substances (it is a functional analogue, not a structural analogue or a model humic acid as claimed frequently in the literature). AQDS reduction can be followed easily by the development of the intense orange color of AHQDS (anthrahydroquinone-2,6-disulfonic acid), the reduced form of AQDS, which can be quantified photometrically at a 450-nm wavelength. After several transfers in liquid medium, dilution series in semisolid agar medium containing AQDS as the electron acceptor give colonies that can be isolated and transferred to liquid medium. Unfortunately, humic acid-reducing bacteria such as fermenting bacteria or Shewanella sp. that depend on more complex substrates as an electron donor (Benz et al., 1998; Lovley et al., 1998) cannot be grown selectively by this approach. They could be included by using their respective substrates, e.g., lactate, but care has to be taken to ensure that the observed growth is not caused by (AQDS-independent) fermentative growth.

Although AQDS has been used in many studies, results have to be interpreted with caution. The redox potential of AQDS ($E_0' = -184$ mV) is at the lower end of the redox potentials determined for humic substances, and bacteria that are able to transfer electrons to humic substances with redox potentials substantially higher than -184 mV can be overlooked when using AQDS. Furthermore, AQDS was shown to be taken up into the cell by a so far unknown mechanism (Shyu et al., 2002). This is hardly possible with humic substances because of their greater size. Consequently, cells able to reduce AQDS do not necessarily reduce humic substances. Furthermore, a potential toxicity of AQDS was reported for pure cultures and indigenous microbial communities (Nevin and Lovley, 2000; Shyu et al., 2002).

To count humic acid-reducing bacteria in sediment or soil, the most-probable number technique (dilution series) can be used. Humic acid concentrations of 0.5 g/liter are used (Kappler et al., 2004). The simultaneous presence of 40 mM ferrihydrite allows reoxidation of the humic acids by electron transfer to ferric iron and thus a much higher turnover of organic substrates, e.g., acetate or lactate. However, this strategy to count humic acid-reducing bacteria in dilution series works only if the number of bacteria reducing ferric iron directly is substantially smaller than the number of humic acid-reducing bacteria; therefore, the number of ferric-iron-reducing bacteria has to be determined in parallel in the absence of humic compounds.

Growth of humic substance-reducing bacteria has to be determined as an increase in biomass/cell numbers. To achieve a measurable amount of substrate oxidation and biomass formation, high concentrations of humic substances have to be used. Unfortunately, because of formation of cell-polymer aggregates accompanied by precipitation processes, direct micro-scopic cell counts are not possible. However, cell counts after the reduction of humic substances have been done if the bacteria can also grow aerobical-ly on plates. After the reduction of humic substances, the cell number in the cultures was determined by plate counts with aerobic heterotrophic medi-um (e.g., *Shewanella* sp.; Lovley *et al.*, 1996), although the authors did not mention if the cells aggregated with humic substances and how they solved this problem. If the bacteria do not grow aerobically on plates, the growth of humic substance-reducing bacteria can be quantified using 5 mM AQDS as the electron acceptor instead of humic substances allowing microscopic cell counting (Lovley *et al.*, 1998). Attempts to measure protein content in humic acid-reducing cultures were not successful because of extensive interference of the protein assay with humic acids (Lovely *et al.*, 1996).

Reduction of humic substances by pure cultures can be shown either by quantifying the redox state of humic substances or by quantifying the oxidation of radiolabeled substrates (e.g., [^{14}C]acetate). When incubating cell suspensions ($\sim$1 mg protein/ml) with [^{14}C]acetate (0.2 mM) and humic acids (2 g/liter), the formation of $^{14}CO_2$ can be monitored over time (Coates *et al.*, 1998; Lovley *et al.*, 1996, 1998). With an electron-accepting capacity of >0.1 mequiv electrons per gram sedimentary or soil humic acid (see earlier discussion), this corresponds to >0.2 mequiv electrons that can be accepted by humic acids per liter of medium. This means that with 2 g/liter humic acids, at least 0.025 mM acetate can be oxidized to CO_2. This method will be useful only in exceptional cases.

Acknowledgments

We thank Andreas Brune for help with redox titrations of humic compounds. This work was supported financially by the European Commission in the PURE project (EVK1-CT-1999–00030).

References

Aiken, G. R., McKnight, D. M., Wershaw, R. L., and Mac Carthy, P. (1985). "Humic Substances in Soil, Sediment, and Water: Geochemistry, Isolation, and Characterization." Wiley, New York.

Bell, P. E., Mills, A. L., and Herman, J. S. (1987). Biogeochemical conditions favoring magnetite formation during anaerobic iron reduction. *Appl. Environ. Microbiol.* **53,** 2610–2616.

Benz, M., Schink, B., and Brune, A. (1998). Humic acid reduction by *Propionibacterium freudenreichii* and other fermenting bacteria. *Appl. Environ. Microbiol.* **64,** 4507–4512.

Brookins, D. G. (1988). "Eh-pH Diagrams for Geochemistry." Springer-Verlag, Berlin.

Cervantes, F. J., de Bok, F. A. M., Duong-Dac, T., Stams, A. J. M., Lettinga, G., and Field, J. A. (2002). Reduction of humic substances by halorespiring, sulphate-reducing and methanogenic microorganisms. *Environ. Microbiol.* **4,** 51–57.

Chao, T. T., and Zhou, L. (1983). Extraction techniques for selective dissolution of amorphous iron oxides from soils and sediments. *Soil Sci. Soc. Am. J.* **47,** 225–232.

Chen, J., Gu, B., Royer, R. A., and Burgos, W. D. (2003). The roles of natural organic matter in chemical and microbial reduction of ferric iron. *Sci. Total Environ.* **307,** 167–178.

Coates, J. D., Ellis, D. J., Blunt-Harris, E. L., Gaw, C. V., Roden, E. E., and Lovley, D. R. (1998). Recovery of humic-reducing bacteria from a diversity of environments. *Appl. Environ. Microbiol.* **64,** 1504–1509.

Cornell, R. M., and Schwertmann, U. (1996). "The Iron Oxides: Structure, Properties, Reactions, Occurrence and Uses." VCH, Weinheim.

Fachgruppe Wasserchemie, G. D. C. (1991). Deutsche Einheitsverfahren zur Wasser-, Abwasser- und Schlammuntersuchung, Bd. II, E1. VCH, Weinheim.

Frimmel, F. H., and Christman, R. F. (1988). "Humic Substances and Their Role in the Environment." Wiley, New York.

Glasauer, S., Weidler, P. G., Langley, S., and Beveridge, T. J. (2003). Controls on Fe reduction and mineral formation by a subsurface bacterium. *Geochim. Cosmochim. Acta* **67,** 1277–1288.

Gu, B., and Chen, J. (2003). Enhanced microbial reduction of Cr(VI) and U(VI) by different natural organic matter fractions. *Geochim. Cosmochim. Acta* **67,** 3575–3582.

Hayes, M. H. B., Mac Carthy, P., Malcolm, R. L., and Swift, R. S. (1989). "Humic Substances II: In Search of Structure." Wiley, Chichester.

Jeon, B. H., Dempsey, B. A., and Burgos, W. D. (2003). Kinetics and mechanisms for reactions of Fe(II) with iron(III) oxides. *Environ. Sci. Technol.* **37,** 3309–3315.

Kappler, A., Benz, M., Brune, A., and Schink, B. (2004). Electron shuttling via humic acids in microbial iron(III) reduction in a freshwater sediment. *FEMS Microbiol. Ecol.* **47,** 85–92.

Kappler, A., and Haderlein, S. B. (2003). Natural organic matter as reductant for chlorinated aliphatic pollutants. *Environ. Sci. Technol.* **37,** 2707–2713.

Kraemer, S. M. (2004). Iron oxide dissolution and solubility in the presence of siderophores. *Aquat. Sci.* **66,** 3–18.

Lovley, D. R. (2000). Dissimilatory Fe(III)- and Mn(IV)-reducing prokaryotes. *In* "The Prokaryotes" (A. Balows, H. G. Trüper, M. Dworkin, W. Harder, and K.-H. Schleifer, eds.), 3rd Ed. Springer-Verlag, Berlin.

Lovley, D. R., and Blunt-Harris, E. L. (1999). Role of humic-bound iron as an electron transfer agent in dissimilatory Fe(III) reduction. *Appl. Environ. Microbiol.* **65,** 4252–4254.

Lovley, D. R., Coates, J. D., Blunt-Harris, E. L., Phillips, E. J. P., and Woodward, J. C. (1996). Humic substances as electron acceptors for microbial respiration. *Nature* **382,** 445–448.

Lovley, D. R., Fraga, J. L., Blunt-Harris, E. L., Hayes, L. A., Phillips, E. J. P., and Coates, J. D. (1998). Humic substances as a mediator for microbially catalyzed metal reduction. *Acta Hydrochim. Hydrobiol.* **26,** 152–157.

Lovley, D. R., and Phillips, E. J. P. (1986). Organic matter mineralization with reduction of ferric iron in anaerobic sediments. *Appl. Environ. Microbiol.* **51,** 683–689.

Malcolm, R. L., and MacCarthy, P. (1986). Limitations in the use of commercial humic acids in water and soil research. *Environ. Sci. Technol.* **20,** 904–911.

Matthiessen, A. (1995). Determining the redox capacity of humic substances as a function of pH. *Vom Wasser* **84,** 229–235.

McKeague, J. A., and Day, J. H. (1966). Dithionite- and oxalate-extractable Fe and Al as aids in differentiating various classes of soils. *Can. J. Soil Sci.* **46,** 13–22.

Nevin, K. P., and Lovley, D. R. (2000). Potential for nonenzymatic reduction of Fe(III) via electron shuttling in subsurface sediments. *Environ. Sci. Technol.* **34,** 2472–2478.

Nurmi, J. T., and Tratnyek, P. G. (2002). Electrochemical properties of natural organic matter (NOM), fractions of NOM, and model biogeochemical electron shuttles. *Environ. Sci. Technol.* **36,** 617–624.

Royer, R. A., Burgos, W. D., Fisher, A. S., Jeon, B.-H., Unz, R. F., and Dempsey, B. A. (2002). Enhancement of hematite bioreduction by natural organic matter. *Environ. Sci. Technol.* **36,** 2897–2904.

Ryden, J. C., McLaughlin, J. R., and Syers, J. K. (1977). Mechanisms of phosphate sorption by soils and hydrous ferric oxide gel. *J. Soil Sci.* **28,** 72–92.

Schnell, S., Ratering, S., and Jansen, K.-H. (1998). Simultaneous determination of iron(III), iron(II), and manganese(II) in environmental samples by ion chromatography. *Environ. Sci. Technol.* **32,** 1530–1537.

Schwertmann, U., and Cornell, R. M. (1991). "Iron Oxides in the Laboratory." VCH, Weinheim.

Schwertmann, U., Schulze, D. G., and Murad, E. (1982). Identification of ferrihydrite in soils by dissolution kinetics, differential x-ray diffraction, and Mössbauer spectroscopy. *Soil Sci. Soc. Am. J.* **46,** 869–875.

Scott, D. T., McKnight, D. M., Blunt-Harris, E. L., Kolesar, S. E., and Lovley, D. R. (1998). Quinone moieties act as electron acceptors in the reduction of humic substances by humics-reducing microorganisms. *Environ. Sci. Technol.* **32,** 2984–2989.

Shyu, B. J. H., Lies, D. P., and Newman, D. K. (2002). Protective role of tolC in efflux of the elctron shuttle anthraquinone-2,6-disulfonic acid. *Appl. Environ. Microbiol.* **184,** 1806–1810.

Stevenson, F. J. (1994). "Humus Chemistry: Genesis, Composition, Reactions." Wiley, New York.

Stookey, L. L. (1970). Ferrozine: A new spectrophotometric reagent for iron. *Anal. Chem.* **42,** 779–781.

Straub, K. L., Benz, M., and Schink, B. (2001). Iron metabolism in anoxic environments at near neutral pH. *FEMS Microbiol. Ecol.* **34,** 181–186.

Straub, K. L., Hanzlik, M., and Buchholz-Cleven, B. E. E. (1998). The use of biologically produced ferrihydrite for the isolation of novel iron-reducing bacteria. *Syst. Appl. Microbiol.* **21,** 442–449.

Straub, K. L., Rainey, F. A., and Widdel, F. (1999). *Rhodovulum iodosum* sp. nov. and *Rhodovulum robiginosum* sp. nov., two new marine phototrophic ferrous-iron-oxidizing purple bacteria. *Int. J. Syst. Bacteriol.* **49,** 729–735.

Straub, K. L., and Schink, B. (2004a). Ferrihydrite reduction by *Geobacter* species is stimulated by secondary bacteria. *Arch. Microbiol.* **182,** 175–181.

Straub, K. L., and Schink, B. (2004b). Ferrihydrite-dependent growth of *Sulfurospirillum deleyianum* by electron transfer via sulfur cycling. *Appl. Environ. Microbiol.* **70,** 5744–5749.

Struyk, Z., and Sposito, G. (2001). Redox properties of standard humic acids. *Geoderma* **102,** 329–346.

Swift, R. S. (1996). Organic matter characterization. *In* "Soil Science Society of America Book Series 5" (D. L. Sparks, ed.), pp. 1018–1020. Soil Science Society of America, Madison, WI.

Szilágyi, M. (1971). Reduction of Fe^{3+} ion by humic acid preparations. *Soil Sci.* **111**, 233–235.

Thamdrup, B. (2000). Bacterial manganese and iron reduction in aquatic sediments. *In* "Advances in Microbial Ecology" (B. Schink, ed.), Vol. 16, pp. 41–84. Kluwer Academic/Plenum, New York.

Visser, S. A. (1964). Oxidation-reduction potentials and capillary activities of humic acids. *Nature* **204**, 581.

Wetzel, R. G. (2001). "Limnology: Lake and River Ecosystems," 3rd Ed. Academic Press, San Diego.

Widdel, F., and Bak, F. (1992). Gram-negative mesophilic sulfate-reducing bacteria. *In* "The Prokaryotes" (A. Balows, H. G. Trüper, M. Dworkin, W. Harder, and K.-H. Schleifer, eds.), 2nd Ed. pp. 3352–3378. Springer-Verlag, Berlin.

Widdel, F., Schnell, S., Heising, S., Ehrenreich, A., Assmus, B., and Schink, B. (1993). Ferrous iron oxidation by anoxygenic phototrophic bacteria. *Nature* **362**, 834–836.

[5] Enrichment, Cultivation, and Detection of Reductively Dechlorinating Bacteria

By Frank E. Löffler, Robert A. Sanford, and Kirsti M. Ritalahti

Abstract

Strategies and procedures for enriching, isolating, and cultivating reductively dechlorinating bacteria that use chloroorganic compounds as metabolic electron acceptors from environmental samples are described. Further, nucleic acid-based approaches used to detect and quantify dechlorinator (i.e., *Dehalococcoides*)-specific genes are presented.

Introduction

A diverse group of bacteria use (poly)chlorinated aromatic and aliphatic compounds as terminal electron acceptors in their energy metabolism (Löffler *et al.*, 2003). This process, termed metabolic reductive dechlorination, involves energy capture from reductive dechlorination reactions. Other terms, including chlororespiration, dehalorespiration, and chloridogenesis, have also been applied to describe growth-coupled reductive dechlorination. Chlororespiration (Löffler *et al.*, 1996) and dehalorespiration

METHODS IN ENZYMOLOGY, VOL. 397   0076-6879/05 $35.00
                   DOI: 10.1016/S0076-6879(05)97005-5

(Maymó-Gatell *et al.*, 1997) were introduced to indicate that energy cap-ture from reductive dechlorination reactions involves respiratory mechan-isms. Chloridogenesis (He *et al.*, 2003b) is a more broadly defined term that encompasses electron transport across a membrane, as well as substrate level phosphorylation, as modes of energy capture. Metabolic reductive dechlorination properly describes these processes, and the abbreviation MRDBs (for metabolic reductively dechlorinating bacteria) is used here.

Metabolic reductive dechlorination as an energy-yielding reaction is intriguing because the anthropogenic release of chlorinated chemicals into the environment started less than a century ago. It remains a common belief that the capability to perform metabolic reductive dechlorination arose within the last century in response to anthropogenic pollution. Al-though the origin and evolution of metabolic reductive dechlorination are currently unclear, it is obvious that biogeochemical chlorine cycles existed for billions of years. A variety of chloroorganic compounds are produced naturally, suggesting metabolic reductive dechlorination represents an ancient lifestyle (Häggblom and Bossert, 2003; Löffler *et al.*, 2003).

Bacteria capable of metabolic reductive dechlorination include mem-bers of numerous genera, including *Desulfomonile*, *Dehalobacter*, *Desulfuromonas*, *Geobacter*, *Sulfurospirillum* (formerly *Dehalospirillum*), *Clostridium*, *Desulfitobacterium*, *Desulfovibrio*, *Anaeromyxobacter*, and *Dehalococcoides* (Bouchard *et al.*, 1996; Chang *et al.*, 2000; DeWeerd *et al.*, 1990; Holliger *et al.*, 1998; Luijten *et al.*, 2003; Sanford *et al.*, 1996, 2002; Scholz-Muramatsu *et al.*, 1995; Smidt and de Vos, 2004; Sun *et al.*, 2000, 2001, 2002; Sung *et al.*, 2003; Utkin *et al.*, 1994). The discovery of these bacteria prompted great interest in the bioremediation community to treat sites contaminated with (poly)chlorinated compounds (Ellis *et al.*, 2000; Lendvay *et al.*, 2003; Major *et al.*, 2002). The spectrum of chlorinated compounds used as growth-supporting electron acceptors by MRDBs is strain specific and spans an array of chloroorganic compounds, including many priority pollutants (Table I).

Some MRDBs exhibit physiological diversity and utilize a variety of electron acceptors, whereas others have very restricted metabolic capabilities. For instance, *Anaeromyxobacter* species exhibit great versa-tility and respire oxygen, nitrate, nitrite, oxidized metal species, fuma-rate, and halophenols (He and Sanford, 2002, 2003; Sanford *et al.*, 2002). *Dehalococcoides* and *Dehalobacter* species, in contrast, require the pres-ence of very specific chloroorganic compounds for energy generation, and no alternative growth-supporting electron acceptors have been shown to suffice (Adrian *et al.*, 2000; He *et al.*, 2003b; Holliger *et al.*, 1998; Maymó-Gatell *et al.*, 1997, 1999; Sun *et al.*, 2002). Similar diversity is seen regarding electron donor requirements. *Sulfurospirillum* and

TABLE I

EXAMPLES OF CHLOROORGANIC PRIORITY POLLUTANTS THAT SERVE AS METABOLIC ELECTRON ACCEPTORS FOR MRDBs[a]

Group	Chloroorganic electron acceptors	Reduced product(s)	Electron donor(s)[b]	Reference
Anaeromyxobacter	2-Chlorophenol	Phenol	Ac	Sanford et al. (2002)
Clostridium	2,6-Dichlorophenol	Phenol	H_2	
	PCE, TCE	cis-DCE	H_2	Chang et al. (2000)
Dehalobacter	PCE, TCE	cis-DCE	nt[c]	Holliger et al. (1998); Schumacher et al. (1997); Sun et al. (2002)
	1,1,1-Trichloroethane	Chloroethane	H_2	
Dehalococcoides	Chloroethenes	DCEs, VC, ethene	—	Adrian et al. (2000); Bunge et al. (2003); Fennell et al. (2004);
	Chloropropanes	Propene	H_2	Griffin et al. (2004); He et al. (2003b, 2005);
	Chlorobenzenes	Chlorobenzenes		Maymó-Gatell et al. (1997); Ritalahti and
	PCBs, PCDDs[d]	PCBs, PCDDs[d]		Löffler (2004); Wu et al. (2002)
Desulfitobacterium	PCE, TCE	TCE, cis-DCE	H_2	Bouchard et al. (1996); De Wildeman et al. (2003);
	Chlorophenols	Phenol, chlorophenols		Neumann et al. (1994, 2004);
	1,2-Dichloroethane	Ethene	—	Sanford et al. (1996); Smidt et al. (2000); Utkin et al. (1994)
Desulfomonile	3-Chlorobenzoate	Benzoate	H_2	DeWeerd et al. (1991); Sun et al. (2001)
Desulfovibrio	2-Chlorophenol	Phenol		Sun et al. (2000)
Desulfuromonas	2-Chlorophenol	Phenol	Ac[e]	Krumholz (1997); Sung et al. (2003)
	PCE, TCE	cis-DCE	Ac	De Wever et al. (2000);
Geobacter	PCE, TCE	cis-DCE	H_2[f]	Sung and Löffler (2004)
	Trichloroacetate	Dichloroacetate	Ac	Luijten (2003);
Sulfurospirillum	PCE, TCE	cis-DCE	H_2	Scholz-Muramatsu et al. (1995)

[a] Also indicated is the ability to derive reducing equivalents for reductive dechlorination from acetate and/or H_2 oxidation.
[b] Note that some populations are more versatile and use a variety of organic compounds as electron donors.
[c] Not tested.
[d] Only a limited number of congeners have been tested, and growth-linked dechlorination has not been always unequivocally established.
[e] *Desulfovibrio dechloracetivorans* couples acetate oxidation to 2-chlorophenol reductive dechlorination but not to sulfate reduction.
[f] *Geobacter* (formerly *Trichlorobacter*) *thiogenes* does not use H_2 as an electron donor.

Anaeromyxobacter species, for instance, utilize a variety of organic electron donors and H_2 (Luijten *et al.*, 2003; Sanford *et al.*, 2002; Scholz-Muramatsu *et al.*, 1995), whereas *Dehalococcoides* populations are strictly hydrogeno-trophic (i.e., require H_2 as an electron donor) (Adrian *et al.*, 2000; He *et al.*, 2003b; Maymó-Gatell *et al.*, 1997). Importantly, a shared characteristic among known MRDBs is the ability to derive reducing equivalents from H_2, acetate, or both substrates to reduce the chlorinated electron acceptor (Table I). The physiological properties of reductively dechlorinating popu-lations have important implications for the design of efficient enrichment and isolation strategies.

Sample Collection

The majority of reductively dechlorinating populations have been isolated from anaerobic environments, including pristine and contaminated sediments, soils, aquifer materials, and digester sludge. Collecting solid sam-ples from contaminated aquifers typically targets the saturated zone, which often involves direct push or hand auger techniques. No standard protocols for sample handling, transportation, and storage exist. Because many de-chlorinators are strict anaerobes, procedures should obviously limit exposure of the sample materials to air. It is common practice to ship samples on blue ice, although the effects of shipping and storage conditions on the survival of dechlorinating populations are currently unclear. In any event, microcosm setup should commence as soon as possible. All laboratory procedures should be performed in the absence of air (i.e., O_2) inside an anoxic chamber or under a stream of sterile, O_2-free gas (e.g., N_2 or Ar). O_2 is removed by passing the gas stream through a heated (350–400°) glass column filled with copper filings (Hungate, 1969). Sample materials are transferred to auto-claved, appropriately sized (e.g., 1 liter), wide-mouth Mason jars that are filled no more than half-way to allow efficient homogenization. Mason jars are also suitable for direct sampling if an auger is used or for collecting samples from river sediments. The sterile sampling container is submerged below the water–sediment interface, opened, completely filled with sedi-ment, and closed before bringing it back to the surface. For collecting small samples or when traveling, 50-ml Falcon plastic tubes work well.

Microcosm Setup

Microcosms are established inside an anoxic chamber or under a stream of O_2-free gas (Breznak and Costilow, 1994; Fennell and Gossett, 2003). An aseptic technique is applied to the extent possible (Fennell *et al.*, 2001; He *et al.*, 2002). Note that flame sterilization is not possible inside most anoxic chambers. Glass vessels (20- to 160-ml nominal capacity, 20-mm opening such as VWR # 66064–348 and #16171–385) that can be closed with Teflon-lined septa (West Pharmaceutical Services, Lionville, PA) or thick butyl rubber stoppers are suited for microcosm setup. While thick butyl rubber stoppers maintain a tight seal even when punctured many times, Teflon-lined septa should be new and punctured with 25 or greater gauge needles only. The physical characteristics of the sample dictate the technique used for introducing the material into the culture vessels. If appropriate, a homogeneous slurry is prepared with anoxic (ground)water collected at the sampling location, with an anoxic phosphate buffer solution (1–10 mM, pH 7–7.5) or quarter- to full-strength saline (5 mM potassium phosphate, 0.85% NaCl, pH 7.2) (Löffler *et al.*, 1997a). The slurry is mixed constantly on a magnetic stirrer and is transferred into the culture vessels with a plastic pipette with the tip cut off or an inverted 10- or 25-ml glass pipette. Coarse or heterogeneous sampling material is transferred using sterile spatulas. The total amount of solids added ranges between 5 and 50 weight percent of the final liquid volume in each vessel. Medium (see later) is added to yield liquid to headspace ratios of 1–1.6:1, and the vessels are sealed. The microcosms are removed from the anoxic chamber and purged for several minutes with a sterile stream of O_2-free N_2/CO_2 (80/20, v/v) using a bent cannula and are resealed. Alternatively, the vessels are purged with the septum in place. The septum is punctured with two 25-gauge needles, one of which supplies a sterile stream of O_2-free purge gas, while the other needle serves as a vent. The pH is typically circumneutral and is adjusted by changing the CO_2 mixing ratio in the purge gas between pH 6.8 and 8. A medium-containing bottle fitted with a thin pH electrode is used to adjust the CO_2 mixing ratio to yield the desired pH in the microcosms. Purging may be omitted if the pH is in the desired range and if the presence of some H_2 (typically present in mixing ratios of around 3% in the anoxic chamber atmosphere) in the vessels is acceptable. If the pH conditions desired fall out of the range of the bicarbonate buffer system (i.e., pH 6.8–8), Good buffer systems are an alternative (Good *et al.*, 1966). In this case, however, one has to consider that the organic buffering substances themselves may serve as metabolic substrates and that some dechlorinators grow better in the presence of bicarbonate (Müller *et al.*, 2004). Microcosms may also be established using groundwater collected from the same sampling locations to confirm that native conditions are conducive for reductive dechlorination activity.

Electron Donors in Microcoms

All known MRDBs use acetate, H_2, or both compounds as electron donors (see later). Hence, media containing H_2 and acetate will provide the electron requirements for the majority of MRDBs. For microcosm setup, acetate is added directly to the medium before dispensing it to individual culture vessels or is added to individual microcosms from a sterile, 1 M, anoxic aqueous stock solution. Acetate is added in two- to fivefold excess of the theoretically needed amount to achieve complete reductive dechlorination. H_2 (5–30% of the headspace volume of a vessel) is added by syringe. Maintaining a positive pressure in the vessels avoids air introduction during sampling events. Because hydrogenotrophic processes will lower the pressure in the culture vessels, additional gas (e.g., N_2/CO_2 or H_2/CO_2) should be added to maintain a pressure that equals or slightly exceeds atmospheric pressure at all times. This is particularly important when Teflon-lined rubber septa are used. Black (Geo-Microbial Technologies, Inc. Ochelata, OK) or blue (Bellco Glass, Inc., Vineland, NJ) butyl rubber stoppers provide a better seal and avoid air contamination, but might not always be appropriate due to their tendency to absorb many chlorinated compounds. Another precaution against gas leaks through the septum is to incubate vessels in the inverted position.

An alternate strategy to provide the key electron donors (i.e., acetate and H_2) to the dechlorinating populations is the addition of 2–10 mM lactate as a substrate. The fermentation of lactate typically yields acetate and H_2, the key electron donors for MRDBs (Fennell *et al.*, 1997). Lactate is also used directly as an electron donor by certain MRDBs, such as *Desulfitobacterium* and *Sulfurospirillum* species. Lactate can be added to the medium and autoclaved before dispensing to individual culture vessels.

Addition of Chlorinated Compounds

The time when the chlorinated electron acceptor is added depends on the physical–chemical properties of the chloroorganic compound(s) of interest. Water-soluble and nonvolatile chlorinated compounds such as chlorobenzoates are added directly to the bulk medium or are added by syringe to sealed microcosms from anoxic, neutralized, sterile stock solutions. Volatile compounds, such as chlorinated solvents, are added undiluted with a pipette immediately prior to sealing the microcosms or to sealed microcosms using gas-tight syringes (e.g., 1800 series from Hamilton Company, Reno, NV). For improved accuracy and reproducibility, the use of a Chaney adapter is recommended (Hamilton Company). Hydrophobic, nonvolatile, chlorinated compounds are dissolved in an organic solvent and added to sterile culture vessels. The solvent is evaporated under a stream of sterile N_2 prior to the addition of the sample material and microcosm setup (Bedard *et al.*, 1996; Fennell *et al.*, 2004). Because many chloroorganic compounds are toxic, the initial strategy is to add the minimum amount of chlorinated electron acceptor that can be quantified readily with the available analytical instrumentation. It is important, however, to provide sufficient chlorinated substrate to support growth during the enrichment process, as the chloroorganic compounds serve as energy-yielding electron acceptors for the target organisms. In the case of chlorinated solvents, inhibitory effects can be avoided by adding the chlorinated electron acceptor dissolved in an inert, hydrophobic carrier such as hexadecane (Krumholz et al., 1996; Sung et al., 2003). Alternatively, halogenated volatile compounds can be sorbed to a Tenax-TA (a solid sorbent, Alltech, Deerfield, IL) bed and delivered to the cultures via the headspace (Brennan and Sanford, 2002). This method is more tedious, as the Tenax is physically separated from the culture medium; however, any interference between carrier material and microbial populations (i.e., the dechlorinators) is avoided. The enrichment and isolation procedures described later focus on the enrichment and isolation of MRDBs that capture energy from chloroethene reductive dechlorination, but similar strategies are applicable to isolating organisms with other electron acceptor utilization patterns.

Analytical Procedures

Chloroorganic compounds are analyzed in the headspace, the aqueous phase, or following extraction of an aqueous sample or the entire microcosm/culture with an organic solvent, as is appropriate for the chloroorganic electron acceptor(s) added. Pressurized systems may interfere with the sampling of volatile analytes (e.g., chlorinated solvents) from the headspace, and appropriate precautions should be taken using gas-tight Hamilton syringes equipped with a valve (e.g., Hamilton SampleLock syringes with a removable male Luer Lock adapter; part #35083). Acetate is measured by high-performance liquid chromatography (Sanford et al., 1996) or gas chromatography (Stansbridge et al., 1993; Vera et al., 2001). H_2 partial pressures are monitored by gas chromatography using a thermal conductivity detector for concentrations above 50 Pa and a reduction gas (AMETEK Process and Analytical Instruments [formerly Trace Analytical], Newark, DE) detector for H_2 concentrations below 50 Pa (Löffler et al., 1999). Chloride release is monitored using the mercury(II) thiocyanate-iron(III) method (Bergmann and Sanik, 1957; Sung et al., 2003) or by ion chromatography (Dionex Corporation, Sunnyvale, CA). The detection of changes in chloride concentration requires a low chloride background and is often not possible in microcosms. Chloride measurements in enrichment cultures may require the preparation of medium with reduced chloride content (Sung et al., 2003). Substrates are replenished when depleted, or the cultures are transferred to fresh medium before the chlorinated electron acceptor(s) has been completely consumed.

Medium

A variety of similar medium recipes exist (Atlas, 2004), and a standard synthetic basal salts medium used successfully for the isolation of reductively dechlorinating microorganisms is described. Media are prepared using the Hungate technique (Balch *et al.*, 1979; Breznak and Costilow, 1994; Hungate, 1969; Miller and Wolin, 1974; Sowers and Noll, 1995).

The following procedure is used to prepare 1 liter of medium.

1. Add 900 ml of distilled water to a 2-liter Erlenmeyer flask and mix on a magnetic stir plate.
2. Add 10 ml of the 100-fold concentrated salts stock solution.
3. Add 1 ml of each trace element solution.
4. Add 0.25 ml of a 0.1% (w/v) resazurin stock solution.
5. Add acetate or the electron donor of choice (1–10 mM). Heat-labile electron donors are added from anoxic, filter-sterilized stock solutions after autoclaving.
6. Add distilled water to a total volume of 1 liter.
7. Transfer the medium to a 2-liter round flask (Pyrex).
8. Bring medium to a boil and boil for 10 min. Use a reflux cooler to avoid evaporative water loss.
9. Transfer flask to an ice bath (careful, hot liquids!) and cool the medium to room temperature under a stream of O_2-free N_2. Remove the reflux cooler when no more steam is formed to conserve cooling water.
10. Add 30 mM $NaHCO_3$, 0.2 mM L-cysteine, and 0.2 mM $Na_2S \times 9H_2O$.

H_2S is a diprotic acid and the following pH-dependent equilibrium exists:

$$H_2S_{(g)} \leftrightarrow H_2S_{(aq)} \leftrightarrow HS^-_{(aq)} + H^+_{(aq)} \leftrightarrow S^{2-}_{(aq)} + H^+_{(aq)}$$

To avoid loss of *H₂S* from the medium while the flask is flushed with purge gas, quick manipulation of the medium following H_2S addition is recommended. In addition, sulfide "poisons" the glass pH electrode, and exposure to sulfide-containing solutions should be limited. No toxic effects of <0.5 mM sulfide on reductively dechlorinating populations have been observed, although sulfide may become inhibitory at higher concentrations. Sulfide not only serves as a reductant. but also is a required sulfur source for some dechlorinators (Sung *et al.*, 2003).

11. Adjust the pH to 7–7.5 by varying the CO_2 content of the purge gas.
12. Dispense medium into appropriately sized vessels that have been purged of air. Do not fill with more than 75% of the nominal capacity of the vessel.

13. Close the vessels without allowing air to enter and secure the stoppers with aluminum crimps.
14. Autoclave the vessels in inverted position after the pink medium turns clear (i.e., the redox indicator resazurin is reduced).
15. Remove the vessels from the autoclave when the medium has reached temperatures below 60°. Handling closed containers with hot liquids is dangerous and must be avoided.
16. Store the medium in the dark at room temperature or 4°. If 160-ml serum bottles are used, this procedure yields 10 bottles with 100 ml of medium. 100-fold concentrated salts stock solution contains per liter: NaCl, 100 g; $MgCl_2 \times 6H_2O$, 50g; KH_2PO_4, 20g; NH_4Cl, 30g; KCl, 30g; $CaCl_2 \times 2H_2O$, 1.5g

Trace element solution A contains per liter: HCl (25% solution, w/w), 10 ml; $FeCl_2 \times 4H_2O$, 1.5 g; $CoCl_2 \times 6H_2O$, 0.19 g; $MnCl_2 \times 4H2O$, 0.1 g; $ZnCl_2$, 70 mg; H_3BO_3, 6 mg; $Na_2MoO_4 \times 2H_2O$, 36 mg; $NiCl_2 \times 6H_2O$, 24 mg; $CuCl_2 \times 2H_2O$, 2 mg

Trace element solution B contains per liter: $Na_2SeO_3 \times 5H_2O$, 6 mg; $Na_2WO_4 \times 2H_2O$, 8 mg; NaOH, 0.5 g

The vitamin solution is prepared as a 1000-fold stock solution and contains per liter (Wolin *et al.*, 1963): biotin, 20 mg; folic acid, 20 mg; pyridoxine hydrochloride, 100 mg; riboflavin, 50 mg; thiamine, 50 mg; nicotinic acid, 50 mg; pantothenic acid, 50 mg; vitamin B_{12}, 1 mg; *p*-aminobenzoic acid, 50 mg; thioctic acid, 50 mg. The pH is adjusted to 7.5 with NaOH. The suspension is stirred rigorously, and aliquots are frozen and stored in the dark. A 200-fold concentrated working solution is prepared by diluting the 1000× stock 1:5 in distilled water. The solution is purged with N_2 to remove O_2, filter sterilized, and stored in the dark at 4°.

Some MRDBs have specific nutritional requirements. *Desulfomonile tiedjei* requires additional vitamin supplements (i.e., 1,4-naphthoquinone, lipoic acid, and hemin) (DeWeerd *et al.*, 1990), and growth of *Dehalobacter restrictus* depends on the amino acids arginine, histidine, and threonine (Holliger *et al.*, 1998). Undefined sources of vitamins and growth factors, such as clarified rumen fluid (Shelton and Tiedje, 1984) or fermented yeast extract (Fennell *et al.*, 1997), can be supplemented to the anaerobic medium. Vitamins are added to the autoclaved medium from sterile, anoxic stock solutions.

Transfers and Enrichment Cultures

Transfers to fresh medium should occur during the active phase of growth (i.e., dechlorination) and before all the chlorinated electron acceptor has been consumed. A decline in cell numbers was reported in

Dehalococcoides-containing cultures once the metabolic electron acceptor had been consumed (He *et al.*, 2003b).

Sediment-Free Cultures

The first goal is to establish sediment-free cultures from microcosms that exhibit reductive dechlorination activity. Dechlorinating microcosms are shaken vigorously, and the solids are allowed to settle. Transfers (1–2%, v/v) are made by syringe using 21-gauge needles. Tilt the culture vessel at an angle to remove supernatant and avoid blocking the needle with solid particles.

Selective Enrichment

Dechlorinating consortia at this stage of enrichment often contain multiple dechlorinating populations, and replicate cultures maintained under different conditions may yield different isolates with distinct physiological properties. The physiological properties of the dechlorinating population(s) of interest determine subsequent enrichment strategies. Hence, a first step involves comparing dechlorination rates and end points in response to different treatment regimes (e.g., electron donors). For instance, if acetate alone supports the dechlorination process of interest, a productive enrichment strategy would use basal salts medium with acetate (electron donor) and the chloroorganic compound (electron acceptor) as energy yielding redox couple.

Nonmethanogenic Cultures

Typically, methanogenesis will occur in the microcosms and enrichment cultures exhibiting reductive dechlorination activity. To eliminate methanogens from the enrichment cultures, the following strategies have been employed successfully. (i) Inhibition of methanogenic activity in the presence of high concentrations of chlorinated solvents has been reported (DiStefano *et al.*, 1992; He *et al.*, 2003a). Hence, continued transfers to medium with high concentrations of chlorinated electron acceptor(s) and regular transfers to fresh medium following the onset of reductive dechlorination activity may result in nonmethanogenic cultures. (ii) The thermodynamics of reductive dechlorination are considerably more favorable than methane formation (Dolfing, 1999, 2003; Dolfing and Beurskens, 1995; Löffler *et al.*, 1999), and reductive dechlorination occurs at electron donor (i.e., H_2, acetate) concentrations that limit or prohibit methanogenic activity. Hence, maintenance and transfer of cultures in medium with low electron donor concentrations (H_2 <100 nM, acetate <500 μM) (He and

Sanford, 2004; Löffler *et al.*, 1999; Yang and McCarty, 1998) can be a successful strategy to select against methanogens. (iii) A third strategy to inhibit methanogens is the addition of 2 mM 2-bromoethanesulfonate (BES). BES, however, is not a selective inhibitor of methanogenic activity and may also affect some dechlorinators (Löffler *et al.*, 1997b). Hence, prolonged enrichment in the presence of BES may eliminate some MRDBs. A limited number (<3) of transfers in the presence of BES was used successfully to eliminate methanogens without losing the initial dechlorinating activity (Löffler *et al.*, 1997b).

Electron Donors for Enriching and Isolating MRDBs

Known MRDBs use acetate and/or H_2 as electron donors. Populations such as *Dehalococcoides* and *Dehalobacter* are strictly hydrogenotrophic, *Desulfuromonas* species use acetate but not H_2 as electron donor, whereas others exhibit versatility and couple reductive dechlorination to the oxidation of a variety of substrates, including acetate and H_2. Most hydrogenotrophic MRDBs grow best when acetate is supplemented as a carbon source (Adrian *et al.*, 2000; He *et al.*, 2003b; Maymó-Gatell *et al.*, 1997). The electron donor requirements of the dechlorinating population(s) of interest should be evaluated, as this information will assist in designing efficient enrichment strategies. For instance, comparing dechlorination rates and end points in response to the electron donor(s) (i.e., acetate or H_2) added to the microcosms will determine whether the process of interest requires H_2 or acetate, or is supported by both electron donors. Note that the preparation of H_2-free medium is not trivial because commercially available gases contain some H_2. This must be taken into account when reductive dechlorination activity is evaluated in vessels purged with "H_2-free" gases. If residual H_2 concentrations are high enough to permit significant dehalogenation, H_2 can be removed from all gas streams using a catalytic H_2 converter available from AMETEK Process & Analytical Instruments (formerly Trace Analytical)(Newark, DE). Electron donors are supplied in at least twofold excess to achieve complete reductive dechlorination of the amount of chlorinated electron acceptor provided. Each reductive dechlorination reaction consumes two reducing equivalents, $2H^+ + 2e^-$, which can also be expressed as 2[H]. One 2[H] is available from H_2 oxidation ($H_2 \rightarrow 2H^+ + 2e^-$). The oxidation of acetate (acetate$^- + 4H_2O \rightarrow 2HCO_3^- + 4 H_2 + H^+$) formally yields four 2[H] or eight reducing equivalents. Because most enrichment cultures contain H_2/CO_2 acetogens, acetate formation will occur in microcosms and enrichment cultures amended with H_2, and acetogenesis will compete with hydrogenotrophic reductive dechlorination for H_2 as the electron donor.

Most microcosms and enrichment cultures will readily oxidize formate to H_2, and much higher H_2 mass loadings can be achieved to support hydrogenotrophic reductive dechlorination (Sanford, 1996). Some MRDBs, including *Desulfitobacterium* and *Sulfurospirillum,* use formate directly as a source of reducing equivalents for reductive dechlorination.

Selecting against H_2/CO_2 Acetogens

Eliminating H_2/CO_2 acetogens from H_2-fed, dechlorinating enrichment cultures is a major obstacle and is often not achieved in liquid cultures. Replacing the commonly used bicarbonate buffer with an alternative buffer system [e.g., morpholinepropanesulfonic acid (MOPS)] (Good *et al.,* 1966) to achieve a constant medium pH is one possible approach to select against H_2/CO_2 acetogens. Reductively dechlorinating populations exhibit significantly lower H_2 consumption threshold concentrations than H_2/CO_2 acetogens (Löffler *et al.,* 1999). Hence, a promising approach to counterselect H_2-utilizing populations not involved in dechlorination is transferring repeatedly under conditions of controlled H_2 flux. The H_2 concentration should be kept below the threshold concentration supporting growth of H_2/CO_2 acetogens (i.e., <100 nM). Ideally, this is achieved using a fatty acid-oxidizing partner population that forms a syntrophic relationship with the hydrogenotrophic dechlorinating population(s) of interest. In such systems, the thermodynamics of fatty acid oxidation control the maximum H_2 concentration in the system, and H_2/CO_2 acetogens can be diluted out following repeated transfers. Such a syntrophic association between an acetate-oxidizing population and a hydrogenotrophic, vinyl chloride-respiring population was a critical step in the isolation of *Dehalococcoides* sp. strain BAV1 (He *et al.,* 2002, 2003a,b). Not all enrichment cultures contain fatty acid-oxidizing populations; however, propionate and butyrate oxidizers are available in type culture collections (e.g., www.atcc.org/; www.dsmz.de). A drawback of exploiting syntrophic processes as an enrichment step, in particular those involving acetate oxidation, is the very long incubation periods lasting many weeks to months. Nevertheless, this approach can be productive in eliminating H_2/CO_2 acetogens from dechlorinating cultures.

Pasteurization

Some dechlorinators are spore formers, and dechlorination activity resumes following pasteurization. For instance, pasteurization was a productive approach for isolating *Desulfitobacterium chlororespirans* Co23 (Sanford *et al.,* 1996), as nonspore-forming vegetative cells are killed at 80°. Once spore-like structures have been observed microscopically (typically in stationary phase cultures), their ability to resist heat treatment can be tested. Culture vessels containing fresh anaerobic medium amended with the electron donor(s) and the chloroorganic electron acceptor(s) are equilibrated to 80° in a water bath. A 1.0% (v/v) inoculum from an actively dechlorinating culture and a stationary phase culture (i.e., all chlorinated electron acceptor reduced) are added to duplicate bottles, respectively. The culture vessels are immediately placed back in an 80° water bath for 10 min. The water bath temperature should be monitored carefully to verify that the temperature does not exceed 80°. The culture vessels are then shaken gently in an ice-water bath and cooled to room temperature, and dechlorination activity is monitored in the pasteurized cultures.

Resistance to Antibiotics

Known *Dehalococcoides* isolates are resistant to penicillin-type antibiotics such as ampicillin, and at least some strains grow in the presence of the glycopeptide antibiotic vancomycin. Antibiotics interfering with cell wall biosynthesis inhibit cell division and growth but do not kill nondividing cells. To ensure that sensitive populations are eliminated, antibiotics should be added to at least three consecutive 1–2% (v/v) transfers at concentrations of 50 to 100 μg/ml. Ampicillin is added from aqueous, anoxic, filter-sterilized stock solutions (50 mg/ml).

Isolation

Metabolic reductive dechlorination is a strictly anaerobic process, although some MRDBs are facultative aerobes or tolerate oxygen exposure. Most MRDBs, however, are strict anaerobes and are very sensitive to oxygen exposure. Hence, successful enrichment and isolation procedures involve strict anaerobic techniques (Balch *et al.*, 1979; Hungate, 1969; Miller and Wolin, 1974). Most dechlorinating populations are relatively easy to propagate in mixed culture but may require extreme care during the isolation process and when maintained in pure culture. *Dehalococcoides* populations are very sensitive to air (i.e., oxygen) exposure and require reduced medium conditions. *Dehalococcoides* organisms will not grow in medium with resazurin in its oxidized (i.e., pink) form. Transfers are conveniently performed with sterile plastic syringes and hypodermic needles that have been flushed with O_2-free nitrogen gas. To remove any residual oxygen contained in the plastic material, the syringes are filled with a freshly prepared, filter-sterilized, 1 mM Na$_2$S $\times$ 9H$_2$O solution prepared in the culture medium. This solution is kept in the syringe barrel for 5–10 min before the solution is discarded, and the syringe is used for inoculum transfer.

Once stable climax consortia are obtained, and no further enrichment can be achieved under the medium and incubation conditions applied, isolating the dechlorinating organisms can be accomplished in liquid medium and/or soft agar cultures. Both strategies apply the dilution-to-extinction principle. Although most MRDBs do not grow on solid surfaces and cannot be propagated using agar slants, many dechlorinating populations do form colonies in semisolid medium prepared with 0.5–0.7% low-melting agarose (Adrian *et al.*, 2000; He *et al.*, 2003b; Krumholz *et al.*, 1996; Sung *et al.*, 2003).

The following procedure is used for dilution-to-extinction in soft agar tubes.

1. To prepare soft agar cultures, weigh 0.6% (weight by volume) of low melting agarose (gelling temperature $<30°$) into 20-ml glass autosampler headspace vials (e.g., VWR) or 24-ml Balch tubes (Bellco).

2. Prepare anaerobic medium amended with acetate (1–10 mM).

3. Purge vials with a sterile stream of N_2/CO_2 (80/20, v/v), being careful not to disturb the agarose powder with the N_2 stream.

4. Dispense 9 ml of medium into 20-ml glass vials, close the vials with Teflon-lined rubber stoppers, and apply aluminum crimps. Alternatively, dispense the medium inside an anaerobic chamber. Cover the vials with aluminum foil to prevent the agarose powder from escaping during transfer to the chamber. Note that most anaerobic chambers contain N_2 and H_2, and quick handling of open medium vessels is needed to minimize loss of CO_2 from the bicarbonate-buffered medium and to maintain the desired pH.

5. Prepare an equal number of vials (dilution tubes) containing 9 ml of medium without low-melting agarose.

6. After autoclaving, cool the medium to around $50°$ and shake the vials gently to distribute the molten agarose (avoid forming gas bubbles). Transfer the vials to a $35°$ water bath.

7. Make additions (e.g., vitamins) as needed. Return the vials to a $35°$ water bath immediately to prevent premature gelling of the agarose.

8. Label all vials and inoculate the 10^{-1} to 10^{-10} liquid dilution vials (the same syringe can be used for the entire dilution series) (Fig. 1). Shake the inoculated vial with the syringe in place, and rinse the needle and the syringe barrel once before withdrawing another milliliter for each subsequent transfer.

9. Remove the agarose-containing vials from the water bath. Inoculate the vials with 1 ml from the liquid dilution-to-extinction series vials. The same syringe can be used if the series is started with the most diluted culture (i.e., 10^{-10}) (Fig. 1).

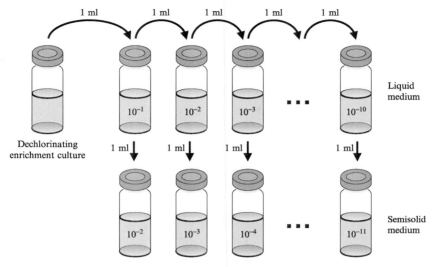

FIG. 1. Preparation of dilution-to-extinction series in liquid and semisolid medium for obtaining isolates.

10. Mix gently immediately after inoculation (avoid forming gas bubbles) and place the vials in upright position in an ice bath until the agarose is solidified (<5 min).

11. If needed, add H_2 (1–5 ml) by syringe to support hydrogenotrophic dechlorinators.

12. Add the chloroorganic electron acceptor directly to the bulk medium, or to individual vials before or after autoclaving, as appropriate. Volatile compounds may be supplied via the gas phase after the agarose solidifies.

13. Incubate the vials in inverted position. Note that some water will form during the gelling process, which will collect on top of the septum. Do not invert the vials, as the liquid may contain organisms and contaminate the surface of the agarose plug.

14. Monitor colony formation biweekly. It will take several weeks to months before small, typically whitish to opaque colonies form. Use a magnifying glass and proper lighting to identify colonies. Not all *Dehalococcoides* populations form visible colonies; if colonies do form, they are tiny and successful observation and transfer require practice and diligence.

Colony Transfer

The vial of the highest dilution with clearly separated colonies is used for transfers to fresh liquid medium to test for dechlorinating activity. Isolating a pure culture often requires repeated dilution-to-extinction series, and it is recommended that a subsequent dilution series is initiated immediately with a single colony picked from an agar shake culture. The colony material is transferred to a vial containing 1.5 ml of anaerobic medium and a small stir bar. The cells are suspended carefully, and 1 ml is used to start another dilution-to-extinction series in liquid and semisolid medium as depicted in Fig. 1.

The following procedure is used for transferring colonies from soft agar tubes.

1. Agar shake cultures are always kept in an inverted position. Mount a vial with visible, clearly separated colonies on a stand with a clamp. Install proper lighting and use a second stand to mount and position a magnifying glass.

2. Open the vial. Note that because vials may be under positive pressure, take appropriate precautions (i.e., safety goggles). A cannula to provide a sterile stream of nitrogen to the vial and thus limit air intrusion may be installed, although this may not be necessary if the procedure is performed quickly and the vial is discarded.

3. Use a 1-ml plastic syringe with a 5- to 7-cm-long needle that has been flushed with O_2-free N_2 and rinsed with a 1 mM Na_2S solution. Fill the needle and syringe with anaerobic medium (0.1–0.2 ml).

4. Target a well-isolated colony without touching another colony in the agarose plug. Transfer the colony material into the syringe by pulling the plunger very carefully. Visually inspect that colony material has been transferred into the syringe barrel.

5. Transfer the colony material to a new vial to test for dechlorinating activity and/or initiate another dilution-to-extinction series. If the colony material has been visibly transferred to the syringe barrel, a small gauge needle may be mounted before puncturing the septum.

Should no colonies form, or if the procedure is not productive, repeated dilution-to-extinction series in liquid medium initiated with the highest dilution that exhibits dechlorinating activity may result in further enrichment and isolation of pure cultures. Dechlorinating activity in liquid cultures is determined with the routine analysis designed for the compound of interest. Note, however, that measurements in agar shake cultures may not always indicate activity, even though colonies of the dechlorinating organism of interest developed. Also consider that alternate substrates may be

TABLE II
Examples of 16S rRNA Gene-Targeted Primers Used in Direct and Nested PCR Approaches to Detect MRDBs[a]

Primer[b]	Sequence (5'-3')	Target	Annealing temperature (°)[c]	Amplicon length (bp)	Reference
Deb 179F Deb 1007R	TGT ATT GTC CGA GAG GCA ACT CCC ATA TCT CTA CGG	Dehalobacter	53	828	Schlötelburg et al. (2002)
Dhc730F Dhc1350R	GCG GTT TTC TAG GTT GTC CAC CTT GCT GAT ATG CGG	Dehalococcoides	58	620	Bunge et al. (2003); He et al. (2003a); Löffler et al. (2000)
FpDhc1 RpDhc1377	GAT GAA CGC TAG CGG CG GTT GGC ACA TCG ACT TCA A	Dehalococcoides	55	1377	Hendrickson et al. (2002)
Dsb 174F Dsb 1373R	AATACCGNATAAGCTTATCCC TAGCGATTCCGACTTCATGTTC	Desulfitobacterium	55*	1199	El Fantroussi et al. (1997)
Dsb 460F Dsb 1084R	TCTTCAGGGACGAACGGCAG CATGCACCACCTGTCTCAT	Desulfitobacterium	55*	624	El Fantroussi et al. (1997)
Dsm 59F Dsm 1054R	CAAGTCGTACGAGAAACATATC GAAGAGGATCGTGTTTCCACGA	Desulfomonile	55*	995	El Fantroussi et al. (1997)
Dsm 205F Dsm 628R	GGGTCAAAGTGCGGCCTCTCGACG GCTTTCACATTCGACTTATCG	Desulfomonile	55*	423	El Fantroussi et al. (1997)
Dsf 205F Dsf p671R	AACCTTCGGGTCCTACTGTC GCCGAACTGACCCCTATGTT	Desulfuromonas	58	466	Löffler et al. (2000); Ritalahti and Löffler (2004)
Geo 63F Geo 418R	CAGGCCTAACACATGCAAGT CCGACCATTCCTTAGGAC	Geobacteriaceae	62	1443	Dennis et al. (2003)
8F 1541R	AGA GTT TGA TCC TGG CTC AG AAG GAG GTG ATC CAG CCG CA	Bacteria	55	1533[d]	Zhou et al. (1997)

[a] Also indicated are amplicon lengths and annealing temperatures.
[b] F, forward primer; R, reverse primer.
[c] Denaturing, annealing, and elongations times are 45 s at the respective temperatures (60 s when noted with an asterisk).
[d] The lengths of 16S rRNA genes from different organisms may vary slightly.

productive in isolating dechlorinating populations. For instance, *Desulfuromonas, Anaeromyxobacter,* and *Geobacter* species use several electron acceptors, and agar shake cultures with fumarate (5–20 mM) as the electron acceptor and acetate (5 mM) as the electron donor produce much larger, reddish colonies after shorter incubation periods than when chlorinated substrates are used. Isolated colonies are then tested for dechlorinating activity in liquid medium.

Nucleic Acid-Based Approaches

Culture-independent, nucleic acid-based approaches are useful to detect, monitor, and quantify dechlorinator-specific genes in environmental samples, laboratory microcosms, and enrichment cultures. A common target is the 16S rRNA gene, and primers specifically targeting individual groups of MRDBs have been designed (Table II). Direct and nested polymerase chain reaction (PCR) approaches have been used successfully to assess the presence of specific dechlorinators in environmental samples (Fennell *et al.*, 2001; Hendrickson *et al.*, 2002; Löffler *et al.*, 2000) and dechlorinating bioreactors (Schlötelburg *et al.*, 2000, 2002). Nested PCR offers unsurpassed sensitivity and, hence, has the least probability of yielding false-negative results (El Fantroussi *et al.*, 1997; Löffler *et al.*, 2000). Quantitative Real-Time (qRTm) PCR has become a standard assessment and monitoring tool, as this technique offers high sensitivity and has the added advantage of providing quantitative information. Quantitative approaches were instrumental in advancing molecular diagnostics in a clinical setting and are now routinely employed in many fields, including environmental microbiology (Mackay, 2004; Sharkey *et al.*, 2004; Walker, 2002). For instance, qRTm PCR has been used to demonstrate growth-linked reductive dechlorination (He *et al.*, 2003a,b; Smits *et al.*, 2004), to establish cause-and-effect relationships in bioremediation applications (Lendvay *et al.*, 2003), to monitor nitrification in wastewater treatment plants (Harms *et al.*, 2003), to examine anaerobic hydrocarbon degradation based on a catabolic gene (Beller *et al.*, 2002), and numerous other environmental applications. The following section describes PCR-based procedures for the qualitative and quantitative assessment of selected MRDBs.

Extracting PCR-Amplifiable DNA

A variety of DNA extraction protocols and commercially available kits exist and have been used to obtain quality DNA. Not all procedures, however, work equally well with all sample materials. Environmental samples may contain compounds that interfere with subsequent PCR amplification, such as humic acids, and DNA isolation from soils and sediments with high clay or organic content may require method optimization (Hurt *et al.*, 2001; Yeates *et al.*, 1998). Preference should be given to environmentally friendly procedures that avoid hazardous phenol/chloroform waste. The initial goal of all procedures is to collect as many bacterial cells as possible from the sample material and to subsequently obtain their DNA. Optimization aims at enhancing cell recovery and cell lysis, as well as purifying high-quality DNA. The following section describes procedures that were employed successfully to extract PCR-amplifiable DNA from a variety of sample materials.

DNA Isolation from Bacterial Cultures

Extracting DNA from cell suspensions typically involves centrifugation, washing and suspending the pellet in buffer, cell lysis, and DNA separation. If column-based kits are used, the binding capacity of the column material determines how much DNA can be applied and, hence, determines the optimal quantity of cells desired for DNA extraction. The following procedure is designed for use of column-based extraction using the QIAamp DNA mini kit (Qiagen, Valencia, CA) or similar, commercially available kits.

Procedure.

1. Collect 10^7–10^8 cells by centrifugation (4°, 30 min, 7500g). Note that some cells are difficult to spin down and that other centrifugation conditions (i.e., longer time and greater g force) may be required. Light microscopic examination of the supernatant following centrifugation is recommended to ensure that no cells remain in suspension.

2. Freezing the cell pellet is recommended to improve subsequent cell lysis.

3. If the Qiagen tissue kit is used, DNA is extracted following the manufacturer's protocol for bacterial cell pellets with the following modifications.

   a. For improved cell lysis, 20 μl of lysozyme (100 mg/ml) (Sigma, St. Louis, MO) and 10 μl of achromopeptidase (25 mg/ml) (Sigma) are added, and the mixture is incubated at 37° for 1 h.

   b. Then, 45 μl of proteinase K (20 mg/ml) (Invitrogen) is added, and the mixture is incubated at 55° for at least 1 h.

   c. Following the lysis steps, and before removing any precipitates, a 5-μl aliquot is analyzed microscopically to verify complete cell lysis.

4. The remainder of the procedure is performed according to the kit manufacturer's protocol. The DNA is eluted in 50–200 μl of 10 mM Tris-HCl buffer (pH 8.5) or nuclease-free water and is quantified.

DNA Isolation from Solids

For small sediment, aquifer, and soil sample sizes (1 g), the MoBio soil extraction kit (MoBio Laboratories Inc., Carlsbad, CA) or similar products have been used successfully. Use of the inhibitor removal solution provided with the MoBio soil extraction kit and the protocol for "maximum yields" is recommended. Depending on the material type and the associated amount of biomass, it may be necessary to perform replicate extractions. Kits that accommodate larger sample masses (10 g) (e.g., MoBio Ultra-Clean Mega Soil DNA kit) or alternative extraction procedures (Hurt et al., 2001; Yeates et al., 1998) are available. A new PowerSoil DNA kit by MoBio may be useful if soils contain high amounts of humic acids. In each of these extraction procedures, the DNA is eluted in final volumes of 50–100 μl of 10 mM Tris-HCl buffer (pH 8.5) and is quantified.

DNA Isolation from Groundwater

The following describes the MoBio Water DNA kit for extracting DNA from water samples. One to 2 liters of groundwater is filtered through the polyether sulfone membrane filters (0.2 μm pore size) supplied with the kit. The volume filtered depends largely on the amount of biomass and particulate matter in the groundwater sample. Use of a vacuum manifold allows filtering multiple samples simultaneously. Membrane filters containing the solids (including bacterial cells) are removed from the filter unit with flame-sterilized forceps, transferred aseptically to a sterile petri dish, and frozen at −20°. After freezing, each filter is quickly cut into strips with a flame-sterilized razor blade and is transferred aseptically to the bead tubes supplied with the kit. The MoBio Water DNA protocol is followed as recommended by the manufacturer, and the DNA is eluted from the column in 3 ml of 10 mM Tris-HCl buffer (pH 8.5). The DNA is concentrated further by adding 0.3 ml of sterile 5 M NaCl solution and two starting volumes (6 ml) of ice-cold, absolute, molecular biology-grade ethanol. Collecting the final DNA is easier if the contents are first transferred from the 50-ml tube to a sterile 15-ml conical plastic centrifuge tube. The tube is refrigerated (4°) for 1 h, and then the DNA is collected by centrifugation (45 min at 7500g, 4°). The pellet is washed with 2 ml of ice-cold, 70% ethanol, collected by centrifugation, and the liquid is decanted. The pellet is air dried and suspended in 50 to 100 μl of 10 mM Tris-HCl buffer (pH 8.5) and is quantified. If the groundwater contains high amounts of particulate matter, the membrane filter with the sediment/bacterial biomass can be processed using a soil DNA kit, as described earlier.

DNA Quantification

DNA is quantified with a UV-VIS spectrophotometer at a wavelength of 260 nm. Readings should fall between 0.05 and 0.5 absorption units, or dilution with distilled water is necessary. For reference, 25 μg/ml of dsDNA will give an A_{260} reading of 0.5 when quartz cuvettes with a 1-cm light path are used. The concentration of DNA is calculated according to the formula ng of DNA/μl = $OD_{260\ nm} \times 50 \times$ dilution factor (Ausubel *et al.*, 1997). For quantifying dilute DNA solutions, a fluorometer is recommended. A fluorescent dye, such as PicoGreen, binds to double-stranded DNA, and the fluorometer allows the detection of as little as 25 pg/ml concentrations (Sambrook and Russell, 2001).

PCR for the Detection and Quantification of Dechlorinating Populations

PCR is a very sensitive technique, and all PCR procedures should include appropriate positive and negative controls to ensure that amplification of template DNA was specific. It is also advisable to perform a dilution series with the control DNA to directly observe the detection limit for the procedure as it is performed. Note that typically much less pure culture DNA is needed to obtain a positive signal than environmental sample DNA. All PCR ingredients and PCR tubes are kept on ice until transfer to the heating/cooling block of the thermocycler.

16S rRNA Gene-Based Qualitative Detection

Direct PCR utilizes isolated DNA directly as a template with MRDB-targeted primers (Table II). This requires a fairly high number of targets in the extracted DNA (>500 copies/μl of DNA). Hence, false-negative results can occur if organisms present in low population sizes are being sought, and data must be interpreted accordingly. To increase sensitivity and to detect just a few target copies per microliters of original DNA sample, a nested PCR approach is used. Fewer than 10 16S rRNA gene target copies per microliter of DNA can be detected; however, two separate rounds of amplification are required. An initial amplification of the communities' 16S rRNA genes is performed using universal bacterial primers 8F and 1541R (Table II) and 20–100 ng of community DNA as template as described (Zhou *et al.*, 1997) prior to amplification with MRDB-specific primer pairs. All PCR protocols are performed similarly and vary with regard to primers used and annealing temperatures. A master mix that contains all ingredients, except template DNA, is prepared according to Table III. To calculate the amount of each reagent to add to the master mix, multiply the number of microliters needed for one reaction by the number of samples to be analyzed plus one extra mixture per 10 reactions to account for pipetting losses (e.g., for 10 PCR tubes, prepare enough mixture for 11 tubes). The reaction mixture is blended by vortexing briefly and is collected at the bottom of the tube by centrifugation. If the volume of any reagent is increased or decreased, the amount of water must be adjusted accordingly to maintain the same total volume. In a 100-μl reaction volume, the master mix (95 μl) is placed into each PCR tube, and 5 μl of template DNA is added and mixed by pipetting or tapping the tube gently. Cycle conditions generally consist of an initial denaturation at 94° for 2 min 10 s, followed by 30 cycles of amplification (94°, 30s; 55°, 45s; 72°, 2min 10s) and a final elongation at 72° for 6 min. Following amplification, 3 μl of PCR product is analyzed on a 1% agarose gel run in 1× TAE buffer at 120 V for 30 min. DNA fragments are visualized by ethidium bromide staining (10 μl of a 10-mg/ml stock solution per 100 ml of distilled water) for 30 min, destaining in distilled water for 10 min, and viewing under UV light. Amplified community 16S rRNA genes of approximately 1500 bp in size are expected with the universal bacterial primers. Smaller reaction volumes (e.g., 25 or 50 μl) may be used to conserve reagents, although the reaction volume is driven by the type of analyses to be performed with the amplicons. For nested PCR, 1:10 and 1:50 dilutions of the 16S rRNA gene amplicons obtained with the universal primers are used as templates for PCR with dechlorinator-targeted primer pairs (Table II). The PCR reaction mix composition is essentially the same as for the universal primer

TABLE III
GENERAL PCR REACTION MIX FOR A 100-μL REACTION VOLUME

Component	Final concentration	Volume (μl)
Sterile, DNA-free water	(add to 100 μl)	61.3
10× PCR buffer	1×	10
MgCl$_2$ (25 mM)	2.5 mM	10
Bovine serum albumin (20 mg/ml)	0.13 mg/ml	0.7
dNTP mix (1:1:1:1, 2.5 mM each)	0.25 mM each	10
Primer 1 (8F, 5 μM)	0.05 μM	1
Primer 2 (1541R, 5 μM)	0.05 μM	1
Taq DNA polymerase	4 U/μl	1
Template DNA	20 pg–100 ng	5

pair, except that the reaction volume is reduced to 25 μl. Again, a master mix is prepared, and 23.75 μl is dispensed into individual PCR tubes. Template DNA (1.25 μl of diluted 16S rRNA gene amplicons generated with the universal primers) is added to each tube and mixed. PCR thermocycler conditions are as described earlier, except that the annealing temperatures for any individual primer pair may vary. Annealing temperatures and expected amplicon sizes for some MRDB-specific PCR primers are shown in Table II, and any exceptions to the PCR procedure for an individual primer pair are also noted. The nested PCR approach requires only 1–10 16S rRNA gene copies in the initial round of community 16S rRNA gene amplification to produce a visible band on an ethidium bromide-stained agarose gel after the MRDB-targeted PCR. However, as all analytical assays, PCR has an inherent detection limit, and false-negative results remain possible.

16S rRNA Gene-Based Detection and Enumeration of Dehalococcoides sp.

Dehalococcoides species represent a relevant group of MRDBs (Adrian *et al.*, 1998; Cupples *et al.*, 2003; He *et al.*, 2003b; Hendrickson *et al.*, 2002; Maymó-Gatell *et al.*, 1997) and received considerable attention for bioremediation applications (Bunge *et al.*, 2003; Ellis *et al.*, 2000; Fennell *et al.*, 2004; Lendvay *et al.*, 2003; Major *et al.*, 2002). The following describes a 16S rRNA gene-targeted qRTm PCR approach to detect and quantify members of the *Dehalococcoides* group. A number of chemistries exist to detect the amplicons formed during each amplification cycle (Mackay, 2004). Described here is a procedure that uses a TaqMan probe (i.e., a linear hybridization probe with a 5′ fluorescent reporter and a 3′ quencher molecule) (Heid *et al.*, 1996) to detect *Dehalococcoides* 16S rRNA genes (He *et al.*, 2003a,b). While not discussed here in detail, SYBR Green, a DNA double strand-specific fluorescent dye, has also been applied successfully in qRTm PCR approaches to quantify *Dehalococcoides*, *Desulfitobacterium,* and *Dehalobacter* populations (Smits *et al.*, 2004). TaqMan PCR reactions contain a forward primer, a reverse primer, the TaqMan probe that utilizes 6-carboxy-fluorescein (FAM) as a reporter fluorochrome on the 5′ end, and *N,N,N′,N′*-tetramethyl-6-carboxy-rhodamine (TAMRA) as quencher on the 3′ end (Table IV). The TaqMan probe binds specifically to the target sequence located between both primer-binding sites. Fluorescent light emission is blocked by the quencher attached to the molecule as long as the probe remains intact. During amplification, the probe is cleaved by the 5′ exonuclease activity of *Taq* DNA polymerase as it extends the sequence from the upstream primer. Thus, a probe is cleaved each time a primer is

TABLE IV

TAQMAN PROBE AND PRIMERS USED TO QUANTIFY *DEHALOCOCCOIDES* 16S rRNA GENES AND PROCESS-SPECIFIC RDase GENES

Primer	Sequence	Target	Reference
RTm*Dhc*F	CTG GAG CTA ATC CCC AAA GCT	*Dehalococcoides* spp.	He *et al.* (2003a,b)
RTm*Dhc*R	CAA CTT CAT GCA GGC GGG	*Dehalococcoides* spp.	He *et al.* (2003a,b)
RTm*Dhc*P	FAM-TCC TCA GTT CGG ATT GCA GGC TGA A-TAMRA	*Dehalococcoides* spp.	He *et al.* (2003a,b)
Bvc925F	AAA AGC ACT TGG CTA TCA AGG AC	*bvcA* gene	Krajmalnik-Brown *et al.* (2004); Aiello (2003)
Bvc1017R	CCA AAA GCA CCA CCA GGT C	*bvcA* gene	Krajmalnik-Brown *et al.* (2004); Aiello (2003)
Bvc977Probe	FAM-TGG TGG CGA CGT GGC TAT GTG G-TAMRA	*bvcA* gene	Krajmalnik-Brown *et al.* (2004); Aiello (2003)
TceA1270F	ATC CAG ATT ATG ACC CTG GTG AA	*tceA* gene	Krajmalnik-Brown *et al.* (2004); Aiello (2003)
TceA1336R	GCG GCA TAT ATT AGG GCA TCT T	*tceA* gene	Krajmalnik-Brown *et al.* (2004); Aiello (2003)
TceA probe	FAM-TGG GCT ATG GCG ACC GCA GG-TAMRA	*tceA* gene	Krajmalnik-Brown *et al.* (2004); Aiello (2003)

extended, and the fluorescent signal increases in proportion to the number of PCR-generated amplicons. The spectrofluorimetric thermal cycler (e.g., the ABI Prism 7700 Sequence Detection System, Applied Biosystems) records the emitted FAM fluorescence intensities of each reaction at regular intervals. A twofold concentrated qRTm PCR master mix is available from ABI and includes *Taq* DNA polymerase, deoxynucleoside triphosphates, and $MgCl_2$. After preparing a master mix with all the components other than template DNA (Table V), 27-μl aliquots are transferred to each ABI optical tube (ABI) in an ice block. Following the addition of 3 μl template DNA, each tube is gently mixed by tapping with a gloved fingertip and is spun in a microcentrifuge to collect the content at the bottom of the tube and to eliminate any trapped air bubbles. Groups of eight tubes are sealed with an optical cap strip (ABI). PCR cycle parameters are as follows: 2 min at 50°, 10 min at 95° followed by 40 cycles of 15 s at 95°, and 1 min at 60°.

qRTm PCR Calibration Curves

qRTm PCR calibration curves are required to estimate the number of target gene copies in a given sample. To translate 16S rRNA gene copy numbers to estimates of cell numbers, knowledge of the 16S rRNA gene copy number of the target organism is required (Crosby and Criddle, 2003; Klappenbach *et al.*, 2001). Information on the 16S rRNA gene copy

TABLE V
qRTm PCR Reaction Mix Composition

Component	Stock solution (mM)	Final concentration (nM)	μl per 30 μl reaction mixture
Water[a]	—	—	10.11
Master mix[b]	2×	1×	15
Probe [c]	100	300	0.09
fwd primer [c]	10	300	0.9
rev primer [c]	10	300	0.9
DNA		(25–250 ng/rxn)	3
		Total volume	30

[a] Nuclease-free, molecular biology grade water (www.idtdna.com).
[b] Master mix purchased from ABI.
[c] Probe and primers available from commercial sources such as Integrated DNA Technologies (www.idtdna.com).

number is available for sequenced genomes (www.tigr.org, www.jgi.doe. gov) or can be determined experimentally (He *et al.*, 2003a,b; Klappenbach *et al.*, 2001). *Dehalococcoides ethenogenes* strain 195, *Dehalococcoides* sp. strain FL2, and *Dehalococcoides* sp. strain BAV1 possess a single 16S rRNA gene copy (He *et al.*, 2003b). Calculations to estimate the number of *Dehalococcoides*-derived 16S rRNA gene copies in a sample assume an average molecular weight of 660 for a base pair in dsDNA, one 16S rRNA gene operon per *Dehalococcoides* genome, and a genome size of 1.5 Mbp (www.tigr.org). The following equation is used to calculate the number of *Dehalococcoides*-derived 16S rRNA gene copies per PCR reaction mix (He *et al.*, 2003a).

16S rRNA gene copies in *Dehalococcoides*

$$\text{chromosomal DNA per reaction mix} = \frac{(\text{DNA ng}/\mu\text{l}) \times (3 \ \mu\text{l/rxn}) \times (6.02 \times 10^{23})}{(1.5 \times 10^6) \times (660) \times (10^9)}$$

Multiplying the 16S rRNA copy number per PCR tube by the volume of DNA obtained from the sample and dividing the resulting value by the total volume or weight of starting sample material yield an estimate of the total number of *Dehalococcoides* 16S rRNA genes per volume unit or weight unit of the sample material. A calibration curve (log DNA concentration versus the cycle number at which fluorescence exceeds the cycle threshold value, C_T) is obtained using serial dilutions of *Dehalococcoides* pure culture genomic DNA or plasmid DNA carrying a cloned *Dehalococcoides* 16S rRNA gene (He *et al.*, 2003a). Equation (1) is also used to calculate the 16S rRNA gene copies/ml of plasmid DNA using a value of 5.5×10^3 (He *et al.*, 2003a) instead of 1.5×10^6 to reflect the size of the plasmid, rather than the *Dehalococcoides* chromosome. The standard curve is prepared by plotting the C_T values obtained for each standard dilution (*y* axis) against the log of the number of 16S rRNA gene copies introduced as template (*x* axis) (Fig. 2).

The equation of the regression line is used to determine the log number of *Dehalococcoides* 16S rRNA gene copies in the unknown sample based on its C_T value. Analyses should be performed in triplicate, using at least two replicate, independent DNA extractions from the same sample material. The qRTm PCR assay is more sensitive than direct PCR, but not as sensitive as the nested PCR approach. Therefore, it is possible that samples test positive with the nested PCR approach but do not contain enough *Dehalococcoides* cells to yield quantitative data with the qRTm PCR approach. In such samples, *Dehalococcoides* cells are detectable but not quantifiable.

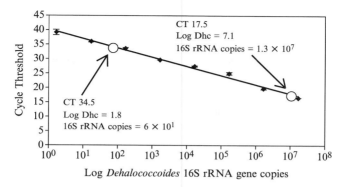

FIG. 2. Standard curve used to estimate the gene copy number in an unknown sample. A standard curve based on 2×10^1 to 2×10^8 16S rRNA gene copies was prepared using *Dehalococcoides*-targeted TaqMan probe and primers. The x axis displays the log *Dehalococcoides* 16S rRNA gene copy number, and the y axis shows the cycle number where fluorescence surpasses the cycle threshold value. In the example shown, R^2 is 0.9922 ($n = 6$, error bars are not shown when within the symbols). The number of *Dehalococcoides*-derived 16S rRNA gene copies in environmental samples and enrichment cultures can be calculated from their C_T values. qRTm PCR has a detection limit of 5–10 16S rRNA gene copies/reaction. With a detection limit of 5 copies/μl of template DNA and a total template DNA volume of 100 μl, this calculates to about 500 (100 μl $\times$ 5 copies/μl DNA) gene copies of interest present in the original sample. The known *Dehalococcoides* spp. possess a single 16S rRNA gene copy, and if the original sample volume was 1000 ml of groundwater, this calculates to less than one *Dehalococcoides* cell per ml (500 copies/1000 ml = 0.5 copies/ml groundwater).

qRTm PCR Quantification of Reductive Dehalogenase (RDase) Genes

While the 16S rRNA gene provides information on the phylogenetic affiliations of reductively dechlorinating populations, it does not always provide information on the dechlorinating activity and the range of chloroorganic compounds used as electron acceptors. Several RDase gene sequences have been retrieved from several MRDBs (summarized in Smidt and de Vos, 2004). Alignment of RDase genes identified conserved regions suitable for primer and amplification of putative RDase gene fragments from pure cultures, consortia, and environmental samples (Regeard *et al.*, 2004; Rhee *et al.*, 2003; von Wintzingerode *et al.*, 2001).

Applying an RDase-specific approach for assigning function (i.e., dechlorinating activity) is of particular interest to distinguish *Dehalococcoides* 16S rRNA gene sequences (He *et al.*, 2003b; Ritalahti and Löffler, 2004). Using the combined knowledge of RDase gene sequences from other MRDBs and genome information of *Dehalococcoides ethenogenes* strain

195 (http://tigrblast.tigr.org/ufmg/; Villemur *et al.*, 2002), degenerate primers sets specifically targeting *Dehalococcoides* RDase genes have been designed (Krajmalnik-Brown *et al.*, 2004). This approach has been applied successfully to retrieve putative RDase genes from *Dehalococcoides* pure cultures (Hölscher *et al.*, 2004; Krajmalnik-Brown *et al.*, 2004) and dechlorinating consortia, although function has been assigned to only few such indicator genes thus far (Krajmalnik-Brown *et al.*, 2004; Magnuson *et al.*, 2000; Müller *et al.*, 2004). Primers specifically targeting the *tceA* gene encoding a TCE reductive dehalogenase identified in *D. ethenogenes* strain 195 and the *vcrA* and *bcvA* genes encoding VC reductive dehalogenases in strain VS and *Dehalococcoides* sp. strain BAV1, respectively, are listed in Table IV. Included in Table IV are TaqMan probes designed for the detection and quantification of *bvcA* and *tceA*. qRTm PCR targeting these indicator genes involves the same protocol as shown in Table V for the quantification of 16S rRNA genes.

The combined application of 16S rRNA gene- and RDase gene-targeted qRTm PCR provides information on the presence and abundance

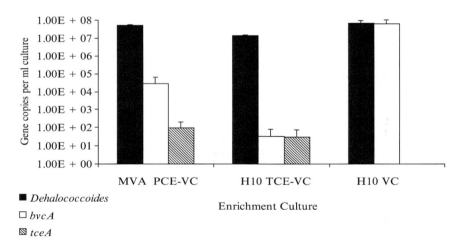

Fig. 3. Molecular tools suggest an untapped diversity of *Dehalococcoides* populations. qRTm PCR analysis of *Dehalococcoides*-containing enrichment cultures that completely detoxify chloroethenes to ethene showed that different *Dehalococcoides* populations contributed to each community. Cultures MVA PCE-VC and H10 TCE-VC were enriched with PCE or TCE (>15 transfers), then VC (>5 transfers), and culture H10 VC was enriched only with VC (>15 transfers). The *y* axis has a logarithmic scale and shows the number of *Dehalococcoides* 16S rRNA gene copies per milliliter of culture fluid, as compared with the number of *bvcA* and *tceA* gene copies. These findings suggest that the majority of *Dehalococcoides* RDase genes involved in chloroethene reductive dechlorination await discovery and that only a fraction of the *Dehalococcoides* community has been described.

of *Dehalococcoides* species as well as process-specific genes. Such information is critical for site assessment and process monitoring at contaminated sites where efficient dechlorination and ethene formation is critical (Lendvay *et al.*, 2003). Both *tceA* and *bvcA* are single copy genes in strain 195 and strain BAV1, respectively (Krajmalnik-Brown *et al.*, 2004), and hence, the combined approach provides an internal control for quantitative assessment. In addition, this two-pronged approach may reveal further information on the dechlorinating *Dehalococcoides* community.

Figure 3 gives quantitative information on three respective chloroethene dechlorinating enrichment cultures enriched with TCE or VC. All of the cultures possessed roughly the same total *Dehalococcoides* cell numbers (2–5 $\times 10^7$ 16S rRNA gene copies/ml of culture liquid). However, the most striking information gleaned from this assay is that our current perception of the dechlorinating *Dehalococcoides* is extremely limited. Only one of the VC-dechlorinating cultures contained predominantly the *bvcA* gene, the *tceA* gene is present in only a minor fraction of the total community, and the majority of the *Dehalococcoides* populations seem to harbor as yet unidentified RDase genes. Hence, the current tools are not comprehensive, and we have only begun to tap into the untold diversity of this largely unexplored bacterial group.

Acknowledgments

We thank all current and previous members of the Löffler lab for valuable contributions and NSF, SERDP, DOE, and EPA for financial support. FEL is indebted to the Alexander von Humboldt Foundation for a Feodor-Lynen fellowship that was critical for initiating this research on MRDBs.

References

Adrian, L., Manz, W., Szewzyk, U., and Görisch, H. (1998). Physiological characterization of a bacterial consortium reductively dechlorinating 1,2,3- and 1,2,4-trichlorobenzene. *Appl. Environ. Microbiol.* **64**, 496–503.

Adrian, L., Szewzyk, U., Wecke, J., and Görisch, H. (2000). Bacterial dehalorespiration with chlorinated benzenes. *Nature* **408**, 580–583.

Aiello, M. R. (2003). Quantitative environmental monitoring of PCE dechlorinators in a contaminated aquifer and PCE-fed reactor. Master's Thesis, Department of Microbiology and Molecular Genetics, Michigan State University.

Atlas, R. M. (2004). "Handbook of Microbiological Media." CRC Press, Boca Raton, FL.

Ausubel, F., Brent, R., Kingston, R. E., Moore, D. D., Seidman, J. G., Smith, J. A., and Struhl, K. (1997). "Short Protocols in Molecular Biology." Wiley, New York.

Balch, W. E., Fox, G. E., Magrum, L. J., Woese, C. R., and Wolfe, R. S. (1979). Methanogens: Reevaluation of a unique biological group. *Microbiol. Rev.* **43**, 260–296.

Bedard, D. L., Bunnell, S. C., and Smullen, L. A. (1996). Stimulation of microbial paradechlorination of polychlorinated biphenys that have persisted in the Housatonic River sediments for decades. *Environ. Sci. Technol.* **30,** 687–694.

Beller, H. R., Kane, S. R., Legler, T. C., and Alvarez, P. J. J. (2002). A real-time polymerase chain reaction method for monitoring anaerobic, hydrocarbon-degrading bacteria based on a catabolic gene. *Environ. Sci. Technol.* **36,** 3977–3984.

Bergmann, J. G., and Sanik, J., Jr. (1957). Determination of trace amounts of chlorine in naphta. *Anal. Chem.* **29,** 241–243.

Bouchard, B., Beaudet, R., Villemur, R., McSween, G., Lépine, F., and Bisaillon, J. G. (1996). Isolation and characterization of *Desulfitobacterium frappieri* sp. nov., an anaerobic bacterium which reductively dechlorinates pentachlorophenol to 3-chlorophenol. *Int. J. Syst. Bacteriol.* **46,** 1010–1015.

Brennan, R. A., and Sanford, R. A. (2002). Continuous steady-state method using tenax for delivering tetrachloroethene to chloro-respiring bacteria. *Appl. Environ. Microbiol.* **68,** 1464–1467.

Breznak, J. A., and Costilow, R. N. (1994). Physicochemical factors in growth. *In* "Methods for General and Molecular Bacteriology" (P. Gerhardt, R. G. E. Murray, W. A. Wood, and N. R. Krieg, eds.), pp. 137–154. American Society for Microbiolgoy, Washington, DC.

Bunge, M., Adrian, L., Kraus, A., Opel, M., Lorenz, W. G., Andreesen, J. R., Görisch, H., and Lechner, U. (2003). Reductive dehalogenation of chlorinated dioxins by an anaerobic bacterium. *Nature* **421,** 357–360.

Chang, Y. C., Hatsu, M., Jung, K., Yoo, Y. S., and Takamizawa, K. (2000). Isolation and characterization of a tetrachloroethylene dechlorinating bacterium, *Clostridium bifermentans* DPH-1. *J. Biosci. Bioeng.* **89,** 489–491.

Crosby, L. D., and Criddle, C. S. (2003). Understanding bias in microbial community analysis techniques due to *rrn* operon copy number heterogeneity. *BioTechniques* **34,** 790–794, 796, 798 passim.

Cupples, A. M., Spormann, A. M., and McCarty, P. L. (2003). Growth of a *Dehalococcoides*-like microorganism on vinyl chloride and *cis*-dichloroethene as electron acceptors as determined by competitive PCR. *Appl. Environ. Microbiol.* **69,** 953–959.

Dennis, P. C., Sleep, B. E., Fulthorpe, R. R., and Liss, S. N. (2003). Phylogenetic analysis of bacterial populations in an anaerobic microbial consortium capable of degrading saturation concentrations of tetrachloroethylene. *Can. J. Microbiol.* **49,** 15–27.

DeWeerd, K. A., Concannon, F., and Sulfita, J. M. (1991). Relationship between hydrogen consumption, dehalogenation, and the reduction of sulfur oxyanions by *Desulfomonile tiedjeii*. *Appl. Environ. Microbiol.* **57,** 1929–1934.

DeWeerd, K. A., Mandelco, L., Tanner, R. S., Woese, C. R., and Suflita, J. M. (1990). *Desulfumonile tiedjei* gen. nov. and sp. nov., a novel anaerobic, dehalogenating, sulfate-reducing bacterium. *Arch. Microbiol.* **154,** 23–30.

De Wever, H., Cole, J. R., Fettig, M. R., Hogan, D. A., and Tiedje, J. R. (2000). Reductive dehalogenation of trichloroacetic acid by *Trichlorobacter thiogenes* gen. nov., sp. nov. *Appl. Environ. Microbiol.* **66,** 2297–2301.

De Wildeman, S., Neumann, A., Diekert, G., and Verstraete, W. (2003). Growth-substrate dependent dechlorination of 1,2-dichloroethane by a homoacetogenic bacterium. *Biodegradation* **14,** 241–247.

Di Stefano, T. D., Gossett, J. M., and Zinder, S. H. (1992). Hydrogen as an electron donor for dechlorination of tetrachloroethene by an anaerobic mixed culture. *Appl. Environ. Microbiol.* **58,** 3622–3629.

Dolfing, J. (1999). Comment on "Competition for hydrogen within a chlorinated solvent dehalogenating anaerobic mixed culture." *Environ. Sci. Technol.* **33,** 2127.

Dolfing, J. (2003). Thermodynamic considerations for dehalogenation. *In* "Dehalogenation: Microbial Processes and Environmental Applications" (M. M. Häggblom and I. D. Bossert, eds.). Kluwer Academic, New York.

Dolfing, J., and Beurskens, J. E. M. (1995). The microbial logic and environmental significance of reductive dehalogenation. "Advances in Microbial Ecology," Vol. 14, pp. 143–206. Plenum Press, New York.

El Fantroussi, S., Mahillon, J., Naveau, H., and Agathos, S. N. (1997). Introduction of anaerobic dechlorinating bacteria into soil slurry microcosms and nested-PCR monitoring. *Appl. Environ. Microbiol.* **63,** 806–811.

Ellis, D. E., Lutz, E. J., Odom, J. M., Buchanan, R. J., Bartlett, C. L., Lee, M. D., Harkness, M. R., and Deweerd, K. A. (2000). Bioaugmentation for accelerated *in situ* anaerobic bioremediation. *Environ. Sci. Technol.* **34,** 2254–2260.

Fennell, D. E., Carroll, A. B., Gossett, J. M., and Zinder, S. H. (2001). Assessment of indigenous reductive dechlorinating potential at a TCE-contaminated site using microcosms, polymerase chain reaction analysis, and site data. *Environ. Sci. Technol.* **35,** 1830–1839.

Fennell, D. E., and Gossett, J. M. (2003). Microcosms for site-specific evaluation of enhanced biological reductive dehalogenation. *In* "Dehalogenation: Microbial Processes and Environmental Applications" (M. M. Häggblom and I. D. Bossert, eds.). Kluwer Academic, New York.

Fennell, D. E., Gossett, J. M., and Zinder, S. H. (1997). Comparison of butyric acid, ethanol, lactic acid, and propionic acid as hydrogen donors for the reductive dechlorination of tetrachloroethene. *Environ. Sci. Technol.* **31,** 918–926.

Fennell, D. E., Nijenhuis, I., Wilson, S. F., Zinder, S. H., and Häggblom, M. M. (2004). *Dehalococcoides ethenogenes* strain 195 reductively dechlorinates diverse chlorinated aromatic pollutants. *Environ. Sci. Technol.* **38,** 2075–2081.

Good, N. E., Winget, G. D., Winter, W., Connolly, T. N., Izawa, S., and Singh, R. M. (1966). Hydrogen ion buffers for biological research. *Biochemistry* **5,** 467–477.

Griffin, B. M., Tiedje, J. M., and Löffler, F. E. (2004). Anaerobic microbial reductive dechlorination of tetrachloroethene (PCE) to predominantly *trans*-1,2 dichloroethene. *Environ. Sci. Technol.* **38,** 4300–4303.

Häggblom, M. M., and Bossert, I. D. (2003). "Dehalogenation: Microbial Processes and Environmental Applications." Kluwer Academic Press, New York.

Harms, G., Layton, A. C., Dionisi, H. M., Gregory, I. R., Garrett, V. M., Hawkins, S. A., Robinson, K. G., and Sayler, G. S. (2003). Real-time PCR quantification of nitrifying bacteria in a municipal wastewater treatment plant. *Environ. Sci. Technol.* **37,** 343–351.

He, J., Ritalahti, K. M., Aiello, M. R., and Löffler, F. E. (2003a). Complete detoxification of vinyl chloride (VC) by an anaerobic enrichment culture and identification of the reductively dechlorinating population as a *Dehalococcoides* species. *Appl. Environ. Microbiol.* **69,** 996–1003.

He, J., Ritalahti, K. M., Yang, K.-L., Koenigsberg, S. S., and Löffler, F. E. (2003b). Detoxification of vinyl chloride to ethene coupled to growth of an anaerobic bacterium. *Nature* **424,** 62–65.

He, J., Sung, Y., Dollhopf, M. E., Fathepure, B. Z., Tiedje, J. M., and Löffler, F. E. (2002). Acetate versus hydrogen as direct electron donors to stimulate the microbial reductive dechlorination process at chloroethene-contaminated sites. *Environ. Sci. Technol.* **36,** 2945–3952.

He, J., Sung, Y., Krajmalnik-Brown, R., Ritalahti, K. M., and Löffler, F. E. (2005). Isolation and characterization of *Dehalococcoides* sp. strain FL2, a trichloroethene (TCE)- and 1,2-dichloroethene-respiring anaerobe. *Environ. Microbiol.* In press.

He, Q., and Sanford, R. A. (2002). Induction characteristics of reductive dehalogenation in the ortho-halophenol-respiring bacterium, *Anaeromyxobacter dehalogenans*. *Biodegradation* **13**, 307–316.

He, Q., and Sanford, R. A. (2003). Characterization of Fe(III) reduction by chlororespiring *Anaeromxyobacter dehalogenans*. *Appl. Environ. Microbiol.* **69**, 2712–2718.

He, Q., and Sanford, R. A. (2004). Acetate threshold concentrations suggest varying energy requirements during anaerobic respiration by *Anaeromyxobacter dehalogenans*. *Appl. Environ. Microbiol.* **70**, 6940–6943.

Heid, C. A., Stevens, J., Livak, K. J., and Williams, P. M. (1996). Real time quantitative PCR. *Genome Res.* **6**, 986–994.

Hendrickson, E. R., Payne, J. A., Young, R. M., Starr, M. G., Perry, M. P., Fahnestock, S., Ellis, D. E., and Ebersole, R. C. (2002). Molecular analysis of *Dehalococcoides*16S ribosomal DNA from chloroethene-contaminated sites throughout North America and Europe. *Appl. Environ. Microbiol.* **68**, 485–495.

Holliger, C., Hahn, D., Harmsen, H., Ludwig, W., Schumacher, W., Tindall, B., Vazquez, F., Weiss, N., and Zehnder, A. J. B. (1998). *Dehalobacter restrictus* gen. nov. and sp. nov., a strictly anaerobic bacterium that reductively dechlorinates tetra- and trichloroethene in an anaerobic respiration. *Arch. Microbiol.* **169**, 313–321.

Hölscher, T., Krajmalnik-Brown, R., Ritalahti, K. M., Wintzingerode, F. V., Görisch, H., Löffler, F. E., and Adrian, L. (2004). Multiple non-identical putative reductive dehalogenase genes are common in "*Dehalococcoides*." *Appl. Environ. Microbiol.* **70**, 5290–5297.

Hungate, R. E. (1969). A roll tube method for cultivation of strict anaerobes. *Methods Microbiol.* **3B**, 117–132.

Hurt, R. A., Qiu, X., Wu, L., Roh, Y., Palumbo, A. V., Tiedje, J. M., and Zhou, J. (2001). Simultaneous recovery of RNA and DNA from soils and sediments. *Appl. Environ. Microbiol.* **67**, 4495–4503.

Klappenbach, J. A., Saxman, P. R., Cole, J. T., and Schmidt, T. M. (2001). rrndb: The Ribosomal RNA operon copy number database. *Nucleic Acids Res.* **29**, 181–184.

Krajmalnik-Brown, R., Hölscher, T., Thomson, I. N. F., Michael Saunders, F. M., Ritalahti, K. M., and Löffler, F. E. (2004). Genetic identification of a putative vinyl chloride reductase in Dehalococcoides sp. strain BAV1. *Appl. Environ. Microbiol.* **70**, 6347–6351.

Krumholz, L. R. (1997). *Desulfuromonas chloroethenica* sp. nov. uses tetrachloroethylene and trichloroethylene as electron acceptors. *Int. J. Syst. Bacteriol.* **47**, 1262–1263.

Krumholz, L. R., Sharp, R., and Fishbain, S. (1996). A freshwater anaerobe coupling acetate oxidation to tetrachloroethene dehalogenation. *Appl. Environ. Microbiol.* **62**, 4108–4113.

Lendvay, J. M., Löffler, F. E., Dollhopf, M., Aiello, M. R., Daniels, G., Fathepure, B. Z., Gebhard, M., Heine, R., Helton, R., Shi, J., Krajmalnik-Brown, R., Jr., C., L. M., Barcelona, M. J., Petrovskis, E., Tiedje, J. M., and Adriaens, P. (2003). Bioreactive barriers: Bioaugmentation and biostimulation for chlorinated solvent remediation. *Environ. Sci. Technol.* **37**, 1422–1431.

Löffler, F. E., Champine, J. E., Ritalahti, K. M., Sprague, S. J., and Tiedje, J. M. (1997a). Complete reductive dechlorination of 1,2-dichloropropane by anaerobic bacteria. *Appl. Environ. Microbiol.* **63**, 2870–2875.

Löffler, F. E., Cole, J. R., Ritalahti, K. M., and Tiedje, J. M. (2003). Diversity of dechlorinating bacteria. *In* "Dehalogenation: Microbial Processes and Environmental Applications" (M. M. Häggblom and I. D. Bossert, eds.), pp. 53–87. Kluwer Academic Press, New York.

Löffler, F. E., Ritalahti, K. M., and Tiedje, J. M. (1997b). Dechlorination of chloroethenes is inhibited by 2-bromoethanesulfonate in the absence of methanogens. *Appl. Environ. Microbiol.* **63**, 4982–4985.

Löffler, F. E., Sanford, R. A., and Tiedje, J. M. (1996). Initial characterization of a reductive dehalogenase from *Desulfitobacterium chlororespirans* Co23. *Appl. Environ. Microbiol.* **62**, 3809–3813.

Löffler, F. E., Sun, Q., Li, J., and Tiedje, J. M. (2000). 16S rRNA gene-based detection of tetrachloroethene (PCE)-dechlorinating *Desulfuromonas* and *Dehalococcoides* species. *Appl. Environ. Microbiol.* **66**, 1369–1374.

Löffler, F. E., Tiedje, J. M., and Sanford, R. A. (1999). Fraction of electrons consumed in electron acceptor reduction (Fe) and hydrogen threshold as indicators of halorespiratory physiology. *Appl. Environ. Microbiol.* **65**, 4049–4056.

Luijten, M. L., de Weert, J., Smidt, H., Boschker, H. T., De Vos, W. M., Schraa, and Stams, A. J. M. (2003). Description of *Sulfurospirillum halorespirans* sp. nov., an anaerobic, tetrachloroethene-respiring bacterium, and transfer of *Dehalospirillum multivorans* to the genus *Sulfurospirillum* as *Sulfurospirillum multivorans* comb. nov. *Int. J. Syst. Evol. Microbiol.* **53**, 787–793.

Mackay, I. M. (2004). Real-time PCR in the microbiology laboratory. *Clin. Microbiol. Infect.* **10**, 190–212.

Magnuson, J. K., Romine, M. F., Burris, D. R., and Kingsley, M. T. (2000). Trichloroethene reductive dehalogenase from *Dehalococcoides ethenogenes*: Sequence of *tce*A and substrate range characterization. *Appl. Environ. Microbiol.* **66**, 5141–5147.

Major, D. W., McMaster, M. L., Cox, E. E., Edwards, E. A., Dworatzek, S. M., Hendrickson, E. R., Starr, M. G., Payne, J. A., and Buonamici, L. W. (2002). Field demonstration of successful bioaugmentation to achieve dechlorination of tetrachloroethene to ethene. *Environ. Sci. Technol.* **36**, 5106–5116.

Maymó-Gatell, X., Anguish, T., and Zinder, S. H. (1999). Reductive dechlorination of chlorinated ethenes and 1,2-dichloroethane by '*Dehalococcoides ethenogenes*' 195. *Appl. Environ. Microbiol.* **65**, 3108–3113.

Maymó-Gatell, X., Chien, Y.-T., Gossett, J. M., and Zinder, S. H. (1997). Isolation of a bacterium that reductively dechlorinates tetrachloroethene to ethene. *Science* **276**, 1568–1571.

Miller, T. L., and Wolin, M. J. (1974). A serum bottle modification of the Hungate technique for cultivating obligate anaerobes. *Appl. Microbiol.* **27**, 985–987.

Müller, J. A., Rosner, B. M., Abendroth, G. V., Meshluham-Simon, G., McCarty, P., and Spormann, A. M. (2004). Molecular identification of the catabolic vinyl chloride reductase from *Dehalococcoides* sp. strain VS and its environmental distribution. *Appl. Environ. Microbiol.* **70**, 4880–4888.

Neumann, A., Engelmann, T., Schmitz, R., Greiser, Y., Orthaus, A., and Diekert, G. (2004). Phenyl methyl ethers: Novel electron donors for respiratory growth of *Desulfitobacterium hafniense* and *Desulfitobacterium* sp. strain PCE-S. *Arch. Microbiol.* **181**, 245–249.

Neumann, A., Scholz-Muramatsu, H., and Diekert, G. (1994). Tetrachloroethene metabolism of *Dehalospirillum multivorans*. *Arch. Microbiol.* **162**, 295–301.

Regeard, C., Maillard, J., and Holliger, C. (2004). Development of degenerate and specific PCR primers for the detection and isolation of known and putative chloroethene reductive dehalogenase genes. *J. Microbiol. Methods* **56**, 107–118.

Rhee, S.-K., Fennell, D. E., Häggblom, M. M., and Kerkhof, L. J. (2003). Detection by PCR of reductive dehalogenase motifs in a sulfidogenic 2-bromophenol-degrading consortium enriched from estuarine sediment. *FEMS Microbiol. Ecol.* **43**, 317–324.

Ritalahti, K. M., and Löffler, F. E. (2004). Populations implicated in the anaerobic reductive dechlorination of 1,2-dichloropropane in highly enriched bacterial communities. *Appl. Environ. Microbiol.* **70,** 4088–4095.

Sambrook, J., and Russell, D. W. (2001). "Molecular Cloning: A Laboratory Manual," 3rd Ed. Cold Spring Harbor Laboratory Press, Cold Spring Harbor, NY.

Sanford, R. A. (1996). *In* "Characterization of Microbial Populations in Anaerobic Food Webs That Reductively Dechlorinate Chlorophenols." Ph.D. thesis. Michigan State University, East Lansing, MI. pp. 1–17.

Sanford, R. A., Cole, J. R., Löffler, F. E., and Tiedje, J. M. (1996). Characterization of *Desulfitobacterium chlororespirans* sp. nov., which grows by coupling the oxidation of lactate to the reductive dechlorination of 3-chloro-4-hydroxybenzoate. *Appl. Environ. Microbiol.* **62,** 3800–3808.

Sanford, R. A., Cole, J. R., and Tiedje, J. M. (2002). Characterization and description of *Anaeromyxobacter dehalogenans* gen. nov., sp. nov., an aryl halorespiring facultative anaerobic Myxobacterium. *Appl. Environ. Microbiol.* **68,** 893–900.

Schlötelburg, C., von Wintzingerode, C., Hauck, R., von Wintzingerode, F., Hegemann, W., and Göbel, U. B. (2002). Microbial structure of an anaerobic bioreactor population that continuously dechlorinates 1,2-D. *FEMS Microbiol. Ecol.* **39,** 229–237.

Schlötelburg, C., von Wintzingerode, F., Hauck, R., Hegemann, W., and Göbel, U. B. (2000). Bacteria of an anaerobic 1,2-dichloropropane dechlorinating mixed culture are phylogenetically related to those of other anaerobic dechlorinating consortia. *Int. J. Syst. Bacteriol.* **50,** 1505–1511.

Scholz-Muramatsu, H., Neumann, A., Meßmer, M., Moore, E., and Diekert, G. (1995). Isolation and characterization of *Dehalospirillum multivorans* gen. nov., sp. nov., a tetrachloroethene-utilizing, strictly anaerobic bacterium. *Arch. Microbiol.* **163,** 48–56.

Schumacher, W. C., Holliger, C., Zehnder, A. J. B., and Hagen, W. R. (1997). Redox chemistry of cobalamin and iron-sulfur cofactors in the tetrachloroethene reductase of *Dehalobacter restrictus*. *FEBS Lett.* **409,** 421–425.

Sharkey, F. H., Banat, I. M., and Marchant, R. (2004). Detection and quantification of gene expression in environmental bacteriology. *Appl. Environ. Microbiol.* **70,** 3795–3806.

Shelton, D. R., and Tiedje, J. M. (1984). Isolation and partial characterization of bacteria in an anaerobic consortium that mineralizes 3-chlorobenzoic acid. *Appl. Environ. Microbiol.* **48,** 840–848.

Smidt, H., Akkermans, A. D. L., van der Oost, J., and de Vos, W. M. (2000). Halorespiring bacteria-molecular characterization and detection. *Enzyme Microb. Technol.* **27,** 812–820.

Smidt, H., and de Vos, W. M. (2004). Anaerobic microbial dehalogenation. *Annu. Rev. Microbiol.* **58,** 43–73.

Smits, T. H. M., Devenoges, C., Szynalski, K., Maillard, J., and Holliger, C. (2004). Development of a real-time PCR method for quantification of the three genera *Dehalobacter*, *Dehalococcoides*, and *Desulfitobacterium* in microbial communities. *J. Microbiol. Methods* **57,** 369–378.

Sowers, K. R., and Noll, K. M. (1995). Techniques for anaerobic growth. *In* "Archaea: A Laboratory Manual: Methanogens" (K. R. Sowers and H. J. Schreier, eds.), p. 15. Cold Spring Harbor Laboratory Press, Cold Spring Harbor, NY.

Stansbridge, E. M., Mills, G. A., and Walker, V. (1993). Automated headspace gas chromatographic analysis of faecal short-chain fatty acids. *J. Chromatogr.* **621,** 7–13.

Sun, B., Cole, J. R., Sanford, R. A., and Tiedje, J. M. (2000). Isolation and characterization of *Desulfovibrio dechloracetivorans* sp. nov., a marine dechlorinating bacterium growing by coupling the oxidation of acetate to the reductive dechlorination of 2-chlorophenol. *Appl. Environ. Microbiol.* **66,** 2408–2413.

Sun, B. L., Cole, J. R., and Tiedje, J. M. (2001). *Desulfomonile limimaris* sp nov., an anaerobic dehalogenating bacterium from marine sediments. *Int. J. Syst. Evol. Microbiol.* **51,** 365–371.

Sun, B. L., Griffin, B. M., Ayala-del-Rio, H. L., Hashasham, S. A., and Tiedje, J. M. (2002). Microbial dehalorespiration with 1,1,1-trichloroethane. *Science* **289,** 1023–1025.

Sung, Y., Ritalahti, K. M., Sanford, R. A., Urbance, J. W., Flynn, S. J., Tiedje, J. M., and Löffler, F. E. (2003). Characterization of two tetrachloroethene (PCE)-reducing, acetate-oxidizing anaerobic bacteria, and their description as *Desulfuromonas michiganensis* sp. nov. *Appl. Environ. Microbiol.* **69,** 2964–2974.

Sung, Y., and Löffler, F. E. (2004). Isolation of a *Geobacter* sp. Strain SZ, an acetate and hydrogen oxidizing dissimilatory Fe(III)- and tetrachloroethene (PCE)- reducing anaerobe, abstr. Q106. *In* Abstracts of the 104th General Meeting of the American Society for Microbiology, New Orleans, LA.

Utkin, I., Woese, C. R., and Wiegel, J. (1994). Isolation and characterization of *Desulfitobacterium dehalogenans* gen. nov., sp. nov., an anaerobic bacterium which reductively dechlorinates chlorophenolic compounds. *Int. J. Syst. Bacteriol.* **44,** 612–619.

Vera, S. M., Werth, C. J., and Sanford, R. A. (2001). Evaluation of different polymeric organic materials for creating conditions that favor reductive processes in groundwater. *Biorem. J.* **5,** 169–181.

Villemur, R., Saucier, M., Gauthier, A., and Beaudet, R. (2002). Ocurrence of several genes encoding putative reductive dehalogenases in *Desulfitobacterium hafniense/frappieri* and *Dehalococcoides ethenogenes. Can. J. Microbiol.* **48,** 697–706.

von Wintzingerode, F., Schlötelburg, C., Hauck, R., Hegemann, W., and Göbel, U. B. (2001). Deveolopment of primers for amplifying genes encoding CprA- and PceA- like reductive dehalogenases in anaerobic microbial consortia, dechlorinating trichlorobenzene and 1,2-dichloropropane. *FEMS Microbiol. Ecol.* **35,** 189–196.

Walker, N. J. (2002). A technique whose time has come. *Science* **296,** 557–559.

Wolin, E. A., Wolin, M. J., and Wolfe, R. S. (1963). Formation of methane by bacterial extracts. *J. Biol. Chem.* **238,** 2882–2886.

Wu, Q., Watts, E. M., Sowers, K. R., and May, H. D. (2002). Identification of a bacterium that specifically catalyzes the reductive dechlorination of polychlorinated biphenyls with doubly flanked chlorines. *Appl. Environ. Microbiol.* **68,** 807–812.

Yang, Y., and McCarty, P. L. (1998). Competition for hydrogen within a chlorinated solvent dehalogenating anaerobic mixed culture. *Environ. Sci. Technol.* **32,** 3591–3597.

Yeates, C., Gillings, M. R., Davison, A. D., Altavilla, N., and Veal, D. A. (1998). Methods for microbial DNA extraction from soil for PCR amplification. *Biol. Proc. Online* **1,** 40–47.

Zhou, J., Davey, M. E., Figueras, J. B., Rivkina, E., Gilichinsky, D., and Tiedje, J. M. (1997). Phylogenetic diversity of a bacterial community determined from Siberian Tundra soil. *Microbiology* **143,** 3913–3919.

[6] Enrichment and Isolation of Iron-Oxidizing Bacteria at Neutral pH

By David Emerson and Melissa Merrill Floyd

Abstract

Methods are provided for the culture of neutrophilic Fe-oxidizing bacteria (FeOB) that grow under microaerobic and anaerobic conditions. For oxygen-requiring lithotrophic Fe oxidizers, it is essential that both O_2 and Fe(II) concentrations are kept low, but that an adequate flux of both O_2 and Fe(II) are provided to support growth. Techniques using opposing gradients of Fe(II) and O_2 are discussed. Preparation of stock solutions of FeS and $FeCl_2$ are described. Methods for the culture of anaerobic FeOB that utilize nitrate or light (anoxygenic photosynthesis) are discussed and presented briefly.

Introduction

The oxidation of ferrous iron to ferric hydrate $Fe(OH)_3$ at circumneutral pH is an exergonic reaction ($\Delta G^{\circ\prime} = -109$ kJ mol^{-1}) using O_2 as the oxidant (Roden *et al.*, 2004). This oxidation supports the growth of a group of microbes uniquely adapted for extracting energy from this reaction. The capacity to grow lithotrophically on Fe(II) is widespread throughout the prokaryotic world, but the majority of isolated microorganisms capable of growing on iron are acidophilic. These organisms usually grow at $<$pH 4 where Fe(II) is stable in the presence of air and can achieve high concentrations in aerated aqueous habitats (Blake and Johnson, 2000).

As the pH changes from acid conditions toward neutrality, the chemical oxidation of Fe(II) becomes much more rapid. As a result, aerobic, neutrophilic Fe-oxidizing bacteria are limited to growing in gradients of O_2 and Fe(II) at oxic–anoxic boundaries (Emerson, 2000). At these interfaces, Fe(II) concentrations may range from tens to several hundreds of micromolar, and O_2 concentrations are typically $<$5% of air-saturated values. At the same time, the flux of both Fe(II) and O_2 may be quite high. Achieving these same conditions in the laboratory presents special challenges for the investigator. Other microbes that oxidize Fe(II) at circumneutral pH avoid the problem of chemical oxidation by living under anaerobic conditions and either carry out anaerobic respiration, coupling Fe-oxidation to nitrate reduction, or utilize anoxygenic photosynthesis with Fe(II) as the electron donor.

METHODS IN ENZYMOLOGY, VOL. 397
Copyright 2005, Elsevier Inc. All rights reserved.
0076-6879/05 $35.00
DOI: 10.1016/S0076-6879(05)97006-7

Microaerophilic Lithotrophs

Gallionella Culture

Most laboratory methods for growing microaerophilic lithotrophs, including *Gallionella* species, can be traced to Kucera and Wolfe (1957). They developed a simple but effective means of creating opposing gradients of O_2 and $Fe(II)$ by adding freshly prepared FeS to glass test tubes containing mineral medium. The original suggestion of using FeS was that of C.B. Van Niel. Hanert (1989, 1992), which made modifications to the liquid medium and preparation of the FeS. For this method, modified Wolfe's mineral medium (MWMM) (see Table I for recipe) can be used. Glass tubes (e.g., 16×120 mm) are filled two-thirds full with sterile, buffered MWMM. Several drops of FeS (see later) are added to the medium; the FeS will settle to the bottom of the tube, simultaneously creating reducing conditions and releasing $Fe(II)$ into the overlying liquid. The tubes are capped with a stopper, leaving a headspace of air, resulting in an oxygen gradient that decreases with depth. In successful enrichments, *Gallionella* will colonize the walls of the tubes and appear as white spots that become increasingly rust colored with age.

Emerson and Moyer (1997) further modified the gradient tube technique by adding a low concentration of agarose to the overlaying mineral salts medium, which allowed unicellular FeOB to stably maintain growth at the interface of O_2 and $Fe(II)$ gradients (see Fig. 1). Addition of the agarose prevented the cells from precipitating out of the oxic–anoxic transition zone as they became mired in Fe-oxide precipitates. This variation has proven useful in isolating new FeOB.

Preparation of Gel-Stabilized Gradient Tubes

Typically 17×60 mm (o.d. $\times$ length) borosilicate glass vials are used as vessels for gradient tube cultures. The top layer, consisting of semisolid mineral media, and the bottom layer, which serves as the iron source, are prepared separately. The bottom layer contains 1% (w/v) high melt agarose and equal volumes of MWMM and FeS. In a separate container, the top layer is prepared by adding 0.15% low melt agarose to MWMM along with 5 mM sodium bicarbonate and 1 μl of trace minerals (www.atcc.org; ATCC MD-TMS) per milliliter of medium. For marine organisms, artificial sea water (ASW) (recipe in Table I) is substituted for MWMM. Both layers are autoclaved. Shortly after autoclaving, 0.75 ml of the bottom layer is pipetted into each culture tube. One-milliliter pipette tips that have been cut to enlarge the opening at the tip to prevent clogging are recommended. The bottom layer is allowed to cool for at least 30 min to ensure it is well

TABLE I

Organism type	Media composition	Iron source
Freshwater microaerophilic lithotrophs	Per liter dH_2O: 1 g NH_4Cl 0.2 g $MgSO_4$ 0.5 g K_2HPO_4 Diluted from 10× stock and sterilized separately	Equal molar amounts of $FeH_8N_2O_8S_2$ and Na_2S. Precipitate washed minimum of four times with boiling dH_2O
Freshwater microaerophilic lithotrophs	Modified Wolfe's mineral medium (MWMM) per liter dH_2O: 1 g NH_4Cl 0.2 g $MgSO_4 \cdot 7H_2O$ 0.1 g $CaCl_2 \cdot 2H_2O$ 0.05 g K_2HPO_4	78 g $FeH_8N_2O_8S_2$ + 44 g Na_2S Prepare in 50° and wash with dH_2O until precipitate is pH neutral
Saltwater microaerophilic lithotrophs	Artificial saltwater (ASW) per liter dH_2O: 27.5 g NaCl 5.38 g $MgCl_2 \cdot 6H_2O$ 6.78 g $MgSO_4 \cdot 7H_2O$ 0.72 g KCl 0.2 g $NaHCO_3$ 1.4 g $CaCl_2 \cdot 2H_2O$ 1 g NH_4Cl 0.05 g K_2HPO_4	See MWMM
Bottle cultures Anaerobic iron-oxidizing nitrate reducers	MWMM or ASW *Ferroglobus* medium per liter dH_2O: 0.34 g KCl 4.30 g $MgCl_2 \cdot 6H_2O$ 0.24 g NH_4Cl 0.14 g $CaCl_2 \cdot 2H_2O$ 18.00 g NaCl 5.00 g $NaHCO_3$ 0.14 g $K_2HPO_4 \cdot 3H_2O$ 1.00 g KNO_3 10.00 ml trace minerals 1.00 mg resazurin 0.50 % $Na_2S.9H_2O$	100 mM $FeCl_2$ stock 2 mM FeS: 0.6 M $NaS_2 \cdot 9H_2O$ + 0.6 M $FeSO_4 \cdot 7 H_2O$ washed with H_2O and dried under N_2/H_2 in anaerobic hood OR 0.25 g $FeCl_2$
Anaerobic iron-oxidizing nitrate reducers	Per liter dH_2O: 0.3 g NH_4Cl 0.05 g $MgSO_4 \cdot 7H_2O$ 0.4 g $MgCl \cdot 6H_2O$ 0.6 g KH_2PO_4 0.1 g $CaCl_2 \cdot H_2O$ 30 ml $NaHCO_3$ Vitamins	1.0 M anoxic stock solution $FeSO_4$

TABLE I (continued)

Organism type	Media composition	Iron source
	Trace elements with EDTA 4 mM NaNO$_3$ buffered to pH 7.0 with NaHCO$_3$	
Photosynthetic lithotrophs	Freshwater medium per liter dH$_2$O: 0.3 g NH$_4$Cl 0.5 g KH$_2$PO$_4$ 0.5 g MgSO$_4$·7H$_2$O (or 0.4 g MCl$_2$·6H$_2$O in sulfate-free medium) 0.1 g CaCl$_2$·2H$_2$O 1 ml trace elements with EDTA 22 ml 1 M NaHCO$_3$ 1 ml filter-sterilized vitamin mixture 1 ml filter-sterilized B$_{12}$	Equal volumes of 0.2 M FeSO$_4$ and 0.2 M Na$_2$S mixed under N$_2$ atmosphere. Wash with sterile medium
Photosynthetic lithotrophs	Marine medium per liter dH$_2$O: to the freshwater medium add 22 g NaCl 3.5 g MgCl$_2$·6H$_2$O 0.5 g KCl and increase CaCl$_2$·2H$_2$O to 0.15 g	See above

set; meanwhile, the top layer is cooled to between 35 and 40°. One μl per milliliter of vitamin solution (www.atcc.org; ATCC MD-VS) is added and the pH is adjusted to within the range of 6.1 to 6.4 by bubbling with sterile CO$_2$ gas. The gassing time is dependent on the volume of medium and the flow rate of the CO$_2$ and should be determined empirically. To make the top layer, 3.75 ml of medium is pipetted over the bottom layer of each tube. The tubes may be capped with either a butyl rubber stopper or a screw cap with a septum to maintain an aerobic headspace. The top layer should be allowed to solidify a minimum of 3 h to a maximum of overnight. Allowing the tubes to sit uninoculated for longer periods will lead to undesirable amounts of abiotic iron oxidation. To inoculate, 10 μl of the desired inoculum is drawn into a pipette tip and inserted just above the FeS layer; the pipette tip is drawn upward as the inoculum is dispensed. Each tube should be stoppered and incubated at the appropriate temperature. Growth is visible as the line of the inoculum spreads into a rust-colored band at the oxic–anoxic interface. It is recommended that the cultures be transferred every 3 to 4 weeks. Refrigerated storage will prolong the useful

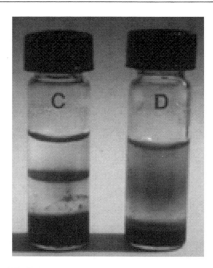

FIG. 1. Illustration of FeOB growing in a gradient tube. The tube on the left has been inoculated with a pure culture of a FeOB; growth is evident as a well-defined band of iron oxides forming in the medium. The tube on the right was uninoculated. Iron oxidation still occurs, but is spread more diffusely through the overlaying gel-stabilized medium.

life of a culture. Long-term storage is achieved by removing the growth band with a pipette and mixing it with an equal volume of sterile 20% glycerol. The glycerol stocks must be immediately frozen in either a $-80°$ freezer or liquid nitrogen.

Gradient Plates

Similarly, microaerophilic FeOB can be cultured using gradient plates (Emerson and Weiss, 2004). These consist of an FeS/agarose solid layer overlaid with a liquid layer (containing no agarose) in standard petri plates. The bottom layer is prepared exactly as described earlier for the gradient tubes. In a separate container the liquid layer is also prepared as described for the gradient tubes except that agarose is omitted. After autoclaving, the FeS/agarose layer is added to petri plates to a depth of 3 to 4 mm; approximately 8 ml per standard 100×15-mm petri plate or 5 ml for small 60×15-mm petri plates. After the FeS/agarose layer has initially solidified, it is allowed to set for approximately 30 min. It is best to do this under anoxic conditions, e.g., by storing the plates in an anaerobic chamber or anaerobe jar that has been flushed with N_2 to prevent chemical oxidation of Fe at the surface. Once the top layer has reached room temperature it is sparged with filter-sterilized CO_2 until the required pH is achieved and then 1 μl/ml of vitamins is added. The entire top layer is inoculated with 2 to 3 ml of the desired organism per 100 ml media. Aliquot 15 ml of the top layer onto each plate (8 ml for the smaller plates). The plates are incubated in a sealed anaerobic jar (www.bd.com; gaspak 150 jars) with a Campy Pak microaerophilic pouch (www.bd.com; BBL campypak). This system produces an atmosphere of approximately 5–10% O_2 and 5 to 12% CO_2 in the headspace. Growth is indicated by accumulated Fe floc, as well as the appearance of an opalescent metallic film on the liquid surface. It is always necessary, however, to confirm the presence of FeOB by epifluoresence microscopy (see later).

Liquid Cultures

Bottle cultures are another option for culturing microaerophilic lithotrophs (Emerson and Moyer, 2002). This method uses serum bottles (typically 125 ml) with butyl rubber stoppers. The MWMM is made up with 20 mM sodium bicarbonate and minerals, and 50 ml is dispensed into each bottle. The liquid in each bottle is gassed with a filter-sterilized N_2:CO_2 (70:30 v/v) gas mix for a minimum of 3 min until the pH has stabilized between 6.2 and 6.4. The bottles are then stoppered with the N_2:CO_2 headspace, crimped with an aluminum seal, and autoclaved. As an alternative buffer, 10 mM MES can be used and the pH adjusted to 6.3 with 1 M sodium bicarbonate. Once the bottles are cool, flame sterilize the stopper with ethanol and inject 50 μl trace vitamins, 0.2 ml 100 mM $FeCl_2$ solution, and 4 ml sterile-filtered air to maintain an O_2 concentration in the headspace of approximately 2%. Typically, an inoculum between 1 and 5% (v/v) is used. The bottles are fed daily or twice daily with $FeCl_2$ and air, 0.2 and 4 ml, respectively. The only way to determine growth in the bottles is to remove samples and check for the presence of cells by epifluorescence microscopy (see later).

Other Methods for Culture of Microaerophilic FeOB

Biphasic Slant Cultures. Verran *et al.* (1995) described a method for preparing biphasic slants for growing *Gallionella*, although this should work for other gradient-requiring Fe oxidizers as well. FeS and 3% agar are combined, autoclaved, and poured into glass test tubes to approximately half full. This layer is bubbled with sterile CO_2 gas for 10 to 15 s, after which the tube is laid at an angle while the agar/FeS solidifies. Hanert's MWMM is prepared and poured into test tubes as the top layer, which is then bubbled with filtered sterilized CO_2 for 5 s, and the tube is stoppered. Growth is visible as a banding pattern on the slope or on the sides of the test tubes.

Bioreactor Methods. Large-scale growth of neutrophilic FeOB is possible using bioreactors that provide control of pH and inputs of O_2 and Fe(II). Neubauer *et al.* (2002) employed a 1.5-liter reactor vessel that contained 1 liter of MWMM. A steady pH was maintained by automated addition of either HCl or $NaHCO_3$, and the O_2 concentration was kept at 0 to 5% air saturation by gassing the vessel with either air or N_2. This system was operated in a fed-batch mode where $FeCl_2$ was pulsed into the system at 4-h intervals, resulting in a concentration of approximately 200 μM immediately after feeding. During the 4-h interval, most of the Fe(II) was oxidized. Specific growth yields in the bioreactor were not as great as in any of the gradient methods described previously, although the much larger volumes afforded by this method can result in good total cell yields.

Hanert (1992) reported growth of lithoautotrophic *Gallionella* in a fermentor culture. This system used 9 liters of filter-sterilized MWMM medium that contained ferrous carbonate and a gas mixture of $N_2/CO_2/O_2$ (94:5:1) at a pressure of 2 atm. The final Fe concentration was approximately 25 ppm. The medium was stirred at 25 rpm and gassed very slowly, approximately 10 ml/min. Growth of *Gallionella* is evidenced by white growth on the walls of the vessel. The ferrous iron will be completely oxidized within 5 days; however, the method is amenable to the development of a flow-through system, thereby extending the life of the culture.

In Situ Growth Methods. Gallionella has also been cultured *in situ* using a bacteriogenic iron oxide reactor apparatus (BRIC) in an underground site known to support the growth of *Gallionella* spp. (Anderson and Pedersen, 2003). Nutrients and ferrous iron were supplied by the local water source via a 6-mm stainless steel pipe. Gravel from local rock provided the base for community development. A manifold system allowed control of water flow through the reactor, as well as oxygen input, and provided sampling ports [see Anderson and Peterson (2003) for details].

Preparation of FeS Stock Solution

FeS is prepared by heating 300 ml of dH_2O to 50° in a 500-ml beaker with a stir bar present. Separately preweigh 46.2 g of ferrous sulfate and 39.6 g of sodium sulfide. While stirring the water rapidly, add the ferrous sulfate followed immediately by the sodium sulfide. A thick black precipitate will form instantly. This mixture is stirred continuously for 2 to 3 min to ensure complete dissolution and mixing of the ferrous sulfate and sodium sulfide. The black FeS sludge is decanted into a narrow-mouthed glass bottle (500 ml) that can be stoppered tightly. The bottle is filled to the top with dH_2O and capped. The FeS is allowed to settle for several hours and then the overlaying water is decanted and replaced. This procedure is repeated at least five times to wash the FeS. After washing, the pH of the FeS solution should be close to neutrality. The FeS layer is normally quite hydrous and can be pipetted with a standard 10-ml pipette.

After removing FeS for use, it is important to top the bottle up with dH_2O and keep it stoppered tightly to limit the influx of oxygen. With limited oxygen exposure, the FeS can be maintained at room temperature for up to 3 months. Even under these conditions the FeS does age, however, and slowly loses its ability to release Fe(II). Each batch of FeS is slightly different. If the FeS smells strongly of sulfide following the washing steps or has a strongly alkaline pH, it should be discarded. Remember that in the presence of acid, sodium sulfide will immediately release hydrogen sulfide, an extremely toxic gas. Preparing FeS in a chemical fume hood is strongly recommended.

Preparation 100 mM FeCl₂ Stock Solution

A 250-ml serum bottle is stoppered, crimped with an aluminum seal, and autoclaved. dH_2O (150 ml) is bubbled with N_2 for approximately 30 min and then 2.16 g ferrous chloride is added. The ferrous chloride must be allowed to dissolve completely. Sparge the serum bottle with filter-sterilized N_2. Using a 60-ml syringe, a syringe filter, and an 18- or 22-gauge needle, the $FeCl_2$ is injected into the serum bottle, with another needle inserted through the stopper as a vent. Gas out the headspace with nitrogen. This stock solution can be kept at room temperature indefinitely; it is preferable to store it in the dark.

Spectrophotometric Analyses of Fe(II) Concentration

In order to determine the concentration of Fe(II) in stock solutions or in an experimental system, a spectrophotometric assay using ferrozine [3-(2-pyridyl)-5,6-diphenyl-1,2,4-triazine disulfonic acid] was developed by Stookey (1970). The ferrozine stock solution is prepared by adding 1 g ferrozine (Sigma Chemical Co.) to 1 liter of 50 mM HEPES that has been adjusted to pH 7.0. Ferrozine reacts with Fe(II) to form a stable, purple, water-soluble complex with a maximum absorbance at 562 nm. Dilutions of 1:10 or 1:100 (sample:ferrozine) are commonly used for environmental samples and experimental systems. Absorbance readings higher than 1.5 are an indication that the sample is too concentrated; samples cannot be diluted after mixing with the reagent.

Visualization of FeOB by Light Microscopy

As FeOB grow, they produce copious amounts of ferric oxyhydroxides that precipitate with the cells; the cells are often difficult to see in the light

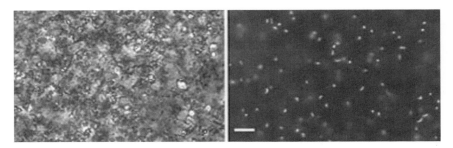

FIG. 2. A photomicrograph showing an FeOB growing in association with Fe oxides. This organism was grown in a gradient tube as illustrated in Fig. 1 and stained with Syto13, a nucleic acid-binding fluorescent dye. The image on the left is viewed by phase contrast and shows the Fe oxides. The same image is shown on the right, but viewed by epifluorescence. The cells are made visible among the Fe oxides. Scale bar: 10 μm.

microscope as they are bound within an Fe-oxide matrix. The best way to see the cells is to stain them with a nucleic acid-binding dye and use epifluorescent microscopy. The stained cells will show up well on the Fe oxides (see Fig. 2).

The most commonly used dyes for this purpose are acridine orange, DAPI, and Syto. Syto 13 (Molecular Probes, Eugene, OR) provides good penetration of the Fe oxides and bright fluorescence. A 0.25 mM stock solution is prepared in sterile dH$_2$O that is then diluted 1:5 with the bacterial culture.

Anaerobic Iron Oxidation Coupled to Nitrate Reduction

The anaerobic coupling of Fe(II) oxidation to nitrate reduction is an energetically favorable process under physiological conditions where ferrous iron normally exists as FeCO$_3$ and the product of the reaction will precipitate as a ferric hydrate (Straub *et al.*, 1996). A number of enrichments and several organisms have been described that appear to grow lithotrophically as a result of this reaction (Benz *et al.*, 1998; Straub and Buchholz-Cleven, 1998; Straub *et al.*, 2004).

The method of Straub and Buchholz-Cleven (1998) is described here for the growth of mesophilic iron-oxidizing nitrate reducers. A basal medium is prepared (see Table I) containing 4 mM NaNO$_3$. The medium is autoclaved and cooled under an atmosphere of filter-sterilized N$_2$:CO$_2$ (90:10, v/v). Sodium bicarbonate is added from a 1 M stock solution (autoclaved under a CO$_2$ atmosphere) to a final concentration of 30 mM; vitamins, trace elements, and a selenite-tungstate solution are also added. The pH is adjusted to 7.0. The medium is dispensed to gas-tight bottles or tubes. $FeSO_4$ is added to growth vessels from a 1 M stock solution to a final concentration of 10 mM; this normally results in a fluffy white precipitate composed primarily of $FeCO_3$. The medium is then ready for inoculation. As growth proceeds, rust-colored Fe oxide precipitates will form. Enrichment and isolation of pure cultures using this medium are generally done by serial dilution to extinction.

The most well-characterized anaerobic Fe oxidizer is *Ferroglobus placidus* (Hafenbrandl *et al.*, 1996). This hyperthermophilic archaeon (optimum growth temperature 85°) was isolated from volcanic sands in Italy and is able to use ferrous iron, H_2, and sulfide as electron donors with NO_3^- as an electron acceptor. It grows at neutral pH. For enrichment and growth of pure cultures on Fe(II), FeS, as prepared below, is used as the energy source. Amorphous FeS (2 mM) is prepared in an anaerobic hood by adding equal molar concentrations of $NaS_2 \cdot 9H_2O$ and $FeO_4S \ M \cdot 7H_2O$. The precipitate is filtered, washed with dH_2O, and dried under N_2/H_2 (95:5, v/v). *Ferroglobus* medium is prepared as described in Table I, and 2 mM FeS is added. The medium should be reduced with 0.5 % $Na_2S \cdot 9H_2O$ prior to autoclaving and the pH adjusted to 7.0. Ten milliliters of media is injected into sterile 120-ml serum bottles that have been sparged under N_2 and stoppered tightly. The gas phase should consist of 300 KPa N_2/CO_2 (80:20, v/v). The medium is inoculated with a 1-ml sample and incubated at 85° while shaking at 200 rpm.

A modified *Ferroglobus* medium that substitutes ferrous carbonate for ferrous sulfide may be used to eliminate the presence of sulfide and demonstrate that growth is truly Fe(II) dependent. To prepare, *Ferroglobus* medium is supplemented with 0.25 g $FeCl_3$ instead of FeS. The medium is boiled and then gassed with N_2 as it cools, and the pH is adjusted to 7.0 with H_2SO_4. The medium is then bottled and stoppered under an atmosphere of pure CO_2. K_2CO_3 is added from a stock solution to a final concentration of 0.27 g/liter in each bottle. Prior to autoclaving, the gas mixture is replaced with $N_2:CO_2$ (80:20; 300 KPa). After autoclaving, $FeCO_3$ should be present as a colorless precipitate, and the medium is ready to inoculate.

Anoxygenic Phototrophs

Some anoxygenic photosynthetic bacteria use ferrous iron as an electron donor (Ehrenreich and Widdel, 1994; Widdel *et al.*, 1993). Medium for culturing these bacteria is prepared (with modifications, from Croal *et al*, 2004) according to the recipe in Table I. The medium should be autoclaved and cooled under filter-sterilized N_2/CO_2 (80:20, v/v). When the medium is cooled, 1 ml trace elements with EDTA, 22 ml sterile 1 M NaHCO$_3$ (autoclaved under CO$_2$), 1 ml filter-sterilized vitamin mixture (4 mg p-aminobenzoate, 1 mg D[+]-biotin, 10 mg nicotinic acid, 5 mg Ca D[+]-pantothenate, 15 mg pyroxidin dihydrochloride, 10 mg thiamine hydrochloride, and 50 mg riboflavin in 1 liter dH$_2$O) and 1 ml filter-sterilized vitamin B$_{12}$ solution (5 mg in 50 ml dH$_2$O) are added. The pH is adjusted to 7.0 with HCl or NaCO$_3$. Using anaerobic techniques the medium is dispensed to 4/5 the volume of the chosen container; sterilized 20-ml tubes or 100-, 250-, or 500-ml flat bottles may be used. The headspace is gassed with N$_2$/CO$_2$ (80:20, v/v) and sealed with a rubber stopper. Fe(II) is added as a 1 M stock solution of FeCl$_2$ (see earlier discussion) to a concentration approximating 10 mM. This may lead to formation of a fluffy white precipitate, most likely vivianite (Fe$_3$[PO$_4$]$_2$·8H$_2$O) or siderite (FeCO$_3$). The precipitate should not affect cell growth. If necessary, it can be removed by filtration (Croal et $al.$, 2004). To determine the soluble Fe(II) concentration accurately, a small aliquot of the medium can be removed with a syringe and passed through a 0.22-μm filter and assayed with ferrozine (see earlier discussion).

Following inoculation, cultures are incubated under light; recommended illumination is from two tungsten lamps (25 W) installed above and below the cultures at a distance of 30 cm. If less intense light levels are desired, the number of bulbs may be reduced and/or the distance from the culture vessels increased. Remember that heat generated by the light bulbs may increase the incubation temperature.

Acknowledgments

We thank Dr. Jeremy Rentz, as well as Laura Croal and Yongqin Jiao, for critically reading over parts of this chapter. Development of some of the procedures described here was supported in part by NSF Grants DEB-9985922 and MCB-0348330 and awards from the NASA Astrobiology Institute, NCC2–1056 and NAG2–1523.

References

Anderson, C. R., and Pedersen, K. (2003). *In situ* growth of *Gallionella* biofilms and partitioning of lanthanides and actinides between biological material and ferric oxyhydroxides. *Geobiology* **1,** 169–178.

Benz, M., Brune, A., and Schink, B. (1998). Anaerobic and aerobic oxidation of ferrous iron at neutral pH by chemoheterotrophic nitrate-reducing bacteria. *Arch. Microbiol.* **169,** 159–165.

Blake, R., II, and Johnson, D. F. (2000). Phylogenetic and biochemical diversity among acidophilic bacteria that respire on iron. *In* "Environmental Microbe-Metal Interactions" (D. R. Lovley, ed.), pp. 53–78. ASM Press, Washington, DC.

Croal, L. R., Johnson, C. M., Beard, B. L., and Newman, D. K. (2004). Iron isotope fractionation by Fe(II)-oxidizing photoautotrophic bacteria. *Geochim. Cosmochim. Acta* **68**, 1227–1242.

Ehrenreich, A., and Widdel, F. (1994). Anaerobic oxidation of ferrous iron by purple bacteria, a new type of phototrophic metabolism. *Appl. Environ. Microbiol.* **60**, 4517–4526.

Emerson, D. (2000). Microbial oxidation of Fe (II) and Mn (IV) at circumneutral pH. *In* "Environmental Microbe-Metal Interactions" (D. R. Lovley, ed.), pp. 31–52. ASM Press, Washington, DC.

Emerson, D., and Moyer, C. (1997). Isolation and characterization of novel iron-oxidizing bacteria that grow at circumneutral pH. *Appl. Environ. Microbiol.* **63**, 4784–4792.

Emerson, D., and Moyer, C. L. (2002). Neutrophilic Fe-oxidizing bacteria are abundant at the Loihi Seamount hydrothermal vents and play a major role in Fe oxide deposition. *Appl. Environ. Microbiol.* **68**, 3085–3093.

Emerson, D., and Weiss, J. V. (2004). Bacterial iron oxidation in circumneutral freshwater habitats: Findings from the field and the laboratory. *Geomicrobiol J.* **21**, 405–414.

Hafenbrandl, D., Keller, M., Dirmeier, R., Rachel, R., Roßnagel, P., Burggraf, S., Huber, H., and Stetter, K. O. (1996). *Ferroglobus placidus* gen. nov., sp. nov., a novel hyperthermophilic archaeum that oxidizes Fe^{2+} at neutral pH under anoxic conditions. *Arch. Microbiol.* **166**, 308–314.

Hanert, H. H. (1989). Nonbudding, stalked bacteria. *In* "Bergey's Manual of Systematic Bacteriology" (J. T. Staley, M. P. Bryant, N. Pfennig, and J. G. Holt, eds.), Vol. 3, pp. 1974–1979. Williams and Wilkins, Baltimore, MD.

Hanert, H. H. (1992). The genus *Gallionella*. *In* "The Prokaryotes" (A. Balows, H. G. Trüper, M. Dworkin, W. Harder, and K. H. Schliefer, eds.), Vol. 4, pp. 4082–4088. Springer-Verlag, New York.

Kucera, S., and Wolfe, R. S. (1957). A selective enrichment method for *Gallionella ferruginea*. *J. Bacteriol.* **74**, 344–349.

Neubauer, S. C., Emerson, D., and Megonigal, J. P. (2002). Life at the energetic edge: Kinetics of circumneutral iron oxidation by lithotrophic iron-oxidizing bacteria isolated from the wetland-plant rhizosphere. *Appl. Environ. Microbiol.* **68**, 3988–3995.

Roden, E. E., Sobolev, D., Glazer, B., and Luther, G. W. (2004). Potential for microscale bacterial Fe redox cycling at the aerobic-anaerobic interface. *Geomicrobiol. J.* **21**, 379–391.

Stookey, L. L. (1970). Ferrozine: A new spectrophotometric reagent for iron. *Anal. Chem.* **42**, 779–781.

Straub, K. L., Benz, M., Schink, B., and Widdel, F. (1996). Anaerobic, nitrate-dependent microbial oxidation of ferrous iron. *Appl. Environ. Microbiol.* **62**, 1458–1460.

Straub, K. L., and Buchholz-Cleven, B. E. (1998). Enumeration and detection of anaerobic ferrous iron-oxidizing, nitrate-reducing bacteria from diverse European sediments. *Appl. Environ. Microbiol.* **64**, 4846–4856.

Straub, K. L., Schonhuber, W. A., Buchholz-Cleven, B. E. E., and Schink, B. (2004). Diversity of ferrous iron-oxidizing, nitrate-reducing bacteria and their involvement in oxygen-independent iron cycling. *Geomicrobiol. J.* **21**, 371–378.

Verran, J., Stott, J. F. D., Quarmby, S. L., and Bedwell, M. (1995). Detection, cultivation and maintenance of *Gallionella* in laboratory microcosms. *Lett. Appl. Microbiol.* **20**, 341–344.

Widdel, F., Schnell, S., Heising, S., Ehrenreich, A., Assmus, B., and Schink, B. (1993). Ferrous oxidation by anoxygenic phototrophic bacteria. *Nature* **362**, 834–836.

[7] High-Throughput Cultivation of Microorganisms Using Microcapsules

By Karsten Zengler, Marion Walcher, Greg Clark, Imke Haller, Gerardo Toledo, Trevin Holland, Eric J. Mathur, Gary Woodnutt, Jay M. Short, and Martin Keller

Abstract

This chapter describes a universal and novel method that provides access to the immense reservoir of untapped microbial diversity by cultivation. This technique uses microcapsules to encapsulate single cells combined with parallel microbial cultivation under low nutrient flux conditions. Under these conditions, single encapsulated cells grow and form microcolonies within the microcapsules. Flow cytometry is used as a sensitive tool to detect growth within the microcapsules. Microcapsules that contain microcolonies (originated from a single encapsulated cell) are sorted individually into microtiter dishes containing organic-rich medium. This high-throughput cultivation can provide more than 10,000 bacterial and fungal isolates per environmental sample.

Introduction

Even though significant advances have been made in understanding microbial diversity, most microorganisms are still only characterized by "molecular fingerprints" and have resisted cultivation. Many different approaches have been developed to overcome the problems associated with the cultivation of microorganisms. One obvious benefit would be the opportunity to investigate the previously inaccessible resources that these microorganisms potentially harbor.

Recombinant DNA approaches for discovering biocatalysts and other molecules directly from the environment without cultivating the host organisms, including functional screening, selection, hybridization-based, and direct DNA-sequencing techniques, were first developed and implemented about a decade ago (Robertson *et al.*, 1996; Short, 1997, 1999, 2000; Stein *et al.*, 1996). These technologies have, to date, yielded many commercially valuable products, such as enzymes (DeSantis *et al.*, 2002; Palackal *et al.*, 2004; Robertson *et al.*, 2004).

To gain a comprehensive understanding of microbial physiology or to access metabolic pathways containing genes dispersed throughout the

METHODS IN ENZYMOLOGY, VOL. 397
Copyright 2005, Elsevier Inc. All rights reserved.

genome, cultivation of microorganisms is advantageous. Only a fraction of all bacterial phyla contain previously cultivated microorganisms (Amann, 2000; Torsvik *et al.*, 2002), with many phyla represented by only a few isolates and some phyla containing only one described species. Half of the bacterial phyla recognized so far consist entirely of as yet uncultured bacteria and have been described solely on the basis of 16S rRNA gene sequences (Keller and Zengler, 2004; Rosselló-Mora and Amann, 2001).

Conventional cultivation of microorganisms is laborious, time-consuming, and, most importantly, selective and biased for the growth of specific microorganisms (Eilers *et al.*, 2000; Ferguson *et al.*, 1984). The majority of cells obtained from nature and visualized by microscopy are viable, but they do not generally form visible colonies on plates (Xu *et al.*, 1982). This may reflect the artificial conditions inherent in most culture media, e.g., extremely high substrate concentrations or the lack of specific nutrients required for growth. In addition, studies using modified media demonstrated the recovery of organisms not previously identified in culture by traditional cultivation methods (Bruns *et al.*, 2002; Janssen *et al.*, 2002; Stevenson *et al.*, 2004). To overcome some of these limitations, we developed a novel high-throughput cultivation method based on the combination of a single cell encapsulation procedure with flow cytometry that enables cells to grow with nutrients that are present at environmental concentrations.

Materials and Methods

Sample Collection and Preparation

Samples are collected fresh and processed immediately in the laboratory in order to keep the microbial population of the sample intact and to prevent any bias based on changes in sample conditions (e.g., temperature and pH). Microbial cells are separated from the sample matrix by repeated sheering cycles followed by density gradient centrifugation (Lindahl and Bakken, 1995; Priemé *et al.*, 1996). A small amount of the sample is resuspended in appropriate buffers (e.g., phosphate-buffered saline, PBS) and homogenized by repeated blending and cooling cycles. Homogenized sample suspensions are separated from soil particles by centrifugation (500 ml tubes 14,000g). The cell pellet is resuspended in 10 ml PBS buffer and is carefully laid over a 60% Nycodenz solution (recommended Sigma-Aldrich), and bacteria are purified by density gradient centrifugation for 2 h. Interphase-containing bacteria, a milky layer on top of a clear Nycodenz cushion, are removed carefully and washed thoroughly with phosphate buffer to remove any residual traces of Nycodenz. The purity of the microbial population is then checked microscopically by staining with SYBR-Green (recommended Molecular Probes) to be able to distinguish between cells and soil debris. This method has been adopted from previously published protocols for the isolation of bacteria from soil samples (Fægri *et al.*, 1977).

Encapsulation of Cells

The concentrated cell suspensions are used for encapsulation. Cells are diluted in appropriate media or buffer (e.g., PBS, pH 7.2) to a final concentration of 10^7 cells/ml. Diluted cell suspensions (0.1 ml) are mixed with 0.5 ml preheated agarose (40°) (OneCell System). Cell agarose mixtures are added to 15 ml of CellMix emulsion matrix (OneCell System) and emulsified at room temperature at 2200 rpm for 1 min using the CellSys 100 microdrop maker (One Cell System) followed by a 1 min emulsifying step (2200 rpm) on ice. The oil-bacterial suspension is then cooled with ice under stirring for 6 min at 1100 rpm. This procedure results in approximately 10^7 microcapsules. Approximately 10% of formed microcapsules are occupied by a single encapsulated cell. Successful encapsulation of single cells is monitored by staining the DNA with a fluorescent dye (e.g., SYBR Green, recommended Molecular Probes) and subsequent microscopy.

Preparation of Soil Extracts

Soil extracts are prepared from the same sample following standard protocols (Vobis, 1992). A fresh sample is mixed with equal amounts of $1\times$PBS (w/v), homogenized, and boiled for 1 h while stirring. After removal of suspended soil particles by centrifugation and filtration, the soil extract is sterilized by filtering through 0.2 μm filters prior to use.

Growth Conditions

The microcapsules are dispensed into sterile chromatography columns XK-16 (recommended Pharmacia Biotech) containing 25 ml of media. Columns are equipped with two sets of filter membranes (0.1 μm at the inlet of the column and 8 μm at the outlet). The filters prevent free-living cells from contaminating the medium reservoir and retain microcapsules in the column while allowing free-living cells to be washed out.

Medium is pumped through the column at a flow rate of 13 ml/h (PharMed tubing, 0.8 mm inside diameter, Cole Parmer). The system can be run in a recirculating or flow-through mode. Soil extract is added to the medium at final concentrations of 0.1 to 1 ml/liter in salt solution (KH_2PO_4, $NaNO_3$, NH_4Cl, $MgSO_4$ in physiological concentrations) and amended with trace minerals and vitamins. Microcapsules are incubated in the columns and the development of microcolonies is monitored microscopically. Microcolonies are sorted individually into 96-well microtiter plates with soil extracts amended with organic carbon sources (e.g., glucose, peptone, and yeast extract).

Flow Cytometry

Microcapsules containing colonies are separated from free-living cells and empty microcapsules by using a flow cytometer (MoFlo, Cytomation). Precise sorting is confirmed by microscopy. On the basis of their distinctive light-scattering signatures, the microcolonies are detected and separated by flow cytometry at a rate of 5000 microcapsules s^{-1}.

Summary and Discussion

Microcolonies containing as few as 20–100 cells are detected (Fig. 1). Nutrient-sparse media are sufficient to support growth, but the carbon content is low enough to prevent microbial "weeds" from "overgrowing" slow-growing microorganisms. The microcapsules allow the reconstituted microbial community members to be simultaneously cultivated "together" and "apart" because each "caged microcolony" can later be separated and analyzed. This method is also applicable for the analysis of interactions between different organisms using *in situ* conditions, e.g., by replacing the

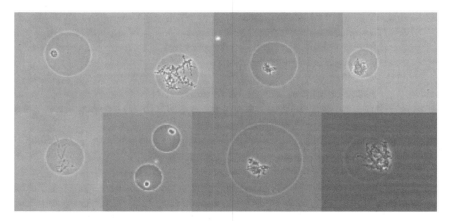

FIG. 1. Photomicrographs of microcapsules containing microcolonies of microorganisms. The size of a microcapsule is between 50 and 80 μm.

encapsulated cells back into the environment. In addition, the ability to reconstitute the community in the column of microcapsules allows for diffusive cross-feeding of metabolites and other molecules (e.g., regulatory and signaling molecules) between members of the community. Microorganisms are grown in an open flowing system where microorganisms are exposed continually to a very low concentration of nutrients. These features simulate the natural environment, thereby preserving some of the community interactions and other specific requirements that might be needed for successful cultivation. After sorting individual microcapsules into microtiter dishes containing organic-rich medium (Fig. 2), most of the cultures grow to a cell density that is greater than 0.1 as measured by optical density at 600 nm (Zengler *et al.*, 2002). This high-throughput cultivation technology is capable of isolating massive numbers of cultures and is therefore sufficient to supply increased diversity for modern high-throughput screening systems in drug discovery (Keller and Zengler, 2004; Zengler *et al.*, 2005).

High-throughput cultivation can provide more than 10,000 bacterial and fungal isolates per environmental sample. However, this increase in throughput requires advanced methods to characterize unique strains rapidly. To deal with this tremendous number of strains, a fully automated high-throughput method to determine the uniqueness of the bacterial

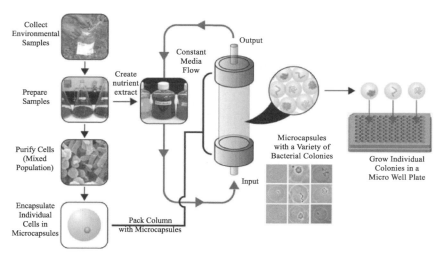

Fig. 2. Flow diagram of Diversa's high-throughput cultivation approach based on the encapsulation of single cells in microcapsules for massively parallel microbial cultivation (Keller and Zengler, 2004). (See color insert.)

and fungal isolates has been developed. Using Fourier transform infrared spectroscopy (Naumann *et al.*, 1991; Wenning *et al.*, 2002) in combination with novel spectra-comparison algorithms, microtiter plate cultures can be analyzed and unique strains are automatically arrayed into new plates (Diversa, unpublished data).

The new technology described in this chapter will continue to advance the transformation of microbiology and natural product discovery, with a focus on new culture-dependent and -independent methods to access microbial diversity.

Acknowledgment

Part of this work was supported by the Office of Science (BER), U.S. Department of Energy, Grant No. DE-FG02-04ER63771.

References

Amann, R. (2000). Who is out there? Microbial aspects of biodiversity. *Syst. Appl. Microbiol.* **23**, 1–8.

Bruns, A., Cypionka, H., and Overmann, J. (2002). Cyclic AMP and acyl homoserine lactones increase the cultivation efficiency of heterotrophic bacteria from the Central Baltic Sea. *Appl. Environ. Microbiol.* **68**, 3978–3987.

DeSantis, G., Zhu, Z., Greenberg, W. A., Wong, K., Chaplin, J., Hanson, S. R., Farwell, B., Nicholson, L. W., Rand, C. L., Weiner, D. P., Robertson, D. E., and Burk, M. J. (2002). An enzyme library approach to biocatalysis: Development of nitrilases for enantioselective production of carboxylic acid derivatives. *J. Am. Chem. Soc.* **124**, 9024–9025.

Eilers, H., Pernthaler, J., Glöckner, F. O., and Amann, R. (2000). Culturability and in situ abundance of pelagic bacteria from the North Sea. *Appl. Environ. Microbiol.* **66**, 3044–3051.

Fægri, A., Torsvik, V. L., and Goksöyr, J. (1977). Bacterial and fungal activities in soil: separation of bacteria and fungi by a rapid fractionated centrifugation technique. *Soil Biol. Biochem.* **9**, 105–112.

Ferguson, R. L., Buckley, E. N., and Palumbo, A. V. (1984). Response of marine bacterioplankton to differential filtration and confinement. *Appl. Environ. Microbiol.* **47**, 49–55.

Janssen, P. H., Yates, P. S., Grinton, B. E., Taylor, P. M., and Sait, M. (2002). Improved culturability of soil bacteria and isolation in pure culture of novel members of the divisions *Acidobacteria, Actinobacteria, Proteobacteria,* and *Verrucomicrobia. Appl. Environ. Microbiol.* **68**, 2391–2396.

Keller, M., and Zengler, K. (2004). Tapping into microbial diversity. *Nature Rev. Microbiol.* **2**, 141–150.

Lindahl, V., and Bakken, L. R. (1995). Evaluation of methods for extraction of bacteria from soil. *FEMS Microbiol. Ecol.* **16**, 135–142.

Naumann, D., Helm, D., and Labischinski, H. (1991). Microbiological characterizations by FT-IR spectroscopy. *Nature* **351**, 81–82.

Palackal, N., Brennan, Y., Callen, W. N., Dupree, P., Frey, G., Goubet, F., Hazlewood, G. P., Healey, S., Kang, Y. E., Kretz, K. A., Lee, E., Tan, X., Tomlinson, G. L., Verruto, J.,

Wong, V. W., Mathur, E. J., Short, J. M., Robertson, D. E., and Steer, B. A. (2004). An evolutionary route to xylanase process fitness. *Protein Sci.* **13**, 494–503.

Priemé, A., Sitaula, J. I. B., Klemedtsson, A. K., and Bakken, L. R. (1996). Extraction of methane-oxidizing bacteria from soil particles. *FEMS Microbiol. Ecol.* **21**, 59–68.

Robertson, D. E., Chaplin, J. A., De Santis, G., Podar, M., Madden, M., Chi, E., Richardson, T., Milan, A., Miller, M., Weiner, D. P., Wong, K., McQuaid, J., Farwell, B., Preston, L. A., Tan, X., Snead, M. A., Keller, M., Mathur, E., Kretz, P. L., Burk, M. J., and Short, J. M. (2004). Exploring nitrilase sequence space for enantioselective catalysis. *Appl. Environ. Microbiol.* **70**, 2429–2436.

Robertson, D. E., Mathur, E. J., Swanson, R. V., Marrs, B. L., and Short, J. M. (1996). The discovery of new biocatalysts from microbial diversity. *Soc. Indust. Microbiol. News* **46**, 3–8.

Rosselló-Mora, R., and Amann, R. (2001). The species concept for prokaryotes. *FEMS Microbiol. Rev.* **25**, 39–67.

Short, J. M. (1997). Recombinant approaches for accessing biodiversity. *Nature Biotechnol.* **15**, 1322–1323.

Short, J. M. (1999). Protein activity of clones having DNA from uncultivated microorganisms. *U.S. Patent* **5,958,672**.

Short, J. M. (2000). Screening for novel bioactivities. *U.S. Patent* **6,030,779**.

Stein, J. L., Marsh, T. L., Wu, K. Y., Shizuya, H., and De Long, E. F. (1996). Characterization of uncultivated prokaryotes: Isolation and analysis of a 40-kilobase-pair genome fragment from a planktonic marine archaeon. *J. Bacteriol.* **178**, 591–599.

Stevenson, B. S., Eichorst, S. A., Wertz, J. T., Schmidt, T. M., and Breznak, J. A. (2004). New strategies for cultivation and detection of previously uncultured microbes. *Appl. Environ. Microbiol.* **70**, 4748–4755.

Torsvik, V., Øvreås, L., and Thingstad, T. F. (2002). Prokaryotic diversity: Magnitude, dynamics, and controlling factors. *Science* **296**, 1064–1066.

Vobis, G. (1992). The genus *Actinoplanes* and related genera. *In* "The Prokaryotes" (A. Balows, H. G. Trüper, M. Dworkin, W. Harder, and K.-H. Schleifer, eds.), Vol. 2, pp. 1029–1060. Springer-Verlag, New York.

Wenning, M., Seiler, H., and Scherer, S. (2002). Fourier-transform infrared microspectroscopy, a novel and rapid tool for identification of yeasts. *Appl. Environ. Microbiol.* **68**, 4717–4721.

Xu, H. S., Roberts, N., Singleton, F. L., Attwell, R. W., Grimes, D. J., and Colwell, R. R. (1982). Survival and viability of nonculturable *Escherichia coli* and *Vibrio cholerae* in the estuarine and marine environment. *Microb. Ecol.* **8**, 313–323.

Zengler, K., Paradkar, A., and Keller, M. (2005). New methods to access microbial diversity for small molecule discovery. *In* "Natural Products: Drug Discovery, Therapeutics and Preventive Medicine" (L. Zhang and A. L. Demain, eds.), pp. 267–285. Humana Press, Totowa, NJ.

Zengler, K., Toledo, G., Rappe, M., Elkins, J., Mathur, E. J., Short, J. M., and Keller, M. (2002). Cultivating the uncultured. *Proc. Natl. Acad. Sci. USA* **99**, 15681–15686.

Section II

Physiological Ecology

[8] Chemotaxis and Behavioral Physiology of Not-Yet-Cultivated Microbes

By Jörg Overmann

Abstract

Chemotaxis assays provide a rapid and efficient means of (1) studying the chemotactic behavior of microorganisms in complex samples and (2) identifying potential growth substrates and generating inocula for subsequent isolation trials. The chemotaxis method thus complements the set of techniques currently available for the investigation of not-yet-cultured microbes. Although restricted to motile and chemotactically active microorganisms, a considerable fraction of species can be covered with this technique, particularly in bacterioplankton communities. Several formats of the chemotaxis assay have been developed. Capillaries are loaded with solutions of test compounds and are inserted in small microscopic chambers, in bottles containing culture suspensions, or incubated directly *in situ*. The latter two techniques are also suitable for experiments with anaerobic bacteria. In flat rectangular glass capillaries, the accumulating microorganisms can be observed directly by light microscopy in a dark field. Afterward, the chemotactically active bacteria can be identified by analyses of their 16S rRNA gene fragments. The method has been used to identify an essential carbon compound required for the growth of previously unculturable phototrophic consortia. This knowledge proved essential for the subsequent successful enrichment of these bacterial associations. Furthermore, it has been shown that different not-yet-cultured members of aerobic lake water bacterioplankton communities are chemotactically active and attracted by different carbon compounds.

Concept

Presently, about 5500 different prokaryotic species are recognized and available in culture (DSMZ, 2005). However, the majority of all species that occur in soil (up to 53,000; Sandaa *et al.*, 1999), sediments (up to 38,000; Torsvik *et al.*, 2002), and water ($\geq$450 species; Acinas *et al.*, 2004; Venter *et al.*, 2004) have so far escaped all cultivation attempts. Only a few notable incursions into this realm have been published (Bruns *et al.*, 2003; Hahn *et al.*, 2003; Janssen *et al.*, 2002; Rappe *et al.*, 2002). Based on culture-independent studies, many of not-yet-cultivated bacteria are only distantly

METHODS IN ENZYMOLOGY, VOL. 397
0076-6879/05 $35.00
DOI: 10.1016/S0076-6879(05)97008-0

related to known species (Rappé and Giovannoni, 2003). Hence, the physiology of novel types of bacteria can only rarely be inferred from their phylogeny, and even then only with a healthy dose of skepticism.

Due to the obvious shortcomings of established culture-based techniques, several culture-independent approaches have been developed in order to study the physiology of not-yet-cultured bacterial phylotypes in the environment. These approaches include the use of

- microautoradiography of single cells combined with fluorescence *in situ* hybridization
- heavy isotope analysis of single cells
- stable isotope probing (Manefield *et al.*, 2002)
- analysis of stable isotope fractionation of highly specific biomarkers (e.g., Glaeser and Overmann, 2003a)
- colocalization of key enzymes with particular types of microorganisms by the enzyme-labeled fluorescence technique (Strojsova *et al.*, 2003)
- analyses of functional genes *in situ* by RING-FISH or CARD-FISH or using extracted mRNA
- labeling of DNA-synthesizing cells by bromodeoxyuridine (Pernthaler *et al.*, 2002)
- monitoring the ribosome content of particular ribotypes

Typically, the aforementioned methods focus on the utilization of a single or very few substrates, on a single metabolic pathway, or yield information on the general physiological state of bacterial cells. However, one prerequisite for the successful cultivation of unknown prokaryotes is an efficient culture-independent screening to determine suitable substrates for growth.

Between 20 and 70% of marine planktonic bacteria are motile. Direct video microscopic observation has revealed that motile bacteria in natural samples are chemotactically active (Fenchel, 2002). Thus, chemotaxis assays potentially provide information on the behavior of a significant portion of a bacterioplankton community, including not-yet-cultured prokaryotes. Although the fraction of motile and chemotactic bacteria in soils or sediments is still unknown, typical soil bacteria have also been shown to react chemotactically. This is particularly true for bacteria associated with plants (e.g., Lopez-de-Victoria and Lovell, 1993). In principle, the chemotaxis method described in this article is also applicable to microbial communities in soils and sediments.

The most instructive way of using chemotaxis assays, however, is to identify potential substrates for subsequent cultivation attempts of not-yet-cultivated microbes. Although exceptions do exist, bacteria typically utilize chemoattractants also as carbon and energy sources for growth (Mesibov and Adler, 1972; Migas *et al.*, 1989; Nealson *et al.*, 1995; Poole *et al.*, 1993). Chemoattractants thus represent potential substrates for cultivation. For example, chemotaxis assays of the previously unculturable phototrophic consortia identified a compound (2-oxoglutarate) that was subsequently used to successfully enrich these bacterial associations for the first time (Fröstl and Overmann, 1998)—more than 90 years after their discovery (Lauterborn, 1906).

As outlined later, particular advantages of the chemotactic assays are speed and simplicity, which permit a parallel testing of a large number of compounds. Second, chemotaxis assays can readily be performed *in situ* in aquatic habitats. Third, chemotaxis is a means to collect live microorganisms as an inoculum for subsequent cultivation attempts (Glaeser and Overmann, 2004). Drawbacks of the method are its selectivity for motile and chemotactically active bacteria and, for certain applications, the rather small sample volumes that can be collected.

Chemotaxis Assays

Principle

In chemotaxis assays, glass capillaries are loaded with defined substrate solutions, inserted in a suspension of motile microorganisms, and the accumulation of cells at the opening of or within the capillary is monitored by direct or indirect methods. The original method of Adler (1973) was modified in order to permit (a) a direct and quantitative microscopic analysis of several samples, (b) parallel testing of a large number of different chemoattractants within the same bacterial sample, and (c) testing of anaerobic bacteria independent of an anaerobic chamber.

Experimental Setup

Short-term experiments are set up in small microscopic chambers (Fig. 1A), which are prepared using small $21 \times 21 \times 0.17$-mm coverslips as spacers between the microscope slide and the lid, which consists of a $60 \times 24 \times 0.17$-mm coverslip. Spacers and the lid are fixed by sealing the two short and one of the long edges of the chamber with a paraffin/mineral oil mixture (4:1, v/v).

Flat rectangular glass capillaries with a length of 50 mm, an inside diameter of 0.1×1.0 mm, and a capacity of 5 μl (Vitrocom, Mountain Lakes, NJ) are used for most applications. These capillaries fit exactly into the opening of the chemotaxis chamber. The specific geometry of these capillaries permits a direct light microscopic examination of their contents.

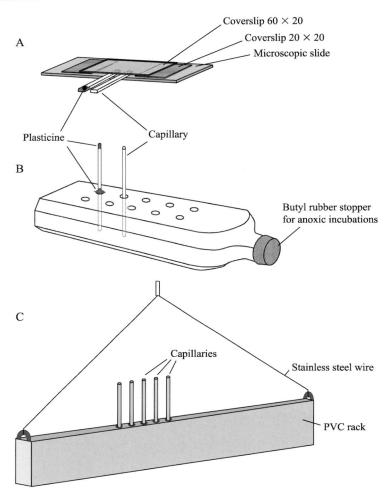

Fig. 1. Schematic view of three different experimental setups for chemotaxis measurements. (A) Laboratory-made microscopic chamber. Two sides and the back of the chamber are sealed with paraffin/mineral oil (not shown). Flat glass capillaries are loaded with substrate solutions, sealed at one end with plasticine, and inserted into the chamber through the opening in the front. (B) A series of chemotaxis capillaries inserted into a 100-ml Meplats bottle. The bottle is filled with a 20-ml sample and is closed with either a screw cap or a butyl rubber stopper for anoxic incubations. (C) Polyvinyl chloride rack for incubation of chemotaxis assays directly in aquatic environments. See text for detailed explanations.

Similarly, larger capillaries with an inner diameter of 0.1 × 2.0 mm (capacity 10 μl; Vitrocom) can be employed in the chemotaxis experiments. Corresponding to their larger volume, these capillaries yield more cells for subsequent analyses. If direct microscopy of the contents is not an issue, conventional round glass capillaries with an inner diameter of 1 mm, a length of 65 mm, and a capacity of 30 μl (e.g., Becton and Dickinson, Orangeburg, NY) may be employed.

Stock solutions of test substrates are prepared at a concentration of 100 mM. The chemoattractants are applied at a final concentration of 0.5–10 mM, which has been found to yield optimum results in 0.1 × 1.0-mm flat glass capillaries (see later). For tests in laboratory cultures, substrates are diluted in a sterile filtered (using disposable cellulose nitrate membrane filters; pore width, 0.2 μm) culture supernatant. The chemotaxis of microorganisms in natural samples is tested after diluting the substrates in sterile filtered aliquots (pore width, 0.1 μm) of the original sample. For tests of anaerobic bacteria, anoxic conditions are maintained by preparing stock solutions anoxically in Hungate or Balch tubes, which are sealed with butyl rubber septa and flushed with N_2. For dilution, culture supernatant or water samples are filled into syringes and then passed, via disposable filters and syringe needles, through septa into sterile glass vials containing an inert N_2 atmosphere. This method is equally well suited for field experiments (Glaeser and Overmann, 2003b, 2004).

In order to start the experiment, the microscopic chamber is tilted to an angle of 45° and is filled slowly with the sample or culture aliquot by means of a Pasteur pipette. Care has to be taken to avoid air bubbles that may be trapped in the inner corners of the microscopic chamber. Capillaries are then filled with diluted test substances by capillary action, sealed at one end with plasticine (Idena, Berlin, Germany), and inserted directly into the microscopic chamber so that their open ends extend well into the sample (Fig. 1A). At least two parallels are used per substrate. Each chemotaxis assay also comprises two control capillaries, which are filled with sterile culture supernatant or sample water. After all capillaries have been inserted, the open side of the microscopic chamber is finally sealed with the paraffin/mineral oil mixture, and the chamber is incubated at an appropriate temperature. Phototrophic microorganisms are incubated in homogeneous dim light (2–10 μmol Quanta · m^{-2} · s^{-1} of daylight fluorescent tubes).

The microscopic chemotaxis chamber is especially suited for short-term chemotaxis measurements of aerobic microorganisms conducted in the laboratory. This technique permits a rapid screening of a limited number of chemoattractants.

Whereas many marine planktonic bacteria are known to move at velocities of up to 400 μm · s^{-1} (Mitchell et al., 1995), therefore accumulating rapidly in chemotaxis experiments, other bacteria, such as the large-celled purple sulfur bacterium *Chromatium okenii*, the sulfate-reducing bacterium *Desulfovibrio desulfuricans*, or phototrophic consortia, swim more slowly ($10–30$ μm · s^{-1}). Analyses of slow-moving microorganisms, analyses of rare microorganisms in natural samples, tests of a large number of chemoattractants under highly defined conditions (i.e., in the same culture aliquot), or collections of larger numbers of bacteria require incubation times of several hours. Samples incubated in custom-made microscopic chambers (Fig. 1A) will usually dry out over such extended incubation periods. Instead, larger volumes of bacterial samples are used and incubated in chemotaxis chambers, which can be produced easily from commercially available glass bottles.

Meplats bottles (volume, 100 ml) are modified for chemotaxis experiments by drilling 10 holes of 3 mm diameter through one of the large side walls (Fig. 1B; Fröstl and Overmann, 1998). The bottles are autoclaved, positioned sideways, and then filled aseptically with 20 ml of liquid sample. For the study of chemotactic bacteria in sediments, slurries are prepared by diluting samples fourfold (v/v) prior to loading of the Meplats bottle (Fröstl and Overmann, 1998). During manipulation of the sample, anoxic conditions can be maintained by gassing the headspace aseptically through the opening of the bottle at a low flow rate with N_2 or N_2/CO_2. After loading with test compounds, each capillary is then inserted into a hole until its end extends into the culture liquid. The capillary is then fixed in the hole using plasticine. The Meplats bottle setup has been used successfully over incubation times of up to 12 h (Kanzler, 2004).

The chemotaxis method has been modified further to study target bacteria directly within aquatic ecosystems. Compared to the two methods described in the preceding sections, the *in situ* method is superior, especially in those cases where only low numbers of target bacteria occur in the natural microbial community, such that prolonged incubation times at constant environmental conditions are necessary, or where environmental conditions cannot be reproduced easily in the laboratory. For direct incubation in the environment, capillaries are fixed vertically in a solid $50 \times 5 \times 3$-cm polyvinyl chloride incubation rack (Fig. 1C, Glaeser and Overmann 2003b), which has 15 bore holes (4 mm diameter, 5 mm depth) drilled in its top. The holes are spaced 3 cm apart in order to avoid interference of individual chemotaxis assays. Before inserting the capillaries, each hole is filled entirely with plasticine. After loading with test substrate, each capillary is inserted into a separate hole, such that its open end faces upward. One-meter steel wires are fixed to both ends of the rack, with their ends connected to a calibrated steel wire, which in turn is connected to a float at the lake surface. With this setup, chemotaxis assays can be positioned at the appropriate water depths with a vertical precision better than 5 cm. Incubation times of up to 5 h have been tested with this setup.

After an appropriate time period, the incubation is terminated by removing the capillaries from the respective chemotaxis setup. For direct microscopy of the accumulated microorganisms, the open end of each capillary is immediately sealed with plasticine. Using a 10, 20, or 40× objective lens and dark-field illumination, bacteria can be visualized within the flat capillaries (Fig. 2A) and their numbers determined with a counting grid (Fig. 2B). Certain bacteria may accumulate by a factor of almost 1000 in the capillaries (Fig. 2B). If the numbers of microorganisms are too high for direct counting or if bacteria are too small or too fast, as is often the case in natural bacterioplankton, counting can be performed by epifluorescence microscopy. The contents of each capillary are collected using small disposable capillary plastic pipette tips (10 μl, Biozym PreCision No. 729009; Biozym, Diagnostik, Hessisch Oldendorf, Germany) and are mixed with 0.5 ml staining solution containing glutaraldehyde at a final

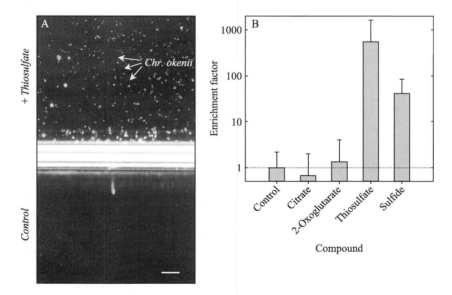

FIG. 2. (A) Dark-field photomicrograph of the chemotactic accumulation of *Chromatium okenii* from a small freshwater lake (Lake Dagow, Brandenburg, Germany). The upper part depicts a section of a capillary loaded with 1 m*M* thiosulfate, and the bottom section shows a control capillary without substrate. Bar: 50 μm. (B) Quantitative analysis of the chemotactic response of *C. okenii* toward different attractants. Values around 1 indicate that no chemotactic enrichment occurred. Vertical bars indicate one standard deviation (J. Fröstl and J. Overmann, unpublished result).

concentration of 2% and the fluorescent DNA stain 4',6-diamidino-2-phenylindole (DAPI) at a final concentration of 0.3 μg · ml^{-1}. After staining the samples for 10 min in the dark, they are filtered through 0.5-cm^2 large defined areas of polycarbonate filters (Isopore, type VCTP, Millipore; pore width, 0.1 μm).

In addition to these quantitative measurements, chemotactically active microorganisms can be identified by a combination of molecular methods (see later).

A specific application of the chemotaxis technique is the aerotaxis assay in large flat glass capillaries described by Eschemann *et al.* (1999). Capillary microslides (Vitro Dynamics, Rockaway, NJ) with inner dimensions of $100 \times 10 \times 1$ mm are filled with the bacterial suspension. A 5-μl gas bubble with a defined partial pressure of oxygen is placed in the middle of the capillary using a Hamilton glass syringe. If required, the capillaries may be sealed with paraffin/mineral oil. Aerotactic bacteria form ring-like bands around the oxygen bubble, with the diameter of the ring depending on the oxygen concentration, cell density, and the respiration rate of the bacteria. Oxygen gradients in the capillary may be analyzed with high precision by the use of oxygen microelectrodes with a tip diameter <10 μm, and the density of the cells by OD measurements across the capillary employing a small (50-μm diameter) optical fiber coupled to a diode array spectrometer (Eschemann *et al.*, 1999).

Variables

Size and Geometry of Capillaries. Experiments (Kanzler, 2004) have revealed that some bacteria do not accumulate in round capillaries of 1 mm diameter, whereas bacteria from the same culture readily enter rectangular capillaries of 0.1×2.0 mm filled with the same chemoattractant solution. Hence, the gradient that develops at the opening and in the first millimeters of rectangular capillaries must differ significantly from that at round capillaries, thereby affecting the chemotactic accumulation of microorganisms. Furthermore, secondary bands developed in 0.1×2.0-mm as opposed to 0.1×1.0-mm rectangular capillaries. These secondary bands were positioned further inside the capillary and were strongly dominated by only one type of bacterium as opposed to the primary bacterial accumulation close to the capillary mouth (Kanzler, 2004). Depending on the microorganism investigated, the geometry of the capillaries employed hence is critical for the degree of enrichment.

Substrate Concentration. The concentration of attractants is a parameter that has been found to be critical for observing a chemotactic response of bacteria (Jaspers, 2000). When series of 10-fold dilutions of yeast extract and amino acid mixtures were tested with natural bacterioplankton, chemotactic accumulations of the bacteria were only detected in capillaries containing $\geq 0.375\%$ (w/v) yeast extract or a highly concentrated amino acid mixture (10 mM each). Steeper gradients at the capillary opening thus result in a more efficient accumulation of chemotactic bacteria.

However, the swimming velocity of marine bacteria has been shown to be inversely related to substrate concentration (Mitchell *et al.*, 1995), such that bacteria moved most rapidly at concentrations of 0.001% tryptic soy broth, but stopped completely at 1.0%. The combined effect of the two opposing effects of substrate concentration in the capillary—more efficient accumulation at the opening and a decrease of swimming velocity at the high substrate concentrations encountered by the cells when entering the capillary—would be expected to lead to an accumulation of immotile cells inside the capillary close to its open end. Indeed, clouds of slow-moving or immotile bacteria were observed repeatedly shortly behind the open-end inside capillaries filled with strong chemoattractants (J. Overmann, J. Fröstl, and B. Kanzler; unpublished data).

Phylogenetic Identification of Chemotactically Active Bacteria

The capillary technique has also been used successfully to identify chemotactically active bacteria by a combination of molecular phylogenetic methods (Fig. 3). In order to collect a sufficiently large number of cells for molecular analyses, at least 10 capillaries need to be incubated in parallel for each compound tested. Prior to sampling of their contents, the capillaries are sterilized on the outside by wiping with 70% (v/v) ethanol. The contents of the capillaries are then harvested by thin disposable plastic pipette tips (10 μl, Biozym PreCision No. 729009; Biozym, Diagnostik, Hessisch Oldendorf, Germany), and the samples are combined in a sterile Eppendorf centrifuge tube. Cells are then pelleted by centrifugation for 20 min at 50,000g and 4°, and the pellet is stored at $-20°$ until further analysis.

Prior to cell lysis, 20 μl of double distilled water or Tris-HCl buffer (10 mM, pH 8.0) is added to each pellet and the cells are resuspended. Cell lysis is performed through five consecutive freeze–thaw cycles, with each cycle comprising a heating step at 100° for 3 min and a freezing step at $-80°$ for 3 min. Two microliters of the resulting cell lysate is used for polymerase chain reaction (PCR) amplification of up to 600-bp-long fragments of the 16S rRNA genes using primers GC341f and 907r (Muyzer *et al.*, 1993). For subsequent separation by denaturing gradient gel electrophoresis (DGGE; Muyzer *et al.*, 1993), primer GC341f carries a 40- bp-long GC fragment at its 5′ end. Amplification products are separated on gradient gels containing a linear gradient of

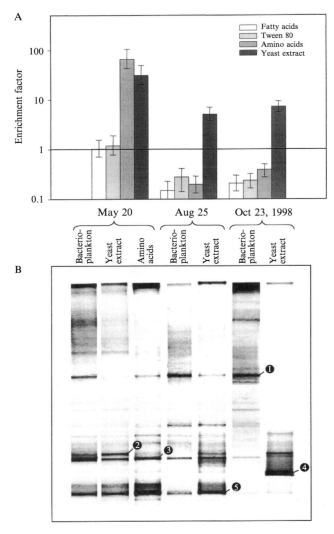

FIG. 3. (A) Chemotactic response of aerobic chemohetrotrophic bacterioplankton in eutrophic Zwischenahner Meer in May, August, and October 1998. A mixture of fatty acids (formate, acetate, propionate; 1 mM each), Tween 80 (0.005%, v/v), an amino acid mixture containing L-cysteine, L-alanine, L-isoleucine, L-leucine, L-asparagine, and L-arginine (1 mM each), and yeast extract (0.375%, w/v) were used as test substrates. Enrichment factors significantly >1 indicate a positive chemotactic response, whereas substrates leading to values <1 act as repellents. (B) Comparison of phylogenetic fingerprints of chemotactically active bacteria with those of the natural bacterioplankton community. Phylotype 1 represents uncultured freshwater *Thermodesulfobacterium* clone LD28; phylotype 2 is related to the β-*Proteobacterium Vogesella indigofera* ATCC19706[T] (96.7% sequence similarity) and was

urea and formamide. After staining with ethidium bromide, individual bands are cut out and sequenced (Glaeser and Overmann, 2004).

The aforementioned approach was used successfully to investigate the chemotactic response of planktonic bacteria in a shallow eutrophic lake (Zwischenahner Meer, Northern Germany) during different seasons (Fig. 3; Jaspers, 2000). Of the four substrates tested, the yeast extract consistently elicited a positive response. In spring, a mixture of amino acids yielded similarly high numbers (Fig. 3A). Phylogenetic DGGE fingerprinting and sequencing of 16S rRNA gene fragments clearly demonstrated, however, that distinctly different types of bacteria responded to the two substrate mixes (Fig. 3B). A mixture of short-chain fatty acids and Tween 80 acted as repellents on two of the three sampling dates (Fig. 3A).

The resolution, i.e., the number of phylotypes detected by PCR-DGGE, is increased significantly when group-specific oligonucleotide primers are used (Gich *et al.*, 2005; Overmann *et al.*, 1999). Aliquots of the cell lysate are amplified in parallel with primers specific for individual phylogenetic groups, such as the *β-Proteobacteria*, or members of the phyla *Bacteroidetes* or *Firmicutes*, and the amplification products are separated in parallel lanes by DGGE.

Generating Inocula for Subsequent Cultivation Attempts

The chemotaxis approach can be used to selectively enrich particular types of bacteria from complex samples. As exemplified by the chemotaxis study of bacterioplankton in Zwischenahner Meer (Fig. 3), phylotypes 2 and 4 were not detected in the natural bacterioplankton community and hence must constitute a minor fraction. However, both could be significantly enriched in capillaries containing yeast extract. The response of phylotype 2 was substrate specific, as this phylotype was not observed in parallel chemotaxis assays containing amino acids (Fig. 3B). In addition, so-far-uncultured members of the freshwater *Actinobacteria* were detected in capillaries containing yeast extract. Information on potential growth substrates can be used to design suitable cultivation media for the targeted isolation of such novel types of bacteria. At the same time, the chemotactic enrichments generated can be used as inoculum for these cultivation attempts.

---

selectively enriched from the May and August samples by chemotaxis toward the yeast extract. Phylotypes 3 and 5 represent two distinct relatives of an uncultured actinomycete clone ACK-M1 (94.2 and 95.3% sequence similarity). Phylotype 4 shows 100% sequence similarity to the *β*-Proteobacterium *Pseudomonas spinosa* ATCC14606. This latter phylotype dominated in the chemotaxis enrichment, although it could not be detected in the natural bacterioplankton community. Modified from Jaspers (2000).

Physiological Measurements Based on the Scotophobic
Response: The Bacteriospectrogramm

Principle

A related, and somewhat more specific, way of exploiting the tactic
behavior of microorganisms is the scotophobic response. Phototrophic

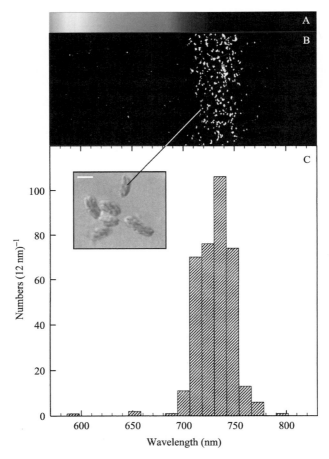

FIG. 4. Bacteriospectrogramm of the phototrophic consortium "*Chlorochromatium
aggregatum.*" (A) Schematic representation of the continuous spectrum focused in the
optical plane of the coverslip of the microscopic chamber. (B) Computerized dark-field
photomicrograph of the chamber. Each dot represents a single consortium. (C) Integrated
number of accumulated consortia per 12-nm interval. (Insert) Differential interference
contrast photomicrograph of six "*C. aggregatum*" consortia. Bar: 5 μm. Modified from Fröstl
and Overmann (1998). (See color insert.)

bacteria as well as phototrophic consortia exhibit a phobic reaction ("Schreckbewegung") when suddenly illuminated with light of high intensity (Buder, 1914). This leads to a wavelength-dependent accumulation, which can be exploited to identify the corresponding light receptor (Fröstl and Overmann, 1998; Glaeser and Overmann, 1999). In addition, discrimination thresholds of light intensities can be determined by exploiting the scotophobic response (Hustede et al., 1989).

Experimental Setup

The wavelength dependence of the scotophobic response of phototrophic bacteria can be determined by illuminating a laboratory-made microscopic chamber (as depicted in Fig. 1A) through a small interference filter with a spectral range of 570–1100 nm (Oriel, Langenberg, Germany). The filter is positioned directly above the field stop of a light microscope, and the spectrum generated is focused in the optical plane of the coverslip of the microscopic chamber. After 1 h of incubation, the interference filter is quickly removed from the light path and photomicrographs are immediately recorded at low magnification (e.g., 120×) in dark field (Fig. 4). Most samples contain immotile bacterial cells attached to the coverslip or bottom of the microscopic chamber. The bacteriospectrogramm can therefore be improved by obtaining an initial image prior to the incubation, which is subtracted from the image taken at the end of the exposure.

References

Acinas, S. G., Klepac-Ceraj, V., Hunt, D. E., Pharino, C., Ceraj, I., Distel, D. L., and Polz, M. F. (2004). Fine-scale phylogenetic architecture of a complex bacterial community. *Nature* **430**, 551–554.

Adler, J. (1973). A method for measuring chemotaxis and use of the method to determine optimum conditions for chemotaxis by *Escherichia coli*. *J. Gen. Microbiol.* **74**, 77–91.

Bruns, A., Nübel, U., Cypionka, H., and Overmann, J. (2003). Effect of signal compounds and incubation conditions on the culturability of freshwater bacterioplankton. *Appl. Environ. Microbiol.* **69**, 1980–1989.

Buder, J. (1914). Chloronium mirabile. *Berichte Dtsch. Bot. Ges.* **31**, 80–97.

Deutsche, Sammlung für Mikroorganismen und Zellkulturen (DSMZ) (2005). Bacterial nomenclature up-to-date. http://www.dsmz.de/bactnom/bactname.htm.

Eschemann, A., Kühl, M., and Cypionka, H. (1999). Aerotaxis in *Desulfovibrio*. *Environ. Microbiol.* **1**, 489–494.

Fenchel, T. (2002). Microbial behavior in a heterogenous world. *Science* **296**, 1068–1071.

Fröstl, J., and Overmann, J. (1998). Physiology and tactic response of "*Chlorochromatium aggregatum.*" *Arch. Microbiol.* **169**, 129–135.

Gich, F., Schubert, K., Bruns, A., Hoffelner, H., and Overmann, J. (2005). Specific detection, isolation and characterization of not-yet-cultured members of freshwater bacterioplankton. Submitted for publication.

Glaeser, J., and Overmann, J. (1999). Selective enrichment and characterization of *Roseospirillum parvum*, gen. nov. and sp. nov., a new purple nonsulfur bacterium with unusual light absorption properties. *Arch. Microbiol.* **171**, 405–416.

Glaeser, J., and Overmann, J. (2003a). Characterization and *in situ* carbon metabolism of phototrophic consortia. *Appl. Environ. Microbiol.* **69**, 3739–3750.

Glaeser, J., and Overmann, J. (2003b). The significance of organic carbon compounds for *in situ* metabolism and chemotaxis of phototrophic consortia. *Environ. Microbiol.* **5**, 1053–1063.

Glaeser, J., and Overmann, J. (2004). Biogeography, evolution and diversity of the epibionts in phototrophic consortia. *Appl. Environ. Microbiol.* **70**, 4821–4830.

Hahn, M. W., Lünsdorf, H., Wu, Q., Schauer, M., Höfle, M. G., Boenigk, J., and Stadler, P. (2003). Isolation of novel ultramicrobacteria classified as *Actinobacteria* from five freshwater habitats in Europe and Asia. *Appl. Environ. Microbiol.* **69**, 1442–1451.

Hustede, E., Liebergesell, M., and Schlegel, H. G. (1989). The photophobic response of various sulfur and nonsulfur purple bacteria. *Photochem. Photobiol.* **50**, 809–815.

Janssen, P. H., Yates, P. S., Grinton, B. E., Taylor, P. M., and Sait, M. (2002). Improved culturability of soil bacteria and isolation in pure culture of novel members of the divisions *Acidobacteria*, *Actinobacteria*, *Proteobacteria*, and *Verrucomicrobia*. *Appl. Environ. Microbiol.* **68**, 2391–2396.

Jaspers, E. (2000). "Zur ökologischen Bedeutung der Diversität planktischer Bakterien: Erkenntnisse aus der Analyse von Reinkulturen." Ph.D. Dissertation, University of Oldenburg.

Kanzler, B. (2004). "Molekularbiologische und physiologische Untersuchungen an phototrophen Kosnortien." Diploma thesis, Ludwig-Maximilians-Universität München.

Lauterborn, R. (1906). Zur Kenntnis der sapropelischen Flora. *Allg. Bot. Z.* **12**, 196–197.

Lopez-de-Victoria, G., and Lovell, C. R. (1993). Chemotaxis of *Azospirillum* species to aromatic compounds. *Appl. Environ. Microbiol.* **59**, 2951–2955.

Manefield, M., Whiteley, A. S., Griffiths, R. I., and Bailey, M. J. (2002). RNA stable isotope probing, a novel means of linking microbial community function to phylogeny. *Appl. Environ. Microbiol.* **68**, 5367–5373.

Mesibov, R., and Adler, J. (1972). Chemotaxis toward amino acids in *Escherichia coli*. *J. Bacteriol.* **112**, 315–326.

Migas, J., Anderson, K. L., Cruden, D. L., and Markovetz, A. J. (1989). Chemotaxis in *Methanospirillum hungatei*. *Appl. Environ. Microbiol.* **55**, 264–265.

Mitchell, J. G., Pearson, L., Bonanzinga, A., Dillon, S., Khouri, H., and Paxinos, R. (1995). Long lag times and high velocities in the motility of natural assemblages of marine bacteria. *Appl. Environ. Microbiol.* **61**, 877–882.

Muyzer, G., de Waal, E. C., and Uitterlinden, A. G. (1993). Profiling of complex microbial populations by denaturing gradient gel electrophoresis analysis of polymerase chain reaction amplified genes coding for 16S rRNA. *Appl. Environ. Microbiol.* **59**, 695–700.

Nealson, K. H., Moser, D. P., and Saffarini, D. A. (1995). Anaerobic electron acceptor chemotaxis in *Shewanella putrefaciens*. *Appl. Environ. Microbiol.* **61**, 1551–1554.

Overmann, J., Coolen, M. J. L., and Tuschak, C. (1999). Specific detection of different phylogenetic groups of chemocline bacteria based on PCR and denaturing gradient gel electrophoresis of 16S rRNA gene fragments. *Arch. Microbiol.* **172**, 83–94.

Pernthaler, A., Pernthaler, J., Schattenhofer, M., and Amann, R. (2002). Identification of DNA-synthesizing bacterial cells in coastal North Sea plankton. *Appl. Environ. Microbiol.* **68**, 5728–5736.

Poole, P. S., Smith, M. J., and Armitage, J. P. (1993). Chemotactic signalling in *Rhodobacter spheroides* requires metabolism of attractants. *J. Bacteriol.* **175**, 291–294.

Rappé, M. S., and Giovannoni, S. J. (2003). The uncultured microbial majority. *Annu. Rev. Microbiol.* **57,** 369–394.

Sandaa, R. A., Torsvik, V., Enger, O., Daae, F. L., Castberg, T., and Hahn, D. (1999). Analysis of bacterial communities in heavy metal-contaminated soils at different levels of resolution. *FEMS Microbiol. Ecol.* **30,** 237–251.

Strojsova, A., Vrba, J., Nedoma, J., Komarkova, J., and Znachor, P (2003). Seasonal study of extracellular phosphatase expression in the phytoplankton of a eutrophic reservoir. *Eur. J. Phycol.* **38,** 295–306.

Torsvik, V., Ovreås, L., and Thingstad, T. F. (2002). Prokaryotic diversity: Magnitude, dynamics, and controlling factors. *Science* **296,** 1064–1066.

Venter, J. C., Remington, K., Heidelberg, J. F., Halpern, A. L., Rusch, D., Eisen, J. A., Wu, D., Paulsen, I., Nelson, K. E., Nelson, W., Fouts, D. E., Levy, S., Knap, A. H., Lomas, M. W., Nealson, K., White, O., Peterson, J., Hoffman, J., Parsons, R., Baden-Tillson, H., Pfannkoch, C., Rogers, Y.-H., and Smith, H. O. (2004). Environmental genome shotgun sequencing of the Sargasso Sea. *Science* **304,** 66–74.

Further Reading

Rappé, M. S., Connon, S. A., Vergin, K. L., and Giovannoni, S. J. (2002). Cultivation of the ubiquitous SAR11 marine bacterioplankton clade. *Nature* **418,** 630–633.

[9] Analysis of Microbial Communities with Electrochemical Microsensors and Microscale Biosensors

By NIELS PETER REVSBECH

Abstract

Electrochemical microsensors for O_2, pH, H_2S, H_2, and N_2O are now available commercially, thus it has become a relatively simple task to analyze the microenvironment in stratified microbial communities for several chemical species. In addition, sensors are available for the physical parameters diffusivity and flow, and based on knowledge about both transport processes and microdistribution of chemistry, it becomes possible to calculate the spatial distribution and local rates of transformations, such as aerobic respiration or denitrification. As compared to other advanced techniques, microsensor equipment is inexpensive. For example, it is possible to start working with oxygen microsensors with an investment of only about $5000. Construction of one's own microsensors is only recommended for the very dedicated user, but the investment here is mainly in terms of man-hours as the equipment is simple and inexpensive. By establishing a microsensor construction facility, it is possible to work with short-lived

METHODS IN ENZYMOLOGY, VOL. 397   0076-6879/05 $35.00
DOI: 10.1016/S0076-6879(05)97009-2

sensors such as ion-selective microsensors for H^+, NO_2^-, NO_3^-, Ca^{2+}, and CO_3^- based on ion exchangers and with microscale biosensors for NO_x^-, NO_2^-, CH_4, and volatile fatty acids based on immobilized bacteria.

Introduction

By the introduction of few micrometer-thick microsensors for analysis of a wide range of chemical species in microbial ecology, it became possible to study small-scale chemical gradients and metabolism in stratified microbial communities (e.g., Revsbech and Jørgensen, 1986). It is even possible to analyze gradients and metabolism associated with individual larger microorganisms such as large colorless sulfur bacteria (Schulz and de Beer, 2002). A simple technique allowed for high-resolution determinations of oxygenic photosynthetic rates (Lassen et al., 1998; Revsbech and Jørgensen, 1986), and diffusion-reaction modeling of concentration profiles allows for the determination of respiration rates (Berg et al., 1998; Epping et al., 1999; Revsbech et al., 1986). Examples of stratified microbial communities that have been studied extensively using microsensors include hypersaline microbial mats (Canfield and DesMarais, 1991; Grotzschel and de Beer, 2002; Revsbech et al., 1983), hot spring microbial mats (Revsbech and Ward, 1984), deep-sea sediments (Glud et al., 1994; Gundersen et al., 1992; Reimers, 1987), soils (Meyer et al., 2002), the termite gut (Schmitt-Wagner and Brune, 1999), and various symbiotic associations such as corals (Kühl et al., 1995). Use of microsensors in microbial ecology requires some laboratory infrastructure, but if the sensors are obtained commercially, the infrastructure requirements are low and also low priced as compared to most other sophisticated analytical techniques. It is the author's experience that having an infrastructure of bench experience is much more essential than sophisticated equipment for the successful application of microsensors. It seems to be difficult for most scientists to start thinking in terms of small-scale diffusion gradients and diffusional transport and therefore also to design optimal experiments. For beginners in the field it is thus often worthwhile to visit a laboratory where microsensors are used routinely to learn about fundamental microsensor handling and to discuss experimental design.

It is the intention of this chapter to give the reader an impression about what sensors may be used for by the nonspecialist, and there is therefore relatively little emphasis on new and still experimental techniques that may only function in the hands of the specialist. Several types of microsensors only work for a few days or even shorter, and although some of these sensors have been in use for decades, they may be of limited interest as a routine tool. If the use of such techniques is crucial, it is usually worthwhile to contact a specialist about joint experiments instead of allocating months to the learning of microsensor construction.

Commercially Available Microsensors and Their Characteristics

The nonfiber optic microsensors that, at present (2005) and to the author's knowledge, are available commercially for use in microbial ecology are listed in Table I, and this section gives a short description of each of them. They are mentioned in the sequence of estimated present use. Not all technical details, such as polarization voltage, are given, but the reader is given an impression about what these sensors may be used for. Technical details are given by the commercial supplier, and even more detail can be found in primary publications and extensive reviews listed later. All of the sensors for chemical analysis listed in Table I exhibit no or almost no sensitivity to flow in the medium if they are constructed correctly. Pressure-compensated versions for use in the deep sea may also be purchased.

Oxygen Microsensor

The oxygen microsensors of interest for microbial ecology are of the micro-Clark design (Clark *et al.*, 1953; Revsbech, 1989) (Fig. 1). For a very detailed description of the functioning of an oxygen microsensor, consult

TABLE I
COMMERCIALLY AVAILABLE MICROSENSORS OF HIGH IMPORTANCE TO MICROBIAL ECOLOGY

Sensor type	Minimum tip diameter	Serious interference	Meter type	Supplier[a]
O_2 micro-Clark	2 μm	—	Ammeter	Diamond, Unisense, AMT
pH glass	10 μm thick 100 μm long	—	Voltmeter	Unisense
pH exchanger	1 μm	—	Voltmeter	Diamond
H_2S micro-Clark	10 μm	—	Ammeter	AMT, Unisense
H_2 micro-Clark	2 μm	H_2S	Ammeter	AMT, Unisense
N_2O, O_2 insensitive	20 μm	H_2S	Ammeter	Unisense
Redox	20 μm	Drifting response	Voltmeter	Unisense
Diffusivity	20 μm	See text	Ammeter	Unisense
Flow	10 μm	See text	Ammeter	Unisense

[a] Diamond: *www.DiamondGeneral.com*; Unisense: *www.Unisense.com*; and AMT: *www.AMT-GMBH.com*.

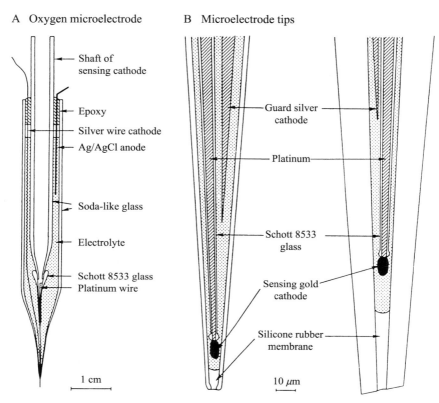

A Oxygen microelectrode

Shaft of
sensing cathode

Epoxy

Silver wire cathode

Ag/AgCl anode

Soda-like glass

Electrolyte

Schott 8533 glass
Platinum wire

1 cm

B Microelectrode tips

Guard silver
cathode

Platinum

Schott 8533
glass

Sensing gold
cathode

Silicone rubber
membrane

10 μm

FIG. 1. Oxygen microsensor of the micro-Clark type shown with both a thin tip, which can be used for analysis where minimum sample disturbance is essential, and a sturdy tip for use in media that might otherwise cause sensor damage. The internal guard cathode removes oxygen from the internal electrolyte that might otherwise interfere. From Revsbech (1989), with permission from Limnology and Oceanography.

Glud *et al.* (2000). They can be made with diameters as small as 1–2 μm, but these are fragile (Fig. 1A), or they can be made with very thick glass walls (Fig. 1B) with total diameters up to about 1 mm, in which case they are extremely robust. The oxygen microsensor is at present the microsensor with the overall best performance. It is sensitive with a detection limit $<1\ \mu M$, fast responding with 90% response times down to 0.1 s, long-term stable with minimal drift during a day, and it does not suffer from pronounced interferences from environmentally relevant chemical species. Due to the availability of very thin sensors, it is possible to perform experiments that cause minimal disturbance to the analyzed medium. The very fast response time makes it possible to record detailed concentration profiles over distances of millimeters within a few minutes, and it also makes it possible to record detailed photosynthesis profiles by the light/dark shift technique (Lassen et al., 1998; Revsbech and Jørgensen, 1986).

Usually there are no problems with interfering agents. Short-term exposure to low concentrations of hydrogen sulfide has little or no effect, but exposure to high concentrations should be avoided, as the zero current of the sensor may be affected. It is possible to block the entrance of oxygen into the microsensor by gas-impermeable matter, resulting in erratic low readings. Usually, such effects are not observed by analysis in microbial ecology, but problems have been encountered by analysis of animal tissue. Plant waxes may also be very efficient in blocking the sensor tip. Where such physical clogging of the sensor tips takes place it may be advantageous to use oxygen optodes (e.g., Kühl and Revsbech, 2001), which are also described elsewhere in this volume. For most profiling purposes, however, electrochemical oxygen sensors are superior to optodes because of the rapid response and extreme range in total diameter from the very thin and delicate to the thick and extremely sturdy. If the tip of an oxygen microsensor or any other electrochemical gas sensor gets coated with sticky material, it is possible to clean it by deep introduction (using a micromanipulator) into, for example, 4% agar gel, and washing in nonpolar and polar solvents such as hexane and ethanol is also possible. Oxygen microsensors of the Clark type copied from the design of Revsbech (1989) are available from Diamond General (DiamondGeneral.com) and Unisense (Unisense.com). The average lifetime of these oxygen microsensors is about 1 year. There is, however, a pronounced tendency to a change in the calibration curve against lower currents by the aging of oxygen microsensors. Apparently, the membrane gets less and less permeable to oxygen. Some sensors may end up with such a low current (e.g., less than 20 pA difference between anoxic and fully aerobic conditions) that electrical noise starts to be a problem, and they must then be discarded. A similar oxygen microsensor is sold by the German company AMT (AMT-GMBH. com). A limited amount of detail is available about this microsensor, and the author has not been able to get any information from the company about prices on sensors and meters. According to the published material, it has an internal reference element (probably lead in an alkaline electrolyte) that causes a self-polarization of the sensor so that no external polarization source is needed. It should have an operational lifetime of about 2 months.

pH Microsensor

The glass pH microsensor (Fig. 2) is, in principle, identical to larger commercial pH sensors, and except for its small size and need for a higher resistance voltmeter, it also has the same basic characteristics. Due to high

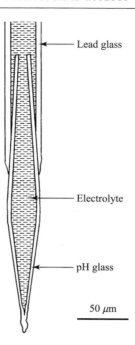

FIG. 2. Tip of glass pH microsensor. The cone made from pH-sensitive (i.e., hydrogen ion exchange) glass will integrate ambient pH over the entire length.

resistance in the circuit of about 10^{10} Ω, the 90% response time is often in the 1-min range when ordinary coaxial shielding of electrical connections is used. Faster responses may be obtained by placing the amplifier in close proximity to the sensor, and presumably also by using a voltmeter with triaxial input wire (Keithley, www.Keithley.com), which alleviates capacity problems associated with the charging of electrical shielding. Because of the very thin glass layers, there is a more pronounced tendency to over-hydration as compared to macroscale pH sensors, causing a lower than Nernstian response and interference from sodium ions. However, a good glass pH microsensor exhibits a response of close to 59 mV per pH unit and very little sodium interference. The minimum dimensions for routinely well-functioning glass pH sensors are, however, diameters of about 10 μm and lengths of exposed pH-sensitive glass of about 100 μm. Such sensors thus integrate pH values over a length of minimum 100 μm, whereas it is possible to get a spatial resolution down in the 1- to 2-μm range by an oxygen microsensor. One problem with microscale glass pH sensors is fragility. It is possible to protect the tip by building the sensors into hypodermic needles (available commercially from Unisense), but this results in a very slow response. Due to high electrical resistance in a small pH sensor, the measurement may be disturbed easily by external magnetic fields. It is therefore highly recommendable to use coaxially shielded sensors; such shielded sensors are sold by the commercial supplier. The lifetime of microscale glass pH sensors is usually about 6 months, and up to 1 year if some non-Nernstin (i.e., less slopes of less than 59 mV per pH unit) response can be accepted. Very small sensors have a shorter lifetime than larger (e.g., 250-μm exposed tip) sensors due to the very thin pH-sensitive glass layer.

LIX-type pH sensors (LIX sensors are described in more general terms later) may have tips in the 1-μm range, and such sensors should be used when the ~100-μm resolution of glass microsensors is insufficient. Diamond General sells down to 1-μm-thick pH sensors that seem to be LIX based. The shelf-life of these sensors by dry storage is probably about 1 month (Cai and Reimers, 2000), which is far more than for most other LIX-type sensors. It has, however, not been possible for the author to get information about the actual lifetime from the company.

H$_2$S Sensor

The Clark-type H$_2$S microsensors (Fig. 3) sold by AMT and Unisense are, in appearance, almost identical to the Clark O$_2$ microsensor. However, H$_2$S is oxidized indirectly through a ferri-ferrocyanide shunt (Jeroschewski et al., 1996), and they thus contain an oxidizing measuring electrode (an anode), where the measuring electrode in an O$_2$ sensor is a cathode. H$_2$S sensors are usually painted black to avoid destruction of the inner electrolyte by light. Light may also result in erratic readings, as the outermost 1 mm or so will be uncoated. The main characteristics for a H$_2$S microsensor are almost the same as for the O$_2$ microsensor, and the detection limit may be <1 μM H$_2$S. There are apparently no major interfering agents in natural environments. The response is usually linear between 0 and 500–2000 μM, but this should be tested for each sensor. The linear range is affected by sensor age, and linearity should thus be tested repeatedly for the same sensor. The shelf-life is usually about 1 year, but prolonged exposure to high sulfide lowers the lifetime, probably by precipitation of elemental sulfur within the sensor. As the fraction of dissolved sulfide present as H$_2$S varies with pH, it is necessary to run parallel pH determinations with pH microsensors. A procedure for calibration and a very detailed description of the H$_2$S sensor in general are found in Kühl and Steuckart (2000). Only about 2% of the dissolved sulfide is found as H$_2$S at pH 8.5, and determinations of low sulfide concentrations are generally not possible at pH values above 8–8.5.

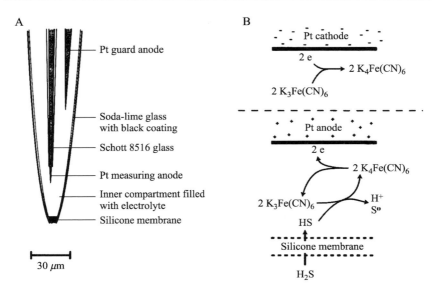

FIG. 3. Tip of H_2S microsensor. Hydrogen sulfide is oxidized by ferricyanide, which is subsequently reoxidized by the platinum anode. An internal guard anode keeps the ferricyanide along the shaft of the measuring anode fully oxidized, although ferrocyanide is produced continuously by the reference cathode (shown to the right). From Kühl et al. (1998), with permission from Aquatic Microbial Ecology.

The first environmental sensor determinations of dissolved sulfide were done with Ag/Ag_2S electrodes (Berner, 1963; Kühl and Steuckart, 2000; Revsbech et al., 1983). The relatively recently developed H_2S sensor is much more reliable and easier to use and, in contrast to the solid-state Ag/Ag_2S electrodes, it does not suffer from a pronounced O_2 interference observed with Ag/Ag_2S microelectrodes.

H_2 Microsensor

A Clark-type H_2 microsensor was first described by Witty (1991). It was basically identical to the oxygen microsensor, but operated with a platinum anode in an acidic electrolyte instead of a gold cathode in an alkaline electrolyte. These sensors were, however, extremely unstable, and the sensitivity to H_2 could disappear within hours. A new version of a Clark-type H_2 microsensor with a different chemistry has been marketed by Unisense. They do not disclose how it is made, but it has long-term stability. However, like the previously developed H_2 sensors, it is extremely sensitive to H_2S, and there may also be other interfering agents. The detection limit is about 0.2 μM. The lifetime is supposed to be about 1 year. A hydrogen microsensor is also sold by AMT, which should have a similar stability as the one from Unisense. The drift during an operational period of 1 week is thus negligible. The AMT sensor also exhibits a pronounced interference from H_2S.

N_2O Microsensor

A Clark-type, oxygen-insensitive N_2O microsensor was developed by Revsbech *et al.* (1988). It only had a lifetime of a few days, and only a small fraction of the sensors manufactured worked satisfactorily. Lately a long-term stable N_2O sensor chemistry has been developed by Unisense, and this new sensor chemistry has been the basis for a new type of oxygen-insensitive N_2O microsensor (Andersen *et al.* 2001) (Fig. 4). The tip diameter of these oxygen-insensitive sensors may be down to about 20 μm, and the detection limit is <1 μM. The linear range is very large (up to several mM), but prolonged exposure of the sensor to only moderately high N_2O may lead to sensor drift. The sensor should thus not be used for long-term monitoring of high N_2O. These sensors are quite slow responding, with 90% response times of about 10–20 s. The most disturbing interference is H_2S, which reversibly inactivates the sensor. N_2O sensors without a front O_2-removing compartment are much simpler to make and are much faster responding and more sensitive to N_2O. Sensitivities approaching 10 nM N_2O may be obtained by such sensors, but oxygen should then be absolutely absent. Most experimental conditions do, however, not allow use of such oxygen-sensitive N_2O sensors.

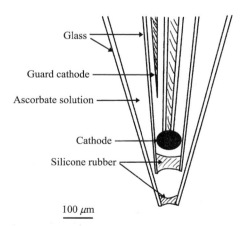

Glass

Guard cathode

Ascorbate solution

Cathode

Silicone rubber

100 μm

FIG. 4. Tip of oxygen-insensitive N_2O sensor. The front compartment with 1–2 M alkaline ascorbate removes all oxygen so that only N_2O reaches the electrochemical N_2O sensor.

Redox Microsensors

Redox potential is, in principle, a well-defined parameter, but it is difficult to perform accurate experimental determinations in microbiological media. This will not be gone into detail here, as these problems have been discussed elsewhere (e.g., Kühl and Revsbech, 2001). However, the reading as such is very simple to perform, and sensors with tips down to 10 μm are available commercially from Unisense.

Diffusivity Sensor

A diffusivity sensor based on the diffusion of a gas away from a sensor tip was described by Revsbech *et al.* (1998) (Fig. 5). A built-in gas sensor monitors the gas concentration near the tip of the sensor, and this concentration is a function of the relative resistance against diffusion away from the sensor tip versus the diffusional resistance inside the sensor. The diffusional resistance inside the sensor is constant, and the signal from the sensor is therefore a simple function of the outside diffusivity. The diffusivity sensors currently sold by Unisense use H_2 as tracer gas, but the author personally prefers the less reactive N_2O. Such sensors may be used to determine diffusivity in biofilms, tissue, sediments, and so on, but

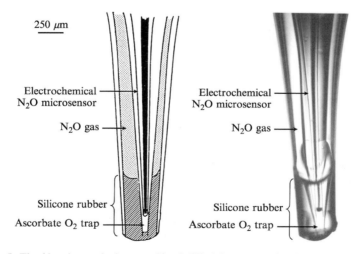

250 μm

Electrochemical N_2O microsensor

N_2O gas

Silicone rubber

Ascorbate O_2 trap

Electrochemical N_2O microsensor

N_2O gas

Silicone rubber

Ascorbate O_2 trap

FIG. 5. Tip (drawing and photograph) of diffusivity sensor showing the internal N_2O sensor, the N_2O reservoir, and the thick silicone plug that constitutes the internal diffusional barrier for gas diffusion. The N_2O concentration, and thus the sensor signal, is a function of the quotient between internal and external diffusional barriers. As the internal barrier is constant, the sensor signal is a function of the external diffusivity. The internal oxygen-insensitive N_2O sensor is shown in more detail in Fig. 4.

inhibitors should be added to prevent consumption of the tracer gas in the medium. In the case of H_2, all biological activity should be stopped by addition of a heavy metal salt such as Hg^{2+}, whereas addition of high concentrations of acetylene should be sufficient by use of N_2O. The diffusivity given by the signal of the sensor corresponds to an integrated value close to the sensor tip. The strongest contribution by far comes from the volume within about two sensor diameters from the tip. This also means that there is a limit to how thin diffusivity sensors should be, as a very local measurement may not be representative for the whole sediment (e.g., a measurement between two sand grains with a sensor so thin that the sand grains themselves are not included in the resulting diffusivity estimate). As a rule of thumb, the sensor should have a diameter of at least twice the average grain diameter in the investigated sediment, and preferably three to four times as large. There is, however, a problem with the response time of very thick sensors, as even a sensor with a diameter of only 300 μm needs a response time of about 10 min to approach a stable signal. It is, however, possible to use nonsteady state signals for determination of diffusivity, in which case even sensors of 1-mm diameter may be read after less than 10 min (N.P. Revsbech, unpublished results).

Flow Microsensors

The diffusivity sensors described earlier may also be used as flow sensors, and they can then be made with tip diameters down to about 10 μm when using H_2 as tracer gas (Unisense). With calibration, such sensors may exhibit an extreme sensitivity in the detection of low rates of flow— down to a few micrometers per second. It is, however, not possible to make flow sensors with an ideal symmetrical sensor tip, and flow sensors may therefore show different responses to different directions of flow. Such sensors could, however, be helpful by investigation of flow fields around small objects, such as meiofauna.

General Equipment for Microsensor Analysis

There are three major suppliers of equipment for microsensor use: Diamond General, World Precision Instruments (WPI), and Unisense. AMT sells a combined volt/ammeter for microsensor use. For use of commercially available microsensors in the environment, Unisense offers the most complete package of equipment, whereas WPI offers a wide variety of equipment for microsensor construction and also a variety of meters, micromanipulators, wires, etc. The equipment essential for use of microsensors is basically a few sensors (the first one may break), a meter (for ammeters with a built-in polarization source), a micromanipulator with a sturdy stand, and either a computer with built-in AD converter or a strip-chart recorder to record data. A dissection microscope on a boom stand (an inexpensive one is sold by WPI) is usually also essential so that the microenvironment in which the microsensor is inserted can be observed. The total cost (2005 prices according to Unisense and Diamond General price lists) of starting to work with microsensors (assuming that a computer with AD converter and a dissection microscope were already present in the laboratory) amounts to about $5000. The more advanced user may prefer to have a motor-controlled micromanipulator and a data acquisition program that can both control the position of the micromanipulator and read the AD converter. The cost of getting started may then increase to about $12,000. The author also advises the advanced user to invest a few hours of work time in the construction of a vibration-protected workplace. The easiest way to do this is to make a table plate of plywood attached to a ca. 5-cm-deep frame that is subsequently filled with concrete. Weighing about 100–150 kg, if the table plate is rested on strips of sturdy foam rubber, most vibrations originating from elevators in the building, outside traffic, and so on will be dampened.

Oxygen microsensors are often used for the determination of photosynthesis in sediments, biofilms, etc. In such work it is necessary to monitor the rate of decrease in oxygen concentration at any particular depth within the first few seconds after darkening. Unisense supplies a version of their data acquisition and micromanipulator control program that can do such operations automatically.

The most common factor causing problems during microsensor analyses is air humidity. In most cases the microsensors sold commercially are sealed so that humidity does not cause problems with short circuits. The user should, however, still be aware that high and variable zero currents from amperometric sensors and calibrations curves exhibiting non-Nernstian responses from potentiometric sensors such as the pH sensors may be due to humidity-caused short circuits. The glass surfaces of the microsensors are especially subject to electrical leaks. At very high air humidities there may also be problems with leaks within the electronics. Problems with static electricity may be a problem at low air humidities, such as those indoors during a cold winter day. The most disturbing effect of static electricity is the puncture of silicone membranes in Clark-type gas sensors. A person can easily get charged with several thousand volts when walking on a synthetic floor covering, and when subsequently touching the equipment, the effect may be a similar potential difference between the sensor interior and the analyzed medium. It is no wonder that about 10 μm of silicone can be punctured by 10,000 V! The best way to avoid ruining sensors by static electricity is to have a ground-connected metal plate under the setup and to touch this plate before touching the equipment. It is also recommended to insert a ground-connected reference electrode into the analyzed medium.

Many *in situ* measurements can be made with the equipment listed previously. The majority of the work the author has done at the hot springs of Yellowstone has been carried out using standard battery-operated equipment and ordinary humidity-sealed microsensors. The instruments have been enclosed in transparent polyethylene plastic bags together with some silica gel to avoid problems with humidity. By *in situ* work at water depths exceeding about 50 cm it is, however, best to have dedicated underwater equipment. Unisense supplies benthic landers for use with microsensors that can be used at water depths up to about 5000 m, as well as pressure-compensated sensors for use with these landers, In addition to the relatively expensive deep-water equipment with built-in micromanipulation and data acquisition, Unisense also supplies shallow-water profilers that can be used down to water depths of about 50 m. These profilers are relatively simple and need cable connection to an above-water unit. Unisense also provides less expensive equipment for scuba-diver operation.

Microsensor Construction

General Procedures and Equipment

Construction of microsensors for one's own use can, as also mentioned earlier, only be recommended for those users willing to spend considerable time and effort on the project. For most short-term applications of noncommercial sensors, collaboration with one of the laboratories that master the construction would be more productive. A possible exception of this is Liquid Ion eXchanger (LIX) sensors, which are based on commercially available ion exchangers from companies such as Fluka and WPI. The construction of these sensors is quite simple and is described in detail by de Beer (2000). It should be emphasized that new and better ion-exchange compounds continue to be synthesized and that there thus is reason to expect more and better LIX sensors in the future. Good LIXes for Mn^{2+} and Fe^{2+} would, for example, be highly welcome! Table II gives a list of sensors that may be relevant for microbial ecology . An in-depth description of the different types of sensors and also references to primary literature can be found in Kühl and Revsbech (2001), de Beer (2000), Revsbech *et al.* (2000), Glud *et al.* (2000), and Cai and Reimers (2000). In addition, general concepts of metal ion detection by *in situ* voltametry are described in Buffle and Tercier-Waeber (2000).

The equipment necessary for the construction of microsensors is actually not very sophisticated. It was described in some detail by Revsbech and

TABLE II
Microsensors with Electrochemical Signal Generation of Potential Interest
to Microbial Ecology[a]

Microsensor type	Minimum tip	Lifetime	Interference	Recent review or primary reference
O_2, Clark	1 μm	1 year	H_2S	Glud *et al.* (2000)
H_2, Clark	1 μm	1 year	H_2S	*www.Unisense.com*, www.AMT-gmbh. com
N_2O, Clark	1 μm	1 year	H_2S	www.unisense.com
N_2O, Clark, O_2 insensitive	20 μm	1 year	H_2S	Andersen *et al.* (2001)
H_2S, Clark	2 μm	1 year	Light	Kühl and Steuckart (2000)
CO_2, Severinghaus	10 μm	Days	H_2S	Cai and Reimers (2000)
pH, glass	10 μm	1 year	No serious	Revsbech and Jørgensen (1986)
pH, LIX	1 μm	Weeks	No serious	de Beer (2000)
NO_2^-, LIX	10 μm	Weeks (dry storage)	H_2S, high Cl^-	de Beer (2000)
NO_3^-, LIX	1 μm	Hours	H_2S, HCO_3^-, Cl^-	de Beer (2000)
CO_3^{2-}, LIX	1 μm	Days	No serious	Choi *et al.* (2002)
NH_4^+, LIX	1 μm	Days	K^+, Na^+	de Beer (2000)
Ca^{2+}, LIX	1 μm	Days	No serious	de Beer (2000)
Redox, bare metal	10 μm	Years	See text	Kühl and Revsbech (2001)
Fe^{2+}, Mn^{2+}, Hg^{2+}, etc., voltametry	100 μm	?	Humic matter	Luther *et al.* (1998)
Diffusivity	20 μm	1 year	See text	Revsbech *et al.* (1998)
Flow	10 μm	1 year	See text	*www.Unisense.com*
Temperature	20 μm	Years	—	www.Unisense.com
NO_2^-, biosensor	20 μm	Days	H_2S, NO, N_2O	Nielsen *et al.* (2004)
NO_x^-, biosensor	20 μm	Days–weeks	H_2S, NO, N_2O	Revsbech *et al.* (2000)
CH_4, biosensor	20 μm	Weeks	H_2, H_2S	Revsbech *et al.* (2000)
VFA, biosensor	20 μm	Days	H_2S, ethanol	Meyer *et al.* (2002)
Glucose, enzyme	10 μm	Days	H_2S	Cronenberg (1991)

[a] For sensor types, "Clark" (Clark *et al.*, 1953) designates amperometric gas sensors with a ion-impermeable membrane; "Severinghaus" (Severinghaus and Bradley, 1958) designates gas sensors based on pH change in a buffer behind an ion-impermeable membrane; "LIX" designates sensors based on a liquid ion exchanger; "biosensor" designates sensors based on immobilized microorganisms; and "enzyme" designates sensors based on immobilized enzyme. The lifetimes indicated are for microsensors; macroscale analogs of, for example, biosensors and sensors based on ion exchangers may have much longer lifetimes. Sensors with minimum tip diameters are usually less sensitive and/or less durable than somewhat larger sensors.

Jørgensen (1986), where the key steps in construction are also described. A microscope with a 10× objective and also, preferably, a higher magnification (20–40×) long-working distance objective is essential, as is a dissection microscope on a boom stand. The inexpensive microscopes sold by WPI are fine for construction work but should be fitted with a graded ocular. In addition to the micromanipulator essential for analysis with microsensors, one more manually operated micromanipulator should be acquired. The only other essential piece of larger equipment is a variable transformer that can supply an output of 0–20 V and 20 A. In addition to the equipment, materials for microsensor construction should, of course, also be bought: glasses, epoxy cement, platinum and silver wires, etc. Again, this will not be gone into detail, as such information can be obtained from the reviews mentioned earlier and with more detail in the publications describing each specific sensor.

Construction of Microscale Biosensors Based on Immobilized Bacteria

Microscale sensors containing immobilized bacteria offer some possibilities for environmental analysis not offered by other classes of sensors, but due to their short lifetimes of days to weeks, these sensors are not sold commercially. The methane biosensor (Damgaard and Revsbech, 1997) is the only sensor currently available for methane, although the possibility of membrane-inlet mass spectrometry (Lloyd et al., 1996) might be investigated as a future alternative. The microscale biosensors for NO_x^- (Larsen et al., 1997) and NO_2^- (Nielsen et al., 2004) based on the design shown in Fig. 6 allow for sensitive analysis of marine systems. The microscale NO_x^- and NO_2^- biosensors are short-lived, but their macroscale analogs (Unisense) have lifetimes of months. The biosensor for volatile fatty acids (VFAs), published by Meyer et al. (2002), is currently the only microsensor for VFAs, which are key compounds in anaerobic degradation, so use of such a sensor may be attractive in many systems.

Construction of the biosensors is based on the same techniques as described by Revsbech and Jørgensen (1986), but for those sensors based on N_2O detection it is probably worthwhile to buy the N_2O sensors from Unisense, as the published N_2O sensors are inferior to those with new sensor chemistry that Unisense has developed. Additionally, the membrane solution for casting ion-permeable membranes should be obtained from Unisense, as the author's own experiences with alternatives has been negative. Although the microscale biosensors based on immobilized bacteria have been developed in the author's laboratory, development has been a difficult learning process. There are two main problems with bacteria-based biosensors: to keep the bacterial population dense and active and for

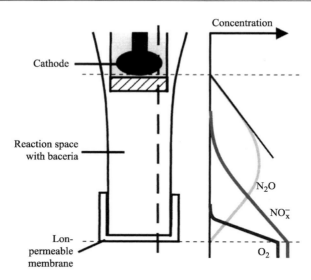

FIG. 6. Microscale biosensor for NO_2^- or NO_x^-. Bacteria in the tip convert incoming NO_2^- or nitrite plus nitrate (NO_x^-) to N_2O, which is subsequently monitored by the internal N_2O microsensor.

the ion sensors to cast ion-permeable membranes that are sufficiently permeable. The first of these problems has been solved by injecting the bacterial biomass obtained by centrifugation or from young agar plate colonies directly into the sensor tips, i.e., introduced via thin glass capillaries, subsequently flushing away excess bacterial biomass with a 0.4% agar growth medium so that only an about 300-μm length of the tip is filled with bacteria. The agar efficiently locks the bacteria in place while at the same time allowing growth of the bacterial population. The membrane is made permeable by resolubilization of the cast membrane in tetrahydrofurane vapor while the inner side of the membrane is in contact with water. This has to be done under the microscope in a humid atmosphere (e.g., in a cold-room), and solubilization has to continue until the membrane almost collapses.

Construction of microscale biosensors so definitely requires specialist instruction. However, the procedures can be demonstrated within a few hours, and new users should, for example, visit the author's laboratory for a day or two to avoid too many frustrations. An illustration of the potential offered by microscale biosensors is shown in Fig. 7, which shows oxygen and NO_x^- profiles analyzed in a diatom-covered sediment during dark and light incubation regimes. Also shown in Fig. 7 are reaction rates calculated

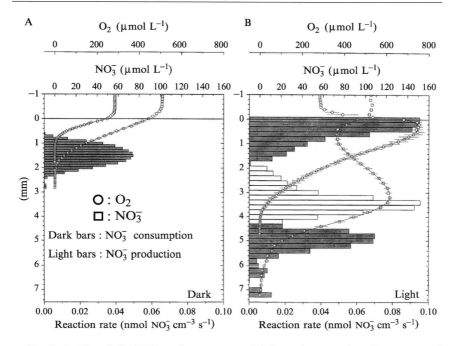

FIG. 7. Profiles of O_2, NO_3^-, and computer-modeled reaction rates in a diatom-covered sediment. The sensor used for analysis was a NO_x^- biosensor, but for simplicity, this is shown as NO_3^-. (A) Data recorded during darkness where the only NO_3^- transforming process was nitrate reduction (presumably denitrification) in the anoxic layers. (B) Data during illumination when algal photosynthesis caused a peak of oxygen in the surface layer and an oxygen penetration down to about 4.3 mm. Nitrate was assimilated in the 0- to 2-mm surface layer, causing a local minimum in NO_3^- concentration, followed by a subsurface peak at the 3-mm depth due to nitrification in the 2- to 4-mm layer. In the anoxic layers below 4.3 mm, NO_3^- was reduced by respiratory processes, and all NO_3^- was depleted at the 6- to 7-mm depth. The slow response of the sensor used (about 1 min for 90% response) causes some inaccuracy in the estimation of exactly where NO_3^- goes to zero.

from the concentration profiles by a diffusion-reaction model. An extensive discussion of similar data and the methods used can be found in Lorenzen *et al.* (1998).

References

Andersen, K., Kjær, T., and Revsbech, N. P. (2001). An oxygen insensitive microsensor for nitrous oxide. *Sensors Actuators B* **81,** 42–48.

Berg, P., Risgaard-Petersen, N., and Rysgaard, S. (1998). Interpretation of measured concentration profiles in sediment pore water. *Limnol. Oceanogr.* **7,** 1500–1510.

Berner, R. A. (1963). Electrode studies of hydrogen sulfide in marine sediments. *Geochim. Cosmochim. Acta* **267,** 563–575.

Buffle, J., and Tercier-Waeber, M.-L. (2000). Biosensors for analysis of water, sludge, and sediments with emphasis on microscale biosensors. In "*In Situ* Monitoring of Aquatic Systems: Chemical Analysis and Speciation" (J. Buffle and G. Horvai, eds.), pp. 279–406. Wiley, New York.

Cai, W.-J., and Reimers, C. E. (2000). Sensors for *in situ* pH and pCO_2 measurements in seawater and at the sediment-water interface. In "*In Situ* Monitoring of Aquatic Systems: Chemical Analysis and Speciation" (J. Buffle and G. Horvai, eds.), pp. 75–119. Wiley, New York.

Canfield, D. E., and Des Marais, D. J. (1991). Aerobic sulfate reduction in microbial mats. *Science* **251,** 1471–1473.

Choi, Y. S., Lvova, L., Shin, J. H., Oh, S. H., Lee, C. S., Kim, B. H., Cha, S. G., and Nam, H. (2002). Determination of oceanic carbon dioxide using a carbonate-selective electrode. *Anal. Chem.* **74,** 2435–2440.

Clark, L. C., Wulf, R., Granger, D., and Taylor, A. (1953). Continuous recording of blood oxygen tension by polarography. *J. Appl. Physiol.* **6,** 189–193.

Cronenberg, C. C. H., Van Groen, H., De Beer, D., and Van den Heuvel, J. C. (1991). Oxygen-independent glucose microsensor based on glucose oxidase. *Anal. Chim. Acta* **242,** 275–278.

de Beer, D. (2000). Potentiometric microsensors for *in situ* measurements in aquatic environments. In "*In Situ* Monitoring of Aquatic Systems: Chemical Analysis and Speciation" (J. Buffle and G. Horvai, eds.), pp. 161–194. Wiley, New York.

Epping, E. H. G., Khalili, A., and Thar, R. (1999). Photosynthesis and the dynamics of oxygen consumption in a microbial mat as calculated from transient oxygen microprofiles. *Limnol. Oceanogr.* **44,** 1936–1948.

Glud, R. N., Gundersen, J. K., Jørgensen, B. B., Revsbech, N. P., and Schulz, H. D. (1994). Diffusive and total oxygen uptake of deep-sea sediments in the south east Atlantic Ocean: *In situ* and laboratory measurements. *Deep-Sea Res.* **41,** 1767–1788.

Glud, R. N., Gundersen, J. K., and Ramsing, N. B. (2000). Electrochemical and optical oxygen microsensors for *in situ* measurements. In "*In Situ* Monitoring of Aquatic Systems: Chemical Analysis and Speciation" (J. Buffle and G. Horvai, eds.), pp. 19–73. Wiley, New York.

Grotzschel, S., and de Beer, D. (2002). Effect of oxygen concentration on photosynthesis and respiration in two hypersaline microbial mats. *Microb. Ecol.* **44,** 208–216.

Gundersen, J. K., Jørgensen, B. B., Larsen, E., and Jannasch, H. W. (1992). Mats of giant sulfur bacteria on deep-sea sediments due to fluctuating hydrothermal flow. *Nature* **360,** 454–456.

Jeroschewski, P., Steuckart, C., and Kühl, M. (1996). An amperometric microsensor for the determination of H_2S in aquatic environments. *Anal. Chem.* **68,** 4351–4357.

Kühl, M., Cohen, Y., Dalsgaard, T., Jørgensen, B. B., and Revsbech, N. P. (1995). The microenvironment and photosynthesis of zooxanthellae in scleractinian corals studied with microsensors for O2, pH, and light. *Mar. Ecol. Prog. Ser.* **117,** 159–172.

Kühl, M., and Revsbech, N. P. (2001). Biogeochemical microsensors for boundary layer studies. In "The Benthic Boundary Layer: Transport Processes and Biogeo-chemistry" (P. Boudreau and B. B. Jørgensen, eds.), pp. 180–210. Oxford Univ. Press, Oxford.

Kühl, M., and Steuckart, C. (2000). Sensors for *in situ* analysis of sulfide in aquatic systems. In "*In Situ* Monitoring of Aquatic Systems: Chemical Analysis and Speciation" (J. Buffle and G. Horvai, eds.), pp. 121–159. Wiley, New York.

Kühl, M., Steuckart, C., Eickert, G., and Jeroschewski, P. (1998). A H_2S microsensor for profiling biofilms and sediments: Application in an acidic lake sediment. *Aquat. Microb. Ecol.* **15,** 201–209.

Larsen, L. H., Kjær, T., and Revsbech, N. P. (1997). A microscale NO3- biosensor for environmental applications. *Anal. Chem.* **69,** 3527–3531.

Lassen, C., Glud, R. N., Ramsing, N. B., and Revsbech, N. P. (1998). An alternative method for measuring benthic microalgal photosynthesis with oxygen microelectrodes. *J. Phycol.* **34,** 89–93.

Lloyd, D., Thomas, K., Price, D., O'Neil, B., Oliiver, K., and Williams, T. N. (1996). A membrane-inlet mass spectrometer miniprobe for the direct simultaneous measurement of multiple gas species with spatial resolution of 1 mm. *J. Microbiol. Meth.,* **25,** 145–151.

Lorenzen, J., Larsen, L. H., Kjær, T., and Revsbech, N. P. (1998). Biosensor determination of the microscale distribution of nitrate, nitrate assimilation, nitrification, and denitrification in a diatom inhabited freshwater sediment. *Appl. Environ. Microbiol.* **64,** 3264–3269.

Luther, G. W., Brendel, P. J., Lewis, B. L., Sundby, B., Lefrancois, L., Silverberg, N., and Nuzzio, D. B. (1998). Simultaneous measurement of O_2, Mn, Fe, I^-, and S(II-) in marine porewaters with a solid-state voltametric microelectrode. *Limnol. Oceanogr.* **43,** 325–333.

Meyer, R. L., Kjær, T., and Revsbech, N. P. (2002). Nitrification and denitrification near a soil-manure interface studied with a nitrate-nitrite biosensor. *Soil Sci. Soc. Am. J.* **66,** 498–506.

Meyer, R. L., Larsen, L. H., and Revsbech, N. P. (2002). A microscale biosensor for measurement of volatile fatty acids (VFA) in anoxic environments. *Appl. Environ. Microbiol.* **68,** 1482–1501.

Nielsen, M., Larsen, L. H., Jetten, M., and Revsbech, N. P. (2004). A bacterium-based NO_2^- biosensor for environmental application. *Appl. Environ. Microbiol.* **70,** 6551–6558.

Reimers, C. E. (1987). An *in situ* microprofiling instrument for measuring interfacial pore water gradients: Methods and oxygen profiles from the North Pacific Ocean. *Deep-Sea Res.* **34,** 2019–2035.

Revsbech, N. P. (1989). An oxygen microelectrode with a guard cathode. *Limnol. Oceanogr.* **34,** 472–476.

Revsbech, N. P., and Jørgensen, B. B. (1986). Microelectrodes: Their use in microbial ecology. *In* "Advances in Microbial Ecology" (K. C. Marshall, ed.), Vol. 9, pp. 293–352. Plenum Press, New York.

Revsbech, N. P., Jørgensen, B. B., Blackburn, T. H., and Cohen, Y. (1983). Microelectrode studies of the photosynthesis and O2, H2S and pH profiles of a microbial mat. *Limnol. Oceanogr.* **28,** 1062–1074.

Revsbech, N. P., Kjær, T., Damgaard, L., Lorenzen, J., and Larsen, L. H. (2000). Biosensors for analysis of water, sludge, and sediments with emphasis on microscale biosensors. *In* "*In Situ* Monitoring of Aquatic Systems: Chemical Analysis and Speciation" (J. Buffle and G. Horvai, eds.), pp. 195–222. Wiley, New York.

Revsbech, N. P., Madsen, B., and Jørgensen, B. B. (1986). Oxygen production and consumption in sediments determined at high spatial resolution by computer simulation of oxygen microelectrode data. *Limnol. Oceanogr.* **31,** 293–304.

Revsbech, N. P., Nielsen, L. P., Christensen, P. B., and Sørensen, J. (1988). Combined oxygen and nitrous oxide microsensor for denitrification studies. *Appl. Environ. Microbiol.* **54,** 2245–2249.

Revsbech, N. P., Nielsen, L. P., and Ramsing, N. B. (1998). A novel microsensor for the determination of diffusivity in sediments and biofilms. *Limnol. Oceanogr.* **45,** 986–992.

Revsbech, N. P., and Ward, D. M. (1984). Microelectrode studies of interstitial water chemistry and photosynthetic activity in a hot spring microbial mat. *Appl. Environ. Microbiol.* **48,** 270–275.

Schmitt-Wagner, D., and Brune, A. (1999). Hydrogen profiles and localization of methanogenic activities in the highly compartmentalized hindgut of soil-feeding higher termites (*Cubitermes* spp.). *Appl. Environ. Microbiol.* **65,** 4490–4496.

Schulz, H. N., and de Beer, D. (2002). Uptake rates of oxygen and sulfide measured with individual *Thiomargarita namibiensis* cells by using microelectrodes. *Appl. Environ. Microbiol.* **68,** 5746–5749.

Severinghaus, J., and Bradley, A. F. (1958). Electrodes for blood pO_2 and pCO_2 determination. *J. Appl. Physiol.* **13,** 515–520.

Further Readings

Damgaard, L. R., and Revsbech, N. P. (1997). A microscale biosensor for methane. *Anal. Chem.* **69,** 2262–2267.

Witty, J. F. (1991). Microelectrode measurements of hydrogen concentrations and gradients in legume nodules. *J. Exp. Bot.* **42,** 20–36.

[10] Optical Microsensors for Analysis of Microbial Communities

By Michael Kühl

Abstract

Fiber-optic microprobes connected to sensitive light meters are ideal tools to resolve the steep gradients of light intensity and spectral composition that prevail in aggregates and surface-associated microbial communities in sediments, biofilms, and microbial mats. They allow for a detailed mapping of light fields and enable insights to the complex optical properties of such highly light-scattering and -absorbing microbial systems. Used in combination with microsensors for chemical species, fiber-optic irradiance microprobes allow for detailed studies of photosynthesis regulation and of the photobiology of microbial phototrophs in intact samples under ambient microenvironmental conditions of the natural habitat. Fiber-optic microprobes connected to sensitive fluorometers enable micro-scale fluorescence measurements, which can be used to map (i) diffusivity and flow; (ii) distribution of photosynthetic microbes, via their photopigment autofluorescence; and (iii) activity of oxygenic photosynthesis via variable chlorophyll fluorescence measurements. Furthermore, by immobilizing optical indicator dyes on the end of optical fibers, fiber-optic

METHODS IN ENZYMOLOGY, VOL. 397   0076-6879/05 $35.00
Copyright 2005, Elsevier Inc. All rights reserved.   DOI: 10.1016/S0076-6879(05)97010-9

microsensors for temperature, salinity, and chemical species such as oxygen, pH, and CO_2 can be realized.

Introduction

Niels Peter Revsbech introduced electrochemical microsensors to microbial ecology in the 1970s. Numerous applications (e.g., reviewed in Revsbech and Jørgensen, 1986) and the introduction of new types of microsensors (reviewed in Kühl and Revsbech, 2001) have since revolutionized our understanding of microenvironments and microenvironmental controls in microbial communities. However, not all relevant environmental variables can be measured with electrochemical measuring principles. Fiber-optic microsensors measure characteristics of the light field (e.g., irradiance or fluorescence) in front of the fiber tip (so-called microprobes) or quantify the amount of an analyte in the tip surroundings via a change in the optical properties of an indicator dye immobilized onto the fiber tip (so-called microoptodes). Fiber-optic microsensors take advantage of the inherent light-guiding capability of optical fibers. The light-collecting and guiding properties of optical fibers depend on the fiber materials used. Light is guided through the central core of the fiber via internal reflection at the core-cladding interface caused by a slightly higher refractive index in the core than in the surrounding cladding material. Both glass fibers and plastic fibers are suitable for sensor fabrication, but plastic fibers show a higher attenuation of ultraviolet (UV) and blue light and are more difficult to cut flat or taper in a controlled way (but see Merchant et al., 1999). For applications involving light guiding of UV radiation, fused silica fibers with a high amount of OH^- are preferable, whereas fibers for VIS-NIR applications have a low amount of OH^- groups in the glass material. Both single mode and multimode fibers can be used for making fiber-optic microsensors. However, the very small core diameter of single mode fibers makes optical alignment difficult and efficient light transmission is best achieved with coherent laser radiation.

The microsensors mentioned in this chapter are all based on multimode optical fibers. There are many good sources for optical fibers, but some companies, such as Polymicro Technologies, USA, and Ceramoptec, Germany, focus on many specialized types of glass fibers, which are excellent for sensor fabrication. This chapter gives an overview of fiber-optic microsensors, which have been applied in environmental microbiology or which have a large potential for application in this field. The development and application of these sensors in microbiology are still limited to a few groups and most examples given in this chapter are from the author's own work. However, microoptodes for oxygen and pH are now available commercially and are used more frequently in environmental studies.

General Construction Procedures for Simple Fiber-Optic Microprobes

Basic Properties and Handling of Optical Fibers

Simple glass fiber-optic microprobes that exhibit defined directional light-emitting or light-collecting properties can be constructed easily from commercially available optical fiber cables, so-called patch cords, that come mounted with standardized fiber connectors; an excellent and more detailed introduction to the construction and characterization of fiber-optic microprobes is given in Vogelmann *et al.* (1991). The simplest microprobe consists of a single-strand multimode fiber-optical cable with the light-collecting end of the optical fiber exposed and flat cut. Such a fiber accepts light within a narrow field of view around its longitudinal axis as given by the acceptance angle:

$$\Theta_a = \sin^{-1}(NA) \tag{1}$$

where the numerical aperture, $NA = \sqrt{(n_1^2 - n_2^2)}/n_0$, is defined by the refractive index of the fiber core, n_1, of the surrounding cladding material, n_2, and of the surrounding medium, n_0, respectively. The acceptance angle gives the maximal angle of incident light, which will be guided by internal reflection through the optical fiber. Note that the acceptance angle of optical fibers varies with the refractive index of the medium surrounding the fiber tip, n_0. Thus, a fiber microprobe with a given acceptance angle in air ($n_0 = 1$) will have a smaller acceptance angle when its tip is immersed in water ($n_0 \sim 1.33$). Typically, such microprobes range from 50 to 200 μm in diameter, depending on which fiber type is used. The construction is as follows (Holst *et al.*, 2000; Kühl and Jørgensen, 1992).

First the protective coating is removed from the fiber cable, exposing the bare optical fiber over a length of $\sim$5–7 cm; this can be done easily with commercial fiber-stripping tools. The fiber end is cut flat with a commercial fiber-cleaving tool and the exposed fiber is then carefully inserted through and fixed (with fast-curing epoxy resin or UV-curing adhesive) in a hypodermic needle, which is mounted on a 1-ml plastic or glass syringe with the plunger removed (Fig. 1A). Alternatively, the fiber end can be cut after fixation in the needle by use of a diamond knife. Best support is obtained with a needle gauge that fits the fiber diameter closely. Fixation can be done with fast-curing epoxy resin or UV-curing glue. The latter is faster and causes no sliding of the fiber during the curing process.

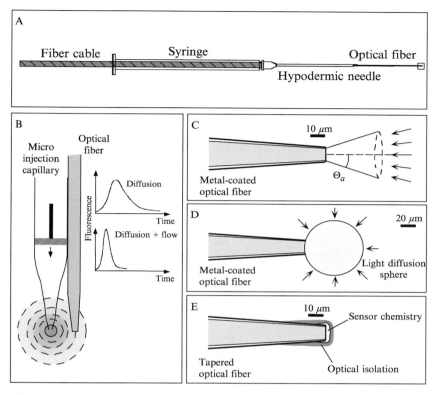

FIG. 1. Overview of different types of fiber-optic microsensors used in environmental microbiology. (A) Schematic drawing of how single-strand fibers can be fixed in a syringe that can be mounted on a micromanipulator. (B) A combined microinjection–microfluorescence sensor for optical measurement of diffusivity and flow. The capillary injects a small plume of a fluorescent dye, the concentration dynamics of which (small curves beside the sensor) is determined with a tapered optical fiber fixed at a known distance to the injection point. (C) Simple tapered fiber-optic microprobe, which is coated on the tapered sides with a metal coating. Such microprobes can be used for field radiance measurements or for microscale fluorescence measurements. (D) A scalar irradiance microprobe consisting of a small light-scattering sphere fixed onto the end of a tapered microprobe. The probe has an isotropic response to incoming light and is used to quantify the total radiative flux in light-scattering media. (E) A microoptode consisting of a tapered microprobe with a layer of optical indicator chemistry immobilized onto the tip (and, in some cases, with an additional optical isolation over the indicator layer). The indicator changes absorbance or luminescence as a function of an analyte in the tip surroundings. From Kühl and Revsbech (2001), with permission of the publisher.

Tapering Optical Fibers and Shaping of the Tapered Tip

By tapering the fiber tip, fiber-optic microprobes with tip diameters of <5 μm can be made. Most sensors are made of fused silica fibers, which soften at high temperatures. Tapering can be done in a small acetylene-oxygen or propane-oxygen flame; the author has had good experience with a hobby welding kit (Rothenberger GmbH, Germany). Alternatively, tapering of fibers can be done in an electric arc, as described in Grunwald and Holst (2004). With advanced (and expensive) laser pullers, even sub-micrometer tip diameters can be made reproducibly. A simple tapering procedure is as follows.

The syringe and needle with the mounted fiber are mounted vertically in a micromanipulator, with the fiber kept straight by a small weight. While observing it under a dissection scope, the fiber is carefully advanced toward the small flame until it softens and is stretched by gravity due to the attached weight. Slow softening and use of less weight make slender tapers, whereas shorter and faster softening and a heavier pulling weight make more conical tapers. Microprobes with short conical tapers exhibit best light transmission to/from the measuring tip and such geometry is advantageous for microscale fluorescence measurements and construction of microoptodes (Kohls et al., 1998). After tapering, the fiber tip is cut flat at an appropriate diameter with a diamond knife or a commercial ceramic fiber-scratching tool. With some training, it is also possible to make a perfect flat cut fiber tip by careful scratching and breaking of the fiber with a sharpened watch-makers forceps while observing the fiber tip on a compound microscope at 10× magnification.

Tapered fiber tips show a narrow light acceptance/emission angle, which is, however, masked by light coupling across the tapered sides of the tip. To avoid this leakage of light, the fiber tip can be coated with an optical isolation such as black enamel paint. The most durable isolation is, however, obtained by applying a metal coating on the fiber taper (Grunwald and Holst, 2004; Vogelmann et al., 1991) (Fig. 1C). Careful heating of the tapered fiber tip in a small flame results in a widening of the acceptance angle. For this, the fiber tip is positioned toward the flame from below while inspected under a dissection microscope. As the fiber tip approaches the flame, the glass softens and forms a rounded tip on the taper. Fiber-optic microprobes with rounded tips show wider acceptance/emission angles than flat cut microprobes with the same diameter and taper geometry. Microprobes with a rounded tip are well suited for optimizing the spatial resolution of microscale fluorescence measurements with microprobes (see later).

Optoelectronic Detection Systems, Positioning, and Data Acquisition

Initially, the fabrication and use of fiber-optic microsensors relied on rather special materials and custom-made equipment, but the extremely rapid development in telecommunications and photonics has since enabled the use of standardized and readily available optical components and materials for sensor fabrication. The same holds true for the required optoelectronic detection systems. The author's initial work with fiber-optic microprobes in the late 1980s was done with a sensitive state-of-the-art diode array system that required water cooling and constant nitrogen flushing and filled a complete laboratory bench (Kühl and Jørgensen, 1992). In recent years, similar sensitivity has been approached with pocket-sized CCD-based spectrometers that can be run directly with a portable PC. The development of high-intensity LEDs (e.g., Luxeon, USA, and Nichia, Japan) and small photomultiplier modules (e.g., Hamamatsu Photonics, Japan/USA) has been another crucial development that has driven the development of fiber-optic microfluorometers. A detailed overview of optoelectronical instrumentation and components used with fiber-optic microsensors is not within the scope of this chapter but can be found in, for example, Holst et al. (2000).

Two types of light meters are commonly used with fiber-optic microsensors in environmental microbiology. An integral measurement of light over a certain wavelength range is done with sensitive photomultiplier tubes in combination with optical filters or a monochromator. Such systems have been developed for measuring photosynthetically active radiation for oxygenic photosynthesis (400–700 nm) (Kühl et al., 1997) or for measuring fluorescence intensity or lifetime with fiber-optic microsensors (Holst et al., 2000; Klimant et al., 1995a,b; Thar et al., 2001). Spectral light measurements are done with sensitive diode array or CCD-based fiber-optic spectrometers. Several fiber-optic spectrometers with suitable sensitivity for UV-NIR radiation (and a relatively low price) are now available commercially (e.g., PMA-11 from Hamamatsu Photonics, USA/Japan, the MMS-series from Carl Zeiss GmbH, Germany, and the USB2000 and QE65000 systems from Ocean Optics, USA).

The positioning of fiber-optic microsensors is done by help of a micromanipulator. The most common model used is a MM33 from Märzhäuser GmbH (Wetzlar, Germany). This is a manually operated micromanipulator, but it can be motorized and interfaced to a computer in various ways. A complete package for motorized positioning and simultaneous data acquisition with microsensors is available from Unisense A/S, Denmark. The system, which allows for autonomous profiling, is developed for use with electrochemical microsensors but is also very useful for measurements with fiber-optic microsensors. Data acquisition (but no positioning control) is also provided with the microooptode-sensing systems from Presens GmbH, Germany.

Special measuring platforms, so-called benthic landers for autonomous *in situ* profiling with microooptodes in sediments, have also been developed for use at water depths up to 6000 m (e.g., Glud *et al.*, 1999a; Wenzhöfer *et al.*, 2001a,b), but none of these are currently available commercially.

Fiber-Optic Microsensors and Their Application

Surface Detection

A general problem when working with microsensors is to align measurements with the surface of the system investigated, e.g., the sediment or biofilm surface. While this can often be accomplished in the laboratory by visual inspection of the sample surface with a dissection microscope, very heterogeneous samples or measurements *in situ* request other methods for surface detection. Optical surface detection is possible by use of a tapered optical fiber connected to a simple modulated light meter equipped with a beam-splitting device (Klimant *et al.*, 1997a). Near-infrared radiation is emitted from the fiber tip, and the reflected NIR radiation from it is collected and guided to a photodiode via the same fiber-optic microprobe. As the microprobe approaches the sediment/biofilm surface, an increasing amount of backscattered radiation is coupled into the fiber tip (Fig. 2A). When the microprobe tip touches the sediment/biofilm interface, the largest increase in backscattered radiation is detected, which can be used to define the surface. With fiber tip diameters of 20–30 μm, the spatial resolution of the optical surface detection is better than 50 μm. When the microprobe tip is fixed to the tip of other microsensors, optical surface detection can be combined easily, e.g., with microscale oxygen measurements (Fig. 2B).

Fiber-Optic Diffusion and Flow Sensors

The mass transfer properties of surface-associated microbial communities are major determinants of microbial activity and zonations. Knowledge about diffusive and advective transport in sediments and biofilms is thus crucial for a quantitative interpretation of microsensor data, e.g., calculations of solute fluxes and reaction rates from concentration microprofiles. Electrochemical microsensors for flow and diffusivity have been developed (see Kühl and Revsbech, 2001), and other techniques, such as high-resolution nuclear magnetic resonance imaging approaches, have

A

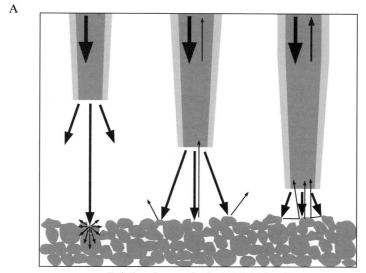

B

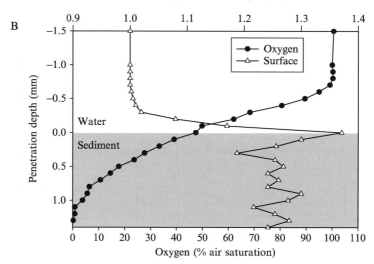

FIG. 2. (A) Principle of optical surface detection with a fiber-optic microprobe. Modulated NIR light is emitted from the fiber tip and is backscattered from the surface. As the microprobe tip approaches the interface, an increasing amount of backscattered light is collected by the microprobe. At the interface the largest relative change in reflectance is measured. (B) Combined microprofiling and surface detection with oxygen microsensors glued together with a fiber-optic microprobe at the measuring tip. After Klimant *et al.* (1997) and Holst *et al.* (2000), with permission of the publishers.

been used to map diffusivity in microbial mats at high spatial resolution (e.g., Wieland *et al.*, 2001). An optical approach for measuring diffusivity and flow was developed by DeBeer (1997) by combining a glass capillary microinjection system with a fiber-optic microprobe connected to a microfluorometer (Fig. 1B).

After injection of a small (picoliter) volume of a fluorescent dye, the plume will equilibrate with the surroundings at a rate that depends on the diffusive and advective transport properties in the matrix. The dye concentration in front of the fiber-optic microprobe tip, which is quantified via the dye fluorescence, will vary over time as the plume is equilibrating. The concentration field of the dye can be described as a function of time by (pg. 29 in Crank, 1975):

$$
\begin{aligned}
C = \frac{1}{2}C_0 \cdot & \left[erf\left(\frac{a+r}{2\sqrt{Dt}}\right) + erf\left(\frac{a-r}{2\sqrt{Dt}}\right) \right] \\
& - \frac{C_0}{r}\sqrt{\frac{Dt}{\pi}} \cdot \left[e^{\left(-(a-r)^2/4Dt\right)} - e^{\left(-(a+r)^2/4Dt\right)} \right]
\end{aligned}
\tag{2}
$$

where t is elapsed time, r is the radius of the plume, a is the initial radius, and C_0 is the initial dye concentration in the injected plume; $erf()$ is the error function. The local dye concentration, C, is approximated by the fluorescence intensity measured by the microprobe. If the distance between the point of injection and the fiber tip is known and the time between injection and fluorescence detection is determined precisely, then the diffusion coefficient, D, can be obtained by iterative fitting of the time-dependent fluorescence signal with Eq. (2). Furthermore, once the diffusion coefficient of the dye has been determined by measuring in stagnant conditions, Eq. (2) can be used to quantify flow velocity, v, by replacing r with $(r + vt)$.

This combined approach has been used to describe the transition from advective transport in the outer layers of biofilms to pure diffusive transport in the biofilm interior (De Beer, 1997). The properties of the fluorochrome are critical for the method. The dye solution must have the same viscosity and density as the liquid in the sediment porewater or biofilm, and the dye must not exhibit strong surface binding or change fluorescence in response to porewater chemistry. In some systems, clogging of the microinjection capillary is a problem.

Fiber-Optic Refractive Index Microsensor

The refractive index is a key factor affecting the optical properties of sediments and biofilms (Kühl and Jørgensen, 1994). The presence of microbial exopolymers with a slightly higher refractive index than water can especially affect the light propagation and it is speculated that this is a key mechanism causing light-trapping effects in the surface layers of sediments and biofilms (Decho *et al.*, 2003; Kühl and Jørgensen, 1994). However, a detailed investigation of the refractive index in surface-associated microbial communities is still missing.

By coating tapered fiber-optic microprobe tips completely with thin metal layers, microsensors for refractive index measurements have been developed (Grunwald and Holst, 2004). The sensing principle is based on the surface plasmon resonance effect: White light sent to the fiber tip causes electron waves (surface plasma waves) at the interface between the metal layer and the surrounding medium that resonate with the incident light, causing a characteristic spectral minimum in the light that is guided back from the fiber tip to a spectral detector. The resonance, and therefore the value of the spectral minimum, is strongly dependent on the refractive index of the medium in immediate contact with the metal-coated fiber tip. Such sensors can be calibrated in commercially available solutions with a precisely known refractive index (e.g., from Cargille Labs, USA; www.cargille.com).

The sensors can be used for measurements of refractive indices from 1.300 to 1.380 at an accuracy of $\sim$0.001 and at a spatial resolution of about 100 μm (Grunwald and Holst, 2004). Examination of the tissues of ascidians containing symbiotic cyanobacteria indeed showed a higher refractive index in the gelatinous test material, where light trapping was observed (Kühl and Larkum, 2002), as compared to the overlaying seawater. With these new sensors it should now be possible to perform the first detailed studies of refractive index in microbial exopolymers, which are often providing the embedding matrix of microbial communities. Such measurements will be important for advancing our understanding of the optical properties of surface-associated microbial communities (Kühl and Jørgensen, 1994).

Fiber-Optic Microprobes for Mapping Light Fields

Surface-associated microbial communities in biofilms and sediments exhibit a strong attenuation of light due to absorption and intense scattering in the matrix of pigmented cells, exopolymers, and mineral grains. The first microscale light measurements in the environment started with the pioneering work of Vogelmann (e.g., Vogelmann and Bjørn, 1984; Vogelmann, 1991) and Jørgensen (e.g., Jørgensen and Des Marais, 1986, 1988), who used tapered fibers to map light fields in plant tissue and microbial mats, respectively. Their work was inspired by developments in biomedical optics, where researchers had developed and used similar techniques for studying the optical properties of animal tissue in connection with photodynamic cancer therapy (reviewed in Star, 1997). Previous reviews of microscale light measurements in environmental microbiology can be found in Jørgensen (1989) and Kühl *et al.* (1994a). A comprehensive introduction to the optics of sediments and biofilms as measured with fiber-optic microprobes is given by Kühl and Jørgensen (1994). Figure 3 gives an overview of central optical parameters relevant for characterizing light fields in microbial communities. Each of the parameters can be measured by a fiber-optic microprobe as described in the following.

Field-Radiance Microprobes. Simple tapered and flat cut fiber microprobes (Fig. 1C) have a narrow acceptance angle for light and thus are ideal tools for mapping the field radiance, $L(\theta, \phi)$, i.e., the incident quantum flux per unit solid angle and area from a certain direction specified by the zenith (θ) and the azimuth (ϕ) angle in a spherical coordinate system. The field radiance is the fundamental parameter for describing the spatial light field and, depending on the actual measuring angle relative to the incident light, different components of the light field can be measured with field radiance

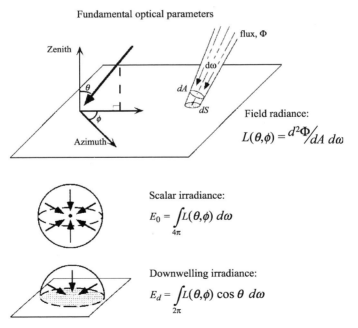

Fundamental optical parameters

Field radiance:
$$L(\theta,\phi) = \frac{d^2\Phi}{dA \, d\omega}$$

Scalar irradiance:
$$E_0 = \int_{4\pi} L(\theta,\phi) \, d\omega$$

Downwelling irradiance:
$$E_d = \int_{2\pi} L(\theta,\phi) \cos\theta \, d\omega$$

FIG. 3. The three basic optical parameters used to characterize light fields in environmental microbiology. Each of the parameters can be measured with a fiber-optic microprobe with light collection properties according to the definition of the parameter (see also Fig. 5). From Kühl and Jørgensen (1994), with permission of the publisher.

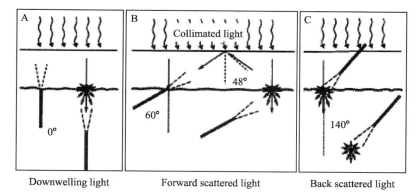

Fig. 4. Fiber-optic microprobes for field radiance measurements can be used to map various components of the light field, such as downwelling light (A), forward scattered light (B), and backscattered light (C) in microbial communities. From Kühl and Jørgensen (1994), with permission of the publisher.

microprobes (Fig. 4). Backscattered light is measured by approaching the biofilm or sediment sample from above, while downwelling and forward scattered light is measured by penetrating the sample from below. The latter requires the sample to be mounted in a core sealed at the bottom with a plug of agar or similar soft material that the microprobe can penetrate.

Numerous field radiance measurements in a sample under multiple different insertion angles allow a very detailed mapping of the light field from which a multitude of optical parameters and information on the radiative transfer can be derived after some mathematical treatment of data (Fukshansky-Kazarinova et al., 1996, 1997, 1998; Kühl and Jørgensen, 1994). However, such measurements are time-consuming, and complete data sets can only be obtained in samples such as sediments and some microbial mats that do not exhibit structural changes over time due to migrating microbes (Kühl et al., 1994a).

Spectral field radiance measurements of downwelling or backscattered light can be used to describe the zonation and migration of differently pigmented photosynthetic microbes at high spatial resolution (e.g., Jørgensen and Des Marais, 1986; Kühl et al., 1994a; Bebout and Garcia-Pichel, 1995; Pringault and Garcia-Pichel, 2004) and have also been used to study the microscale distribution of special protective host pigments in symbiont-bearing corals (Salih et al., 2000); such measurements complement similar studies based on microscale fluorescence measurements with tapered microprobes (see later). However, it is important to realize that field radiance measurements only map the radiant flux from a narrow solid angle and therefore do not quantify the total amount of light available at a certain position in a microbial community.

Scalar Irradiance Microprobes. Optical studies with field radiance microprobes show that diffuse scattered light is a major source of energy for photosynthesis in biofilms and sediments (Jørgensen and Des Marais, 1988; Kühl and Jørgensen, 1994). Photosynthetic microbes in such systems harvest light from all directions, and it is most relevant to relate microbial photosynthesis at a given point to the total incident quantum flux from all directions, i.e., the spherical integral of the radiance distribution, which is the so-called quantum scalar irradiance,

$$E_0 = \int\limits_{4\pi} L(\theta, \phi)$$

This parameter can be measured with a fiber-optic scalar irradiance microprobe consisting of a 80- to 100-μm-wide integrating sphere immobilized onto the tip of a 10- to 15-μm-wide tapered field radiance microprobe (Fig. 1D). Two fabrication methods have been described.

One method is based on dip coating the radiance microprobe in a viscous white mixture of methacrylate and titanium dioxide crystals (Lassen *et al.*, 1992a); best dispersion of the light-scattering particles in methacrylate is obtained with organically coated titanium dioxides often used in the painting industry. Due to the surface tension of the viscous material, a small droplet adhering to the fiber tip will tend to form a sphere, which solidifies as the solvent evaporates. Another method is based on tapering the fiber tip into a long thin filament, which is dipped in magnesium oxide and then melted back and shaped into a scattering vitro-ceramic spherical tip by careful heating in a small flame (Garcia-Pichel, 1995). The latter method allows construction of scalar irradiance microprobes for UV radiation. This is not possible with the dip-coating procedure, as UV is absorbed in the methacrylate matrix. With some training, both types of sensors can be made with an excellent isotropic response, i.e., light from all directions is channeled equally effectively by into the fiber and guided to the detector (Fig. 5). It is, however, crucial that the tapered sides of the fiber tip are coated with an optical isolation (see earlier discussion) in order to avoid nonideal light collection at oblique angles.

Scalar irradiance microprobes are easy to use, even in the field, as they can be positioned into the sample from above. The "blind spot" where the optical fiber connects with the integrating sphere only amounts to about 6% of the spherical surface area, and when inserted at a zenith angle of 135–145° relative to the incident light, self-shadowing is insignificant.

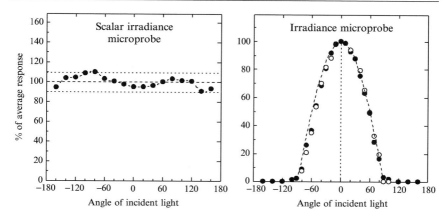

FIG. 5. Angular light collection properties of fiber-optic scalar irradiance and irradiance microprobes. The ideal angular response is indicated with dashed lines. From Kühl *et al.* (1994a), with permission of the publisher.

Scalar irradiance microprobes are relatively sturdy and can last for several years; they have been used in all kinds of systems, ranging from quartz sand to very cohesive microbial mats, plant, and animal tissues.

Calibration is often done by normalizing measurements obtained in a sample to measurements of the downwelling scalar irradiance (=downwelling irradiance in a collimated vertically incident light beam) measured with the microprobe tip positioned, at identical distance and orientation to the light source, over a black light well with minimal reflection of the downwelling light; a small petri dish or beaker coated with thick black velvet or with a matte black enamel paint has proven an excellent material for this. If the calibration is done with a calibrated lamp having a known absolute intensity for a given wavelength and distance from the light source, readings of the scalar irradiance microprobe can also be calibrated in absolute units of μmol photons m^{-2} s^{-1} (see, e.g., Kühl *et al.*, 1997).

Spectral attenuation coefficients of scalar irradiance, $K_0(\lambda)$, with depth, z, can be calculated as

$$K_0(\lambda) = -\frac{d \ln[E_0(\lambda)]}{dz} = -\frac{\ln[E_0(\lambda)_1/E_0(\lambda)_2]}{z_2 - z_1} \tag{3}$$

where $E_0(\lambda)_1$ and $E_0(\lambda)_2$ are the spectral scalar irradiance measured at depths z_1 and z_2, respectively. Equation (3) can also be used to calculate attenuation coefficients of radiance and irradiance.

Scalar irradiance microprobes are ideal tools for optical studies in surface-associated communities (e.g., Kühl *et al.*, 1994a; Garcia-Pichel, 1995). In combination with oxygen microsensor measurements, detailed photobiological studies can be performed on intact samples, where regulation of microbial photosynthesis can be studied under natural conditions as a function of spectral composition or intensity (e.g., Ploug *et al.*, 1993; Lassen *et al.*, 1992b; Kühl *et al.*, 1996, 1998). The intensity and spectral characteristics of available light in different depths can be mapped and the presence of different photopigments can be determined (Fig. 6). Light availability in undisturbed samples can be precisely aligned with the distribution of oxygen and oxygenic photosynthesis (Fig. 7). This enables studies of how photosynthesis is regulated by light intensity in undisturbed samples with steep light gradients (Fig. 8), and spectral attenuation and photosynthetic action spectra can be determined for 0.1-mm-thin layers (Fig. 9). Probably the most surprising finding when scalar irradiance microprobes were first used in environmental microbiology studies was

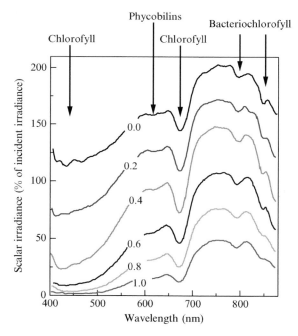

FIG. 6. Spectral measurements of scalar irradiance in a coastal sediment covered by a stratified microbial mat containing diatoms, cyanobacteria, and purple sulfur bacteria. Characteristic minima in the spectra correspond to absorption maxima of different photopigments as indicated with arrows.

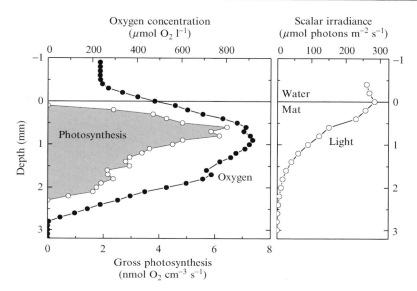

FIG. 7. Combined measurements of oxygen, oxygenic photosynthesis, and quantum scalar irradiance of photosynthetic active radiation (PAR, 400–700 nm) in a coastal sediment with a cyanobacterial mat. After Kühl *et al.* (1992), with permission of the publisher.

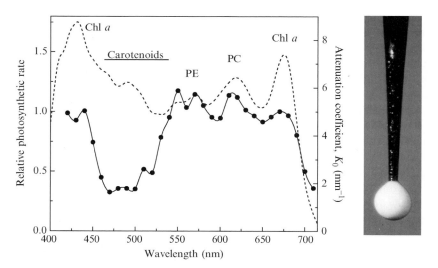

FIG. 8. Scalar irradiance attenuation spectrum (dashed line) and photosynthetic action spectrum (solid line) measured in the upper 0.1 mm of an intact epilithic cyanobacterial biofilm. The absorption regions of major photosynthetic pigments [chlorophyll *a*, phycocyanin (PC), and phycoerythrin (PE)] are indicated. After Kühl *et al.* (1996), with permission of the publisher.

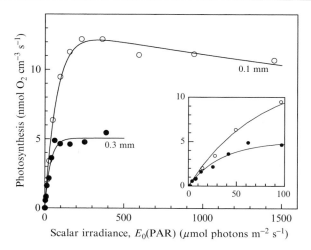

FIG. 9. Gross photosynthesis vs scalar irradiance measured at two different depths in a hypersaline cyaobacterial mat. The upper mat layer gets photoinhibited at high irradiance, while the strong light attenuation in the mat only results in saturation of photosynthesis at a 0.3-mm depth. After Kühl *et al.* (1997), with permission of the publisher.

that the scalar irradiance exhibits a surface or near-surface maximum (Kühl and Jørgensen, 1992, 1994; Lassen *et al.*, 1992) in many sediments and microbial mats (Figs. 6, 7, and 11). However, such light-trapping effects have also been observed in other scattering media and tissues (e.g., Vogelman and Björn, 1986) and the phenomenon is not a violation of thermodynamics but a consequence of photon statistics and path length distribution. Incident light is strongly scattered in the matrix of cells, mineral grains, and exopolymers, causing photons to travel an increased path length per vertical distance traversed. This increases the residence time of photons near the sediment/biofilm surface, which add onto incident photons and causes a local transient enhancement of scalar irradiance. The upper layers of strongly scattering sediments and mats thus act as a kind of "bottleneck" for photons on their way into deeper layers.

Irradiance Microprobes. The most commonly used quantification of light intensity in environmental microbiology is done in terms of the quantum irradiance, i.e., the integral quantum flux per unit horizontal surface area from above, i.e., the downwelling irradiance,

$$E_d = \int_{2\pi} L(\theta, \phi)\cos(\theta)$$

or from below, the upwelling irradiance, E_u. Note that the incident field radiance is weighted by the cosine of its zenith angle, and scattered light traveling at oblique angles is thus weighted less than vertically incident light. The ratio between upwelling and downwelling irradiance is called the irradiance reflectance,

$$R = E_u/E_d$$

which is a key parameter in remote-sensing studies of soils, sediment and mats. The difference between downwelling and upwelling irradiance is the vector irradiance, $\vec{E} = E_d - E_u$, which is a measure of the net radiant energy input to a system.

Irradiance microprobes with a good angular cosine response are difficult to make and have not found widespread application; a practical problem is also that microprofiling of downwelling irradiance requires penetration of the sample from below. The fabrication of irradiance microprobes involves the formation of a small light-scattering disk on the end of a tapered or untapered optical fiber, which is then coated on the sides with black enamel paint (Kühl *et al.*, 1994b; Lassen and Jørgensen, 1994). In practice, this is done by first applying a small droplet of methacrylate doped with TiO_2 (the same material used for making scalar irradiance microprobes) onto the fiber tip. After complete hardening, the white droplet is coated with black enamel paint. Subsequently, the now black droplet is carefully ground down with a small diamond drill bit to form a white disk surrounded by a black rim. With some training, such irradiance microprobes can be made with an almost ideal angular cosine response to incident light (Fig. 5).

As indicated earlier, neither microscale radiance nor irradiance measurements are sufficient for quantifying the available light for photosynthesis in sediments, biofilms, and other surface-associated microbial communities. However, the combined use of irradiance and scalar irradiance sensors is a strong tool for microscale determination of spectral absorption coefficients, $a(\lambda)$, in such intensely light-scattering systems by using Gershuns equation (Kühl and Jørgensen, 1994; Lassen and Jørgensen, 1994):

$$a(\lambda) = K_{\vec{E}}(\lambda) \cdot \frac{\vec{E}(\lambda)}{E_0(\lambda)} \tag{4}$$

where $K_{\vec{E}}$ is the vertical attenuation coefficient [see Eq. (3)] of the vector irradiance, $\vec{E}$, and E_0 is the scalar irradiance. Absorption coefficients are important for calculating quantum yields of photosynthesis.

Microscale Mapping of Pigment Fluorescence and Photosynthetic Activity

While microbial activity can be studied in great detail and almost noninvasively with microsensors, the distribution of microorganisms is mainly determined with various semidestructive microscopic techniques in combination with specific *in situ* hybridization or staining protocols (Amann and Kühl, 1998). The distribution of different photosynthetic microorganisms can, however, be detected and discriminated in intact samples by microscale mapping of their characteristic pigment fluorescence with a tapered fiber-optic microprobe (Kühl and Fenchel, 2000; Thar *et al.*, 2001) (Fig. 10). By fixing fiber-optic microprobes to the measuring tip of an oxygen microsensor, the distribution of oxygen and oxygenic photosynthesis can be mapped simultaneously with the distribution of chlorophyll (Fig. 11).

Using the same kind of microprobes with a more advanced pulse–amplitude-modulated fluorescence detector system (Schreiber *et al.*, 1996), the photosynthetic activity of microbes with oxygenic photosynthesis can be mapped. This method relies on measurements of variable

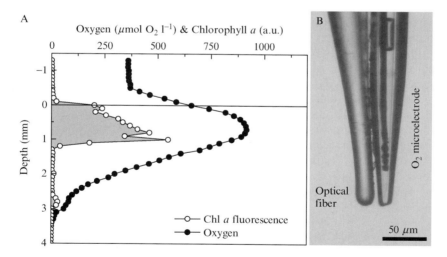

FIG. 10. Combined measurements of oxygen and chlorophyll *a* fluorescence (A) with a Clark-type oxygen microsensor glued together with the tapered tip of a fiber-optic microprobe (combined diameter $\sim$40 μm, B). Measurements were done at a temperature of 0.5° and a downwelling irradiance of 130 μmol photons m^{-2} s^{-1} in an Arctic sediment covered with a dense biofilm of benthic diatoms (Young Sound, northeast Greenland) (Kühl *et al.*, unpublished data).

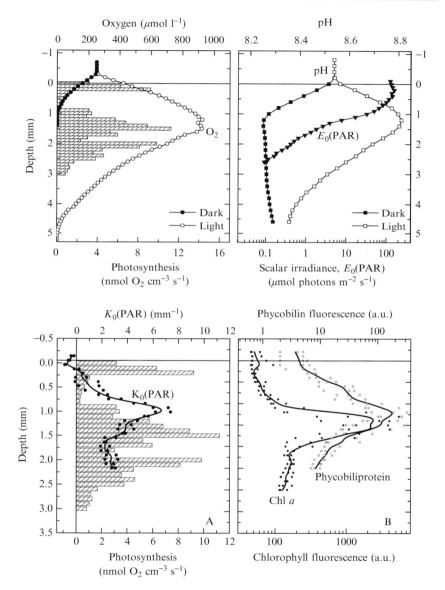

FIG. 11. Combined measurements of oxygen, pH, oxygenic photosynthesis, scalar irradiance (top), light attenuation, and photopigment fluorescence (bottom) in a cyanobacterial mat. From Kühl and Fenchel (2000), with permission of the publisher.

chlorophyll fluorescence from PSII in oxygenic phototrophs with the so-called saturation-pulse method (Schreiber, 2004). The method, which is not described in detail here, allows the determination of the effective quantum yield of PSII-related electron transport, ϕ_{II} (in units of mole electrons transported per mole quanta causing charge separation in PSII), from measurements of the chlorophyll fluorescence yield under ambient light, F, and during an intense saturation pulse that fully closes the reaction center, F_m':

$$\phi_{II} = \frac{F_m' - F}{F_m'} \tag{5}$$

When the sample is fully dark adapted, Eq. (3) is used to calculate the maximal PSII quantum yield, $\phi_{II(max)}$, from the minimal fluorescence signal in the dark, F_0, and the maximum fluorescence yield in the saturation pulse, F_m. As the actinic light intensity increases, the reaction center approaches full closure and the effective quantum yield decreases. Relative rates of the photosynthetic activity (in terms of a relative electron transport rate, $rETR$) can be estimated as $rETR = \phi_{II} \cdot E_0$, where E_0 is the scalar irradiance of actinic light.

Much more information on the status and function of the photosynthetic apparatus can be obtained from variable chlorophyll fluorescence measurements [a comprehensive treatment is given in Papageorgiou and Govindjee (2004)]. A simple measure of the nonphotosynthetic energy dissipation, i.e., the nonphotochemical quenching, NPQ, can be calculated as

$$NPQ = \frac{F_m - F_m'}{F_m'} \tag{6}$$

Variable fluorescence measurements are rapid, and measurements of rETR and energy dissipation as a function of different irradiances can be obtained repeatedly over time scales ranging from seconds to hours, allowing for detailed studies of photoacclimation and photoinhibition. The first microscale measurements of variable chlorophyll fluorescence were done in terrestrial leafs at a spatial resolution of 20–30 μm with a custom-made detection system (Schreiber *et al.*, 1996). A commercial version is now available (Microfiber-PAM, Walz GmbH, Germany), and this system has been used to map photosynthesis in sediments, biofilms, and corals (Ralph *et al.*, 2002). A combined microsensor, where the tip of a microfiber probe is fixed to the measuring tip of an oxygen microsensor (Fig. 10A), can be used to obtain both oxygen- and fluorescence-based measurements of photosynthesis at $\sim$100-μm spatial resolution. Combined measurements variable chlorophyll fluorescence and oxygen with fiber-optic of microsensors (see later) were used to study the oxygen regulation of photosynthesis in a cyanobacterial mat (Schreiber *et al.*, 2002). Variable chlorophyll fluorescence-imaging systems have been developed and applied for studying photosynthesis at similar high spatiotemporal resolution in various surface-associated microbial communities (e.g., Grunwald and Kühl, 2004; Kühl *et al.*, 2005; Ralph *et al.*, 2005).

Microoptodes

Microooptodes are tapered fiber-optic microsensors with one or more optical indicators immobilized at the measuring tip of the fiber (Holst *et al.*, 2000; Klimant *et al.*, 1997) (Fig. 1E). The indicator (also called the sensor chemistry) changes its optical properties (e.g., absorbance or fluorescence) reversibly as a function of the local level of the analyte to be measured. Miniature optical chemosensors were first developed for blood-gas analysis in the biomedical field, and the basic-sensing principles (see, e.g., Wolfbeis, 1991, 2003) have largely been adopted and optimized in the development of microooptodes for use in environmental microbiology. This section focuses mainly on microooptodes for oxygen, which are now applied frequently in various fields of environmental microbiology. In addition, it comments briefly on microooptodes for pH, CO_2, and temperature and their potential for application in environmental microbiology. Several types of microooptodes are available commercially (see, e.g., the Web pages of Presense GmbH, World Precision Instruments Inc., and Ocean Optics Inc.).

Oxygen Microooptodes

OPTICAL SENSING OF OXYGEN AND CALIBRATION. Oxygen can be measured optically by different luminescent indicator dyes that exhibit a dynamic quenching of their luminescence intensity, I, and life-time (=decay time), τ, for a given oxygen concentration, c, according to the well-known Stern–Volmer relation (Wolfbeis, 1991, 2003):

$$\frac{I}{I_0} = \frac{\tau}{\tau_0} = \frac{1}{1 + K_{SV} \cdot c} \tag{7}$$

where I_0 and τ_0 are the luminescence intensity and life-time, respectively, of the indicator in the absence of oxygen, and K_{SV} is a characteristic quenching coefficient of the indicator. Energy from excited dye molecules is transferred to molecular oxygen, whereby singlet oxygen is formed, which is rapidly relaxing under emission of heat. Note that oxygen optodes exhibit the highest signal and signal changes at low oxygen concentrations.

When indicators are immobilized in a matrix material, their lumines-cence as a function of oxygen often exhibits a nonideal Stern–Volmer behavior, which can be described by the modified relation (Klimant *et al.*, 1995a,b) (Fig. 12):

$$\frac{I}{I_0} = \frac{\tau}{\tau_0} = \frac{1-\alpha}{1 + K_{SV} \cdot c} + \alpha \tag{8}$$

where α is a nonquenchable fraction of the indicator luminescence. The oxygen concentration, c, can thus be expressed as

$$c = \frac{(I_0 - I)}{K_{SV} \cdot (I - I_0\alpha)} = \frac{(\tau_0 - \tau)}{K_{SV} \cdot (\tau - \tau_0\alpha)} \tag{9}$$

For a given mixture of indicator and matrix material, α is usually constant over the dynamic range, and Eq. (8) accurately describes the nonlinear behavior of I or τ vs oxygen concentration (Fig. 12). α can be determined from measurements of luminescence from at least three de-fined oxygen concentrations (e.g. 0, $c_1 = 20\%$ oxygen saturation, and $c_2 = 100\%$ oxygen saturation) as follows (Glud *et al.*, 1996).

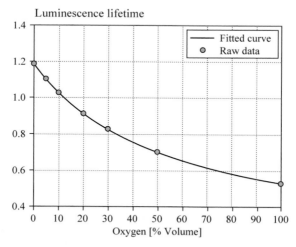

FIG. 12. Calibration curve of an oxygen microoptode based on the oxygen-dependent luminescence lifetime of an indicator dye immobilized at the end of a tapered microprobe. Solid symbols represent actual measurements, and the solid line was fitted to data with Eq. (8). From Holst *et al.* (2000), with permission of the publisher.

First the quenching constant, K_{SV}, can be determined as

$$K_{SV} = \frac{I_0(c_2 - c_1) - (I_1 c_2 - I_2 c_1)}{(I_1 - I_2) \cdot c_1 \cdot c_2} \tag{10}$$

The nonquenchable fraction, α, can then be determined as

$$\alpha = \frac{I_1(1 + K_{SV} c_1) - I_0}{I_0 \cdot K_{SV} \cdot c_1} \tag{11}$$

Once α has been determined, calibration of optical oxygen sensors can thus be done by a simple two-point calibration, i.e., measurements at zero oxygen and at a known oxygen concentration. In many aquatic applications of oxygen microsensors, a reading for zero oxygen is reached when profiling into deeper layers of sediments and biofilms, and a reading in the overlaying water can be related to the oxygen content determined by a Winkler titration or from tables of oxygen solubility in water if the medium is air-flushed and has a known temperature and salinity (such tables can be downloaded freely from *www.unisense.com* and *www.presense.de*). If a zero oxygen reading is not reached in the measurement, a zero oxygen calibration point can be measured by immersing the microoptode tip into a small subsample of the overlaying water made anoxic by the addition of small amounts of sodium dithionite.

FABRICATION OF OXYGEN MICROOPTODES. The first oxygen microoptodes were developed by Klimant *et al.* (1995) and have since found increasing application in various fields of environmental microbiology. Oxygen microoptodes are made from tapered fiber-optic microprobes (see earlier discussion), where the sensing tip is coated with an optical indicator dissolved in a hydrophobic and oxygen-permeable immobilization matrix (Fig. 1E) [see Holst *et al.* (2000) for more details on microoptode construction]. A small droplet of this so-called sensor cocktail is applied to the tip of the microprobe and will form a solid sensor layer after the solvent is evaporated. The thin sensor layer does not absorb all excitation light, which can lead to local stimulation of oxygen production when microoptodes are used for measurements in dense photosynthetic samples such as planktonic aggregates or biofilms and microbial mats. Also, strong intrinsic fluorescence in the sample, e.g., high chlorophyll concentrations, can interfere with the measurements. In order to avoid such effects, the sensor layer is coated (by dip coating) with a thin black layer of an oxygen-permeable optical isolation, i.e., black silicone or a black solution of the Teflon AF polymer (Dupont, USA).

The operational range and sensitivity of oxygen microoptodes can be optimized by the choice of indicator type and immobilization matrix. Two types of indicator material are used commonly for oxygen microoptodes: metalloorganic complexes of ruthenium(II) (excited by blue light and emitting orange light) (Klimant et al., 1995, 1999) and porphyrines with platinum or palladium as the central atom (excited by UV or blue-green light and emitting red or NIR light) (Papkovsky, 2004). Tris(4,7-diphenyl-1,10-phenantroline)-ruthenium(II) (RuDPP) is the most frequently used ruthenium-based indicator dye with oxygen microoptodes. RuDPP-based microoptodes exhibit very good photostability, as well as good and homogeneous sensitivity over a broad dynamic range from 0 to >500% air saturation. Porphyrine-based microoptodes have a more narrow dynamic range (platinum–porphyrines: 0–250%, and palladium-porphyrines: 0–~10% air saturation) but exhibit higher sensitivity (Fig. 13).

The oxygen solubility and permeability of the immobilization matrix also affect the sensitivity of oxygen microoptodes. For a given ambient oxygen level, a higher solubility results in a higher amount of quenching

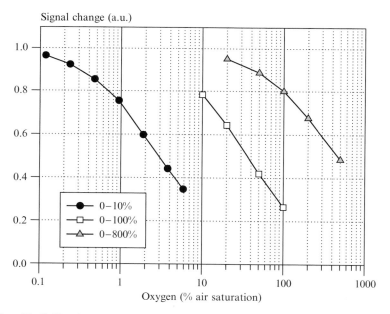

FIG. 13. Calibration curves of oxygen optodes made of three different materials: palladium–porphyrin (●), platinum–porphyrin (□), and ruthenium (△). The oxygen zero point of all curves is suppressed because of the logarithmic scale. From Holst et al. (2000), with permission of the publisher.

oxygen in the sensor layer, causing a higher signal change. In practice, the need for good mechanical stability of the matrix is limiting the choice of material. Although silicone has high oxygen solubility, it is not a good matrix for oxygen microoptodes, which are used for profiling in solid or semisolid substrates. Instead, polystyrene, which has significantly lower oxygen solubility, good mechanical stability and adhesion to the fiber glass, is often used for immobilizing the indicator layer to the fiber tip. Organically modified sol-gels (so-called *Ormosils*) can also be used as an immobilization matrix with oxygen microoptodes, and their oxygen permeability can be optimized by the proper choice of formation conditions and amounts of precursor material (Klimant *et al.*, 1999).

PERFORMANCE AND APPLICATION OF OXYGEN MICROOPTODES. First applications of oxygen microoptodes relied on the use of luminescence intensity-based measurements. However, intensity-based luminescence measurements are subject to interferences from ambient light, changes of the optical properties in the surroundings of the measuring tip, e.g., when entering from water into a scattering and/or fluorescent matrix, and microbending of the optical fiber, causing changes in light transmission. While most of these interferences can be overcome or minimized by coating the measuring tip with an optical isolation (see earlier discussion), they become irrelevant with luminescence lifetime-based measurements, which rely on time-dependent changes in signal, which are essentially independent on absolute signal intensity (as long as there is enough signal at all times) (Holst *et al.*, 2000).

Oxygen microoptodes and instrumentation for lifetime-based oxygen measurements are available commercially (from Presense GmbH, Germany), and since their development in 1995 they have been used in many different aquatic and terrestrial habitats and systems (e.g., Schreiber *et al.*, 2002; Mock *et al.*, 2002). For most applications in environmental microbiology oxygen microoptodes and microelectrodes are equally suitable and give identical results. However, for special applications needing a very fast response time (<0.5 s), such as measurements of gross photosynthesis with the light–dark shift method (see Revsbech and Jørgensen, 1986), oxygen microelectrodes remain the best alternative; oxygen microoptodes currently have response times down to about 1 s in aqueous media.

Oxygen microoptodes (and oxygen optodes in general) have no inherent oxygen consumption in contrast to oxygen microelectrodes that generate a current in proportion to the amount of oxygen consumed by the measuring cathode (see Kühl and Revsbech, 2001). Microoptodes therefore show the same signal in a solution kept at constant oxygen levels whether it is stagnant or not, whereas oxygen microelectrodes typically show a 1–3% lower signal under stagnant conditions (Klimant *et al.*,

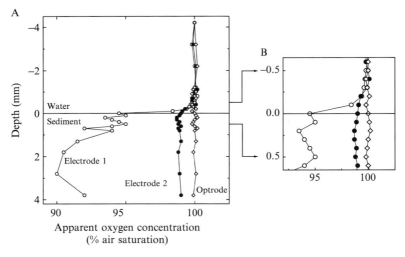

FIG. 14. Stirring sensitivity of microsensors can lead to artifacts when measuring oxygen microprofiles in surface-associated microbial communities. Microprofiles of apparent O_2 concentration in an artificial sediment with a stirred water phase above, as measured by an O_2 microoptode and Clark-type O_2 microelectrodes with a high (electrode 1) and a low (electrode 2) stirring sensitivity, respectively. There was no real gradient of O_2 present in the system. From Klimant et al. (1995), with permission of the publisher.

1995a; Fig. 14). Reduced gases such as hydrogen sulfide can poison oxygen microelectrodes, which is not the case with oxygen microooptodes.

By using porphyrine indicator dyes, oxygen microooptodes can be made with a very high sensitivity toward low oxygen concentrations (down to 1 ppb), where oxygen microelectrodes have a limited sensitivity. Another advantage of oxygen microooptodes in comparison to electrochemical oxygen microsensors is that they exhibit a relatively small-pressure dependency of the signal. This is an advantage for *in situ* applications where microooptodes are used for oxygen profiling in the deep sea (Glud et al., 1999a; Wenzhöfer et al., 2001a,b). Oxygen microooptodes are solid-state sensors, which perform well even at subzero temperatures, and have been used for oxygen measurements in sea ice (Mock et al., 2002), where the sensors were frozen into the ice matrix during formation, and for profiling the sea ice–water interface (Mock et al., 2003).

Other Microooptodes for Use in Environmental Microbiology. Several other types of microooptodes have been described (Wolfbeis, 2004), and several of these have found application in environmental microbiology. This section gives a brief account of the most relevant types. With the accelerating development of optical sensor technology, the author foresees that many new types of microoptodes will be developed and will find increasing application in environmental microbiology. An interesting topic not touched upon in this chapter is the construction of fiber-optic biosensors (e.g., Marazuela and Moreno-Bondi, 2002; Wolfbeis, 2004), where enzymes, antibodies, fluorescent molecular beacons, or whole cells are immobilized onto fiber tips, where they induce an optical change when reacting (or binding) to a specific compound in the tip surroundings. The field is under rapid development, specifically the use of whole cells containing reporter gene fusions (Hansen and Sørensen, 2001), which will allow the construction of new types of fiber-optic microsensors for use in environmental microbiology.

TEMPERATURE MICROOPTODES. Luminescence dyes exhibit a temperature dependency, which can be used to fabricate temperature microoptodes. A simple way to accomplish this is to immobilize luminescent oxygen-sensitive dyes in a matrix impermeable of oxygen or, alternatively, to place an oxygen microoptode inside a sealed capillary (Holst et al., 2000). Such temperature microoptodes exhibit a linear response of the luminescence lifetime vs temperature over a range from 0 to 45° and with a precision of about 0.1–0.2° (Klimant et al., 1997). In comparison to microthermocouples, which are affected by heat conduction to or from the measuring tip via the wires when measuring in a temperature gradient, temperature microoptodes give a better spatial resolution, as heat conduction in the glass material is much less than in metal wires (Holst et al., 1997b). Temperature microoptodes can be used to map steep temperature gradients in hot spring environments or for studying temperature dynamics in strongly absorbing microbial communities such as microbial mats (Klimant et al., 1997).

pH MICROOPTODES. Like oxygen, pH is a key variable in most biological studies and many optical measuring schemes have been developed (see e.g., Wolfbeis, 1991), where pH is evaluated via a change in absorbance or fluorescence of a pH-sensitive indicator. Such optical pH sensors respond according to the mass acting law and do not show a log-linear response over a wide pH range such as potentiometric pH electrodes. Optical pH sensors thus have a limited dynamic range of approximately 3 to 4 pH units, i.e., ±1.5–2 pH units around the pK value of the pH indicator. However, the resolution can be very high in the optimal range. While electromagnetic fields and sulfide or flow velocity effects on the reference electrode influence electrochemical pH measurement, the main interference on optical pH sensors is the ionic strength. However, this problem can be overcome using calibration buffers of similar ionic strength as the sample. Typically, calibration is performed in three to four buffer solutions covering the pH range of the sensor. There are many different pH indicators available for optical sensors, but most indicators have a pK value toward the neutral or acidic range, whereas sensitive indicators with a pK around the pH of seawater are rare.

A simple pH microsensor was developed by Kohls *et al.* (1997) and was based on the immobilization of a commercial pH indicator (N9, Merck) in a sol-gel matrix onto the tip of a fiber-optic microprobe. The indicator changes absorbance as a function of pH, and this change is monitored by measuring the reflectance from the fiber tip when feeding two different wavelengths, i.e., light from a blue and a yellow LED, into the microfiber. The indicator exhibits a pK of ~8.2, and the pH microoptode exhibits a good accuracy of ± 0.05 pH units over a range of pH 7–9 (the total useful range being pH 6.5–10). This microsensor thus covers the most relevant pH range for most applications in environmental microbiology.

Klimant and co-workers (2001; Kosch, 1999) have described a more advanced measuring scheme for pH based on fluorescent decay time measurements, and such pH microoptodes are now available commercially (from Presense GmbH). In these microsensors, the sensor chemistry consists of an inert long decay time reference dye and a short decay time indicator dye, which changes its fluorescence intensity due to the pH value. It can be shown (Klimant *et al.*, 2001) that the average decay time represents the ratio of the two fluorescence intensities. Therefore, the signal of such pH microoptodes is referenced internally and is thus insensitive to fluctuations of the light source. Commercial pH microoptodes have a dynamic range of pH 5.5–8.5 at a resolution of 0.01 pH units and with a response time <30 s.

CO$_2$ MICROOPTODES. Carbon dioxide levels in microbial communities are affected directly by autotrophic and heterotrophic microbes and indirectly by the interplay of microbial metabolism with the carbonate system in aquatic environments. In surface-associated communities, steep concentration gradients of CO_2 and pH exist and microscale measurements of CO_2 are thus highly relevant. Hitherto, electrochemical CO_2 microsensors have been applied mostly in environmental microbiology (e.g., DeBeer *et al.*, 1997, 2000; Köhler-Rink and Kühl, 2001), but *in situ* profiling with a CO_2 microoptode in the deep sea has been reported (Wenzhöfer *et al.*, 2002).

The latter study was done with a microoptode, which was based on the so-called Severinghaus principle (see Holst *et al.*, 2000), where CO_2 diffuses across a gas-permeable membrane into an internal buffer solution and induces a pH shift that can be monitored with a pH sensor. The microoptode consisted of a microcapillary (tip diameter 20 μm) sealed with a silicone plug and filled with a pH-sensitive luminescent dye (HPTS, Fluka).

The pH of the internal solution was monitored just behind the membrane by a tapered microprobe inserted into the outer casing and connected to a microfluorometer. Such CO_2 microoptodes have a response time of <2 min and a detection limit <5 μmol CO_2 liter^{-1}.

More recently, a CO_2 microoptode based on a tapered fiber coated with a nonaqueous hydrophobic CO_2-sensitive layer was described (Neurauter et al., 2000). This sensor has a detection limit of <1 μmol CO_2 liter^{-1} and a response time <1 min. To our knowledge, this sensor has not yet been applied in environmental microbiology, although it has excellent measuring characteristics for studying the carbonate system in aquatic microbial communities.

Planar Optodes. An attractive feature of optical chemical sensors is the possibility of scaling up from microscale point measurements to simultaneous high-resolution imaging of larger areas by immobilizing the optical indicator on sensor foils instead of at the tip of an optical fiber. Using the same measuring and calibration principles, such sensor foils (so-called planar optodes) can be monitored with fluorescence intensity or lifetime imaging systems (e.g., Holst et al., 1998; Glud et al., 2001) for imaging two-dimensional oxygen (e.g., Glud et al., 1996, 1999b; König et al., 2001; Grunwald and Holst, 2001), pH (Hulth et al., 2002; Liebsch et al., 2000), and CO_2 (Liebsch et al., 2000) dynamics in microbial communities.

Acknowledgments

I thank colleagues, postdocs, and students with whom I have had the pleasure to develop and apply various fiber-optic microsensor techniques. I especially mention Bo Barker Jørgensen, Carsten Lassen, Ingo Klimant, and Gerhard Holst. The work has relied on the excellent technical assistance of Anni Glud. The Danish Natural Science Research Council, the Carlsberg Foundation (Denmark), the Max-Planck Society (Germany), and the European Commission (contracts: MAS3-CT-950029, MAS3-CT-970078, EVK3-CT-1999-00010, QLK3-CT-2002-01938) have strongly supported my own work with fiber-optic microsensors over the past 15 years.

References

Bebout, B., and Garcia-Pichel, F. (1995). UV B-induced vertical migrations of cyanobacteria in a microbial mat. *Appl. Environ. Microbiol.* **61,** 4215–4222.

Beyenal, H., Lewandowski, Z., Yakymyshyn, C., Lemley, B., and Wehri, J. (2000). Fiber-optic microsensors to measure backscattered light intensity in biofilms. *Appl. Opt.* **39,** 3408–3412.

Crank, J. (1975). "The Mathematics of Diffusion," 2nd Ed. Oxford Univ. Press, New York.

De Beer, D. (1997). Microenvironments and mass transfer phenomena in biofilms and activated sludge studied with microsensors. *In* "Proceedings of the International Symposium on Environmental Biotechnology" (H. Verachtert and W. Verstraete, eds.), pp. 217–224. Koninklijke Vlaamse Ingenieurs Verininging, Antwerpe.

Decho, A. W., Kawaguchi, T., Allison, M. A., Louchard, E., Reid, R. P., Stephens, C., Voss, K., Wheatcroft, R., and Taylor, B. B. (2003). Sediment properties influencing up-welling spectral reflectance signatures: The biofilm gel effect. *Limnol. Oceanogr.* **48,** 431–443.

Fukshansky-Kazarinova, N., Fukshansky, L., Kühl, M., and Jørgensen, B. B. (1996). Theory of equidistant three-dimensional radiance measurements with optical microprobes. *Appl. Opt.* **35,** 65–73.

Fukshansky-Kazarinova, N., Fukshansky, L., Kühl, M., and Jørgensen, B. B. (1997). General theory of three-dimensional radiance measurements with optical microprobes. *Appl. Opt.* **36,** 6520–6528.

Fukshansky-Kazarinova, N., Fukshansky, L., Kühl, M., and Jørgensen, B. B. (1998). Solution of the inverse problem of radiative transfer on the basis of measured internal fluxes. *J. Quant. Spectr. Rad. Trans.* **59,** 77–89.

Glud, R. N., Klimant, I., Holst, G., Kohls, O., Meyer, V., Kühl, M., and Gundersen, J. K. (1999a). Adaptation, test and *in-situ* measurements with O_2 microoptodes on benthic landers. *Deep-Sea Res. A* **46,** 171–183.

Glud, R. N., Kühl, M., Kohls, O., and Ramsing, N. B. (1999b). Heterogeneity of oxygen production and consumption in a photosynthetic microbial mat as studied by planar optodes. *J. Phycol.* **35,** 270–279.

Glud, R. N., Ramsing, N. B., Gundersen, J. K., and Klimant, I. (1996). Planar optrodes, a new tool for fine scale measurements of two dimensional O_2 distribution in benthic communities. *Mar. Ecol. Progr. Ser.* **140,** 217–226.

Glud, R. N., Tengberg, A., Kühl, M., Hall, P., Klimant, I., and Holst, G. (2001). An *in situ* instrument for planar O_2 optode measurements at benthic interfaces. *Limnol. Oceanogr.* **46,** 2073–2080.

Grunwald, B., and Kühl, M. (2004). A system for imaging variable chlorophyll fluorescence of aquatic phototrophs. *Ophelia* **58,** 79–89.

Grunwald, B., and Holst, G. (2004). Fibre optic refractive index microsensor based on white-light SPR excitation. *Sens. Act. A* **113,** 174–180.

Hansen, L. H., and Sørensen, S. J. (2001). The use of whole-cell biosensors to detect and quantify compounds or conditions affecting biological systems. *Microb. Ecol.* **42,** 483–494.

Holst, G., Klimant, I., Kohls, O., and Kühl, M. (2000). Optical microsensors and microprobes. *In* "Chemical Sensors in Oceanography" (M. Varney, ed.), pp. 143–188. Gordon & Breach.

Holst, G., Kohls, O., Klimant, I., König, B., Richter, T., and Kühl, M. (1998). A modular luminescence lifetime imaging system for mapping oxygen distribution in biological samples. *Sens. Act. B* **51,** 163–170.

Holst, G., Kühl, M., Klimant, I., Liebsch, G., and Kohls, O. (1997b). Characterization and application of temperature micro-optodes for use in aquatic biology. *SPIE Proc.* **2980,** 164–171.

Hulth, S., Aller, R. C., and Engstrom, P. (2002). A pH plate fluorosensor (optode) for early diagenetic studies of marine sediments. *Limnol. Oceanogr.* **47,** 212–220.

Jørgensen, B. B. (1989). Light penetration, absorption and action spectra in cyanobacterial mats. *In* "Microbial Mats: Physiological Ecology of Benthic Microbial Communities" (Y. Cohen and E. Rosenberg, eds.), pp. 123–137. *Am. Soc. Microbiol.*, Washington, DC.

Jørgensen, B. B., and Des Marais, D. J. (1986). A simple fiber-optic microprobe for high resolution light measurements: Application in marine sediment. *Limnol. Oceanogr.* **31,** 1376–1383.

Jørgensen, B. B., and Des Marais, D. J. (1988). Optical properties of benthic photosynthetic communities: Fiber-optic studies of cyanobacterial mats. *Limnol. Oceanogr.* **33,** 99–113.

Klimant, I., Holst, G., and Kühl, M. (1995b). Oxygen micro-optrodes and their application in aquatic environments. *SPIE Proc.* **2508,** 375–386.

Klimant, I., Holst, G., and Kühl, M. (1997). A simple fiber-optic sensor to detect the penetration of microsensors into sediments and other biological materials. *Limnol. Oceanogr.* **42,** 1638–1643.

Klimant, I., Huber, C., Liebsch, G., Neurauter, G., Stangelmayer, A., and Wolfbeis, O. S. (2001). Dual lifetime referencing (DLR): A new scheme for converting fluorescence intensity into a frequency-domain or time-domain information. *In* "New Trends in Fluorescence Spectroscopy: Application to Chemical and Life Sciences" (B. Valeur and J. C. Brochon, eds.), pp. 257–275. Springer Verlag, Berlin.

Klimant, I., Meyer, V., and Kühl, M. (1995a). Fiber-optic oxygen microsensors, a new tool in aquatic biology. *Limnol. Oceanogr.* **40,** 1159–1165.

Klimant, I., Ruckruh, F., Liebsch, G., Stangelmayer, A., and Wolfbeis, O. S. (1999). Fast response oxygen micro-optodes based on novel soluble ormosil glasses. *Mikrochim. Acta* **131,** 35–46.

Kohls, O., Holst, G., and Kühl, M. (1998). Micro-optodes: The role of fibre tip geometry for sensor performance. *SPIE Proc.* **3483,** 106–108.

Kohls, O., Klimant, I., Holst, G., and Kühl, M. (1997). Development and comparison of pH microoptodes for use in marine systems. *SPIE Proc.* **2978,** 82–94.

Köhler-Rink, S., and Kühl, M. (2001). Microsensor studies of photosynthesis and respiration in the larger foraminifera *Amphistegina lobifera* and *Amphisorus hemprichii. Ophelia* **55,** 111–122.

König, B., Holst, G., Kohls, O., Richter, T., Glud, R. N., and Kühl, M. (2001). Imaging of oxygen distribution at benthic interfaces: A brief review. *In* "Organism-Sediment Interactions" (J. Y. Aller, S. A. Woodin, and R. C. Aller, eds.), pp. 63–73. Univ. South Carolina Press, Columbia.

Kühl, M., Chen, M., Scheiber, U., Ralph, P. J., and Larkum, A. W. D. (2005). A niche for cyanobacteria with chlorophyll. *d. Nature* **433,** 820.

Kühl, M., and Fenchel, T. (2000). Bio-optical characteristics and the vertical distribution of photosynthetic pigments and photosynthesis in an artificial cyanobacterial mat. *Microb. Ecol.* **40,** 94–103.

Kühl, M., Glud, R. N., Ploug, H., and Ramsing, N. B. (1996). Microenvironmental control of photosynthesis and photosynthesis-coupled respiration in an epilithic cyanobacterial biofilm. *J. Phycol.* **32,** 799–812.

Kühl, M., and Jørgensen, B. B. (1992). Spectral light measurements in microbenthic phototrophic communities with a fiber-optic microprobe coupled to a sensitive diode array detector. *Limnol. Oceanogr.* **37,** 1813–1823.

Kühl, M., and Jørgensen, B. B. (1994). The light field of micro-benthic communities: Radiance distribution and microscale optics of sandy coastal sediments. *Limnol. Oceanogr.* **39,** 1368–1398.

Kühl, M., and Larkum, A. W. D. (2002). The microenvironment and photosynthetic performance of Prochloron sp. in symbiosis with didemnid ascidians. *In* "Cellular Origin and Life in Extreme Habitats" (J. Seckbach, ed.), Vol. 3, pp. 273–290. Kluwer, Dordrecht.

Kühl, M., Lassen, C., and Jørgensen, B. B. (1994a). Optical properties of microbial mats: Light measurements with fiber-optic microprobes. *In* "Microbial Mats: Structure, Development and Environmental Significance" (L. J. Stal and P. Caumette, eds.), pp. 149–167. Springer, Berlin.

Kühl, M., Lassen, C., and Jørgensen, B. B. (1994b). Light penetration and light intensity in sandy sediments measured with irradiance and scalar irradiance fiber-optic microprobes. *Mar. Ecol. Progr. Ser.* **105,** 139–148.

Kühl, M., Lassen, C., and Revsbech, N. P. (1997). A simple light meter for measurements of PAR (400–700 nm) with fiber-optic microprobes: Application for P vs. I measurements in microbenthic communities. *Aq. Microb. Ecol.* **13,** 197–207.

Lassen, C., and Jørgensen, B. B. (1994). A fiber-optic irradiance microsensor (cosine collector): Application for *in situ* measurements of absorption coefficients in sediments and microbial mats. *FEMS Microbiol. Ecol.* **15,** 321–336.

Lassen, C., Ploug, H., and Jørgensen, B. B. (1992a). A fibre-optic scalar irradiance microsensor: Application for spectral light measurements in sediments. *FEMS Microbiol. Ecol.* **86,** 247–254.

Lassen, C., Ploug, H., and Jørgensen, B. B. (1992b). Microalgal photosynthesis and spectral scalar irradiance in coastal marine sediments of Limfjorden, Denmark. *Limnol. Oceanogr.* **37,** 1813–1823.

Liebsch, G., Klimant, I., Frank, B., Holst, G., and Wolfbeis, O. S. (2000). Luminescence lifetime imaging of oxygen, pH, and carbon dioxide distribution using optical sensors. *Appl. Spectrosc.* **54,** 548–559.

Marazuela, M. D., and Moreno-Bondi, M. C. (2002). Fiber-optic biosensors: An overview. *Anal. Bioanal. Chem.* **372,** 664–682.

Merchant, D. F., Scully, P. J., and Schmitt, N. F. (1999). Chemical tapering of polymer optical fibre. *Sens. Act. A* **76,** 365–371.

Mock, T., Dieckmann, G., Haas, C., Krell, A., Tison, J. L., Belem, A., Papadimitriou, S., and Thomas, D. N. (2002). Micro-optodes in sea ice: A new approach to investigate oxygen dynamics during sea ice formation. *Aq. Microb. Ecol.* **29,** 297–306.

Mock, T., Kruse, M., and Dieckmann, G. (2003). A new microcosm to investigate the oxygen dynamics of the sea-ice water interface. *Aq. Microb. Ecol.* **30,** 197–205.

Neurauter, G., Klimant, I., and Wolfbeis, O. S. (2000). Fiber-optic microsensor for high-resolution pCO_2 sensing in marine environment. *Fresen. J. Anal. Chem.* **366,** 481–487.

Papageorgiou, G., and Govindjee, C. (2004). "Chlorophyll Fluorescence: A Signature of Photosynthesis." Kluwer, Dordrecht.

Papkovsky, D. B. (2004). Methods in optical oxygen sensing: Protocols and critical analyses. *Methods Enzymol.* **383,** 715–734.

Ploug, H., Lassen, C., and Jørgensen, B. B. (1993). Action spectra of microalgal photosynthesis and depth distribution of spectral scalar irradiance in a coastal marine sediment of Limfjorden, Denmark. *FEMS Microbiol. Ecol.* **102,** 261–270.

Pringault, O., and Garcia-Pichel, F. (2004). Hydrotaxis of cyanobacteria in desert soils. *Microb. Ecol.* **47,** 366–373.

Ralph, P., Gademann, R., Larkum, A. W. D., and Kühl, M. (2002). Spatial heterogeneity in active fluorescence and PSII activity of coral tissues. *Mar. Biol.* **141,** 639–646.

Ralph, P. J., Schreiber, U., Gademann, R., Kühl, M., and Larkum, A. W. D. (2005). Coral photobiology studied with a new imaging PAM fluorometer. *J. Phycol.* **41,** 335–342.

Revsbech, N. P., and Jørgensen, B. B. (1986). Microelectrodes: Their use in microbial ecology. *Adv. Microb. Ecol.* **9,** 293–352.

Salih, A., Larkum, A. W. D., Cox, G., Kühl, M., and Hoegh-Guldberg, O. (2000). Fluorescent pigments in corals are photoprotective. *Nature* **408,** 850–853.

Schreiber, U. (2004). Pulse-amplitude-modulation (PAM) fluorometry and saturation pulse method: An overview. *In* "Chlorophyll Fluorescence: A Signature of Photosynthesis" (G. Papageorgiou and C. Govindjee, eds.), pp. 279–319. Kluwer, Dordrecht.

Schreiber, U., Gademann, R., Bird, P., Ralph, P., Larkum, A. W. D., and Kühl, M. (2002). Apparent light requirement for activation of photosynthesis upon rehydration of desiccated beachrock microbial mats. *J. Phycol.* **38,** 125–134.

Schreiber, U., Kühl, M., Klimant, I., and Reising, H. (1996). Measurement of chlorophyll fluorescence within leaves using a modified PAM fluorometer with a fiber-optic microprobe. *Photosynth. Res.* **47**, 103–109.

Star, W. M. (1997). Light dosimetry *in vivo. Phys. Med. Biol.* **42**, 763–787.

Thar, R., Kühl, M., and Holst, G. (2001). A fiber-optic fluorometer for microscale mapping of photosynthetic pigments in microbial communities. *Appl. Environ. Microbiol.* **67**, 2823–2828.

Vogelmann, T. C., and Björn, L. O. (1984). Measurements of light gradients and spectral regime in plant tissue with a fibre optic probe. *Photochem. Photobiol.* **41**, 569–576.

Vogelmann, T. C., Martin, G., Chen, G., and Buttry, D. (1991). Fibre optic microprobes and measurement of the light microenvironment within plant tissues. *Adv. Bot. Res.* **18**, 256–295.

Wenzhöfer, F., Adler, M., Kohls, O., Hensen, C., Strotmann, B., Boehme, S., and Schulz, H. D. (2001b). Calcite dissolution driven by benthic mineralization in the deep-sea: *In situ* measurements of Ca^{2+}, pH, pCO_2 and O_2. Geochim. *Cosmochim. Acta* **65**, 2677–2690.

Wenzhöfer, F., Holby, O., and Kohls, O. (2001a). Deep penetrating benthic oxygen profiles measured *in situ* by oxygen optodes. *Deep Sea Res. A* **48**, 1741–1755.

Wieland, A., van Dusschoten, D., Damgaard, L. R., de Beer, D., Kühl, M., and Van As, H. (2001). Fine-scale measurement of diffusivity in a microbial mat with NMR imaging. *Limnol. Oceanogr.* **46**, 248–259.

Wolfbeis, O. S. (1991). "Fiber Optic Chemical Sensors and Biosensors." CRC Press, Boca Raton, FL.

Wolfbeis, O. S. (2003). A review on milestones in opt(r)ode technology unitil the year 2000. *In* "Optical Sensors for Industrial, Environmental and Clinical Applications" (R. Narayanaswamy and O. S. Wolfbeis, eds.), pp. 1–34. Springer, Berlin.

Wolfbeis, O. S. (2004). Fiber optic chemical sensors and biosensors. *Anal. Chem.* **76**, 3269–3283.

Further Readings

Amman, R., and Kühl, M. (1998). *In situ* methods for assessment of microorganisms and their activities. *Curr. Opin. Microbiol.* **1**, 352–358.

De Beer, D., Glud, A., Epping, E., and Kühl, M. (1997). A fast responding CO_2 microelectrode for profiling in sediments, microbial mats and biofilms. *Limnol. Oceanogr.* **42**, 1590–1600.

De Beer, D., Kühl, M., Stambler, N., and Vaki, L. (2000). A microsensor study of light enhanced Ca^{2+} uptake and photosynthesis in the reef-building coral *Favia* sp. *Mar. Ecol. Progr. Ser.* **194**, 75–85.

Holst, G., Glud, R. N., Kühl, M., and Klimant, I. (1997a). A microoptode array for fine scale measurements of oxygen distribution. *Sens. Act. B* **38–39**, 122–129.

Holst, G., Kühl, M., and Klimant, I. (1995). A novel measuring system for oxygen microooptodes based on a phase modulation technique. *SPIE Proc.* **2508**, 387–398.

Klimant, I., Kühl, M., Glud, R. N., and Holst, G. (1997b). Optical measurement of oxygen and temperature in microscale: Strategies and biological applications. *Sens. Act. B* **38**, 29–37.

Kosch, U., Klimant, I., Werner, T., and Wolfbeis, O. S. (1998). Strategies to design pH optodes with luminescence decay times in the μsecond time regime. *Anal. Chem.* **362**, 73–78.

Kühl, M., Fenchel, T., and Kazmierczak, J. (2003). Growth, structure and calcification potential of an artificial cyanobacterial mat. *In* "Fossil and Recent Biofilms, a Natural History of Life on Earth" (W. E. Krumbein, D. Paterson, and G. Zavarzin, eds.), pp. 77–102. Kluwer, Dordrecht.

Vogelmann, T. C., and Björn, L. O. (1986). Plants as light traps. *Physiol. Plant.* **68**, 704–708.

[11]   *In Situ* Measurements of Metabolite Fluxes: Microinjection of Radiotracers into Insect Guts and Other Small Compartments

By ANDREAS BRUNE and MICHAEL PESTER

Abstract

In microbial ecology, it is of great interest to determine metabolic activities under *in situ* conditions, i.e., without disturbing the structure of the community and the spatial arrangement of individual populations by experimental manipulation. Microinjection of radiotracers and subsequent analysis using the isotope dilution technique has proven to be a powerful method to measure metabolic fluxes in small biological systems, e.g., the intestinal tract of termites. The large variety of commercially available radiolabeled substrates and the identification and quantitation of radiolabeled products by chromatographic methods allow for investigation of the complete metabolic network in a given system.

Introduction

Most microbial habitats are not homogeneous, but are rather characterized by physicochemical gradients of various metabolites. Soils, sediments, microbial mats, and the intestinal tract of insects are prominent examples of highly structured biological systems where the spatial separation of microbial populations gives rise to steep metabolite gradients and also controls the fluxes between sources and sinks of the respective metabolites (see Kühl and Jørgensen, 1992; Brune *et al.*, 2000).

In the past, metabolic activities in such habitats were often studied with homogenized samples (for examples, see Breznak and Switzer, 1986; Skyring, 1987; Brauman *et al.*, 1992 and references therein). However, homogenization can severely bias the results because it destroys the structure of the original system, increases the distance between microbial cells, and introduces buffer components and substrate concentrations different from those in the natural habitat. Therefore, it is not surprising that metabolic activities obtained with homogenized material often do not mirror the metabolic activities *in situ* (Jørgensen, 1978; Kühl and Jørgensen, 1992; Tholen and Brune, 1999).

Introduction of minute amounts and volumes of radiolabeled substrates into structurally intact systems, e.g., by microinjection, circumvents such

METHODS IN ENZYMOLOGY, VOL. 397   0076-6879/05 $35.00

experimental constraints. The minimal mechanical disturbance preserves the integrity of the system and ensures that physicochemical gradients and spatial arrangement of microbial populations are maintained. Because radiolabeled metabolites can be detected with high sensitivity, only minute amounts and volumes of the respective metabolites need to be injected, which avoids undesirable changes in size and concentration of the metabolite pools *in situ*.

There are several prerequisites for the applicability of this method. Most importantly, the system has to be in a steady state, i.e., net changes of metabolite pools and metabolic fluxes during a given time interval should be negligible. Furthermore, the formation rate of a given product can be analyzed only if the label in the product is trapped by dilution in a relatively large product pool so that the turnover of labeled product is negligible.

Another important consideration concerns distribution of the label after injection. For systems where the spatial distribution of microbial populations in the system can be considered homogeneous at the scale of label distribution, as in the case of a horizontally stratified sediment, integrated rates can be determined even if the injected label is distributed unevenly within each layer—unless the label is carried out of the system during the incubation. However, in systems in which the microbial populations are distributed heterogeneously within the range of the injected label, as in the case of insect guts, a fast and even distribution of the injected radiolabel by diffusion and/or advective mixing is required to calculate rates based on the isotope dilution technique.

Microinjection of radiotracers has been applied successfully to determine metabolic rates in various gradient habitats. For example, depth profiles of *in situ* sulfate reduction rates in sediment cores (Moeslund *et al.*, 1994) and cyanobacterial mats (Jørgensen, 1994) were determined by injection of $^{35}SO_4^{2-}$ into different layers. Another example is the measurement of metabolic rates and the respective carbon flux within intact hindguts of various termite species by microinjection of ^{14}C-labeled substrates (Tholen and Brune, 1999, 2000). This article focuses on the latter example—the experimental procedure and possible pitfalls are discussed in detail.

Theoretical Background

In small systems with a heterogeneous spatial arrangement of the microbial populations, it is important that the time required for an even distribution of the injected radiolabel is considerably shorter than the time interval used for the determination of the metabolic activity. At short distances, diffusion is the most important transport mechanism, and the time (t) required for a molecule to travel by diffusion a certain distance (s) in space is described by the Einstein–Smoluchowski equation:

$$t = \frac{s^2}{6 \cdot D} \tag{1}$$

where D is the apparent diffusion coefficient of the injected substrate in a given environment. Based on Eq. (1), and assuming a diffusion coefficient for small molecules in aqueous medium of approximately 10^{-5} cm^2 s^{-1} (Ambrose *et al.*, 1999), it can be estimated that an injected substance travels approximately 1 mm within 60 s after injection. In closed biological systems, the time needed for total mixing is often reduced further by advection, e.g., caused by peristalsis and movements of protists in gut environments. This allows the study of systems even larger than those controlled solely by diffusion. The mathematics of diffusion is covered in detail by Crank (1975).

The determination of metabolic rates in a small, enclosed, and well-mixed system requires knowledge of the size of the respective substrate and product pools. Depending on the size of the substrate pool, two situations can be distinguished.

i. The pool size (N) is small and is therefore increased considerably by the injected substrate. In this case, the observed decrease of radiolabel in the substrate pool over time (dX_S/dt) is directly proportional to the turnover rate of the substrate (R_S):

$$\frac{dX_S(t)}{dt} = R_S \cdot A \tag{2}$$

where A is the specific radioactivity of the labeled element in the substrate pool. The increased pool size, however, may remove kinetic constraints of the system, and the resulting potential rates may exceed the *in situ* rates considerably.

ii. The amount of injected label is small and does not increase the size of the substrate pool greatly. In the ideal case, the pool size can be considered constant, and the decrease of radioactivity in the substrate pool (X_S) reflects the dilution of the initially injected label (X_0) by nonlabeled substrate entering the pool and labeled substrate leaving the pool (Fig. 1):

$$X_S(t) = X_0 \cdot e^{\mu \cdot t} \tag{3}$$

where μ is the turnover rate constant. In its logarithmic form, Eq. (3) yields a linear function and the turnover rate constant can be determined from the slope of the linear regression of the experimental data:

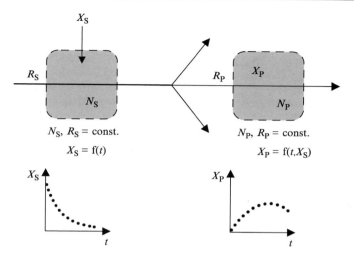

FIG. 1. Principle of flux measurements using the isotope dilution technique. At steady-state conditions, the pools of substrate (N_S) and the product(s) (N_P) and the corresponding fluxes through the respective pools (R_S, R_P) are constant parameters. The hypothetical time courses for the washout of injected radiolabeled substrate (X_S) and the accumulation of radiolabeled product (X_P) are indicated; they can be used to calculate turnover rates and metabolic fluxes (see text).

$$\ln X_S(t) = \mu \cdot t + \ln X_0 \tag{4}$$

The substrate turnover rate can then be calculated using the experimentally determined pool size and the turnover rate constant:

$$R_S = \mu \cdot N \tag{5}$$

Calculation of product formation rates is slightly more complex. The specific radioactivity of the substrate pool is determined by the amount of radiolabel injected into the pool:

$$A = \frac{X}{N} \tag{6}$$

Due to the washout of radioactivity from the substrate pool [Eq. (3)], the specific radioactivity of the pool decreases over time:

$$A(t) = A_0 \cdot e^{\mu \cdot t} \tag{7}$$

Despite a constant rate of substrate-to-product conversion (steady state), this decrease in the specific radioactivity of the substrate pool also causes a decrease in the formation rate of labeled product [dX_P/dt] (Fig. 1). dX_P/dt is obtained by substituting the specific radioactivity in Eq. (2) by Eq. (7). The resulting Eq. (8) describes formation of a labeled product as a function of time, turnover-rate constant of the substrate pool, and the turnover rate of this particular product (R_P).

$$\frac{dX_P(t)}{dt} = R_P \cdot A_0 \cdot e^{\mu \cdot t} \tag{8}$$

Integration of Eq. (8) over time ($t_0 = 0$; $X_0 = 0$) yields a new exponential function (Eq. 9), which allows the formation rate of a given product to be calculated from the experimentally obtained data points [X_P; t], the turnover-rate constant, and the initial specific radioactivity of the substrate pool.

$$X_P(t) = \frac{R_P \cdot A_0}{\mu}(e^{\mu \cdot t} - 1) \tag{9}$$

Practical Example

The model just described has been applied successfully in a study of metabolite fluxes in the hindgut of the termite *Reticulitermes flavipes*, employing ^{14}C-labeled metabolites (Tholen and Brune, 2000). The hindgut paunch of this termite is characterized by steep gradients of O_2 and H_2 (Brune *et al.*, 1995; Ebert and Brune, 1997; Fig. 2) and by a heterogeneous distribution of the microbial community within the gut (Yang *et al.*, 2005). The lactate pool in the hindgut fluid was rather small (ca. 1 mM), but the rapid washout of injected label from the substrate pool indicates a rapid

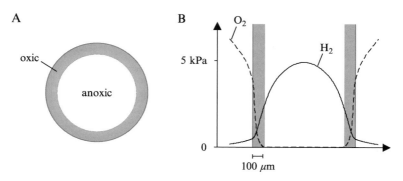

FIG. 2. Oxygen status and important metabolite gradients in the agarose-embedded hindgut of the termite *Reticulitermes flavipes*: (A) radial section of the hindgut and (B) oxygen and hydrogen gradients in a radial section. Scheme from Brune and Friedrich (2000).

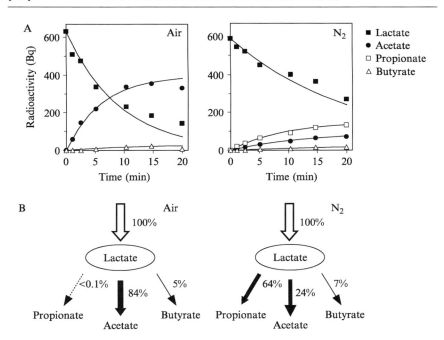

FIG. 3. Original data from a microinjection study of lactate turnover in the hindgut of the termite *Reticulitermes flavipes* (Tholen and Brune, 2000). (A) Time course of label distribution among different metabolites after microinjection of [^{14}C]lactate into agarose-embedded guts incubated under air or under a nitrogen atmosphere. Values represent the mean of four injections into separate guts. Predicted curves for radioactive substrate depletion and product formation (solid lines) are based on the kinetic model described in the text [Eqs. (4) and (9)]. (B) Impact of oxygen on the relative rates of product formation from lactate produced in the primary fermentations. The slight imbalance is most likely due to the formation of CO_2 not being accounted for.

turnover of lactate (Fig. 3A). The experimentally determined turnover rate constant [Eq. (4)], which gave a good fit with experimental data when inserted into Eq. (3), yielded a high lactate turnover rate [Eq. (5)], representing about one-third of the total carbon flux through the animal. Acetate was the major product of lactate turnover, and the experimentally determined product formation rates also gave good fits with experimental data when inserted into Eq. (9). When the experiment was repeated under a nitrogen atmosphere, the lactate turnover rate decreased, and the product formation rates changed fundamentally when compared to the situation under air (Fig. 3B), which provides a perfect example underlining the necessity to determine metabolic fluxes under *in situ* conditions.

Experimental Setup

Microinjection Apparatus

The introduction of minute volumes of radiotracer (down to submicroliter amounts) into the sample is accomplished by means of microinjection. The microinjection apparatus is a hydraulic system consisting of a fine-tipped glass capillary filled with light mineral oil ($d_{25°} \approx 0.8$ g ml^{-1}) connected to a microliter syringe. The syringe can be actuated either mechanically or by a motorized mechanism. The positioning of the capillary tip is controlled by a manual micromanipulator (MM 33, Märzhäuser, Wetzlar, Germany) and is monitored with the help of a stereomicroscope (Fig. 4A).

A critical point of the microinjection is the accuracy of the volume delivery. During method development, we tested three different variants of the microinjection apparatus. Initially, we used a Microdispenser (Drummond Scientific Company, Broomall, PA), which resembles an automatic pipette with a steel piston that inserts directly into the capillary tip, thus creating a direct connection between capillary tip and syringe. However, the volume delivery of the Microdispenser was not reproducible at volumes <100 nl. In addition, the transmission of mechanical disturbances resulting from the manual operation of the instrument led to uncontrolled movements of the capillary tip, which affected the tip position and the integrity of the sample.

To alleviate these problems, the tip and the actuator were separated by connecting the capillary via hydraulic tubing to a microliter syringe, which was operated manually using a micrometer drive. This variant was applied successfully in several studies (Tholen and Brune, 1999, 2000), but the setup was inferior to the first variant because it was difficult to remove air bubbles introduced into the hydraulic tubing during instrument setup and capillary changes.

For this reason, we have developed a third setup in which the tip is again connected directly to the syringe, but the manual control of the syringe piston is replaced by a microprocessor-controlled stepping motor (Fig. 4A). The current system consists of a 50-μl syringe with a luer-lock connection (Hamilton Company, Reno, NV) actuated by a Micro Pump II operated via a Micro4 microprocessor controller (World Precision Instruments, Inc., Sarasota, FL). This setup is the most accurate and convenient solution to date and allows reproducible injection in the desired volume range (30–50 nl; E. Miambi, A. Fujita, M. Pester, and A. Brune, unpublished results).

For certain applications, microinjection and subsequent incubation must be carried out under a defined atmosphere. For this purpose, samples are

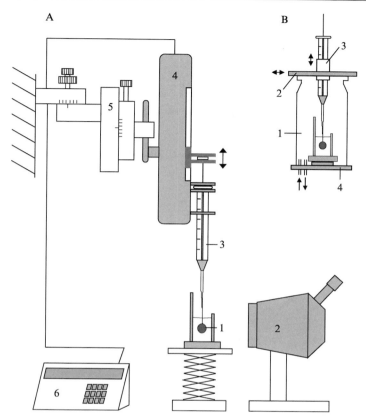

FIG. 4. (A) Schematic view of the most advanced setup used for microinjection of radiotracers into insect guts: (1) isolated gut embedded in agarose, placed on a scissor-lift table, (2) stereomicroscope, (3) oil-filled microliter syringe with a fine-tipped glass capillary, (4) motor drive, (5) manual micromanipulator controlling tip position, fixed to a solid steel support, and (6) microprocessor controller for motor drive. (B) Schematic view of the modifications for microinjection under a defined headspace: (1) glass bell, (2) PVC cover plate and (3) shaft, which allow lateral and vertical positioning, and (4) base plate with gas inlet and outlet for continuous flushing.

placed into a glass bell (inverted membrane filtration funnel; Sartorius, Göttingen, Germany) covered with a round PVC plate (Fig. 4B). The micropipette is inserted through a small vertical tube glued into a central hole of the PVC plate; the inner diameter of the tube is only slightly larger than the outer diameter of the microliter syringe. The setup is sealed with a thin film of silicone oil applied to all moving parts. With this device, the micropipette tip can be positioned vertically and laterally within the bell, while the headspace of the bell is flushed continuously with the desired gas mixture. At flow rates of 2 ml s^{-1}, the backpressure at the outlet is negligible.

Preparation of Fine-Tipped Capillaries

In addition to the reproducibility of the hydraulic system, the tip of the glass capillary is decisive for a successful microinjection. Prefabricated tips are available commercially (World Precision Instruments, Inc.), but are expensive and do not have an optimal shape.

Custom-made capillaries with the desired tip diameter and length of the taper are constructed by pulling a suitable glass capillary (50 μl; Duran glass, 1.0 mm i.d, 1.5 mm o.d.), either manually or in a commercial capillary puller. For manual pulling, it is best to first reduce the original diameter of the capillary by gentle, manual pulling after heating the capillary locally with a small gas flame (e.g., of a small soldering torch). The final tip is pulled in a second step using a platinum heating loop and gravitational force. A commercial capillary puller allows tip production in a one-step process and yields more reproducible tapers, but is not essential. Tip diameter and taper length can be varied by changing the heat and the pulling force. The final tip diameter is adjusted by breaking off the tip under a microscope using precision steel tweezers (e.g., Dumont No. 5, Dumont, Montignez, Switzerland).

The connection to the syringe is made by fitting the blunt end of the capillary into a plastic luer-lock adapter. The luer-lock adapter is removed from a disposable syringe needle by briefly heating the steel next to the adaptor (e.g., with a cigarette lighter or the pilot flame of a Bunsen burner) and pulling the adapter and needle apart. The opening in the plastic adaptor is enlarged to the diameter of the glass capillary with a suitable drill bit. The blunt end of the capillary is introduced into the adaptor and is connected by melting the plastic carefully with a small flame. A tight seal is crucial and can be improved by applying a fast-curing epoxy glue. Before connection to the syringe, the capillary is filled with the same oil as the syringe.

System Check and Calibration

The most common problem is an air bubble in the hydraulic system, which is usually introduced when the different parts of the hydraulic system are connected. Because of the large compressibility of air, movements of the piston are not exactly translated into movement of the liquid in the capillary tip, resulting in both inaccurate and poorly reproducible injections.

Therefore, the hydraulic system should be scrutinized for air bubbles by eye and under a stereomicroscope. Air bubbles are trapped most commonly at the luer-lock connection between the capillary tip and the syringe. Even if no air bubbles are visible, the reproducibility and accuracy of the injections have to be tested. This is done by injecting the desired volume of radiotracer of known volume activity directly into a series of scintillation vials filled with scintillation cocktail and determining the radioactivity using a liquid scintillation counter.

To avoid evaporation of the aqueous sample and to avoid blocking, the tip should be kept immersed in oil between injections.

Radiolabel Concentration and Injection Volumes

To avoid artifacts, the injection should not increase pools sizes or dilute pool concentrations of the metabolites substantially (see earlier discussion). Nevertheless, the label should be sufficient to detect the injected compound and its metabolic products with the analytical equipment available. While the sensitivity of liquid scintillation counting depends mainly on the measuring time, identification and quantitation of metabolites by chromatographic techniques can easily become a limiting factor.

To address both constraints, it is advisable to use the label as supplied by the vendor, i.e., undiluted, and at the highest available specific radioactivity. This is especially important in the case of ^{14}C-labeled compounds. Because of the long half-life time of the ^{14}C isotope, commercial preparations are actually provided at a specific radioactivity that already comes close to the theoretically possible value, where all carbon atoms are labeled (2.3 MBq μmol^{-1} or 62.4 mCi mmol^{-1} per μmol^{-1} C). Because microinjection will always involve only minute amounts of a given compound, cost is usually not an issue.

In the example mentioned earlier, we injected 50 nl [^{14}C]lactate (600 Bq) into the hindgut paunch of *R. flavipes* (total volume 0.72 μl). Due to the high specific radioactivity of the lactate preparation (6.1 MBq μmol^{-1}), the amount of lactate injected (0.1 nmol) caused only a relatively small dilution of the intestinal lactate pool (1 nmol). The detection limit of the individual radiolabeled metabolites by high-performance liquid chromatography (HPLC) was in the range of 2–5 Bq per peak (Tholen and Brune, 2000).

Microinjection into Termite Guts

Ideally, metabolite fluxes should be studied by introducing radiotracer into the gut *in vivo*, i.e., still within the insect body. However, even with immobilized termites, it is impossible to position the capillary tip into a specific gut region. Moreover, the fragile tips break easily when penetrating the cuticle. Therefore, guts have to be isolated from the termites and embedded in agarose made up in Ringer's solution to prevent desiccation and to keep the guts physiologically active. The gut is removed easily from the agarose at the end of the incubation. Both the gut contents and any radiolabeled metabolites that diffused out of the gut and were trapped in the agarose are available for subsequent analyses.

Sample Preparation

Termites are dissected, and guts are embedded in agarose in Ringer's solution (7.5 g NaCl liter^{-1}, 0.35 g KCl liter^{-1}, and 0.21 g CaCl$_2$ liter^{-1}) in a suitable chamber. The bottom and side walls of the chamber are constructed of a U-shaped PVC spacer; front and back walls are made of large microscope coverslips. The termite gut is placed flat and fully extended onto a thin layer of 2% agarose at the bottom of the microchamber and is covered quickly with a layer of molten 0.5% agarose (45°), which should cool and solidify immediately. Ideally, the overlay should not be thicker than 1–2 mm. Several guts can be prepared in parallel and stored in a moist chamber to avoid desiccation.

Microinjection

A chamber with an embedded gut is placed onto a small scissor-lift table and, if indicated, covered by a glass bell to control the gas headspace (Fig. 4B). Using the micromanipulator, the capillary tip containing the radiotracer is positioned outside of the agarose directly above the gut. Then, the tip is slowly inserted through the agarose into the gut and positioned at the gut center. Usually, the gut wall is indented upon contact with the capillary tip and relaxes after penetration. This visual control is important to avoid injecting the radiotracer into the agarose. With the motorized setup, the injection can be performed at a high rate of volume delivery (e.g., 1000 nl s^{-1}); in that case, also a sudden, barely perceptible increase of the gut volume indicates a successful injection. After injection, the capillary tip is retracted and the gut is placed in a moist chamber for the desired incubation time.

For the interpretation of the results, it is important to determine the exact amount of radiotracer delivered into the gut at the time of injection and the amount of injected tracer that leaks from the gut into the surrounding agarose because of possible damage to the gut caused by the injection procedure. For this purpose, it is convenient to use a compound that does not diffuse across the gut wall and is not metabolized within the gut (e.g., [^{14}C]polyethylene glycol 4000).

Analysis of Metabolites

At the end of the incubation, guts are removed from the agarose and placed immediately into a defined volume of ice-cold stopping solution (0.2 M NaOH) and homogenized by sonication with an ultrasonic probe (e.g., Ultraschallprozessor 50 H equipped with a microtip, Dr. Hielscher GmbH, Teltow, Germany). After centrifugation (10 min, 20,000g, 4°), the supernatant is analyzed for total radioactivity and individual metabolites. The pellet is washed with stopping solution and then analyzed for residual radioactivity.

The agarose block is placed into an Eppendorf tube containing a defined volume of the stopping solution and stored for a few days at 4° to allow metabolites in the agarose to equilibrate with the solution. Cutting the agarose into smaller pieces will speed up the process. After equilibration, the liquid phase is analyzed for total radioactivity and individual metabolites. Volume determination of the remaining agarose block by weighing or analyzing the remaining agarose for total radioactivity allows accurate radioactivity recoveries for the substrate and the single products to be calculated.

The alkaline character of the stopping solution prevents the loss of volatile substances, such as CO_2 and short-chain fatty acids. It is advisable to include any metabolites of interest (nonlabeled) in the stopping solution to act as an internal standard and as a carrier to prevent unspecific loss of the radiotracer.

The overall recovery of radioactivity after microinjection and incubation is determined by liquid scintillation counting (LSC) of aliquots of the gut and agarose samples. Incomplete recovery of the injected radioactivity usually indicates loss of volatile metabolites (e.g., CO_2 or CH_4) during incubation or sample preparation. Individual metabolites are identified and quantitated by HPLC or gas chromatography coupled with radioactivity detection. Detection efficiencies and possible quenching effects of the samples have to be considered. Radiolabeled CO_2 is separated from the samples by a flow injection method described by Hall and Aller (1992) and quantitated by LSC.

References

Ambrose, D., Berger, L. I., Clay, R. W., Covington, A. K., and Eckerman, K. F. and 23 other authors (1999). "Handbook of Chemistry and Physics." CRC Press, London.

Brauman, A., Kane, M. D., Labat, M., and Breznak, J. A. (1992). Genesis of acetate and methane by gut bacteria of nutritionally diverse termites. *Science* **257,** 1384–1387.

Breznak, J. A., and Switzer, J. M. (1986). Acetate synthesis from H_2 plus CO_2 by termite gut microbes. *Appl. Environ. Microbiol.* **52,** 623–630.

Brune, A., Emerson, D., and Breznak, J. A. (1995). The termite gut microflora as an oxygen sink: Microelectrode determination of oxygen and pH gradients in guts of lower and higher termites. *Appl. Environ. Microbiol.* **61,** 2681–2687.

Brune, A., Frenzel, P., and Cypionka, H. (2000). Life at the oxic-anoxic interface: Microbial activities and adaptations. *FEMS Microbiol. Rev.* **24,** 691–710.

Brune, A., and Friedrich, M. (2000). Microecology of the termite gut: Structure and function on a microscale. *Curr. Opin. Microbiol.* **3,** 263–269.

Crank, J. (1975). "The Mathematics of Diffusion," 2nd Ed. Clarendon Press, Oxford.

Ebert, A., and Brune, A. (1997). Hydrogen concentration profiles at the oxic-anoxic interface: A microsensor study of the hindgut of the wood-feeding lower termite *Reticulitermes flavipes* (Kollar). *Appl. Environ. Microbiol.* **63,** 4039–4046.

Hall, P. O. J., and Aller, R. C. (1992). Rapid, small volume, flow injection analysis of ΣCO_2 and NH_4^+ in marine and freshwaters. *Limnol. Oceanogr.* **37,** 1113–1119.

Jørgensen, B. B. (1978). A comparison of methods for the quantification of bacterial sulfate reduction in coastal marine sediments. *Geomicrobiol. J.* **1,** 11–27.

Jørgensen, B. B. (1994). Sulfate reduction and thiosulfate transformations in a cyanobacterial mat during a diel oxygen cycle. *FEMS Microbiol. Ecol.* **13,** 303–312.

Kühl, M., and Jørgensen, B. B. (1992). Microsensor measurements of sulfate reduction and sulfide oxidation in compact microbial communities of aerobic biofilms. *Appl. Environ. Microbiol.* **58,** 1164–1174.

Moeslund, L., Thamdrup, B., and Jorgensen, B. B. (1994). Sulfur and iron cycling in a coastal sediment: Radiotracer studies and seasonal dynamics. *Biogeochemistry* **27,** 129–152.

Skyring, G. W. (1987). Sulfate reduction in coastal ecosystems. *Geomicrobiol. J.* **5,** 295–374.

Tholen, A., and Brune, A. (1999). Localization and *in situ* activities of homoacetogenic bacteria in the highly compartmentalized hindgut of soil-feeding higher termites (*Cubitermes* spp.). *Appl. Environ. Microbiol.* **65,** 4497–4505.

Tholen, A., and Brune, A. (2000). Impact of oxygen on metabolic fluxes and *in situ* rates of reductive acetogenesis in the hindgut of the wood-feeding termite *Reticulitermes flavipes.* *Environ. Microbiol.* **2,** 436–449.

Yang, H., Schmitt-Wagner, D., Stingl, U., and Brune, A. (2005). Niche heterogeneity determines bacterial community structure in the termite gut (*Reticulitermes santonensis*). *Environ. Microbiol.* **7,** 916–932.

[12] Assaying for the 3-Hydroxypropionate Cycle of Carbon Fixation

By Michael Hügler and Georg Fuchs

Abstract

The 3-hydroxypropionate cycle is a novel pathway for autotrophic CO_2 fixation, which has been demonstrated in the thermophilic phototrophic bacterium *Chloroflexus aurantiacus*; a yet to be defined variant of this

METHODS IN ENZYMOLOGY, VOL. 397
Copyright 2005, Elsevier Inc. All rights reserved.

pathway occurs in autotrophic members of the Sulfolobales (Crenarchaeota). The 3-hydroxypropionate cycle consists of the conversion of acetyl-CoA to succinyl-CoA, via malonyl-CoA, 3-hydroxypropionate, propionyl-CoA, and methylmalonyl-CoA. Carboxylation of acetyl-CoA and propionyl-CoA by acetyl-CoA/propionyl-CoA carboxylase are the CO_2 fixation reactions. Succinyl-CoA serves as a precursor of cell carbon and also as a precursor of the starting compound acetyl-CoA. In *C. aurantiacus*, the cycle is completed by converting succinyl-CoA to malyl-CoA and cleaving malyl-CoA to acetyl-CoA and glyoxylate. Glyoxylate is then converted in a second cyclic pathway to pyruvate, which serves as a universal cell carbon precursor. The fate of succinyl-CoA in Sulfolobales is at issue. Assays used to study the characteristic enzymes of this novel pathway in *C. aurantiacus* are reported.

Introduction

The 3-hydroxypropionate cycle is a novel pathway for autotrophic CO_2 fixation that has been studied in the thermophilic phototrophic bacterium *Chloroflexus aurantiacus* (Eisenreich *et al.*, 1993; Herter *et al.*, 2001, 2002b; Holo, 1989; Holo and Sirevag, 1986; Strauss and Fuchs, 1993; Strauss *et al.*, 1992). So far *C. aurantiacus* is the only member of the bacteria known to fix CO_2 via the 3-hydroxypropionate cycle, but a modified version of this pathway seems to be present within members of the Sulfolobaceae in the domain of Archaea. Key enzyme activities of the 3-hydroxypropionate cycle have been found in *Acidianus ambivalens*, *Acidianus brierleyi*, *Metallosphaera sedula*, *Sulfolobus metallicus*, and *Sulfolobus* sp. strain VE6, all thermophilic or hyperthermophilic and acidophilic members of the Crenarchaeota (Burton *et al.*, 1999; Hügler *et al.*, 2003a; Ishii *et al.*, 1997; Menedez *et al.*, 1999; Norris *et al.*, 1989). While some features of the cycle in Sulfolobaceae are still unknown, some enzymes of *C. aurantiacus* have been studied in more detail.

In this bicyclic pathway (Fig.1), three molecules of bicarbonate are fixed into one molecule of pyruvate. In a first cycle, glyoxylate is formed from two molecules of bicarbonate, which in a second cycle is converted with another molecule of bicarbonate to pyruvate. Both cycles regenerate acetyl-CoA, which is the starting molecule. The ATP-dependent acetyl-CoA carboxylase, a biotin enzyme, is one of two CO_2-fixing enzymes of the pathway and forms malonyl-CoA from acetyl-CoA and bicarbonate. Malonyl-CoA is reduced to 3-hydroxypropionate in a NADPH-dependent reaction, which is catalyzed by the bifunctional enzyme malonyl-CoA reductase. This key enzyme comprises malonate

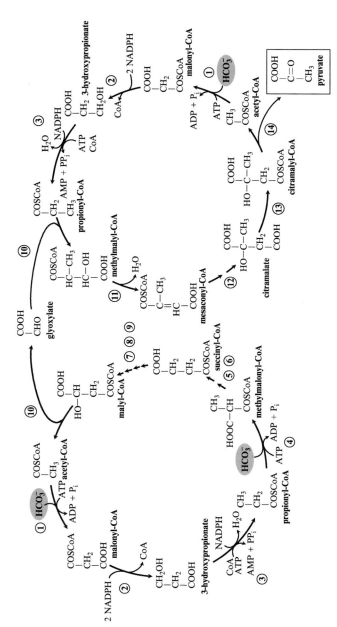

Fig. 1. The proposed bicyclic 3-hydroxypropionate cycle in *C. aurantiacus*. The enzymatic reactions involved are (1) acetyl-CoA carboxylase (E.C. 6.4.1.2), (2) malonyl-CoA reductase (E.C. 1.1.1.- and E.C. 1.2.1.-), (3) propionyl-CoA synthase (E.C. 6.2.1.-; E.C. 4.2.1.-; and E.C. 1.3.1.-), (4) propionyl-CoA carboxylase (E.C. 6.4.1.3), (5) methylmalonyl-CoA epimerase (E.C. 5.1.99.1), (6) methylmalonyl-CoA mutase (E.C. 5.4.99.2), (7) succinyl-CoA:L-malate CoA transferase, (8) succinate dehydrogenase (E.C. 1.3.5.1), (9) fumarate hydratase (E.C. 4.2.1.2), (10) L-malyl-CoA lyase (E.C. 4.1.3.24), (11) β-methylmalyl-CoA lyase (E.C. 4.1.3.24), (12) unknown reactions transforming methylmalyl-CoA to citramalate, (13) succinyl-CoA:citramalate CoA transferase, and (14) citramalyl-CoA lyase.

semialdehyde and 3-hydroxypropionate dehydrogenase activity (Hügler *et al.*, 2002). 3-Hydroxypropionate is further converted to propionyl-CoA by a trifunctional enzyme, propionyl-CoA synthase, the second key enzyme of the cycle. This large enzyme comprises 3-hydroxypropionate: CoA ligase, 3-hydroxypropionyl-CoA dehydratase, and propionyl-CoA dehydrogenase activities (Alber and Fuchs, 2002). A second ATP- and bicarbonate-dependent carboxylation reaction, catalyzed by the biotin enzyme propionyl-CoA carboxylase, leads to the formation of methyl-malonyl-CoA. In Sulfolobaceae, both acetyl-CoA and propionyl-CoA carboxylation reactions are catalyzed by one promiscuous enzyme, acetyl-CoA/propionyl-CoA carboxylase, which has been purified and characterized from *A. brierleyi* and *M. sedula* (Chuakrut *et al.*, 2003; Hügler *et al.*, 2003b). The isomerization of methylmalonyl-CoA to succi-nyl-CoA is catalyzed by the conventional enzymes methylmalonyl-CoA epimerase and methylmalonyl-CoA mutase. Succinyl-CoA is used for the activation of malate to malyl-CoA. The transfer of the CoA moiety is catalyzed by succinyl-CoA:L-malate CoA transferase (Herter *et al.*, 2001). Succinate in turn is oxidized to malate by enzymes of the citric acid cycle, succinate dehydrogenase and fumarate hydratase.

The cleavage of malyl-CoA into acetyl-CoA and glyoxylate (the CO_2 fixation product of the first cycle) is catalyzed by another promiscuous enzyme, L-malyl-CoA/β-methylmalyl-CoA lyase. This enzyme not only catalyzes the cleavage of L-malyl-CoA, but also the condensation of glyoxylate with another molecule of propionyl-CoA forming β-methyl-malyl-CoA. This enzyme connects both cycles and has been purified and characterized from *C. aurantiacus* (Herter *et al.*, 2002a). The en-zymes converting β-methylmalyl-CoA to citramalate have not been studied in detail yet; it has been postulated that this reaction sequence involves hydratase and CoA transferase reactions (Herter *et al.*, 2002b). The subsequent activation of citramalate to citramalyl-CoA is catalyzed by succinyl-CoA:citramalate CoA transferase. In the final reaction of the second cycle, citramalyl-CoA is cleaved by citramalyl-CoA lyase into pyruvate, the CO_2 fixation product, and acetyl-CoA. Acetyl-CoA is converted to propionyl-CoA by acetyl-CoA carboxylase, malonyl-CoA reductase, and propionyl-CoA synthase closing the second cycle. Pyruvate is further metabolized by pyruvate phosphate dikinase forming phosphoenolpyruvate. Phosphoenolpyruvate carboxy-lase acts as anaplerotic enzyme catalyzing the bicarbonate-dependent carboxylation of phosphoenolpyruvate, which results in the formation of oxaloacetate.

Preparation of Cell Extracts

French Press

Wet cells are resuspended in a twofold volume of cold buffer (100 mM Tris-HCl, pH 7.8, 0.1 mg DNase I per 1 ml), the cell suspension is passed through a French pressure cell at 137 MPa, and the lysate is ultracentrifuged (100,000g, 4°, 1 h). The supernatant should be used immediately.

Mixer-Mill

Wet cells (150–200 mg) are resuspended in a 1.5-ml Eppendorf vial in 500–800 μl of cold buffer (100 mM Tris-HCl, pH 7.8, 0.1 mg DNase I per 1 ml). Glass beads (1 g, diameter 0.1–0.25 mm) are added, and the suspension is shaken in a mixer-mill (type MM2, Retsch, Haare, Germany) for 7 min at 100% intensity (30 Hz). Following a centrifugation step (12,000g, 4°, 10 min), the lysate can be used for enzymatic tests. This is the method of choice when only a small amount of cell mass is available.

Synthesis of Coenzyme A Thioesters and 3-Hydroxypropionate

Malonyl-CoA, acetyl-CoA, propionyl-CoA, succinyl-CoA, L-malyl-CoA, β-methylmalyl-CoA, and 3-hydroxypropionate, some of which are not available commercially, are synthesized. Monothiophenylmalonate, the precursor of malonyl-CoA, is synthesized according to published procedures (Imamato et al., 1982; Pollmann and Schramm, 1964). The final synthesis of malonyl-CoA is described in Hügler et al. (2002). The CoA-thioesters of acetate, propionate, and succinate are synthesized according to the method described by Simon (1957) and Stadtman (1957) proceeding from their anhydrides. L-Malyl-CoA is synthesized according to the method of Eggerer et al. (1964). The slightly modified method is described in Herter et al. (2002a). β-Methylmalyl-CoA is synthesized enzymatically using recombinant L-malyl-CoA/β-methylmalyl-CoA lyase as described in Herter et al. (2002a).

Enzymatic Assays

All enzyme assays are performed under aerobic conditions at 55°, the growth temperature of C. aurantiacus. To a 1-ml assay mixture, 10–40 μl cell lysate is added, depending on the reaction.

Acetyl-CoA/Propionyl-CoA Carboxylase (EC 6.4.1.2 and EC 6.4.1.3)

Principle

In a discontinuous assay the fixation of ^{14}C from [^{14}C]bicarbonate into nonvolatile, acid-stabile products is measured. After 1, 5, and 10 min of incubation, 100-μl samples are withdrawn by syringe from the rubber stoppered 1-ml glass vial (0.5 ml assay mixture, 0.5 ml gas phase) and added to scintillation vials containing 20 μl of 6 M HCl (final pH 1–2). Volatile $^{14}CO_2$ (nonfixed) is removed by shaking the vials vigorously for 3 h. Then 3 ml of scintillation cocktail (e.g., Rotiszint 2200, Roth, Karlsruhe, Germany) is added, and the remaining radioactivity in the sample is determined by liquid scintillation counting. The radioactivity in samples of two control experiments, in which the substrate and extract, respectively, are omitted, serve as blanks and controls. The amount of product formed is calculated based on the amount of acid-stabile radioactivity and the specific radioactivity of "CO_2" (added bicarbonate) in the assay.

Reagents

The assay mixture (0.5 ml) contains 100 mM Tris-HCl, pH 7.8, 5 mM MgCl$_2$, 5 mM DTE, 4 mM ATP, 2 mM NADPH, 10 mM NaHCO$_3$, 37 kBq [^{14}C]Na$_2$CO$_3$ (specific radioactivity of total "CO_2" in the assay, 3.7 Bq nmol^{-1}), and 0.4 mM acetyl-CoA or propionyl-CoA. Acetyl-CoA/propionyl-CoA carboxylase is ATP-, NaHCO$_3$- and acetyl-CoA- or propionyl-CoA-dependent. Either of these substrates could be added from stock solutions to start the enzymatic reaction. In routine assays the reaction is started by the addition of acetyl-CoA or propionyl-CoA, respectively. NADPH is not required; however, it stimulates the acetyl-CoA carboxylation reaction by transforming the reaction product malonyl-CoA to 3-hydroxypropionate, which is catalyzed by endogenous malonyl-CoA reductase of the cell extract.

Malonyl-CoA Reductase (EC 1.1.1.- and EC 1.2.1.-)

Principle

The continuous assay is based on the malonyl-CoA-dependent oxidation of two NADPH. The oxidation is measured spectrophotometrically at 365 nm [$\varepsilon_{365\ nm}(NADPH) = 3.4 \times 10^3\ M^{-1}\ cm^{-1}$] in glass cuvettes.

Reagents

The assay mixture (0.5 ml) contains 100 mM Tris-HCl, pH 7.8, 2 mM MgCl$_2$, 0.4 mM NADPH, and 0.5 mM malonyl-CoA. The addition of malonyl-CoA starts the reaction.

Propionyl-CoA Synthase (EC 6.2.1.-, EC 4.2.1.-, and EC 1.3.1.-)

Principle

The oxidation of NADPH is followed spectrophotometrically at 365 nm. The trifunctional enzyme catalyzes the ATP- and CoA-dependent ligation of 3-hydroxypropionate to 3-hydroxypropionyl-CoA forming AMP and pyrophosphate. Dehydration of 3-hydroxypropionyl-CoA leads to acrylyl-CoA, which is reduced in a NADPH-dependent reaction to propionyl-CoA. Therefore, the overall reaction can be measured as the ATP-, CoA-, and 3-hydroxypropionate-dependent oxidation of NADPH.

Reagents

The assay mixture (0.5 ml) contains 100 mM Tris-HCl, pH 7.8, 10 mM KCl, 5 mM MgCl$_2$, 3 mM ATP, 0.5 mM CoA, 0.4 mM NADPH, and 1 mM 3-hydroxypropionate. The addition of 3-hydroxypropionate starts the reaction.

Succinyl-CoA:L-Malate CoA Transferase

Succinyl-CoA:L-malate CoA transferase catalyzes the reversible transfer of the coenzyme A moiety from succinyl-CoA to L-malate forming L-malyl-CoA and succinate. The enzyme could be measured by two different methods. In a coupled spectrophotometric assay, the formation of L-malyl-CoA is followed using the activity of L-malyl-CoA lyase. In a discontinuous assay, coenzyme A esters are analyzed by HPLC.

Coupled Assay

Principle. The reaction of succinyl-CoA:L-malate CoA transferase could be coupled to the cleavage of L-malyl-CoA to glyoxylate and acetyl-CoA by endogenous L-malyl-CoA lyase. The overall reaction could be measured spectrophotometrically at 324 nm by following the succinyl-CoA- and L-malate-dependent formation of glyoxylate phenylhydrazone (see L-malyl-CoA lyase).

Reagents. The assay mixture (0.5 ml) contains 200 mM morpholinopropanesulfonic acid (MOPS)/KOH, pH 7.0, 4 mM MgCl$_2$, 3.5 mM phenylhydrazinium chloride, 1 mM succinyl-CoA, and 5 mM L-malate. Either substrate can be used to start the reaction.

HPLC Assay

Principle. Succinyl-CoA:L-malate CoA tranferase can be measured directly by following the succinyl-CoA-dependent formation of L-malyl-CoA. The CoA-thioesters can be separated by HPLC on a reversed-phase C18 column (LiChrospher 100, 5 μM, 125 × 5 mm; Merck, Germany). A gradient of 1–8% acetonitrile (v/v) in 50 mM potassium-phosphate buffer, pH 6.7, and a flow rate of 1 ml min^{-1} over 30 min are used. CoA-thioesters are detected at 260 nm. The retention times are 2 min (malate, succinate), 10 min (L-malyl-CoA), and 13 min (succinyl-CoA, CoA).

Reagents. The assay mixture (0.5 ml) contains 100 mM potassium-phosphate buffer, pH 6.7, 10 mM KCl, 10 mM MgCl$_2$, 1 mM succinyl-CoA, and 5 mM L-malate. Either substrate can be used to start the reaction. After incubation for 0, 2, 5, and 10 min, samples of 100 μl are taken, and the reaction is stopped by adding 10 μl of 6 M HCl. Protein is removed by centrifugation (12,000g, 4°, 10 min), and the samples are analyzed by HPLC. A control reaction is carried out in which succinyl-CoA is omitted.

L-Malyl-CoA Lyase (EC 4.1.3.24) and β-Methylmalyl-CoA Lyase

Lyase Reaction

Principle. The reversible cleavage of malyl-CoA into acetyl-CoA and glyoxylate is measured in a continuous spectrophotometrical assay by following the formation of glyoxylate-phenylhydrazone. The cleavage of β-methylmalyl-CoA leads to propionyl-CoA and glyoxylate and can be followed in the same way. Phenylhydrazine reacts with free aldehyde groups forming phenylhydrazones, which can be detected at 324 nm ($\varepsilon_{324\ nm} = 17,000$ M^{-1} cm^{-1}).

Reagents. The assay mixture (0.5 ml) contains 200 mM MOPS/KOH, pH 7.7, 4 mM MgCl$_2$, 3.5 mM phenylhydrazinium chloride, and 0.5 mM L-malyl-CoA or β-methylmalyl-CoA. The addition of L-malyl-CoA or β-methylmalyl-CoA starts the reaction.

Condensation Reaction

Principle. The condensation of acetyl-CoA or propionyl-CoA with glyoxylate forms L-malyl-CoA or β-methylmalyl-CoA. This reaction is monitored in a discontinuous assay. After incubation, the reaction is stopped and the mixture is incubated with phenylhydrazinium chloride. Residual glyoxylate reacts with phenylhydrazine forming glyoxylate phenylhydrazone, which can be quantified by measuring the absorption at 324 nm.

Reagents. The assay mixture (0.2 ml) contains 200 mM MOPS/KOH, pH 7.7, 4 mM MgCl$_2$, 2 mM glyoxylate, and 4 mM acetyl-CoA or propionyl-CoA. The reaction is started by the addition of the CoA ester. After incubation (1, 5, 10, 20 min), samples (25 μl) are retrieved and added to 0.975 ml of 200 mM MOPS/KOH, pH 7.4, containing 3.5 mM phenylhydrazinium chloride at room temperature. After 15 min of incubation at room temperature, the formed glyoxylate phenylhydrazone is detected at 324 nm.

Succinyl-CoA:Citramalate CoA Transferase and Citramalyl-CoA Lyase

Coupled Assay

Principle. Succinyl-CoA:citramalate CoA transferase catalyzes the reversible transfer of the coenzyme A moiety from succinyl-CoA to citramalate forming citramalyl-CoA and succinate. Citramalyl-CoA is cleaved into acetyl-CoA and pyruvate by citramalyl-CoA lyase. This overall reaction can be measured spectrophotometrically at 324 nm by following the succinyl-CoA- and citramalate-dependent formation of pyruvate phenylhydrazone (for details, see succinyl-CoA:L-malate CoA transferase).

Reagents. The assay mixture (0.5 ml) contains 200 mM MOPS/KOH, pH 7.0, 4 mM MgCl$_2$, 3.5 mM phenylhydrazinium chloride, 1 mM succinyl-CoA, and 5 mM D/L-citramalate. Either substrate can be used to start the reaction.

References

Alber, B. E., and Fuchs, G. (2002). Propionyl-CoA synthase from *Chloroflexus auranticus*, a key enzyme of the 3-hydroxypropionate cycle for autotrophic CO$_2$ fixation. *J. Biol. Chem.* **277**, 12137–12143.

Burton, N. P., Williams, T. D., and Norris, P. R. (1999). Carboxylase genes of *Sulfolobus metallicus*. *Arch. Microbiol.* **172**, 349–353.

Chuakrut, S., Arai, H., Ishii, M., and Igarashi, Y. (2003). Characterization of a bifunctional archeal acyl-CoA carboxylase. *J. Bacteriol.* **185**, 938–947.

Eggerer, H., Remberger, U., and Grünwald, C. (1964). Zum Mechanismus der biologischen Umwandlung von Citronensäure. V. Citrat-Synthase, eine Hydrolase für Malyl-CoA. *Biochem. Z.* **229**, 436–453.

Eisenreich, W., Strauss, G., Werz, U., Fuchs, G., and Bacher, A. (1993). Retrobiosynthetic analysis of carbon fixation in the phototrophic eubacterium *Chloroflexus auranticus*. *Eur. J. Biochem.* **215**, 619–632.

Herter, S., Busch, A., and Fuchs, G. (2002a). L-Malyl-CoA/β-methylmalyl-CoA lyase from *Chloroflexus auranticus*, a bifunctional enzyme involved in autotrophic CO_2 fixation. *J. Bacteriol.* **184**, 5999–6006.

Herter, S., Farfsing, J., Gad'on, N., Rieder, C., Eisenreich, W., Bacher, A., and Fuchs, G. (2001). Autotrophic CO_2 fixation in *Chloroflexus aurantiacus*: Study of glyoxylate formation and assimilation via the 3-hydroxypropionate cycle. *J. Bacteriol.* **183**, 4305–4316.

Herter, S., Fuchs, G., Bacher, A., and Eisenreich, W. (2002b). A bicyclic autotrophic CO_2 fixation pathway in *Chloroflexus aurantiacus*. *J. Biol. Chem.* **277**, 20277–20283.

Holo, H. (1989). *Chloroflexus aurantiacus* secretes 3-hydroxypropionate, a possible intermediate in the assimilation of CO_2 and acetate. *Arch. Microbiol.* **151**, 252–256.

Holo, H., and Sirevåg, R. (1986). Autotrophic growth and CO_2 fixation in *Chloroflexus aurantiacus*. *Arch. Microbiol.* **145**, 173–180.

Hügler, M., Huber, H., Stetter, K. O., and Fuchs, G. (2003a). Autotrophic CO_2 fixation pathways in Archaea (Crenarchaeota). *Arch. Microbiol.* **179**, 160–173.

Hügler, M., Krieger, R. S., Jahn, M., and Fuchs, G. (2003b). Characterization of acetyl-CoA/propionyl-CoA carboxylase in *Metallosphaera sedula*: Carboxylating enzyme in the 3-hydroxypropionate cycle for autotrophic carbon fixation. *Eur. J. Biochem.* **270**, 736–744.

Hügler, M., Menendez, C., Schägger, H., and Fuchs, G. (2002). Malonyl-coenzyme A reductase from *Chloroflexus aurantiacus*, a key enzyme of the 3-hydroxypropionate cycle for autotrophic CO_2 fixation. *J. Bacteriol.* **184**, 2404–2410.

Imamato, T., Kodera, M., and Yokoyama, M. (1982). A convenient method for the preparation of S-esters of thio analogs of malonic acid. *Bull. Chem. Soc. Jpn.* **55**, 2303–2304.

Ishii, M., Miyake, T., Satoh, T., Sugiyama, H., Oshima, Y., Kodama, T., and Igarashi, Y. (1997). Autotrophic carbon dioxide fixation in *Acidianus brierleyi*. *Arch. Microbiol.* **166**, 368–371.

Norris, P. A., Nixon, A., and Hart, A. (1989). Acidophilic, mineral-oxidizing bacteria: The utilization of carbon dioxide with particular reference to autotrophy in *Sulfolobus*. *In* "Microbiology of Extreme Environments and Its Potential for Biotechnology" (M. S. Da Costa, J. C. Duarte, and R. A. D. Williams, eds.), pp. 14–43. Elsevier, London.

Pollmann, W., and Schramm, G. (1964). Reactivity of metaphosphate esters prepared from P_4O_{10} and ethylether. *Biochim. Biophys. Acta* **80**, 1–7.

Simon, E. (1957). S-Succinyl-CoA. *Biochem. Preps.* **5**, 30.

Stadtman, E. R. (1957). Preparation and assay of acyl coenzyme A and other thiol esters; use of hydroxylamine. *Methods Enzymol.* **3**, 931–941.

Strauss, G., Eisenreich, W., Bacher, A., and Fuchs, G. (1992). ^{13}C-NMR study of autotrophic CO_2 fixation pathways in the sulfur-reducing archaebacterium *Thermoproteus neutrophilus* and in the eubacterium *Chloroflexus aurantiacus*. *Eur. J. Biochem.* **205**, 853–866.

Strauss, G., and Fuchs, G. (1993). Enzymes of a novel autotrophic CO_2 fixation pathway in the phototrophic bacterium *Chloroflexus aurantiacus*, the 3-hydroxypropionate cycle. *Eur. J. Biochem.* **215**, 633–643.

Further Reading

Ménendez, C., Bauer, Z., Huber, H., Gad'on, N., Stetter, K. O., and Fuchs, G. (1999). Presence of acetyl-coenzyme A (CoA) carboxylase and propionyl-CoA carboxylase in autotrophic *Crenarchaeota* and indication for the operation of a 3-hydroxypropionate cycle in autotrophic carbon fixation. *J. Bacteriol.* **181**, 1088–1098.

[13] Analysis of Trace Hydrogen Metabolism

By Frank E. Löffler and Robert A. Sanford

Abstract

H_2 maintains an important role in bacterial metabolism in anaerobic environments. The ability to measure and analyze H_2 concentrations in biological systems is often useful for determining the physiological state of the microbiota. Methods for precisely analyzing H_2 in bacterial cultures and environmental samples are now available. This chapter discusses H_2 measurements from both the theoretical and practical methodology perspective of the analysis and the interpretation of data.

Introduction

H_2 is a simple molecule that plays a central role in microbial metabolism. Many microorganisms not only consume H_2 as an energy source, they also produce H_2 in their metabolism (Zehnder and Stumm, 1988). In anaerobic ecosystems, H_2 has even been referred to as the universal currency that connects populations with distinct physiologies in anaerobic food webs (Wolin, 1982; Wolin and Miller, 1982). The extent of H_2 consumption and generation is governed by the thermodynamics of enzymatically catalyzed reactions (Dolfing and Harrison, 1992; Dolfing and Janssen, 1994; Thauer *et al.*, 1977). Due to the presence of multiple hydrogenases (enzymes involved in H_2 oxidation and proton [H^+] reduction) with distinct properties, both H_2 consumption and H_2 generation can occur simultaneously within a single bacterial population or between populations in a microbial community (Fauque *et al.*, 1988). In natural ecosystems, the flux of reduced organic compounds, H_2-forming fermentation processes, and H_2-consuming (hydrogenotrophic) reactions govern the steady-state H_2 concentrations. In a mixed microbial community, this interplay between H_2 production and H_2 consumption among populations is often referred to as interspecies H_2 transfer (Schink and Friedrich, 1994). Theoretically, bulk H_2 concentrations are controlled by the energetics of the predominant terminal electron-accepting process (TEAP). Hence, H_2 concentration measurements can be used to indicate and delineate the predominant TEAP occurring in a particular environment (Lovley *et al.*, 1994). The ability of many microorganisms to rapidly influence H_2 concentrations through regulating hydrogenase activity also makes H_2 an attractive

METHODS IN ENZYMOLOGY, VOL. 397   0076-6879/05 $35.00
 DOI: 10.1016/S0076-6879(05)97013-4

metabolite to monitor the response of the microbial community to pertur-
bations (e.g., flux of reduced carbon compounds and electron acceptors) in
anaerobic ecosystems. This chapter discusses methods and procedures used
to analyze and interpret H_2 concentrations in laboratory cultures and
environmental systems.

Consumption Threshold versus Compensatory H_2 Concentration

The minimum H_2 concentration obtained during the reduction of a
specific terminal electron acceptor in the absence of other electron donors
is called the H_2 consumption threshold (or just H_2 threshold) (Lovley,
1985; Cord-Ruwisch et al., 1988; Lovley and Goodwin, 1988; Conrad,
1996). The H_2 threshold concentration associated with different TEAPs is
inversely correlated with the Gibbs free energy (ΔG^0) of the redox reaction (Tables I and II) (Breznak, 1994; Cord-Ruwisch et al., 1988; Lovley
and Goodwin, 1988; Löffler et al., 1999; Lu et al., 2001; Luijten et al., 2004).
According to the thermodynamics of the redox reactions, TEAPs will
occur in the following progression: acetogenesis > methanogenesis > sulfate reduction (sulfidogenesis) > ammonification > metabolic reductive
dechlorination (chlororespiration) > Fe(III) reduction > Mn(IV) reduction > denitrification > aerobic respiration. According to this model, H_2
threshold concentrations will follow the same progression (Table I).

When both H_2 generation and consumption are occurring simultaneously, a compensatory steady-state H_2 concentration is observed
(Fig. 1). Compensation occurs due to the kinetic balance between forward
(H_2 production) and reverse (H_2 consumption) reactions. Equal rates in
both directions result in compensatory steady-state concentrations or H_2
compensation concentrations. Compensatory H_2 levels are always higher
than the consumption threshold concentrations for a particular TEAP. For
example, the tropospheric concentration of H_2 of 0.56 ppmv can be viewed
as a global compensatory steady-state concentration, which is well above
the concentration of H_2 achieved in closed aerobic incubations (<0.01
ppmv) (Conrad et al., 1983). In anaerobic environments, however, a thermodynamic relationship between a particular TEAP and the steady-state
H_2 concentration exists. For example, Lovley and Goodwin (1988) demonstrated that steady-state H_2 concentrations are controlled by the physiological state of the H_2-consuming organisms (i.e., what electron acceptors
they are using) and are independent of the kinetics of the H_2-producing or
H_2-consuming reactions. This threshold model, however, does not always
apply and is affected by the specific environmental conditions and the
specific microorganisms present. Hoehler et al. (1998) found no difference
in steady-state H_2 concentrations when sulfate, Fe(III), or Mn(IV) was

TABLE I

H$_2$ CONSUMPTION THRESHOLD CONCENTRATIONS AND GIBBS FREE ENERGY CHANGES (ΔG^0)
OF DIFFERENT REDOX COUPLES WITH H$_2$ AS THE ELECTRON DONOR

TEAP	H$_2$ threshold concentrations[a]		$\Delta G^{0,b}$	Reference
	[ppmv]	[nM][c]	[kJ/mol H$_2$]	
Denitrification ($NO_3^- \to N_2O, N_2$)	<0.06	<0.05	−240	Lovley and Goodwin 1988; Lovley et al., 1994
Mn (IV) reduction ($MnO_2 \to Mn^{2+}$)	<0.06	<0.05	−157	Conrad, 1988; Lovley and Goodwin, 1988; Lovley et al., 1994
Ammonification ($NO_3^- \to NH_4^+$)	0.02–0.03	0.015–0.025	−149.9	Conrad, 1988; Cord-Ruwisch et al., 1988
Metabolic reductive dechlorination (chlororespiration)	<0.1	0.04–0.3	−130 to −187	Löffler et al., 1999; Luijten et al., 2004
Fe(III) reduction ($Fe(OH)_3 \to Fe^{2+}$	0.13–1	0.1–0.8	−108	Conrad, 1988; Lovley and Goodwin, 1988; Lovley et al., 1994
Sulfate reduction ($SO_4^{2-} \to HS^-$)	1.3–19	1–15	−38.0	Conrad, 1988; Cord-Ruwisch et al., 1988; Lovley and Goodwin, 1988
Methanogenesis	6–120	5–95	−33.9	Conrad, 1988; Cord-Ruwisch et al., 1988; Lovley and Goodwin, 1988
Acetogenesis	430–4660	336–3640	−26.1	Breznak, 1994; Cord-Ruwisch et al., 1988

[a] Note that direct environmental measurements often yield values in the lower range compared with threshold concentrations obtained for the same TEAPs in laboratory-based cultures.

[b] The ΔG^0 values had been published previously (Dolfing and Harrison, 1992; Dolfing and Janssen, 1994; Thauer et al., 1977) or were calculated from free energies of formation (ΔG_f).

[c] Concentrations calculated assuming a temperature of 25°.

TABLE II

EQUATIONS USED TO CALCULATE THE FREE ENERGY OF REDOX REACTIONS

1. $aA + bB \rightarrow cC + dD$

2. $\Delta G° = [c\Delta G_f(C) + d\Delta G_f(D)] - [a\Delta G_f(A) + b\Delta G_f(B)]$

Example redox reaction

3. $CO_2 + 4H_2 \rightarrow CH_4 + 2H_2O$

4. $\Delta G°_{rxn} = [\Delta G_f(CH_4) + 2\Delta G_f(H_2O)] - [\Delta G_f(CO_2) + 4\Delta G_f(H_2)]$

Temperature effect on ΔG

5. $\Delta G_f = \Delta H_f - T\Delta S$ $\Delta H_f \rightarrow$ enthalpy of formation
    $\Delta S \rightarrow$ entropy
    $T \rightarrow$ temperature in °Kelvin

6. $\Delta G_{rxn} = \Delta G°_{rxn} + RT \ln Q$

From example:

7. $Q = \dfrac{[C]^c[D]^d}{[A]^a[B]^b}$ $Q = \dfrac{[CH_4]}{[CO_2][H_2]^4}$ $[H_2O] = 1^a$

_____

[a] H_2O is not included in the calculation of Q because its concentration (activity) in dilute aqueous systems is 1.

added as an electron acceptor to sediment microcosms, although the threshold model would predict a higher H_2 concentration in the sulfate-amended microcosms. This was attributed to the lack of metal-reducing microbes in the sediment investigated; however, another explanation is that the H_2 production rate was high and the bioavailability of oxidized metal species was low relative to the available sulfate. Thus, in this example, the compensatory steady-state concentrations observed were not representative of a particular TEAP.

Other environmental factors also impact H_2 concentrations. Electron acceptor concentration, pH, and temperature all influence reaction thermodynamics (Hoehler et al., 1998). For example, temperature affects the free energy, ΔG, in two ways (Table II). First, the ΔG under standard conditions varies with temperature so that a specific $\Delta G°_T$ exists for any temperature (T). Second, the ΔG is influenced by temperature as the term RT lnQ is used to determine its value, where Q is the product to substrate

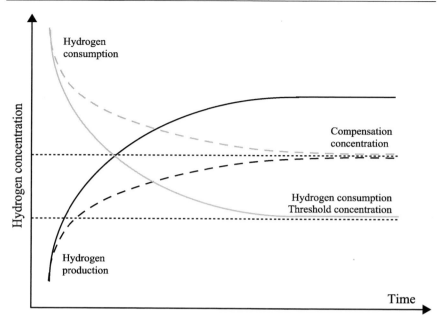

FIG. 1. Illustration of how H_2 production and consumption combine to create a compensatory steady-state concentration (compensation concentration).

ratio of the chemical reaction (Table II). Consequently, as temperature increases, ΔG becomes less negative and less energy is available from a particular TEAP. Several reports have noted that H_2 threshold concentrations follow this temperature dependence (Lee and Zinder, 1988; Zehnder and Stumm, 1988; Zinder, 1994). Lee and Zinder (1988) demonstrated a 250-fold increase in H_2 partial pressures in a thermophilic, methanogenic coculture over a temperature range from 0 to 80°. The electron acceptor concentration (or bioavailability) will also impact the H_2 consumption threshold, as this concentration is included in the term Q and influences the magnitude of ΔG (Table II). In addition, decreasing electron acceptor concentrations will result in lower H_2 oxidation rates if the reaction follows Michaelis–Menten kinetics. Thus, as electron acceptor concentrations decrease, H_2 threshold concentrations will increase. This phenomenon can be demonstrated experimentally by comparing H_2 consumption thresholds with very low and high electron acceptor concentrations (Hoehler et al., 1998).

Obviously, several complicating factors exist, and interpreting H_2 concentration data obtained in experimental, laboratory-based systems, particularly in environmental systems, must be interpreted accordingly. It is generally very difficult to draw any definite conclusions about microbial activity and the dominating TEAP in an environmental system when only H_2 concentrations are measured. In order to draw meaningful conclusions, H_2 measurements should be complemented by chemical and microbiological monitoring and include analyses of pH, concentrations of electron acceptors and reduced product(s), and possibly analysis of the active bacterial populations in the system.

Measuring Microbial H_2 Metabolism

Although microbial metabolism involves dissolved H_2, the analysis targets gaseous (headspace) H_2. Hence, H_2 analysis assumes equilibrium between aqueous phase concentrations (molar) and gas phase concentrations (partial pressure) of the gas. Consequently, great care must be taken to avoid H_2 loss from the sample and contamination of the sample with H_2 present in the surrounding air.

H_2 Sampling from Cultures

In the case of laboratory microcosms, microbial consortia, and pure cultures, H_2 is analyzed in the headspace of the culture vessels (e.g., 160-ml serum bottles). Black butyl rubber stoppers must be used to avoid H_2 exchange with the surrounding atmosphere, and appropriate controls must accompany all experiments to account for H_2 decrease or increase not due to microbial activity. Following the addition of a known amount of H_2, measurements are continued until H_2 reaches a steady-state concentration. The time to reach a constant concentration can be quite variable, ranging from hours to days or weeks, and depends on the electron acceptor utilized, the bacterial population involved in the H_2-consuming process, cell density, and incubation conditions (see case study given later). If H_2 is the only source of reducing equivalents in the culture system, the steady-state condition represents the consumption threshold concentration for the culture and TEAP being evaluated. A compensatory steady-state H_2 concentration is indicated when other organic electron donors are present in the system. H_2 production from an organic electron donor can be monitored by initially removing all H_2 from the culture headspace and measuring the H_2 increase with time. Hoehler et al. (1998) demonstrated this principle using sediments incubated in serum bottles with very low initial H_2 concentrations. The only caveat for delineating H_2 consumption thresholds from compensatory steady-state concentration measurements is that with the energetically most favorable TEAPs, such as O_2 and NO_3^- reduction, H_2 concentrations are likely to be below the detection limit of the currently available H_2 analytical instruments even in the presence of organic electron donors. In such cases, the H_2 concentration cannot be determined accurately. To avoid this type of situation, measurements should be performed in well-defined systems to prevent erroneous data interpretation. Fortunately, most H_2-consuming anaerobic TEAPs have H_2 consumption threshold concentrations well above the detection limit of the currently available analytical systems (Table I). This facilitates the comparison between compensatory steady-state concentrations and H_2 consumption threshold concentrations.

H_2 Sampling from Groundwater

To quantify H_2 concentrations in groundwater, the so-called bubble strip method has been developed (Chapelle et al., 1995). A modified version of this method involves purging a gas sampling bulb (Microseeps, Pittsburgh, PA; www.microseeps.com) with groundwater pumped from the desired depth of a well. Groundwater is pumped at a rate sufficient to flush the volume of the bulb at least one to two times per minute. Flow is maintained through the bulb for several minutes to eliminate air bubbles (McInnes and Kampbell, 2000). The sampling bulb is then inverted and 20 ml of ultrapure, H_2-free N_2 is added to the system by syringe. Dissolved gases are allowed to equilibrate for 30 min. An alternate approach that has also been applied in the field involves the same system but pumping continues during the equilibration phase. Hence, injection of air suffices because the H_2 introduced with the injected air will equilibrate with the groundwater pumped through the bulb. Subsamples (e.g., 15 ml) of gas from the sample bulb can then be transferred to glass vials of known volume (e.g., 20 ml) that have been flushed with H_2-free N_2 and closed with a thick butyl rubber stopper. Samples can be stored in the short term but analysis of H_2 and other gases such as methane should occur as soon as possible. Factors that impact the quality of the H_2 analysis include the pumping rate and the temperature of the water, which can affect the rate of diffusion and consequently the time required to reach equilibrium. Ideally, samples should be taken repeatedly over an extended period of time to verify that H_2 concentrations are stable. Contact to iron metal during sampling must be avoided because abiotic H_2 formation can occur and affect H_2 measurements (Chapelle et al., 1997). Hence, H_2 measurements in newly installed wells with metal casings will not yield meaningful results.

H_2 Analysis

H_2 is typically measured with a gas chromatograph (GC) equipped with a reduction gas detector. Stand-alone systems (e.g., the reduction gas analyzer) are available from Trace Analytical (Sparks, MD). Headspace samples are injected into a gas sampling loop and separated on a molecular sieve analytical column (MS 5A, 0.79 m by 0.3175 cm, mesh 60/80; Trace Analytical) at an oven temperature of 40° (or room temperature). Ultrahigh purity nitrogen (99.999%) should be used as the carrier gas to achieve maximum sensitivity at a typical flow rate of about 30 ml min^{-1}. The carrier gas passes through a catalytic combustion converter (Trace Analytical) to remove traces of H_2 prior to entering the analytical system. The detection limit for H_2 under optimum conditions reaches 5–10 ppbv. To ensure that the detector response remains in the linear portion of the standard curve, samples with greater than 100 ppmv H_2 may require dilution with H_2-free N_2. This can be achieved using a serum bottle that has been flushed with H_2-free N_2 and contains a few glass beads to allow efficient mixing of the gases. Dilutions are done immediately before analysis. A gas sample loop is recommended for injection, as it is difficult to keep atmospheric H_2 from contaminating the sample during direct injection. The volume of the gas sample loop (0.1 to 1.0 ml) can be changed to decrease or increase sensitivity. Samples with H_2 concentrations >100 ppmv can also be analyzed on a GC equipped with a thermal conductivity detector, which has a much broader dynamic range but will not provide the sensitivity of the reduction gas detector. Hence, systems with thermal conductivity detection are not useful to measure H_2 threshold concentration except for some H_2/CO_2 acetogenic populations (Breznak, 1994). A typical linear standard curve for the reduction gas detector has a dynamic range from 10 ppbv to 100 ppmv. Standards are prepared by adding varying volumes of a 100-ppmv standard (Scott Specialty Gases, Scotty II available from Supelco) to 160-ml serum bottles purged with H_2-free N_2. Alternatively, undiluted H_2 can be used to prepare standards covering the desired concentration range as illustrated in Fig. 2. Glass beads are added to each dilution vessel to allow efficient mixing of the gases before further dilution or GC injection. Gastight syringes with a valve should be used whenever handling gaseous samples. A Micro Gas Blender for standard preparation is available from Trace Analytical (Sparks, MD), but the high cost of this instrument ensures that the traditional dilution procedure will remain popular.

The sampling of headspace gas from cultures or from standards also requires some care. When using a 1.0-ml sample loop, at least 3 ml of headspace gas should be withdrawn for each injection. With cultures or microcosms, it is necessary to add an equivalent volume of H_2-free inert gas

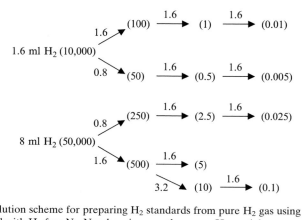

Fig. 2. Dilution scheme for preparing H_2 standards from pure H_2 gas using 160-ml serum bottles purged with H_2-free N_2. Numbers in parentheses are H_2 partial pressures (ppmv), and standards used to make a calibration curve are in bold letters. All other numbers indicate the volume of gas in milliliters added to 160-ml serum bottles fitted with thick butyl rubber stoppers.

to maintain a constant pressure in the culture vessels. This gas addition will have a dilution effect on the headspace H_2 concentration that can be calculated if the headspace volume is known. Obviously, this effect is small when the headspace volume is large. With a 3-ml sample size and five repeated sampling events from a 160-ml serum bottle without medium, the actual H_2 concentration would decrease less than 10%. In liquid/headspace systems, the effect is even smaller on a per milliliter gas basis due to the equilibration between aqueous and headspace H_2 concentrations. When samples are withdrawn, the headspace gas should be flushed several times back and forth through the sampling syringe prior to injection.

Gas leakage through the serum bottle septum is a concern, particularly when the culture H_2 concentrations are below the H_2 concentration of the surrounding air. The average H_2 mixing ratio in the troposphere is 0.56 ppmv (Conrad et al., 1983), and even higher values are generally measured in microbiological laboratories where H_2 is used for media preparation and analytical applications. Therefore, low H_2 concentrations (<0.5 ppmv) measured for denitrification, metabolic reductive dechlorination (chlororespiration), and Fe(III) reduction cannot be explained by H_2 leakage out of the serum bottles. In contrast, one can argue that the experimental threshold concentrations might be even lower for these TEAPs due to the leakage of atmospheric H_2 into the culture bottles. Leakage is prevented by using thick butyl rubber stoppers (Geo-Microbial Technologies, Inc. Ochelata, OK; Bellco Glass, Inc., Vineland, NJ) with low gas diffusion coefficients.

H_2 Data Interpretation

Because H_2 solubility varies with temperature and concentration units are not universally reported, the following section deals with the conversion and interpretation of H_2 analytical data. A conversion equation is presented that allows estimation of the aqueous molar H_2 concentration for any temperature when the equilibrium headspace concentration is known.

H_2 concentrations in microbiological experiments are often expressed in parts per million by volume (ppmv) according to the convention adapted by Conrad (1996). Many studies, however, use other units, such as atmospheres (atm) or Pascal (Pa) to express H_2 partial pressures, or report dissolved molar H_2 concentrations (Table III). To compare headspace (ppmv) to aqueous concentrations, the following equation is used:

$$H_{2(aqueous)} = (L \times P)/R \times T \tag{1}$$

where $H_{2 (aqueous)}$ is the dissolved phase H_2 concentration in nM; L is the Ostwald coefficient for H_2 solubility (Wilhelm et al., 1977) (Table IV); R is the universal gas constant (0.0821 liter $\times$ atm K^{-1} mol^{-1}); P is the pressure (atm); and T is the temperature in degrees Kelvin ($^\circ$K) (Lovley et al., 1994). Table IV shows how both the Ostwald coefficient and the calculated dissolved H_2 concentration vary with temperature. Data are shown for a 1.0-ppmv concentration; however, all conversions are proportional and any gas phase concentration (in ppmv) can be used (e.g., for a 10-ppmv partial pressure, multiply all values by a factor of 10). Because Ostwald coefficients are not reported for all temperatures, the relationship between

TABLE III
CONVERSION BETWEEN UNITS COMMONLY USED TO EXPRESS H_2 CONCENTRATIONS/PARTIAL PRESSURES

% (v/v) in headspace	ppmv	Pa[a]	atm	H_2 dissolved [nM] (25°)[b]
0.000001	0.01	0.001013	10^{-8}	0.0078
0.00001	0.1	0.01013	10^{-7}	0.078
0.0001	1	0.1013	10^{-6}	0.78
0.001	10	1.013	10^{-5}	7.8
0.01	100	10.13	10^{-4}	78.14
0.1	1000	101.3	10^{-3}	781.4
1	10^4	1013	0.01	7814
10	10^5	10,130	0.1	78,140
100	10^6	101,300	1	781,400

[a] Pa = atm $\times$ 101325.
[b] Calculated according to $H_{2(aqueous)} = (LP)/RT \times 10^9$.

TABLE IV

DETERMINATION OF nM H$_2$ CONCENTRATION AT DIFFERENT TEMPERATURES BASED ON GASEOUS
H$_2$ PARTIAL PRESSURES IN ppmv[a]

Temperature °K (°C)	Ostwald coefficient[b]	Concentration [ppmv]	H$_2$ dissolved [nM]
273 (0)	0.02187	1.0	0.976
283 (10)	0.02033	1.0	0.875
288 (15)	0.01981	1.0	0.838
293 (20)	0.01941	1.0	0.807
298 (25)	0.01913	1.0	0.782
303 (30)	0.01895	1.0	0.762
308 (35)	0.01887	1.0	0.746
328 (55)	0.01935	1.0	0.719

[a] Case shown is for 1.0 ppmv. $H_{2(aqueous)} = (LP)/RT) \times 10^9$, where $H_{2(aqueous)}$ is the dissolved concentration in nM; L is the Ostwald coefficient for H$_2$ solubility; R is the universal gas constant (0.0821 liter $\times$ atm K^{-1} mol^{-1}); P is the pressure (atm); and T is the temperature in °K (Wilhelm *et al.*, 1977).
[b] Ostwald coefficients vary with temperature.

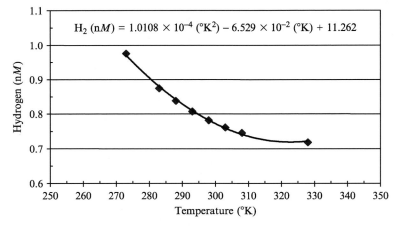

$$H_2 \text{ (nM)} = 1.0108 \times 10^{-4} \text{ (°K}^2) - 6.529 \times 10^{-2} \text{ (°K)} + 11.262$$

FIG. 3. Correlation between temperature (degrees Kelvin; 273°K = 0°) and aqueous H$_2$ concentration at a H$_2$ gas phase concentration of 1.0 ppmv (0.1013 Pa). Data fit was achieved with a second-order polynomial equation with an $r^2 = 0.998$ [Eq. (2)]. Equation (2) can be used to approximate aqueous H$_2$ concentration when the Ostwald coefficient is unknown.

temperature and aqueous H$_2$ concentration with a H$_2$ gas phase partial pressure of 1.0 ppmv is shown in Fig. 3. The following second-order equation derived from the plot of headspace H$_2$ concentration versus T (°K) can be used to approximate aqueous H$_2$ concentrations when the Ostwald coefficient is unknown.

$$[H_2](nM) = [H_2](ppmv) \times [1.0108 \times 10^{-4}(°K^2) \\ -6.52899 \times 10^{-2}(°K) + 11.262] \tag{2}$$

Note that there is an approximate equivalency between 1.0 ppmv (in the gas phase) and 1 nM (in the dissolved phase), particularly at colder temperatures (10–15°), like those typically observed in subsurface (i.e., groundwater) environments.

Case Study: H_2 Consumption Threshold Concentrations Linked to Metabolic Reductive Dechlorination (Chlororespiration)

The threshold model predicts that chlororespiring bacteria consume H_2 below those threshold concentrations observed for acetogens, methanogens, and sulfidogens (Lovley and Goodwin, 1988; Löffler et al., 1999). This model was confirmed for numerous pure and mixed cultures that use chloroorganic compounds as terminal electron acceptors. Figure 4 shows typical H_2 utilization profiles for two different bacterial species: *Anaeromyxobacter dehalogenans* and *Desulfitobacterium chlororespirans* (Sanford et al., 1996, 2002). Although the final threshold concentrations obtained in both cultures (0.1 and 0.5 ppmv, respectively) were similar, the rates of H_2 utilization were very different. In cultures of similar cell density, *D. chlororespirans* required 20 times longer (>200 h) to reach the H_2 consumption threshold concentrations compared with *A. dehalogenans*. These findings indicate that the H_2 consumption kinetics for metabolic reductive dechlorination is species specific and can be quite different. Although the rates of H_2 consumption do not influence the final H_2 threshold concentrations, it is important to verify that the H_2 concentrations measured are the end points of H_2 consumption. Threshold and compensatory steady-state concentrations are approached asymptotically (Fig. 1), and sequential measurements are needed to establish apparent end point concentrations. Thus, any measurements of H_2 consumption threshold or compensatory steady-state concentration should reflect at least three sequential time points with little or no variation so that slow H_2 uptake or H_2 production rates can be delineated.

The fundamental biochemical principle that makes trace H_2 measurements a valuable tool is that the majority of microorganisms studied to date possess hydrogenases that are associated with both consumption (uptake hydrogenases) and generation of H_2. Not all populations, however, possess hydrogenases and consume and/or produce H_2. For instance, *Desulfuromonas michiganensis* strain BB1 cannot utilize H_2 as an electron donor but couples tetrachloroethene (PCE) reductive dechlorination to

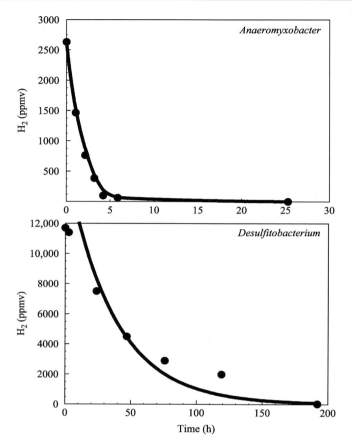

Fig. 4. Comparison of H_2 uptake rates between *Anaeromyxobacter dehalogenans* strain 2CP-C and *Desulfitobacterium chlororespirans* strain Co23 grown with 2-chlorophenol and 3-chloro-4-hydroxybenzoate, respectively, as metabolic electron acceptors. Note that the X axis scale differs between the two graphs.

acetate oxidation (Sung *et al.*, 2003). Strain BB1, however, generates H_2 from acetate oxidation under electron acceptor (i.e., PCE)-limiting conditions (Fig. 5). Similar observations were made with *Geobacter sulfurreducens* when acetate oxidation and H_2 production occurred under iron-limiting conditions (Cord-Ruwisch *et al.*, 1998). Thus, ecosystems dominated by acetotrophic, H_2-forming bacteria may have H_2 concentrations that are higher than predicted by thermodynamics of the dominant TEAP alone.

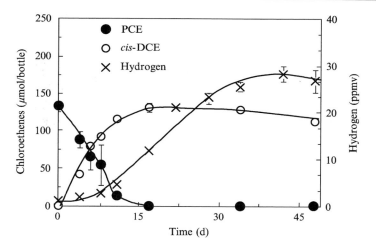

FIG. 5. Generation of H_2 from acetate oxidation by *Desulfuromonas michiganensis* strain BB1 under electron acceptor (PCE-)-limiting conditions. The final H_2 concentration exceeds the expected steady-state level for metabolic reductive dechlorination by two orders of magnitude.

Conclusions

The ability to obtain accurate H_2 concentration measurements from microbial cultures and environmental samples (e.g., groundwater) facilitates analyzing and monitoring fundamental metabolic processes greatly (i.e., TEAPs). Typically, the H_2 concentration reflects the compensatory steady state: a balance between the thermodynamic consumption threshold (rate of utilization) and H_2 generation (rate of production). This complex balance of H_2 consumption and generation produces steady-state concentrations that correlate with the ΔG of the predominant TEAP in most environmental systems. Thermodynamic H_2 consumption threshold concentrations are usually not observed in environmental systems due to the presence of reduced organic compounds, but can be determined in defined systems (i.e., pure cultures).

Acknowledgments

The authors are indebted to the Center for Microbial Ecology at Michigan State University for providing opportunities and inspiration for starting this research on bacterial hydrogen metabolism and acknowledge NSF for receiving continuation funds (CAREER award to FEL).

References

Breznak, J. A. (1994). Acetogenesis from carbon dioxide in termite guts. *In* "Acetogenesis" (H. L. Drake, ed.), pp. 303–330. Chapman & Hall, New York.

Chapelle, F. H., McMahon, P. B., Dubrovsky, N. M., Fujii, R. F., Oaksford, E. T., and Vroblesky, D. A. (1995). Deducing the distribution of terminal electron-accepting processes in hydrologically diverse groundwater systems. *Water Res. Res.* **31,** 359–371.

Chapelle, F. H., Vroblesky, D. A., Woodward, J. C., and Lovley, D. R. (1997). Practical considerations for measuring hydrogen concentrations in groundwater. *Environ. Sci. Tech.* **31,** 2873–2877.

Conrad, R. (1996). Soil microorganisms as controllers of atmospheric trace gases (H_2, CO, CH_4, OCS, N2O, and NO). *Microbiol. Rev.* **60,** 609–640.

Conrad, R., Aragno, M., and Seiler, W. (1983). The inability of hydrogen bacteria to utilize atmospheric hydrogen is due to threshold and affinity for hydrogen. *FEMS Microbiol. Lett.* **18,** 207–210.

Cord-Ruwisch, R., Lovley, D. R., and Schink, B. (1998). Growth of *Geobacter sulfurreducens* with acetate in syntrophic cooperation with hydrogen-oxidizing anaerobic partners. *Appl. Environ. Microbiol.* **6,** 2232–2236.

Cord-Ruwisch, R., Seitz, H.-J., and Conrad, R. (1988). The capacity of hydrogenotrophic anaerobic bacteria to compete for traces of hydrogen depends on the redox potential of the terminal electron acceptor. *Arch. Microbiol.* **149,** 350–357.

Dolfing, J., and Harrison, B. K. (1992). Gibbs free energy of formation of halogenated aromatic compounds and their potential role as electron acceptors in anaerobic environments. *Environ. Sci. Technol.* **26,** 2213–2218.

Dolfing, J., and Janssen, D. B. (1994). Estimates of Gibbs free energies of formation of chlorinated aliphatic compounds. *Biodegradation* **5,** 21–28.

Fauque, G., Peck, H. D., Jr., Moura, J. J. G., Huynh, B. H., Berlier, Y., Der Vartanian, D. V., Teixeira, M., Przybyla, A. E., Lespinat, P. A., Moura, I., and Le Gall, J. (1988). The three classes of hydrogenases from sulfate-reducing bacteria of the genus *Desulfovibrio*. *FEMS Microbiol. Rev.* **54,** 299–344.

Hoehler, T. M., Alperin, M. J., Albert, D. B., and Martens, C. S. (1998). Thermodynamic control of hydrogen concentrations in anoxic sediments. *Geochim. Cosmochim. Acta* **62,** 1745–1756.

Lee, M. J., and Zinder, S. H. (1988). Hydrogen partial pressures in a thermophilic acetate-oxidizing methanogenic coculture. *Appl. Environ. Microbiol.* **54,** 1457–1461.

Löffler, F. E., Tiedje, J. M., and Sanford, R. A. (1999). Fraction of electrons consumed in electron acceptor reduction and hydrogen thresholds as indicators of halorespiratory physiology. *Appl. Environ. Microbiol.* **65,** 4049–4056.

Lovley, D. R. (1985). Minimum threshold for hydrogen metabolism in methanogenic bacteria. *Appl. Environ. Microbiol.* **49,** 1530–1531.

Lovley, D. R., Chapelle, F. H., and Woodward, J. C. (1994). Use of dissolved H_2 concentrations to determine distribution of microbially catalyzed redox reactions in anoxic groundwater. *Environ. Sci. Technol.* **28,** 1205–1210.

Lovley, D. R., and Goodwin, S. (1988). Hydrogen concentrations as an indicator of the predominant terminal electron-accepting reactions in aquatic sediments. *Geochim. Cosmochim. Acta* **52,** 2993–3003.

Lu, X. X., Tao, S., Bosma, T., and Gerritse, J. (2001). Characteristic hydrogen concentrations for various redox processes in batch study. *J. Environ. Sci. Health A Tox. Haz. Subst. Environ. Eng.* **36,** 1725–1734.

Luijten, M. L., Roelofsen, W., Langenhoff, A. A., Schraa, G., and Stams, A. J. (2004). Hydrogen threshold concentrations in pure cultures of halorespiring bacteria and at a site polluted with chlorinated ethenes. *Environ. Microbiol.* **6**, 646–650.

McInnes, D. M., and Kampbell, D. (2000). The bubble stripping method for determining dissolved hydrogen (H_2) in well water. *Field Anal. Chem. Technol.* **4**, 283–296.

Sanford, R. A., Cole, J. R., Loffler, F. E., and Tiedje, J. M. (1996). Characterization of *Desulfitobacterium chlororespirans* sp. nov., which grows by coupling the oxidation of lactate to the reductive dechlorination of 3-chloro-4-hydroxybenzoate. *Appl. Environ. Microbiol.* **62**, 3800–3808.

Sanford, R. A., Cole, J. R., and Tiedje, J. M. (2002). Characterization and description of *Anaeromyxobacter dehalogenans* gen. nov., sp. nov., an aryl-halorespiring facultative anaerobic myxobacterium. *Appl. Environ. Microbiol.* **68**, 893–900.

Schink, B., and Friedrich, M. (1994). Energetics of syntrophic fatty acid oxidation. *FEMS Microbiol. Rev.* **15**, 85–94.

Sung, Y., Ritalahti, K. M., Sanford, R. A., Urbance, J. W., Flynn, S. J., Tiedje, J. M., and Löffler, F. E. (2003). Characterization of two tetrachloroethene-reducing, acetate-oxidizing anaerobic bacteria and their description as *Desulfuromonas michiganensis* sp. nov. *Appl. Environ. Microbiol.* **69**, 2964–2974.

Thauer, R. K., Jungermann, K., and Decker, K. (1977). Energy conservation in chemotrophic anaerobes. *Bacteriol. Rev.* **41**, 100–180.

Wilhelm, E., Battino, R., and Wilcock, R. J. (1977). Low pressure solubility of gases in liquid water. *Chem. Rev.* **77**, 219–262.

Wolin, M. J. (1982). Hydrogen transfer in microbial communities. *In* "Microbial Interactions And Communities" (A. T. Bull and J. H. Slater, eds.), Vol. 1, pp. 323–356. Academic Press, London.

Wolin, M. J., and Miller, T. L. (1982). Interspecies hydrogen transfer: 15 years later. *ASM News* **48**, 561–565.

Zehnder, A. J. B., and Stumm, W. (1988). Geochemistry and biogeochemistry of anaerobic habitats. *In* "Biology of Anaerobic Microorganisms" (A. J. B. Zehnder, ed.), pp. 1–38. Wiley, New York.

Zinder, S. H. (1994). Syntrophic acetate oxidation and "reversible acetogenesis." *In* "Acetogenesis" (H. L. Drake, ed.), pp. 386–415. Chapman & Hall, New York.

[14]   Advances in Microscopy: Microautoradiography of Single Cells

By Jeppe Lund Nielsen and Per Halkjær Nielsen

Abstract

Microautoradiography (MAR) is an efficient method to obtain reliable information about the ecophysiology of microorganisms at the single cell level in mixed communities. Data obtained by the traditional MAR method can now be improved significantly when MAR is combined with fluorescence *in situ* hybridization (FISH) with oligonucleotide probes for the identification of target organisms. This chapter discusses how to use

METHODS IN ENZYMOLOGY, VOL. 397   0076-6879/05 $35.00
DOI: 10.1016/S0076-6879(05)97014-6

MAR-FISH in various ecosystems with emphasis on the type of information to be obtained, incubation conditions, and detailed protocols for the MAR technique. It also discusses new MAR applications and the type of information that can be obtained, e.g., the use of dual labeling to investigate the simultaneous uptake of several substrates by individual prokaryotes.

Introduction

The microbial consortia in natural or engineered systems typically consist of many groups of prokaryotes, many of which are still unidentified. Our understanding of their role and function in the environment is even inferior. We have only a very limited understanding of a few distinct groups, very often those with easily recognizable morphologies or specialized activities. A traditional way of obtaining information on the roles of microbes in the environment is to isolate and cultivate the microorganisms using more or less selective media. However, only a small proportion of the microorganisms may be cultivated, and if they are, their physiology and activity as investigated under culture conditions may not necessarily reflect the activities and functions in the environment. The application of molecular techniques has revealed the identity of many uncultured organisms in complex environmental systems, which cannot be detected by direct cultivation approaches. Therefore, culture-independent techniques are compulsory in order to investigate the physiology and ecology of these uncultivated microorganisms.

A number of new staining methods and molecular methods have been applied to provide information about the activity of prokaryotes directly in complex microbial systems, and microautoradiography is probably the most efficient and most versatile method to date. MAR has been used in microbial ecology for many years (Brock and Brock, 1966, 1968), and since it was first applied in combination with fluorescence *in situ* hybridization for identification of the microorganisms as described by Lee *et al.* (1999) and Ouverney and Fuhrman (1999), MAR has been revived and applied in numerous approaches and combinations. The combined technique has been described with several acronyms, such as MAR-FISH, STARFISH, and MicroFISH (Cottrell and Kirchman, 2000; Lee *et al.*, 1999; Ouverney and Fuhrman, 1999), but the basic principle is identical.

This article describes the microautoradiographic technique and how to apply it for investigating the physiology and ecology of cultured or uncultured single prokaryotic cells directly in complex microbial consortia. We focus on two important aspects: incubation conditions, which are critical for the type of information to be obtained about the physiology of the microorganisms investigated, and protocols for performing MAR-FISH, which has to be done very carefully. The basic theory of MAR has not been included, but it can be found elsewhere (e.g., Carman, 1993; Rogers, 1979).

Type of Information Obtained by MAR-FISH

MAR has been used in combination with various simple staining techniques but without combination with FISH to characterize microbial activity in mixed communities (Andreasen and Nielsen, 1997; Meyer-Reil, 1978; Tabor and Neihof, 1982). However, the information obtained cannot be related to specific microorganisms without them having a very atypical activity or morphology. In some studies the number of active cells assimilating a specific compound (e.g., leucine, thymidine, acetate, glucose, N-acetylglucosamine, cAMP) or a mixture of radiolabeled organic substrates (e.g., amino acids) is obtained. Also, the number of phytoplankton or autotrophic bacteria incorporating labeled bicarbonate can be quantified. By using specific incubation conditions and inhibitors it is possible to enumerate various functional bacterial groups in more detail. In activated sludge, for example, denitrifiers, sulfate reducers, polyphosphate-accumulating bacteria, and methanogens have been enumerated using MAR in combination with 4',6-diamino-2-phenylindole, dihydrochloride (DAPI) staining for all procaryotes (Nielsen and Nielsen, 2002a,b).

When MAR is combined with FISH, the method becomes much more useful for detailed studies of the ecophysiology of probe-defined microorganisms, identification of MAR-positive bacteria, and determination of degradation pathways in microbial communities. Examples of information that can be obtained are presented next.

Ecophysiology of Probe-Defined Microorganisms

Investigations into specific probe-defined microorganisms have provided information about which organic substrates that can be taken up (e.g., Ouverney and Fuhrman, 1999; Kong *et al.*, 2004), autotrophic or mixotrophic activity (e.g., Gray *et al.*, 2000; Daims *et al.*, 2001; Nielsen *et al.*, 2000), activity under various electron acceptor conditions and response to inhibitors (e.g., Ito *et al.*, 2002), and uptake of inorganic phosphate (e.g., Lee *et al.*, 1999; Kong *et al.*, 2004). Also, quantitative kinetic parameters such as cell-specific uptake rates and substrate affinity constants (K_s) can be obtained (Nielsen *et al.*, 2003). Information about the physiology of microorganisms of interest may be used to design efficient isolation strategies.

Partial Identification of MAR-Positive Bacteria

If unknown bacteria degrade a certain labeled compound, e.g., an environmental pollutant (Yang *et al.*, 2003), the identity can be partly or fully resolved by applying various FISH probes. An example is the partial identification of Fe(III) reducers in activated sludge (Nielsen *et al.*, 2002). If the specific identity of the organism is desired, such information can greatly reduce the work with analysis of clone libraries and gene probe design (Kong *et al.*, 2005).

Overview of Degradation Pathways of Organic Matter in a Microbial Community

By investigating an ecosystem with a number of labeled substrates assumed to be central to the degradation of organic matter and combining this with a population structure investigation, it is possible to obtain important information about structure and function of an entire ecosystem. As an example, it was shown that planktonic *Archaea* play an important role in amino acid turnover in marine systems (Overney and Fuhrman, 2000; Teira *et al.*, 2004). Kindaichi *et al.* (2004) studied interactions between nitrifying bacteria and heterotrophic bacteria in autotrophic nitrifying biofilms, where they identified several groups of heterotrophs that could grow on compounds excreted by the autotrophs.

The Microautoradiographic Procedure

Overview of the Method

The original procedure in MAR-FISH for incubations of a microbial sample is outlined in Fig. 1. The sample is incubated under defined conditions for some time, typically a few hours, to ensure that a sufficient amount of radiotracer is taken up by the active cells. Several factors should be considered in the design of a proper sample incubation: type and specific activity of the radiotracer, concentration of unlabeled substrate, concentrations of electron acceptors (oxygen, nitrate, or others), inhibitors, biomass concentration, temperature, and incubation time. It is advisable to measure substrate consumption and incorporation into the biomass to ensure optimal incubation conditions. After incubation, the samples are fixed in paraformaldehyde or ethanol (to allow FISH analysis of gram-negative and gram-positive cells, respectively) and washed to remove surplus radiotracer. Homogenization, cryosectioning, dilution, or concentration of the sample before it is mounted on a cover glass, a glass slide, or a polycarbonate filter may be necessary for optimal visualization of the MAR-FISH signal. For FISH, the standard protocols can be followed using 5(6)-carboxyfluorescein-N-hydroxysuccinimide ester (FLUOS)-, and Cy3-labeled oligonucleotide probes before the radiation-sensitive emulsion is applied. The emulsion on glass slides is exposed some days before being

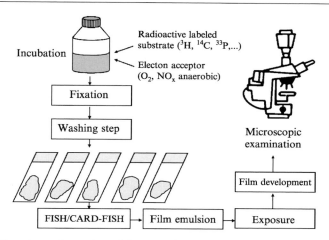

FIG. 1. Experimental overview of the procedure for conducting a microautoradiography experiment on a sample from a complex microbial community. (See color insert.)

developed using standard photographic methods. MAR-positive and -negative bacteria can be examined by combining bright field or phase contrast with epifluorescence microscopy or confocal laser-scanning microscopy.

Sample Pretreatment

The sample should preferably be collected just prior to incubation, as storage may alter the activities of the indigenous organisms in the sample. Depending on the nature of the sample and the subject of investigation, storage for 1–2 days at 4° may be applied. Also, based on the composition and nature of the system investigated, it might be advisable just prior to incubation to aerate the sample to ensure the removal of easily degradable organic matter. When examining anaerobic activity, this preincubation step can be replaced with an anoxic step to remove nitrate or other electron acceptors. In a few special cases, the background level of electron donor or other secondary metabolites might be too high to ensure a proper design of the incubation, and special care, such as washing the sample, might be necessary (Gray *et al.*, 2000). When examining certain systems such as aggregates (e.g., biofilms), physical pretreatment (homogenization and/or solubilization) might be needed to avoid diffusional limitations of substrate or electron acceptor (Nielsen and Nielsen, 2005).

Sample Incubation

Sample incubation has to be in accordance with the type of information expected from the MAR experiments, and there are several important

TABLE I

IMPORTANT PARAMETERS TO CONSIDER FOR OPTIMAL INCUBATION OF SAMPLES FOR
MAR EXPERIMENTS

Parameter	Consider for optimal incubation
Type of radiotracer	Energy of tracer (high energy gives lower resolution of MAR)
	Labeling position in tracer (different atoms in substrate molecule may be transformed differently)
Addition of unlabeled substrate	Final substrate concentration (labeled + unlabeled substrate). If possible, it should be well above half-saturation constant (K_s) during the entire incubation period
Specific activity of the labeled compound in sample (μCi/mol)	High specific activity aims at cells with low activity or low yield (incorporation) and vice versa
Electron acceptor(s)	Presence of oxygen, nitrate, nitrite, sulfate, or others. Avoid toxicities and depletion. Maintain the level during incubation
Incubation time	As short as possible (a few hours) and no longer than the time of depletion of substrate/electron acceptor
Inhibitors	Inhibition of other influencing metabolisms
Volume and biomass concentration	Ensure sufficient number of cells

parameters to consider regarding incubation of the sample with radiotracer (Table I). Different information can be obtained by MAR (see earlier discussion), and it is important to detail this question and determine how to examine a specific question in a given system. The radiotracer is chosen based on the isotope (typically ^{3}H or ^{14}C labeling of substrates) and often on the location of the labeled atom(s) in the organic molecule (e.g., C1 or C2 labeling of acetate). Labeling should preferably be on an atom, which is incorporated directly into the biomass. Because the energy of the isotope is related inversely to the resolution of the resulting MAR image, the lower the energy, the higher the resolution. In contradiction to this, isotopes with higher energy yield higher sensitivity and higher dynamic range valuable to quantitative examinations. Our experience is that tritium is generally preferred due to the higher resolution, better than 1 μm (see example in Fig. 3). If additional studies of transformation of organic matter are carried out by quantitative analyzing CO_2 production and incorporation of carbon into the biomass, ^{14}C labeling is preferable. However, instead of using the same samples for MAR and transformation studies, it is often better to run separate parallel incubations, as lower tracer amounts are needed for transformations studies compared to MAR incubations.

Each radiotracer can be purchased with different specific radioactivities (Ci/mmol) and concentrations (μCi/ml). It is, however, not very important which one is chosen, as it often only constitutes a minor part of the total amount of substrate in the incubation (because of added nonlabeled substrate). Only if the final concentration during the incubation must be very small (only in systems with very low microbial activity, so the substrate is not depleted during incubation), a high specific activity of the radiotracer is needed so that the amount added to the sample can be kept low without losing the wanted level of radioactivity. In other cases, the purchased tracer is adjusted with unlabeled substrate to obtain the specific activity and concentration wanted.

In order to optimize the incubation conditions for each type of microbial sample, it is usually important to test different amounts of the specific activity of the tracer, substrate concentration, biomass concentration, and incubation time. Usually, the first step is to fix the incubation time, which is dependent on the system, but typically is between 1 and 6 h. If the incubation time is too long (too many hours, days), there is a risk of population growth and of uptake of metabolites by other bacteria than the target population.

The second step is to decide on a proper substrate concentration (unlabeled plus labeled substrate). The concentration should be sufficient to avoid substrate depletion during incubation. Levels well above the value of the half-saturation constant (K_s) for the target organisms should usually be selected to get maximum uptake of all target bacteria and are thus highly dependent on the system investigated. However, K_s is hardly known in most systems, so a substrate concentration range based on microbial activity is often used. For highly active systems (biofilms, wastewater treatments systems, etc.), a concentration in the upper range of normal concentrations for the actual environment is selected, typically 0.1–2 mM. For systems with low activity (e.g., many marine environments), 1–50 nM or sometimes lower concentrations are used (e.g., Cottrell and Kirchman, 2000; Ouverney and Fuhrman, 2000). Given this, the third step is to adjust the biomass concentration in such a way that the substrate is not depleted during the incubation period. In high-rate systems we usually apply a biomass concentration of 0.1–1 g suspended solids/liter or 0.06–0.6 g organic matter/liter. This is based on our experience with substrate uptake rates in the range of 0.5–2 g substrate/g organic matter/hour. Dilution of biomass is made by filtered water from the same sample. A high dilution would ensure that the substrate is not depleted during incubation, but means that a larger sample volume is needed in order to get enough biomass to perform the subsequent MAR procedure. Normal incubation volumes are 2- to 5-ml samples. Larger incubation volumes increase the need for radiotracer accordingly.

The tracer must be diluted properly with unlabeled substrate to get an overall specific activity of the tracer, which ensures that only active cells

TABLE II

TYPICAL SETUP FOR TWO MAR-FISH EXPERIMENTS OF A HIGH-ACTIVITY AND A LOW-ACTIVITY
SYSTEM FOR THE USE OF ACETATE UPTAKE UNDER AEROBIC CONDITIONS

Parameter	Low activity	High activity
Type of radiotracer	[^{3}H]acetate (Na)	[^{3}H]acetate (Na)
Specific activity (tracer)	50–64,000 mCi/mmol	50–64,000 mCi/mmol
Addition of unlabeled acetate	To a final concentration of 0.5–50 nM	To a final concentration of 2 mM
Specific activity (overall in sample)	10^4–10^6 μCi/μmol	2 μCi/μmol
Electron acceptor	Oxygen (provided by airspace in bottle)	Oxygen (provided by airspace in bottle)
Incubation time	6–24 h	3 h
Temperature	20°	20°
Inhibitors	None	None
Biomass concentration	10^6–10^7 cells/ml	0.5 g organic matter/liter $\sim$$10^9$ cells/ml
Volume	5 ml	2 ml
Exposure time	3–14 days	3–4 days

take up sufficient substrate during the incubation in order to be visualized as MAR-positive cells. The amount of tracer required to detect one active cell is approximately 10^{-15} Ci ($3 \cdot 10^{-5}$ Bq), if the sample is exposed 4–6 days during the MAR procedure.

Table II shows two typical experimental conditions: a highly active microbial system (activated sludge) and a low activity system (marine sample). The length of the incubation must be adjusted to ensure a measurable uptake and should be based on knowledge obtained from control experiments from the same system. For anaerobic experiments it is important to apply strict anaerobic techniques, e.g., by removing oxygen from the bottles by flushing/evacuation several times with oxygen-free nitrogen. Trace amounts of oxygen in high-rate systems can be removed by introducing a preincubation step with an additional electron donor, but without addition of a radioactive tracer. The substrates and tracers are added using strictly anaerobic techniques for anaerobic experiments. A problem often encountered is the presence of unwanted nitrate or nitrite, which can also be removed similarly to oxygen by introducing a preincubation step. This also works best in high-rate systems. All sample vials should be sufficiently mixed during incubations by horizontal placement on a rotary table.

A preincubation step with an unlabeled substrate should always be applied for all incubations with electron acceptors other than oxygen (e.g., nitrate, nitrite, sulfate). An anaerobic preincubation of 1–2 h is important, as many aerobic bacteria take up labeled substrate the first 0.5–1 hour after changing from aerobic to anoxic conditions—possibly because there were traces of oxygen present—and that they were unable to take up the same substrate after 1 or 2 h preincubation with unlabeled substrate. A preincubation can also be used to determine whether assimilation of a certain substrate is taken up for storage or for further metabolic activity, e.g., growth (e.g., Andreasen and Nielsen, 2000; Kong *et al.*, 2004; Kragelund *et al.*, 2005). For instance, some bacteria have a certain capacity to assimilate and store substrate under anaerobic conditions, but are unable to grow. A preincubation of 3–4 h with an unlabeled substrate saturates the storage capacity, and by adding labeled substrate after this preincubation step, no uptake can be found under anaerobic conditions. The addition of substrate only from the start of the incubation period would give positive uptake (and MAR signal) but not show whether the organisms used the substrate for storage or actually metabolized the substrate.

Under certain, but rare, conditions, chemical or physical factors can cause silver grain formation in the absence of radioactivity, also called chemography, thus a negative control must be included. A pasteurized sample (heated to 80° for 10 min) is suitable and must be prepared just prior to the experiment to avoid germination of bacterial spores. Changes in pH, temperature, and other physicochemical factors should be checked before and after each type of incubation.

Termination of Incubation, Fixation, Wash, and Staining

Incubation of samples is terminated by fixing the biomass by adding freshly prepared 8% paraformaldehyde (PFA) solution in phosphate-buffered saline (PBS) (130 mM sodium chloride, 10 mM sodium phosphate buffer, pH 7.2) directly to the incubation vials (final concentration 4%). For fixation of gram-positive bacteria, a final concentration of 50% EtOH/PBS is used for termination. In anaerobic incubations the fixative is added using a syringe and needle. The samples are fixed for 1–3 h at 4°.

In experiments with sufficient biomass, the samples are pelleted by centrifugation and are washed three times by resuspension and centrifugation using filtered (pore size 0.22 μm) tap water. In cases of limited biomass, the sample can be concentrated by filtering the sample through a polycarbonate filter (pore size 0.22 μm) and subsequently washing by filtering prefiltered water through the filter. Biomass attached to the filter can either be removed from the filter by backwashing or by detachment using 3-aminopropyltriethoxysilane-treated coverslips (see later). Alternatively, the sample can be centrifuged (10,000g for 10 min), and 90–95% of the supernatant is removed carefully by pipetting. This can be repeated a number of times until most unbound tracer is washed away. The advantage of nonfiltered samples is that they can be immobilized easily on cover glass. Cover glass allows a good visualization through the cover glass so any stain or hybridization underneath the MAR film emulsion is still fully visible. If visualizing through the film, the silver grains of MAR-positive cells typically block the staining signal. Filtered samples can be visualized by peeling off the exposed film and then placing it on a glass slide and subsequently staining it. Another approach uses the adhesiveness of an aminoethoxysilane-treated slide and, by assuring close contact between cells and the glass surface with clamps, the cells can be transferred from the filter to a cover glass surface (Cottrell and Kirchman, 2000). Efficient transfer of cells from a filter surface must be examined in every case by performing controls, typically carried out by DAPI or acridine orange staining of the cells on the filter after transfer.

The washed sample biomass can be immobilized on hydrophilic (gelatin-coated) glass slides or cover glasses (see later) and air dried on the bench or in a 60° oven, which in our experience slightly increases the attachment of single cells to the gelatin-coated surface. Gelatin coating is performed using cover glasses washed thoroughly in boiling water containing dish detergent for 5 min and rinsed in distilled water. After drying, the slides are dipped in a jar with a 0.1% gelatin, 0.01% $CrK(SO_4)_2$ suspension heated to 70° in which it is allowed to stand for 5 min before air drying overnight at room temperature in a vertical position.

Most often around 20 μl of the fixed, washed biomass is placed on the slide/cover glass and then dispersed on the glass surface using the side of a pipette tip. The remaining biomass from the sample can be stored at $-20°$ in 50% EtOH/PBS until later use. Stored samples containing PBS should be rinsed carefully in distilled water after immobilization to avoid autofluorescence from precipitated salt crystals and film-damaging effects.

Before immobilization of the biomass to the coverslips, the samples may be treated to improve visualization of single cells. If aggregate-forming bacteria are to be investigated, the sample must often be homogenized or sectioned into thin slices by cryosectioning to be able to distinguish the source of the MAR signal. The sample is embedded in cryosectioning material (Tissue-Tek OCT; Miles, Elkhart, IN) and frozen before slicing it into a suitable thickness (a few micrometers and up to 20 μm) and transferring it to a glass surface. Embedding material can be washed away easily prior to further treatment without any significant loss of cells. Alternatively, gentle use of a glass tissue grinder can often disrupt the microbial aggregate without destroying the microcolonies and without lysing the cells. Planktonic cells and filamentous bacteria outside the aggregates can be investigated without any further treatment. Special staining procedures (such as FISH, DAPI, gram, and Neisser staining) can be made at this point.

Fluorescence in Situ Hybridization

The MAR-FISH combination can be performed by hybridizing the sample directly on the cover glass. Hybridizations of the fixed cells are carried out using fluorescently labeled oligonucleotides targeting specific ribosomal RNA sequences. We recommend the use of probes labeled with the sulfoindocyanine dye Cy3, which usually gives a high fluorescence intensity, but FLUOS can also be used. In our experience, Cy5 (far red) has a too small Stoke shift and can easily be mistaken with the reflection of the emitting wavelength from the silver grains.

A sequential dehydration in series of ethanol solutions (50, 80, and 96% EtOH; 2 min in each) allows a better visualization of thick samples, a better mobilization of the probe through the cell walls, and helps in removing traces of embedding material when cryosectioning is applied. After the cover glasses have dried, 8 μl of hybridization buffer with the required formamide concentration is mixed with 1 μl (50 ng/μl) of each probe in the center of the cover glass. The sample is then hybridized at 46° for 1.5 h in a moisture chamber (e.g., a 50-ml plastic Falcon tube containing a tissue paper moistened with the applied hybridization buffer), taking care to place the cover glass horizontally, maintaining the probe at the center of the cover glass. Unspecifically bound probe is washed away by dipping the cover glass into washing buffer (48° for 15 min). Salt crystals are avoided by gently submerging the cover glass in distilled water prior to air drying. After any desired staining (e.g., DAPI), the glasses are allowed to air dry.

Catalyzed Reporter Deposition–FISH (CARD-FISH)

In many natural ecosystems, the activity level of the microorganisms is so low that the normal FISH procedure provides fluorescence levels too weak to be detected. The identification of such cells can instead be examined by the significantly more sensitive CARD-FISH technique (catalyzed reporter deposition-FISH). The CARD-FISH procedure can, after some modifications, be combined easily by MAR, and such a combination has revealed that MAR can be used to determine the activity of cells with low ribosomal content (J.L. Nielsen et al., unpublished results; Teira et al., 2004).

Modifications of the CARD-FISH procedure can be carried out by immobilizing the cells on a cover glass surface instead of filters. The fixed and washed sample is transferred to a gelatin-coated cover glass and allowed to dry completely at 50°. To avoid loss of cells during the process of hybridization, a thin layer of low-melting agarose gel (0.2%) is placed onto the sample by a pipette and is processed further through this agarose layer. Permeabilization is carried out depending on the nature of the cells and should be optimized in every case. Typically, a treatment in lysozyme (10 g/liter in 50 μM EDTA, 50 μM Tris-HCl, pH 7.5) for 30 min at 37° followed by treatment with achromopeptidase (60 U/ml in 0.01 M NaCl, 0.01 M Tris-HCl) (at 37° for 30 min) is required for the horseradish peroxidase (HRP)-labeled probe to enter the cells. The cover glass is then washed in distilled water and treated with 0.1 M HCl for 10 min and washed again in distilled water. The hybridization is then performed by mixing 2 μl HRP probe (final concentration 0.05 μM) in 200 μl hybridization buffer [0.9 M NaCl, 20 mM Tris-HCl, pH 7.5, 10% (w/v) dextran sulfate, 0.02% (w/v) sodium dodecyl sulfate, 1% blocking reagent (Boehringer Mannheim, Mannheim, Germany)] and formamide according to the probe used. Hybridization takes place at least 2 h at 35° followed by a 10-min washing step at 37° in hybridization buffer. At this step, it is important to keep the slides moist throughout the hybridization and amplification procedure. Amplification of the tyramide signal is carried out by incubating the slide in a 50-ml tube with PBS and 0.05% Triton X-100 for 15 min at room temperature. After removal of excess buffer, the slide is incubated for 30 min at 37° in the dark after the addition of 5 μl Cy3-tyramide (1 mg/ml) and amplification buffer [10% (w/v) dextran sulfate, 2 M NaCl, 0.1% (w/v) blocking reagent, and 0.0015% H_2O_2 in PBS]. The slide is then washed in 50 ml PBS containing 0.05% Triton X-100 and is subsequently rinsed in distilled water, both at room temperature. Following this procedure, the slide is processed with film emulsion as described later.

Autoradiographic Procedure

The following steps are performed in a darkroom with safe light for black and white film development procedures. The film emulsion (LM-1, Amersham Biotech) (usually diluted by the addition of 1 part distilled water and 9 parts film emulsion) is placed in a water bath at 43° and is shaken gently every 2 min for 10 min to avoid air bubbles in the emulsion. The emulsion is poured carefully into a dipping vessel to a level of approximately 4 cm. The glass slides/cover glasses are dipped carefully in the emulsion for 5 s. The slides are then placed vertically on a folded tissue paper for 5 s, and the side without the sample is cleaned with a piece of tissue paper. The slides are placed horizontally on a plastic tray, and the emulsion is allowed to solidify for 2 h in the dark before they are placed in a slide box with water-free silica gel. The box is covered with aluminum foil and exposed at 4°. The exposure time is normally between 2 and 10 days, depending on the sample, labeling, incubation, etc. To ensure sufficient exposure time for both active and less active cells to appear MAR-positive, variable exposure length must be compared regarding the number of MAR-positive cells. For development of the film, the slides are placed in the developer (Kodak D19, 40 g/liter) for 3.0 min. They are drip dried before being placed in distilled water for 1 min. The slides are placed in fixer (Na-thiosulfate, 300 g/liter) for 4 min. To remove the fixer, the slides are washed in distilled water for 3x3 min. The slides must air dry at least 4 h before microscopic examination. Storage in a refrigerator preserves the slides for at least 1 year.

Microscopic Evaluation

The microscopic evaluation of MAR can be performed using either bright-field microscopy (either DIC or phase-contrast microscopy work ideally) or a laser-scanning microscope (LSM) using the transmittent light mode. We use two different systems: an upright microscope stand, Zeiss Axioskop 2 plus and a Zeiss LSM 510 Meta (Carl Zeiss, Oberkochen, Germany) mounted on an inverted Axioplane microscope stand with UV laser (Coherent Enterprise; Coherent, Inc., Santa Clara, CA), an Ar-Kr laser (458 and 488 nm) and two He-Ne lasers (543 and 633 nm). MAR signals are best visualized by the laser-scanning microscope without oil, but in order to avoid fading of the fluorescence signal, we normally add a small drop of antifading agent (e.g., Citifluor or Vectashield) under a cover glass. An inverted microscope stand facilitates visualization, whereas an upright microscope requires mounting and immobilization of the coverslide onto a slide using tape. LSM images can be obtained using the standard software package (Zeiss LSM 510 v. 3.02), and epifluorescence images are captured using a charge-coupled device camera (CoolSnap HQ; Photometrics, Tucson, AZ). Confocal LSM images yield higher intensities of the fluorescence signal and increase the resolution by removing out-of-focus fluorescence. The MAR signal cannot be recorded confocally, but is best visualized by applying the transmittent mode. By using a normal epifluorescence microscope without an inverted stand, the cover glasses are placed upside down on a glass slide coated with the antifading agent. In this way, the silver grains are underneath the cells rather than covering them, making it easier to visualize the FISH signal.

In order to find the optimal exposure times, several extra glass slides should always be included for daily inspection of the silver grain densities. In this case, it is easiest to combine DAPI staining with the inspection of the MAR signal by switching between bright field and fluorescence microscopy. Enumeration of the number of MAR-positive cells and the amount of silver grains should be evaluated in order to ensure a proper exposure time. Too brief an exposure will only yield the most active organisms in the sample, whereas too lengthy exposures will increase background signals, thus decreasing resolution of the MAR technique.

Usually, the amount of silver grains on top of the cells is assessed in qualitative terms as MAR-negative or MAR-positive. A clear MAR-positive result should be easy to distinguish from the background level, and we often use at least four to five silver grains located on top or close to a single cell as a guideline (with background levels below 0.01 silver grains per μm^2). Examples of MAR combined with the normal FISH procedure and the CARD-FISH procedure for identification of the microorganisms are shown in Fig. 2. Furthermore, a real quantification of the MAR signal can give valuable information about the substrate uptake by the individual cells (Nielsen et al., 2000; Pearl and Stull, 1979). A proper quantification, however, requires procedures that ensure prevention of leakage from the cells and use of internal standards. The procedure for quantitative MAR (Q-MAR) is described in detail elsewhere (Nielsen et al., 2003).

Example of Application: Dual Uptake of Substrates

In some studies the simultaneous uptake of two substrates of probe-defined populations has been investigated under in situ conditions (Kong et al., 2004). Such investigations can be carried out by adding both substrates simultaneously, but with only one (and different) compound labeled in parallel incubations. If the probe-defined cells are MAR-positive under both conditions, the cells are able to take up both substrates simultaneously. However, in cases where specific probes are not available or if only a minor part of the population is active, this approach does not work. Instead, it is possible to use dual labeling by applying a weak and a strong isotope for labeling of the substrates. Because the half distance for production of silver grains is different for the two isotopes (less than 1 μm and 2–3 μm for ^3H and ^{14}C, respectively), it is possible to see whether a certain cell has taken up ^3H- or ^{14}C-labeled substrates or if it has taken up both.

Figure 3 shows such an example. Filamentous bacteria and single cells from an activated sludge system were investigated for simultaneous uptake of [^3H]acetate and [^{14}C]propionate. Both filamentous bacteria, hybridizing with the probe for "Candidatus Meganema perideroedes" (belonging to the Alphaproteobacteria, Nielsen et al., 2003), and many unidentified single cells were able to take up both substrates simultaneously. However, several unidentified single cells and some filamentous bacteria hybridizing with a probe for Cytophaga (CF319a) could assimilate only acetate and not propionate. In order to obtain such images as shown in Fig. 3, the specific

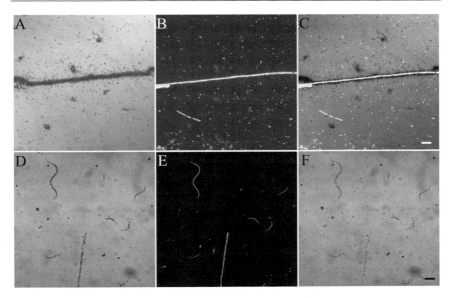

FIG. 2. Microautoradiography (MAR) combined with fluorescence *in situ* hybridization (FISH) or catalyzed reporter deposition FISH (CARD-FISH). Images A–C and D–F show the same microscopic view of an activated sludge sample (Grindsted wastewater treatment plant, Grindsted, Denmark) and a river water sample (North Pine Dam, in southeast Queensland, Australia). (A) MAR image after incubation of the sample with [^{14}C]propionate under aerobic conditions. Note the presence of a MAR-positive filament and a few single cells (seen as areas covered by a thick layer of silver grains). (B) The same microscopic field as A, but recorded for fluorescence after hybridization with a Cy3-labeled probe specific for *Meganema perioderoedes* (probe Gr1-Cy3) and a FLUOS-labeled bacterial probe (probe EUB338-FLUOS). (D) Water sample incubated with [^{3}H]thymidine under aerobic conditions, and (E) hybridization of CARD-FISH after hybridization with a Cy3-labeled probe specific for *Actinomycetes* (probe HG1–654-Cy3). (C and F) Overlay of A-B and D-E, respectively. All images were recorded by laser-scanning microscopy (LSM510 Meta, Carl Zeiss, Jena, Germany). Bar: 10 μm. (See color insert.)

activity of the two labeled compounds and the exposure time must be optimized. In the actual case, the concentrations were 0.7 mM and 20 μCi/ml for acetate and 1.4 mM and 20 μCi/ml for propionate, incubation time was 2 h, and exposure time was 17 h.

Example of Application: Heterotrophic Assimilation of ^{14}CO$_2$

A new variation of the MAR method has been developed, the HetCO$_2$-MAR (Hesselsøe *et al.*, 2005). This approach is based on the fact that most heterotrophic bacteria assimilate CO$_2$ in various carboxylation reactions

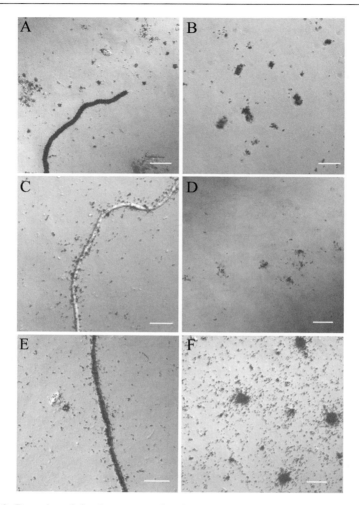

Fig. 3. Detection of simultaneous uptake of two substrates by bacteria from an activated sludge sample (Grindsted wastewater treatment plant, Grindsted, Denmark) using dual labeling of [^{3}H]acetate and [^{14}C]propionate. Images A and B show MAR-positive filaments and single cells, respectively, after incubation with [^{3}H]acetate. Images C and D show MAR-positive cells after uptake of [^{14}C]propionate, whereas images E and F show the uptake of both [^{3}H]acetate and [^{14}C]propionate. Note that silver grains are located up to 2–3 μm from MAR-positive cells with the more energetic isotope ^{14}C, whereas the less energetic ^{3}H isotope only forms silver grains up to 1 μm from the cell and thus yields a higher resolution. By incubating with a relatively higher final specific activity of the ^{3}H-labeled substrate it is possible to see cells taking up both labeled substrates simultaneously. Bar: 20 and 10 μm in A, C, and E and in B, D, and F, respectively.

during biosynthesis. Typically 1–10% of the biomass carbon has been derived from CO_2 fixation. Assimilation of $^{14}CO_2$ by heterotrophic bacteria can be used for isotope labeling of active microorganisms in environmental samples and visualized by MAR-FISH. The MAR signals are comparable with the traditional MAR approach based on the uptake of ^{14}C-labeled organic substrates [see the detailed protocol described by Hesselsøe et al. (2005)]. The new HetCO$_2$-MAR approach appears to differentiate better between substrate uptake (and possibly storage) and substrate metabolism that result in growth. Another obvious advantage is that isotope labeling of heterotrophic microorganisms is no longer restricted to the use of commercially available and often expensive labeled organic substrates, so any organic compound or mixture of organic substrates can, in principle, be investigated for potential uptake and metabolism just by focusing at the uptake and incorporation of $^{14}CO_2$. This novel HetCO$_2$-MAR approach expands significantly the possibility for studying the ecophysiology of uncultivated microorganisms.

Common Methodological Problems

MAR, particularly MAR-FISH, has some drawbacks and limitations. Extensive experience with both MAR and FISH is required, preferably access and experience with a confocal laser microscope, the techniques are time-consuming, and it is relatively expensive due to the cost of radiotracers. In some countries, the safety regulations require separate laboratories and equipment for isotope work. It is important to conduct all experiments in rooms dedicated to isotope work and to follow all local safety instructions. Get all formal permissions from the local safety officer. Also, all involved staff should be properly trained and advised. If all safety instructions are followed, MAR is not a dangerous method.

The availability of suitable radiolabeled substrates can be a problem. Often, the suppliers can only offer a limited selection of either ^3H- or ^{14}C-labeled organic compounds, whereas more rarely sold compounds can only be produced upon request at a special price. However, by screening the suppliers, most common organic compounds can be purchased at a reasonable price. Also, use of the novel HetCO$_2$-MAR approach as described earlier may circumvent these problems.

Interpretation of a positive or negative MAR signal must always be carried out carefully. Controls and duplicates (on independent samples) should always be included. It is also very important to evaluate uptake/no uptake of different substrates only if comparable treatment has been applied, such as same amount of tracer and cold substrate and same incubation time and exposure time. If unexpected results are obtained, these must always be checked for errors in the procedure, particularly related to the incubation conditions, and the experiments should be repeated. This is especially critical for anaerobic conditions.

Conclusion

The method of assessing microbial activity of probe-defined microorganisms under *in situ* conditions as described in this chapter has been used successfully during the last 5–6 years in several ecosystems. A wide range of hitherto uncultured organisms from very diverse environments has been examined and their activities have been established. Microautoradiography is a very versatile technique that can be applied in most microbial environments and yields information on the ecophysiology of individual microorganisms. Recent improvements, such as the combination of MAR with CARD-FISH for microorganisms with low ribosome content, quantitative MAR for obtaining kinetic data, and the detection of dual-labeled cells, offer new options for more detailed studies of microbial consortia. Also, combinations with other *in situ* methods, e.g., formation of internal storage products or measurements of local microenvironments by microsensors, can provide valuable information. Although new improvements are most likely under way, the technique today is fully capable of conducting many detailed and valuable studies of microbial communities.

References

Andreasen, K., and Nielsen, P. H. (1997). Application of microautoradiography for the study of substrate uptake by filamentous microorganisms in activated sludge. *Appl. Environ. Microbiol.* **63,** 3662–3668.

Andreasen, K., and Nielsen, P. H. (2000). Growth of *Microthrix parvicella* in nutrient removal activated sludge plants: Studies of *in situ* physiology. *Wat. Res.* **34,** 1559–1569.

Brock, T. D., and Brock, M. L. (1966). Autoradiography as a tool in microbial ecology. *Nature* **209,** 734–736.

Brock, M. L., and Brock, T. D. (1968). The application of micro-autoradiographic techniques to microbial ecology. *Theor. Angewendte Limnol.* **15,** 1–29.

Carman, K. (1993). Microautoradiographic detection of microbial activity. *In* "Handbook of Methods in Aquatic Microbial Ecology" (P. F. Kemp, B. F. Sherr, E. B. Sherr, and J. J. Cole, eds.), pp. 397–404. Lewis Publishers, London.

Cottrell, M. T., and Kirchman, D. L. (2000). Natural asssemblage of marine proteobacteria and members of the *Cytophaga-flavobacter* cluster consuming low- and high-molecular weight dissolved organic matter. *Appl. Enivron. Microbiol.* **66,** 1360–1363.

Daims, H., Nielsen, J. L., Nielsen, P. H., Schleifer, K.-H., and Wagner, M. (2001). *In situ* characterization of *Nitrospira*-like nitrite-oxidizing bacteria active in wastewater treatment plants. *Appl. Environ. Microbiol.* **67,** 5273–5284.

Gray, N. D., Howarth, R., Pickup, R. W., Jones, J. G., and Head, I. M. (2000). Use of combined microautoradiography and fluorescence *in situ* hybridization to determine carbon metabolism in mixed natural communities of uncultured bacteria from the genus *Acromatium*. *Appl. Environ. Microbiol.* **66,** 4518–4522.

Hesselsøe, M., Nielsen, J. L., Roslev, P., and Nielsen, P. H. (2005). Isotope labeling and microautoradiography of active heterotrophic bacteria based on assimilation of $^{14}CO_2$. *Appl. Environ. Microbiol.* **71,** 646–655.

Ito, T., Nielsen, J. L., Okabe, S., Watanabe, Y., and Nielsen, P. H. (2002). Phylogenetic identification and substrate uptake patterns of sulfate-reducing bacteria inhabiting an aerobic-anaerobic sewer biofilm. *Appl. Environ. Microbiol.* **68,** 356–364.

Kindaichi, T., Ito, T., and Okabe, S (2004). Ecophysiological interaction between nitrifying bacteria and heterotrophic bacteria in autotrophic nitrifying biofilms as determined by microautoradiography by microautoradiography-fluorescence in situ hybridization. *Appl. Environ. Microbiol.* **70,** 1641–1650.

Kragelund, C., Nielsen, J. L., Thomsen, T. R., and Nielsen, P. H. (2005). Ecophysiology of the filamentous *Alphaproteobacterium Meganema perideroedes* in activated sludge. *FEMS Microbiol. Ecol.* (in press).

Kong, Y., Nielsen, J. L., and Nielsen, P. H. (2004). Microautoradiographic study of *Rhodocyclus*-related poly-P accumulating bacteria in full-scale enhanced biological phosphorus removal plants. *Appl. Environ. Microbiol.* **70,** 5383–5390.

Kong, Y., Nielsen, J. L., and Nielsen, P. H. (2005). Identity and ecophysiology of uncultured *Actinobacteria*-related polyphosphate accumulating organisms in full-scale enhanced biological phosphorus removal plants. *Appl. Environ. Microbiol.* **71,** 4076–4085.

Lee, N., Nielsen, P. H., Andreasen, K., Juretschko, S., Nielsen, J. L., Schleifer, K.-H., and Wagner, M. (1999). Combination of fluorescent *in situ* hybridization and microautoradiography: A new tool for structure-function analysis in microbial ecology. *Appl. Environ. Microbiol.* **65,** 1289–1297.

Meyer-Reil, L.-A. (1978). Autoradiography and epifluorescence microscopy combined for the determination of number and spectrum of actively metabolizing bacteria in natural waters. *Appl. Environ. Microbiol.* **36,** 506–512.

Nielsen, J. L., Aquino de Muro, M., and Nielsen, P. H. (2003). Determination of viability of filamentous bacteria in activated sludge by simultaneous use of MAR, FISH and reduction of CTC. *Appl. Environ. Microbiol.* **69,** 641–643.

Nielsen, J. L., Christensen, D., Kloppenborg, M., and Nielsen, P. H. (2003). Quantification of cell-specific substrate uptake by probe-defined bacteria under *in situ* conditions by microautoradiography and fluorescence *in situ* hybridization. *Environ. Microbiol.* **5,** 202–211.

Nielsen, J. L., Juretschko, S., Wagner, M., and Nielsen, P. H. (2002). Abundance and phylogenetic affiliation of iron reducers in activated sludge as assessed by fluorescence *in situ* hybridization and microautoradiography. *Appl. Environ. Microbiol.* **68,** 4629–4636.

Nielsen, J. L., and Nielsen, P. H. (2002a). Quantification of functional groups in activated sludge by microautoradiography. *Wat. Sci. Tech.* **46,** 389–395.

Nielsen, J. L., and Nielsen, P. H. (2002b). Enumeration of acetate-consuming bacteria by microautoradiography under oxygen and nitrate respiring conditions in activated sludge. *Wat. Res.* **36,** 421–428.

Nielsen, P. H., Aquino de Muro, M., and Nielsen, J. L. (2000). Studies on the *in situ* physiology of *Thiothrix* spp. in activated sludge. *Environ. Microbiol.* **2,** 389–398.

Nielsen, P. H., and Nielsen, J. L. (2005). Microautoradiography: Recent advances within the studies of the ecophysiology of bacteria in biofilms. (in press).

Ouverney, C. C., and Fuhrman, J. A. (1999). Combined microautoradiography-16S rRNA probe technique for determination of radioisotope uptake by specific microbial cell type *in situ*. *Appl. Environ. Microbiol.* **65,** 1746–1752.

Ouverney, C. C., and Fuhrman, J. A. (2000). Marine planktonic *Archaea* take up amino acids. *Appl. Enivron. Microbiol.* **66,** 4829–4833.

Pearl, H. W., and Stull, E. A. (1979). In defense of grain density autoradiography. *Limnol. Oceanogr.* **24,** 1166–1169.

Rogers, A. W. (1979). "Techniques of Autoradiography." Elsevier/North-Holland Biochemical, New York.

Tabor, S. P., and Neihof, R. A. (1982). Improved microautoradiographic method to determine individual microorganisms active in substrate uptake in natural waters. *Appl. Environ. Microbiol.* **44,** 945–953.

Teira, E., Reinthaler, T., Pernthaler, A., Pernthaler, J., and Herndl, G. J. (2004). Combining catalyzed reporter deposition-fluorescence in situ hybridization and microautoradiography to detect substrate utilization by *Bacteria* and *Archaea* in deep ocean. *Appl. Environ. Microbiol.* **70,** 4411–4414.

Yang, Y, Zarda, A., and Zeyer, J (2003). Combination of microautoradiography and fluorescence *in situ* hybridization for identification of microorganisms degrading xenobiotic contaminants. *Environ. Technol. Chem.* **22,** 2840–2844.

Further Reading

Ginige, M. P., Hugenholtz, P., Daims, H., Wagner, W., Keller, J., and Blackall, L. L. (2004). Use of stable-isotope probing, full-cycle rRNA analysis, and fluorescence in situ hybridization-microautoradiography to study a methanol-fed denitrifying microbial community. *Appl. Environ. Microbiol.* **70,** 588–596.

[15] Atomic Force Microscopy of Bacterial Communities

By Megan E. Núñez, Mark O. Martin, Phyllis H. Chan, Lin K. Duong, Anil R. Sindhurakar, and Eileen M. Spain

Abstract

This chapter discusses atomic force microscopy (AFM) for the benefit of microbiologists who are interested in using this technique to examine the structures and dynamics of bacteria. AFM is a powerful technique for imaging biological samples at the nanometer to micrometer scale under nondestructive conditions. In order to be imaged with AFM, bacteria must be supported by a surface, which presents challenges because many laboratory strains of bacteria are planktonic. Still, in nature many bacteria live at surfaces and interfaces. This chapter discusses the benefits and difficulties of different methods that have been used to support bacteria on surfaces for AFM imaging and presents two methods in detail used to successfully grow and image bacteria at solid–liquid and solid–air interfaces. Using these methods it is possible to study bacterial morphology and

METHODS IN ENZYMOLOGY, VOL. 397
0076-6879/05 $35.00
DOI: 10.1016/S0076-6879(05)97015-8

interactions in a native state. These explorations by AFM have important applications to the study of different kinds of bacteria, interfacial bacterial communities, and biofilms.

Overview

Since atomic force microscopy was developed in 1986 (Binnig *et al.*, 1986), it has been used to image a variety of materials and surfaces at the nanometer to micrometer scale. Of particular interest to those of us who work with biological systems, AFM has been used to image DNA, proteins, eukaryotic and prokaryotic cells, and cell surfaces. The purpose here is to discuss how AFM can be used to image bacterial cells in as native an environment as possible.

In AFM, a very small sharp tip is scanned back and forth across the surface of a sample, much like a reader of Braille scans his finger across a page. Small vertical tip deflections are used to generate an image of the three-dimensional shape or contours of the image (Fig. 1). On a hard, smooth sample, even small deflections on the order of nanometers can be detected; on softer samples and rougher surfaces the resolution is diminished, but it is still tens or hundreds of nanometers. Furthermore, because the tip touches the surface, the technique can potentially reveal not only information about shape and size, but also about tip–sample interactions involving texture, adhesion, and elasticity of the sample. Although a laser is used to detect the deflection of the tip, it does not penetrate the sample at all, and thus AFM is fundamentally different than light and electron microscopies.

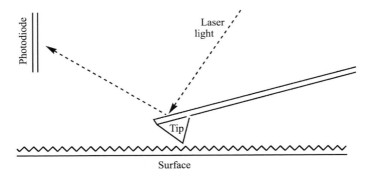

FIG. 1. Scheme of the atomic force microscope. A sharp tip is scanned across a surface, and small deflections in that tip are used to generate an image of the shape and contours of the sample. The laser does not penetrate the sample and is used to detect deflections in the tip.

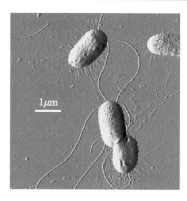

Fig. 2. AFM image of *E. coli* strain ZK1056 grown on a glass surface. Note the long flagella and shorter straight pili surrounding the cells. These cells were imaged in contact mode in air.

The major benefits of AFM compared to many other forms of microscopy are that it provides better resolution than most forms of light microscopy and that samples need not be chemically fixed, stained, sliced, or imaged under vacuum as needed for electron microscopy. For delicate specimens or nanometer-level details, this minimal preparation can be critical. For example, we have used AFM to visualize individual DNA strands, as well as bacterial flagella and pili (Fig. 2). A variety of dynamic biological processes have been imaged by AFM in a native liquid environment, including cells growing and changing and enzymes carrying out their work.

On the flip side, AFM does have a couple of major drawbacks. First, because of the way that the tip moves across the sample, this technique in theory only images the outside surface of a sample providing a topographical map. Surprisingly, we have resolved features *inside* of bacterial cells right through the outer membranes. For example, we have reproducibly imaged the bacterial predator *Bdellovibrio bacteriovorus* growing inside the periplasm of its prey cell, along with the shrunken remains of the cytoplasm and inner membrane of the prey (Fig. 3A) (Núñez *et al.*, 2003). We also clearly resolved rows of magnetosomes, insoluble membrane-encased iron-containing nanoparticles, inside of the magnetotactic bacterium *Magnetospirillum magnetotacticum* (Fig. 3B). Even in the absence of interesting internal structures, the AFM can still provide important information about external structures, including the flagella and pili mentioned earlier. The technique is also particularly sensitive to differences in the texture and properties of the cell membrane because the tip interacts with the surface of the sample. Distinct differences in the texture of the outer

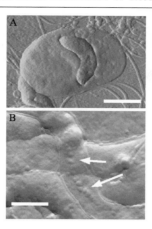

Fig. 3. Images of intracellular structures. In theory, AFM can provide information only about the outside of a cell; in practice, interior features can often be observed as well. (A) Inside the periplasm of an *E. coli* prey cell, a peanut-shaped *Bdellovibrio bacteriovorus* predator grows and digests the prey cytoplasm. Scale bar: 1 μm. (B) A row of small spherical magnetosomes can be seen inside a *Magnetospirillum magnetotacticum* cell. Magnetosomes, membrane-bound iron-containing nanoparticles, are generally around 20–100 nm in diameter. Scale bar: 500 nm.

membrane have been noted between gram-negative and -positive cells viewed in air, such that gram-negative cells generally appear wrinkled and positive cells appear smooth (see later dicussion) (Umeda *et al.*, 1998).

A bigger challenge is that the object to be imaged must be supported by a solid surface. For molecular-scale samples, techniques have been developed for adhering DNA and proteins to hard, smooth surfaces such as mica. For eukaryotic cells, adhesion to a surface is rarely a big problem. For bacteria, however, adhesion to a surface has proved to be surprisingly difficult because the cells tend to float away into the solution where they are effectively invisible. A few techniques have been used to circumvent this problem. Some of these methods have been compared and discussed previously (Camesano *et al.*, 2000).

The easiest way to prepare bacteria for AFM is to spot the liquid bacterial culture onto mica or glass and let it dry (Boyle-Vavra *et al.*, 2000; Robichon *et al.*, 1999; Umeda *et al.*, 1998). Mica and glass are the two most common substrates for biological AFM because they are inexpensive and flat. The surface can be treated with gelatin or poly-L-lysine to improve adhesion (Doktycz *et al.*, 2003). This sample preparation has the advantage that it is quick and easy, but the disadvantage that even when the cells are rinsed first, on the surface they are often embedded in sticky debris and the dried components of the medium (Fig. 4). This sticky debris not

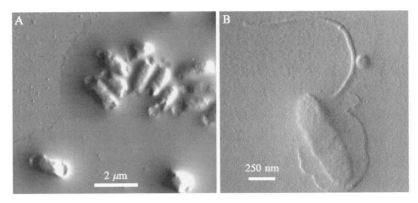

Fig. 4. Images of bacteria deposited onto mica, allowed to dry, and imaged by AFM. The bacteria do not adhere strongly to the surface and thus cannot be rinsed. Cells are surrounded by a layer of sticky residue, probably from the growth medium. In some cases, cells can be completely covered in this material. (A) *E. coli*. Note that the ends of the cells are sunken, probably due to drying. (B) *Bdellovibrio bacteriovorus*. The flagellum at the top is poorly resolved due to contaminants on the surface.

only buries the cells, but can adhere to the tip and generate artifacts in the image. This method also has the disadvantage that bacterial cells cannot be imaged under liquid because they have no real adhesion to the surface; when a drop of liquid is placed on the substrate, the cells float away.

Another approach is to fix the cells with glutaraldehyde (Razatos, 2001; Razatos *et al.*, 1998) or chemically cross-link them to the surface with 1-(3-dimethylaminopropyl)-3-ethylcarbodiimide HCL (EDC) (Camesano *et al.*, 2000). This approach effectively attaches and localizes bacteria to glass surfaces, allowing numerous force measurements to be performed. Contact angle measurements suggest that cells treated with glutaraldehyde remain hydrophobic (Razatos *et al.*, 1998). Thus, it was concluded that the outer membrane charge density, and presumably its structure, remains essentially native. We have not used this method in our laboratories because modifying the outer membranes of our cells would prevent or mask the bacterial interactions that we are trying to study.

A third approach to the problem of keeping bacterial cells on a surface is to use a vacuum to pull the cells through filters, with pores slightly smaller than the cells so that the cells become wedged in the holes. If a cell fits into a hole tightly, it can be imaged in either air or liquid. This method has been used very successfully by Dufrêne's group to acquire images and force data for a variety of microbes (Dufrêne, 2002, 2004). This method does have a couple of limitations. As the authors pointed out (Dufrêne, 2002), this method only works if the cells are spherical. Also, it seems that few cells are effectively captured in the pores, providing a relatively small number of data points. We found that small oblong cells were pulled through the filter and disappeared, whereas larger oblong cells were distorted and torn apart by the vacuum. We also work with fragile cells that are easily burst by mechanical stresses and thus are not amenable to this method.

We have explored another approach using the same type of filters. Instead of using a frit and aspirator to wedge the cells into the filters, we place the filters on top of an agar medium plate and allow the solution of bacteria to wick through slowly. The cells, which are too large to go through the pores, are left on top of the filter. Because this process is much slower and gentler, the cells are not distorted nor wedged into the holes on the filter. The filter can be then removed to the AFM and imaged immediately (Fig. 3B). In addition to its gentleness, this method has the advantage that it also works for different shapes of cells. On the downside, because the cells are not wedged into the filter, they cannot be imaged in liquid because they will float away.

Still, what is unique and valuable about this method is that the cells and filters can be left on top of the agar plate for many days, during which time the cells *continue to grow and multiply*. This hydrated solid–air interface turns out to be a hospitable environment for a variety of bacteria, depending on the choice of plate medium, temperature, and other conditions. We have used this method to individually image different species of bacteria over several days (Fig. 5), as well as to investigate predatory interactions between different species of bacteria (Núñez *et al.*, 2003). This "filter growth method" has also been used by Hansma and colleagues to image growth and change in *Pseudomonas* communities (Auerbach *et al.*, 2000; Steinberger *et al.*, 2002). Details of this method are discussed at length later.

There is yet another way to image bacteria on a surface. Instead of wrestling with the challenges of chemical or physical adhesion to a surface, we turn to the advantages afforded by nature. Many strains of bacteria readily adhere to glass or other solid surfaces and spontaneously form simple biofilms (Beech *et al.*, 2002). With this method, a piece of solid substrate such as glass or metal is placed in a liquid bacterial culture and is later removed after the bacteria have had time to adhere. Tens or hundreds of bacteria can be readily imaged in a short time. Because of their adhesion to the surface, the bacteria can be rinsed clean and imaged well in air or even in liquid (Figs. 2 and 6). These cells are not distorted, shrunken, or swollen.

We allowed a biofilm-forming strain of *Escherichia coli* to adhere to glass and explored the resulting bacterial community by AFM, discerning several features characteristic of biofilms. The *E. coli* cells adhered to the surface in clumps or sheets. Although adhered, the cells continue to divide

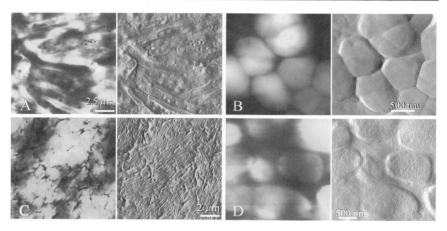

FIG. 5. Bacteria grown on filters over media plates and imaged by AFM. Height images are shown on the left of each pair. Light colors represent high regions, and dark colors represent low regions. Deflection images, which emphasize differences in texture, are shown on the right in each pair. (A) *Aquaspirillum serpens*. These very long cells grow in piles and layers on top of each other. (B) *Micrococcus luteus*. These gram-positive cells appear smooth and show ridges from cell division. (C) Host-independent mutant *Bdellovibrio bacteriovorus*. (D) *Ensifer adhaerens*. These gram-negative cells are notably wrinkled and display multiple flagella.

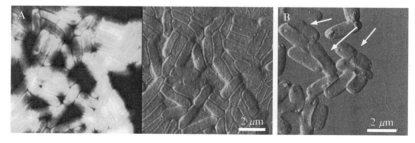

FIG. 6. Simple *E. coli* biofilms grown on glass and imaged by AFM. A fresh culture of *E. coli* strain ZK1056 was added to a 24-well plate with a sterile coverslip partially submerged. Within a day, bacteria adhered to the surface of the glass. The glass could then be removed, rinsed, and viewed by AFM. Although only one strain of bacteria was imaged here, we have mixed different strains of bacteria and observed their interaction by AFM in the same manner. (A) In rich media, bacteria grow to a thick film at the air–liquid interface. The height image at the left shows the location, shape, size, and height of the cells; the deflection image at the right shows textures and smaller structures, such as the many flagella used by the cells to adhere to the glass. (B) In more dilute media, bacteria still adhere to the surface, but do not grow so densely. On the surface they can grow and divide (arrows).

and reproduce *in situ*. Exploiting the high resolution of the AFM, we imaged multiple flagella and pili adhering the cells to the glass (Fig. 2). Despite the light rinsing of cells to remove excess growth medium and loose cells, the bacteria are also often surrounded by a thin layer of extracellular material, which is well documented in biofilms (Danese *et al.*, 2000; Whitchurch *et al.*, 2002). Interestingly, this thin, colorless layer of extracellular material lacks contrast in an optical microscope unless stained, but we can image it by AFM because the tip detects small changes in surface height and texture. AFM is eminently useful to explore other vital biological processes and interactions in these biofilms, e.g., bacterial predation (Núñez *et al.*, 2004). Some details of this sample preparation are discussed later.

It is important to mention that there are other specialized AFM techniques not discussed here that have relevance to bacteriology. In one interesting twist, membrane proteins have been imaged within *pieces* of membranes isolated from cells and transferred to flat surfaces (reviewed in Dufrêne, 2002). In another example, bacteria have been fixed to the AFM *tip* in order to measure adhesion forces between bacteria and a flat surface (Ong *et al.*, 1999).

When bacterial cells can be grown and supported on a surface, many interesting experiments involving AFM become possible. Smaller features, such as flagella and pili, wrinkles and scars, excretions, and cell debris, can be examined at high resolution. Bacteria cannot only be imaged individually, but in groups, populations, and mixed communities, at different concentrations and growth conditions. The elasticity of cells and their adhesion to the tip can be measured quantitatively under the right conditions. Beyond all of these fascinating possibilities, AFM is also straightforward to learn, even for undergraduate students. Some of the most basic details of tips, imaging parameters, and data processing are introduced next.

Experimental Details

The first step in any of these procedures is to grow fresh, dense cultures of bacteria (greater than 10^7 cells/ml is best). We use a variety of media and cells in our experiments, but the details will need to be adapted for specific organisms.

Imaging Bacterial Communities on Filters

A fresh culture of bacteria is grown to obtain a healthy, vigorous, and concentrated population of bacteria. Isopore polycarbonate filters from Millipore are sterilized by autoclaving in a covered glass petri dish. These filters are available with a variety of diameters (including 13 and 25 mm) and pore sizes (0.2, 0.4, 0.8, 1.2, and 2.0 μm and up), making it easy to choose a pore size smaller than the bacteria. We generally use 25-mm-diameter filters with 0.2- or 1.2-μm pores. Once sterilized the filters are removed individually with flame-sterilized forceps and placed on top of solidified sterile media plates. It is important to place the shiny side of the filter facing upward, as the dull side has a rougher texture and more jagged pore edges, neither of which image as nicely as the shiny side. Next, 8–15 μl of a solution containing the bacteria, suspended in either in culture medium or 10 mM MgSO$_4$, is deposited onto the center of the filters. The solution makes a rounded bubble on top of the filter that soaks through into the plates within about 30 min. As long as the pore size is smaller than the bacteria, the bacteria remain on top while the cell debris goes through the filter. At this point, filters coated with bacterial cells can be peeled up carefully from the plates using sterile forceps and fastened to 15-mm round stainless-steel pucks (Ted Pella) with double-sided adhesive tape. If necessary the excess filter can be cut off with a razor or scissors. The microorganisms can be imaged immediately without any further treatment.

Alternatively, instead of peeling the filter up immediately, the bacteria can be incubated on top of the filter for periods ranging from 1 h to several days. During this time, it is hoped that the cells will continue to grow before imaging, as found for several different gram-negative and -positive bacteria (Fig. 5). For some cells, incubation at room temperature is generally sufficient to initiate growth and complex behaviors, but the agar plate can also be placed in an incubator, refrigerator, or other controlled environment at this step to encourage the cells to grow. Mixtures of bacteria can be placed on the filter simultaneously, sequentially, or side by side to examine more complex interactions between different organisms or strains.

Although we initially had some concerns about imaging bacteria in air, the fact that the cells appear healthy and vigorous convinced us that the cells imaged on filters are quite native. In fact, in this hydrated air–solid interface, the cells grow, divide, multiply, move, and spread outward across the filter. The vigor of the cells leads us to conclude that this interface must resemble hospitable surfaces in nature, and this system can mimic biofilms grown at the air–aqueous interface. We find that even after removing the filter from the agar plate and imaging the bacteria for half an hour, a large fraction of the cells can be scraped off and grown in solution. Given that bacteria grow over the filter *and* across the pores and given that they do not seem to prefer particular pore sizes, it seems reasonable to conclude that either there is a layer of liquid above the filter or bacteria are able to absorb liquid directly through the filter. These cells must also quickly adapt to a drier environment because they do not become sunken and dried out when the filter is removed to the AFM.

Bacteria can be imaged for up to several hours after removal from the agar by either contact or tapping-mode AFM in air. Our instrument is a Digital Instruments Multimode SPM with a Nanoscope IIIa or IV controller. Although the tapping mode is generally suggested for imaging soft samples, we find that contact mode also gives clear, sharp images. Furthermore, contact mode tips (oxide-sharpened silicon nitride) are slightly less expensive, and this mode is more compatible with force measurements. Relatively high gains and low tip speeds ensure that the tip has enough time in the feedback loop to track over tall cells, preventing streaking of the images. Choosing small deflection values during engagement prevents impaling the sample with the tip. Bacterial samples contain plentiful sugars, proteins, and cell debris, so as always with AFM it is important to be cautious of artifacts, indicating that debris is stuck to the tip.

In contact mode, samples can be monitored in both height and deflection modes. The height mode reveals the vertical height of objects on the surface, whereas the less-quantitative deflection mode gives images of the finer texture of the cells. The kinds of images generated in each mode are shown in Fig. 5 for comparison. In tapping mode, another choice is to use phase mode. This mode reflects changes in the phase of oscillation of the cantilever due to the tip being stuck or dragged. When imaging cells, which are large and sticky everywhere, the phase signal is not always useful. Instead, phase information may be most useful for revealing sticky regions on the smooth hard substrate around a cell.

When beginning imaging, the tip should be brought down in the center of the filter where the bacteria were placed. Generally, this spot is easy to find as a dull region on the shiny filter. If bacteria are allowed to grow for a day or more, this spot may change color and grow thicker in height or larger in diameter. It is important to move the tip around, taking data from different parts of the filter, because there can be significant differences in the size and shape of cells between locations, especially from the center of the bacterial community to the edge. Begin by scanning a large area and zoom in slowly on areas of interest. The response of the piezoelectric scanner is inherently nonlinear (termed hysteresis), so zooming in to a small area quickly may result in imaging somewhere other than the desired location. In general, a piezoelectric scanner with a 100-μm maximum scan size is useful for viewing bacteria, but for bacteria 5μm or shorter, it is quite possible to take nice data with a 12-μm scanner.

We have attempted to image bacteria under fluid by placing a 40-μl drop of liquid on the filter after taping it to the puck, but the bacteria did not remain adhered to the surface and could not be imaged. Even gentle suction through the filter and some drying prior to the addition of liquid did not keep the cells adhered to the filter, and we have so far been unwilling to try harsher methods to do so, although we believe it possible. Imaging in fluid is desirable primarily because measurements of forces (sample tip adhesion or sample elasticity) *must* be done in liquid, as meniscus forces between the tip and the hydrated surfaces mask any other signal. For more information about measuring the adhesion and elasticity of cells, see Dufrêne (2002) and references therein.

Once image data have been collected, generally only one modification to the images is performed. The vertical offset between scan lines is removed ("flattening"), as provided by Veeco's/Digital Instruments' image software. For a given image, this is accomplished by subtraction of the average vertical value of the scan line from each point in the scan line. Lengths and heights of objects can also be calculated using the software. Although there are many ways to make a given piece of data look more attractive, be wary of overprocessing your data, as it is easy to do in this software package. Images can then be exported to a graphics program (e.g.,, Adobe Photoshop) for final adjustment of dimensions and contrast, cropping, and adding scale bars.

Imaging Bacteria Forming Simple Biofilms on Glass

It has become clear that the preferred growth environment for many bacteria is at an interface, whether it be solid–liquid or liquid–air, despite the fact that planktonic bacteria have been the main subjects of study for more than 100 years. We can take advantage of this proclivity of bacteria to grow at surfaces by placing our substrate into a bacterial culture and allowing the bacteria to adhere naturally.

We have used a strain of *E. coli* (strain ZK1056) that is known to form biofilms (O'Toole *et al.*, 1999). Round 15-mm sterile glass coverslips are used as the solid substrate. The coverslips are placed upright in the wells of a 24-well plate using sterile plastic stands (home-made stands are constructed from the lids of 1.7-ml centrifuge tubes). A 1-ml volume of fresh medium containing *E. coli* (~10^7 cells/ml) is added to each well. The 24-well plate is immediately covered and transferred to a 30° incubator. Over the course of 2–5 days, a thick white line is formed on the glass at the air–liquid interface, indicating the presence of a simple bacterial biofilm. The thickness of the biofilm depends on the medium chosen: a thick biofilm forms in rich medium, a more sparse biofilm forms in dilute medium, and no biofilm forms at all in minimal medium (Núñez *et al.*, 2004).

A different coverslip is removed each day for up to 5 days, rinsed gently three times in a beaker of water, and dried gently. This drying is done by wicking the liquid off with a paper towel or lightly blowing the liquid off with a stream of nitrogen and then allowing the coverslip to sit in a covered clean petri dish for a few minutes until any remaining water has dried. Next, the coverslips are fastened to stainless-steel AFM sample discs (Ted Pella, Inc., Redding, CA) using double-sided adhesive tape. Samples are placed into the AFM and imaged immediately by contact or tapping mode in air without any further treatment. We have seen differences in the biofilm thickness and cell morphology with the location on the slide and the depth in the liquid. The thickest biofilm is formed at the interface between the air and the liquid, so tip placement on the slide is important. When using an AFM with an optical microscope attached, it may be possible to locate this interfacial film directly; otherwise, it is helpful to mark the location of the biofilm on the back side of the glass with a marker to help with tip placement. Imaging of the samples is otherwise the same as described earlier. Because of the ease of preparation and imaging, several different types of samples can be examined in a single day, capturing multiple "snapshots" of bacterial processes and interactions.

Conclusions

By exploiting the natural tendency of many bacteria to thrive on solid surfaces, we have developed two methods for imaging bacteria at high resolution by atomic force microscopy. The methods described here should be widely useful for the examination of bacterial morphology, communities, and interactions.

Acknowledgments

The authors gratefully acknowledge the National Science Foundation (Leveraged Starter Grant, Research Planning Grant, and Presidential Early Career Award for Scientists and Engineers), the Dreyfus Foundation, Research Corporation, ACS Petroleum Research Fund, the James Irvine Foundation, Occidental College, Mount Holyoke College, the Clare Boothe Luce foundation, and the Howard Hughes Medical Institute for generous financial support. The authors also thank Roberto Kolter for his gift of ZK1056.

References

Auerbach, I. D., Sorensen, C., Hansma, H. G., and Holden, P. A. (2000). Physical morphology and surface properties of unsaturated *Pseudomonas putida* biofilms. *J. Bacteriol.* **182,** 3809–3815.

Beech, I. B., Smith, J. R., Steele, A. A., Penegar, I., and Campbell, S. A. (2002). The use of atomic force microscopy for studying interactions of bacterial biofilms with surfaces. *Colloids Surf. B* **23,** 231–247.

Binnig, G., Quate, C. F., and Gerber, C. (1986). Atomic force microscopy. *Phys. Rev. Lett.* **56,** 930–933.

Boyle-Vavra, S., Hahm, J., Sibener, S. J., and Daum, R. S. (2000). Structural and topological differences between a glycopeptide-intermediate clinical strain and glycopeptide-susceptible strains of *Staphylococcus aureus* revealed by atomic force microscopy. *Antimicrob. Agents Chemother.* **44,** 3456–3460.

Camesano, T. A., Natan, M. J., and Logan, B. E. (2000). Observation of changes in bacterial cell morphology using tapping mode atomic force microscopy. *Langmuir* **16,** 4563–4572.

Danese, P., Pratt, L. A., and Kolter, R. (2000). Exopolysaccharide production is required for development of *Escherichia coli* K-12 biofilm architecture. *J. Bacteriol.* **182,** 3593–3596.

Doktycz, M., Sullivan, C., Hoyt, P., Pelletier, D., Wu, S., and Allison, D. (2003). AFM imaging of bacteria in liquid media immobilized on gelatin coated mica surfaces. *Ultramicroscopy* **97,** 209–216.

Dufrêne, Y. F. (2002). Atomic force microscopy, a powerful tool in microbiology. *J. Bacteriol.* **184,** 5205–5213.

Dufrêne, Y. F. (2004). Refining our perception of bacterial surfaces with the atomic force microscope. *J. Bacteriol.* **186,** 3283–3285.

Núñez, M., Martin, M., Chan, P., and Spain, E. (2004). Predation, death, and survival in a biofilm: Investigating *Bdellovibrio* by atomic force microscopy. *Biointerfaces* **42,** 263–271.

Núñez, M., Martin, M., Duong, L., Ly, E., and Spain, E. (2003). Investigations into the life cycle of the bacterial predator *Bdellovibrio bacteriovorus* 109J at an interface by atomic force microscopy. *Biophys. J.* **84,** 3379–3388.

Ong, Y.-L., Razatos, A., Georgiou, G., and Sharma, M. M. (1999). Adhesion forces between *E. coli* bacteria and biomaterial surfaces. *Langmuir* **15,** 2719–2725.

O'Toole, G., Pratt, L. A., Watnick, P., Newman, D. K., Weaver, V. B., and Kolter, R. (1999). Genetic approaches to the study of biofilms. *Methods Enzymol.* **54,** 49–79.

Razatos, A. (2001). Application of atomic force microscopy to study initial events in bacterial adhesion. *Methods Enzymol.* **337,** 276–285.

Razatos, A., Ong, Y.-L., Sharma, M. M., and Georgiou, G. (1998). Molecular determinants of bacterial adhesion monitored by atomic force microscopy. *Proc. Natl. Acad. Sci. USA* **95,** 11059–11064.

Robichon, D., Girard, J.-C., Cenatiempo, Y., and Cavellier, J.-F. (1999). Atomic force microscopy imaging of dried or living bacteria. *C.R. Acad. Sci. III* **322,** 687–693.

Steinberger, R., Allen, A., Hansma, H. G., and Holden, P. A. (2002). Elongation correlates with nutrient deprivation in *Pseudomonas aeruginosa*-unsaturated biofilms. *Microb. Ecol.* **43,** 416–423.

Umeda, A., Saito, M., and Amako, K. (1998). Surface characteristics of gram-negative and gram-positive bacteria in an atomic force microscope image. *Microbiol. Immunol.* **42,** 159–164.

Whitchurch, C., Tolker-Nielson, T., Ragas, P. C., and Mattick, J. (2002). Extracellular DNA required for bacterial biofilm formation. *Science* **259,** 1487.

Section III

Nucleic Acid Techniques

[16]   Nucleic Acid Recovery from Complex Environmental Samples

By KEVIN J. PURDY

Abstract

Effective extraction of nucleic acid from environmental samples is an essential starting point in the molecular analysis of microbial communities in the environment. However, there are many different extraction methods in the literature and deciding which one is best suited to a particular sample is very difficult. This article details the important steps and choices in deciding how to extract nucleic acids from environmental samples and gives specific details of one method that has proven very successful at extracting DNA and RNA from a range of different samples.

Introduction

The application of molecular techniques revolutionized environmental microbiology (Pace *et al.*, 1986) by greatly improving our ability to analyze and investigate natural microbial communities. However, this work is predicated on the effective extraction of nucleic acids from environmental samples, but despite nearly 20 years of research in this area, there is still no consensus on the most effective method of nucleic acid extraction from environmental samples (Forney *et al.*, 2004). None of the available methods are perfect; all are empirically designed methods that are acceptable because they work with the samples and for the purpose at hand. However, it is possible to determine the best available methods by considering what the nucleic acid is needed for, which types of nucleic acid need to be extracted, the quality required, and the number of samples to be processed.

Planning an Effective Extraction Method

There are three questions that must be addressed when planning an extraction protocol:

1. What type of sample is being analyzed?
2. What type of analysis is the nucleic acid being used for?
3. What types of nucleic acid will need to be extracted?

METHODS IN ENZYMOLOGY, VOL. 397
0076-6879/05 $35.00
DOI: 10.1016/S0076-6879(05)97016-X

Sample Type

Each type of sample brings its own problems, which to some extent explains the variety of methods that can be found in the literature. The most problematical samples to work with are organic-rich soils, as these contain high concentrations of organic degradation products such as humic and fulvic acids, which are easily coextracted with nucleic acid and inhibit downstream analytical processes (Tebbe and Vahjen, 1993). Water samples, while usually posing fewer difficulties with contaminating inhibitors, are often low in biomass and so require concentration prior to analysis. Thus, the type of sample, the biomass within the sample, and the potential for contaminants should all be considered when embarking on an extraction. Several published methods that target different environmental samples are used as exemplars in this article.

What Is the Nucleic Acid Being Used For?

The analytical method that is to be utilized in an experiment affects the type of extraction that should be used. For example, standard polymerase chain reaction (PCR) analysis requires relatively clean DNA, whereas accurate real-time PCR demands a greater level of purity. Table I indicates

TABLE I
QUALITY AND TYPE OF NUCLEIC ACID USUALLY REQUIRED FOR A NUMBER OF COMMONLY USED
MOLECULAR METHODS IN ENVIRONMENTAL MICROBIOLOGY

| Method | Nucleic acid | Quality | | Other factors |
		Size	Contamination	
PCR	DNA	>5 kb	Low	Avoid carryover of SDS
RT-PCR	RNA	Not degraded	Very low	Remove all traces of DNA
RT-PCR	DNA	>5 kb	Minimal	Requires strict controls
Hybridization to DNA	DNA	>5 kb	Low	High yields are needed
Hybridization to RNA	RNA	Not degraded	Low	High yields are needed
Gene expression	mRNA	Not degraded	Very low	Remove all traces of DNA
Metagenomic analysis	DNA	>10 kb	Low	Need high molecular weight DNA

the type of nucleic acid and minimum quality in terms of DNA fragment size and levels of contamination required for the application of a number of common molecular analyses. A secondary question to consider here is replication. In order to address ecological questions, it is usually necessary to replicate experiments and samples. While these considerations are outside of the scope of this chapter, the number of both biological and sample replicates could be a major factor when deciding on an extraction method.

What Type of Nucleic Acid is Being Extracted and Analyzed?

Extracellular DNA

Although this type of nucleic acid is rarely studied, a significant proportion of extractable DNA can be extracellular (Dell'Anno and Corinaldesi, 2004; Ogram *et al.*, 1987), which can be important in studies on gene transfer in the environment (Lilley *et al.*, 1994).

Cellular Genomic DNA

This is the commonest target of extraction processes. Effective extraction of cellular genomic DNA is absolutely essential in studies on microbial biodiversity and community structure with the main issue being the lysis of as many members of the community as possible. In addition, environmental genomic studies require high-quality genomic DNA (DeLong, 2004).

RNA

The analysis of activity in microbial communities usually requires the extraction of high-quality RNA. RNA is degraded very rapidly by both intracellular and extracellular RNases and these ubiquitous and extremely resilient proteins represent the major difficulty when extracting RNA. Thus, special steps must be taken to ensure successful extraction of RNA.

Maintaining RNase-Free Workspace. All traces of RNase must be removed from all glassware, metalware, solutions, plasticware, and anything that the RNA extract may come into contact with. Because skin is literally covered in RNases, nothing that will be used with RNA should be touched with bare hands; powderless gloves must be worn at all times.

GLASSWARE, METALWARE, AND PLASTICWARE. All glassware and metalware must be cleaned, wrapped in aluminium foil, and then baked at 260° for at least 3 h. In the past, all plasticware (tips and Eppendorf tubes, etc) were autoclaved at 15 psi and 121° for at least an hour. However, it is now possible to buy ready-to-use plasticware that is certified DNase and RNase free.

SOLUTIONS. Great care must be taken when making solutions for RNA work. Water should be as high purity as possible; it is safer to buy water that is certified RNase free and autoclaved for 1 h as described earlier in baked bottles. Many RNA workers (Blumberg, 1987) recommend that solutions be treated with diethyl pyrocarbonate (DEPC), which is a nonspecific inhibitor of RNases, prior to autoclaving. Autoclaving destroys the DEPC but because DEPC reacts with Tris, Tris-containing solutions should be prepared from DEPC-treated, autoclaved water. However, DEPC treatment is not absolutely necessary if preautoclaved RNase-free water is used to make solutions and all other precautions are strictly adhered to.

RNases can contaminate dry chemicals if they are dispensed with spatulas that are not prebaked to RNA standards. To avoid this problem, all chemicals for RNA work should be kept for RNA work only and dispensed with prebaked spatulas or into prebaked measuring cylinders. All solutions should be dispensed in sensibly sized aliquots into prebaked bottles prior to autoclaving [we use small Duran bottles (Fisher Scientific, UK) with the lids autoclaved separately described earlier]. Solutions are aliquoted so that should a sample degrade, then all the solutions in use can be discarded rather than chasing the cause of the degradation. Certain chemicals cannot be autoclaved; high concentration sodium dodecyl sulfate (SDS) and any guanidium solutions are two examples. In these cases, absolute adherence to maintaining an RNase-free environment is essential or, alternatively, purchase premade solutions that are guaranteed RNase free.

AN RNASE-FREE ZONE. RNase is commonly used in a number of standard molecular techniques, such as plasmid preparation methods, and so can contaminate pipettes and work benches. It is absolutely essential that pipettes and plasticware used with this enzyme are *never* used in the preparation of RNA. Do so by having dedicated pipettes for RNA work and setting aside an area where RNA-only work is performed. Ensure that stocks of RNase are stored in a freezer that is not used to store any RNA-work related chemicals, samples, or stocks. If all these precautions are observed, RNA extractions are no more difficult than DNA extractions.

The Extraction Process

Having defined the sample, the downstream analytical methods, and the type(s) of nucleic acid to be extracted, it is now necessary to determine which is the most effective method for the task in hand. To determine this, it is best to divide the extraction process into several distinct sections: (1) pretreatment of the sample, (2) lysis of the cells, (3) nucleic acid isolation, and (4) nucleic acid purification. This article describes the most effective methods in all of these areas and then details the use of one method that has proven extremely useful in the past in the analysis of a variety of different types of nucleic acid from a variety of sample types.

Sample Pretreatment

Many researchers have found that nucleic acid quality and yield can be improved by pretreating samples. Table II details some examples of sample pretreatment and their consequences for a variety of common environmental samples.

Water Samples

The major issue with many water samples is low biomass, which is usually addressed by sample concentration using filtration (Giovannoni *et al.*, 1990). Filters can be extracted directly or cells are washed off prior to extraction. Generally, these are relatively low contamination samples, as the majority of organic matter collected is cellular. There are potential problems with filtration, particularly if labile organisms are the object of investigation, such as certain protozoa, and in these cases great care must be taken to avoid premature cell lysis.

TABLE II
RESULTS OF DIFFERENT TYPES OF PRETREATMENT ON THE FINAL YIELD AND QUALITY OF EXTRACTED NUCLEIC ACID FROM DIFFERENT ENVIRONMENTAL SAMPLES

			Nucleic acid quality		
Sample	Major issue	Pretreatment	Yield	Size	Contamination
Water	Low biomass	None	Very low	5–20 kb	Very low
		Concentrate	Low/medium	5–20 kb	Very low
Biofilms	High biomass, polysaccharide	None	High	5–20 kb	Very low
Soils/ sediment	Humic contamination	None	High	0–10 kb smear	Method dependent
		Washed sample	High	0–20 kb smear	Method dependent
		Cells extracted	Low	5–20 kb	Very low

Biofilms

Like water samples, biofilms generally represent low contamination samples but are also high biomass systems. Thus, nucleic acid extraction is often very successful from these types of samples. However, the highly structured nature of biofilms can make representative sampling difficult.

Soils and Sediments

Without a doubt the most challenging samples are those from soils and sediments. Both are complex matrices of mineral and organic matter. Mineral matrices can interfere with extractions, as clay particles adsorb organic matter, particularly nucleic acid. However, the major issue is coextraction of recalcitrant phenolic compounds, such as humic and fulvic acids that derive from degrading organic matter. These contaminants are potent inhibitors of many molecular biological enzymes, particularly *Taq* and *Reverse Transcriptase*. Direct extraction exposes cellular nucleic acid to these contaminating compounds. This can be avoided by separating the cells from the matrix prior to extraction but there are problems with such indirect extraction methods.

The issue of whether cells within a soil or sediment sample should be extracted prior to extraction or whether nucleic acid should be extracted directly appeared to be settled in the early days of environmental nucleic acid extraction. It was clear that yields were significantly higher if nucleic acid was extracted directly from the environmental matrix rather than after separation of the cells (Holben *et al.*, 1988; Ogram *et al.*, 1987). However, other reports have revived the idea of cell separation as part of an effective extraction method (Courtois *et al.*, 2001; Gabor *et al.*, 2003). This is because the lower nucleic acid yield from cell separation methods does not seem to adversely affect the diversity of the recovered microbial community; in fact, the recoverable diversity may even be greater if indirect extraction is utilized (Gabor *et al.*, 2003). Methods that can be used to separate cells from the environmental matrix have been reviewed by Bakken and Lindahl (1995). Indirect extractions also produce less contaminated nucleic acid of higher molecular weight and so would be more amenable to metagenomic and real-time PCR analysis.

As stated earlier, extracellular DNA is rarely the target for extraction but can interfere with diversity studies by leading to a signal being detected from cells that had already lysed. Therefore, it is preferable to remove this DNA prior to the extraction of cellular nucleic acid. This can be achieved by washing sediment and soil samples with an alkaline phosphate solution immediately after sampling (Tsai and Olson, 1991). This wash needs to be performed immediately after sampling, as nonfrozen storage of samples leads to changes in the community structure (Rochelle *et al.*, 1995), while washing after frozen storage, which will result in some cell lysis, could change the recovered microbial community.

There are a number of pretreatment methods in the literature that are designed specifically to reduce humic contamination. These include chemical flocculation using $Al(NH_4)(SO_4)_2$ (Braid *et al.*, 2003), pretreatment using Tween 20 and solvents (Purohit *et al.*, 2003), and sample washing with Triton X-100 and EDTA (Fortin *et al.*, 2004; Watson and Blackwell, 2000). All are reported to be effective at reducing the contamination of nucleic acids by humics and other inhibitory substances but have not been widely used to date.

Lysis

The critical step in any nucleic acid extraction method is the lysis step. Lysis methods can be divided into four separate types: (1) chemical, (2) enzymatic, (3) mechanical, and (4) others. In the most widely utilized extraction methods, a mixture of lysis techniques is usually employed.

Chemical Lysis

A number of chemical are used to lyse cells; these include detergents, such as SDS and sarkosyl, phenol, and chaotrophic salts, such a guanidium thiocyanate. A detergent is included to aid cell membrane disruption and often phenol is added as both a lytic agent and to aid nucleic acid purification. Chaotrophic salts, which disrupt protein structure, are used to protect the highly labile RNA from RNase activity.

Enzymatic Lysis

Specific enzymes can be used to break down the cell walls of defined types of microbes: lysozyme is used to lyse gram-negative bacteria, achromopeptidase lyses gram-positive cells, and lyticase lyses fungal cells. Thus, the obvious problem with enzymatic lysis is that it is selective. In addition, extended periods of incubation are often required with resultant degradation of released RNA and increased nucleic acid contamination with humics. A positive side to enzymatic lysis is that it is very gentle and can produce high molecular weight DNA. Proteinase K is also often included in enzymatic extraction methods to degrade proteins in the samples to facilitate nucleic acid release.

Mechanical Lysis

Cells can be lysed by breaking them physically using mechanical forces. Lysis is aided greatly by the addition of glass beads that help create shear forces that rip cells open when samples are shaken at high speed, so-called bead beating. For bacteria and archaea, beads of about 0.1 mm in diameter are recommended, whereas larger beads of 0.2–1 mm diameter are usually more effective with fungi. It has been reported that bead beating is the most effective lysis method presently available (de Lipthay *et al.*, 2004; Moré *et al.*, 1994). Other alternatives are homogenization and French press, but these are only really effective with "clean" samples that do not contain substantial amounts of mineral matrix. In addition, bead-beating equipment is often designed to allow processing of multiple samples, thus facilitating a higher throughput of samples.

Other Methods

This group includes the much-used freeze/thaw and sonication methods. Both have been used extensively and efficiently (Picard *et al.*, 1992; Tsai and Olson, 1991). Another alternative that has been used occasionally is lysis using microwave heating (Picard *et al.*, 1992). However, as stated earlier, mechanical lysis produces greater yields of nucleic acid with equivalent fragment sizes. Most extraction methods incorporate more than one lytic step or component. Table III lists five methods used on different samples to illustrate this point.

There are a large number of lysis methods that have been used to extract nucleic acid from environmental samples, but it is clear that bead beating in the presence of SDS is the most effective in terms of yield (Moré *et al.*, 1994). However, for certain analyses, such as metagenomics, alternative methods may be preferable.

Nucleic Acid Isolation

Classical DNA isolation utilizes an alkaline phosphate buffer (pH 8.0) (Marmur, 1961; Sambrook *et al.*, 1989), which has been used effectively in extractions from environmental samples. Isolation can be aided by the addition of pH-equilibrated phenol or phenol/chloroform/isoamylalcohol (IAA), as addition of these not only helps lysis, but also reduces humic contamination, precipitates proteins at the phenol/aqueous interface, and partitions nucleic acids into the aqueous phase.

There are a number of methods to differentially extract RNA from DNA. A hot acid phenol extraction can be used to differentially extract

TABLE III
DIFFERENT LYSIS METHODS USED IN FIVE EXTRACTION PROCEDURES ON FIVE DIFFERENT
TYPES OF SAMPLES

Sample	Lysis method				Reference
	Chemical	Enzymatic	Mechanical	Other	
Filtered water	SDS	Proteinase K	[a]	[a]	Giovannoni et al. (1996)
Biofilm	SDS	Lysozyme, proteinase K	French press	[a]	Ruff-Roberts et al. (1994)[b]
Soil	SDS	Lysozyme	[a]	Freeze/thaw	Tsai and Olson (1991)
Sediment	SDS/ phenol	[a]	Bead beating	[a]	Purdy et al. (1996)
Plant roots	Guanidium	[a]	[a]	Sonication	Hahn et al. (1990)[c]

[a] Not incorporated in the method.
[b] From Ruff-Roberts et al. (1994).
[c] From Hahn et al. (1990).

RNA (Moran et al., 1993). The phenol is acid equilibrated with ACE buffer (10 mM sodium acetate, 10 mM NaCl, 3 mM EDTA, pH 5.1) and used at 65°. Isopycnic acid density centrifugation has also been used to separate and purify RNA (Giovannoni et al., 1996) but is not rapid and can be technically demanding. Hydroxyapatite can be used to separate DNA and RNA (Bernardi, 1971) and has been used very effectively in a simple and rapid spin-column method. This extraction method is described in more detail later. DNA and RNA can also be separated by enzymatic means. RNA can be digested from a DNA extract by adding RNase to the lysis buffer, as occurs in plasmid preparation methods (Sambrook et al., 1989). Conversely, DNase can be used to remove DNA from an RNA extract after purification (Miskin et al., 1999). However, the DNase must be completely RNase free and then removed after digestion.

Nucleic Acid Purification

After nucleic acid isolation it is usually necessary to purify the product. Contaminants may include protein, carbohydrates, or other compounds such as humic acids. Different methods are required to remove these compounds and different extraction methods have different associated problems.

Removal of Protein

Protein is often coextracted with DNA and can be removed using classical methods of protein separation, such as phenol/chloroform extraction, hydroxyapatite separation, and caesium chloride density centrifugation. Phenol/chloroform/IAA, as described earlier, partitions nucleic acids into the aqueous phase and precipitates proteins at the aqueous/organic interface (Sambrook *et al.*, 1989). Hydroxyapatite has been used to separate DNA from RNA and from proteins in analytical and preparative HPLC (Bernardi, 1971) and has been used to purify extracted DNA in gravity columns (Ogram *et al.*, 1987) and in spin columns (Purdy *et al.*, 1996). Caesium chloride density gradients have been used to purify DNA from sediments (Selenska and Klingmüller, 1991).

Carbohydrates

Coextraction of carbohydrates with nucleic acid can be a problem when extracting from plant material and can be alleviated using hexadecyltrimethyl ammonium bromide (Tsushima *et al.*, 1995).

Humic Contamination

By far the biggest problem in molecular environmental microbiology is the coextraction of humic and fulvic acids. Some approaches attempt to reduce contamination during lysis and isolation, such as adding acid-washed polyvinylpolypyrrolidone [PVPP (Holben *et al.*, 1988)] or phenol to the lysis buffer and reducing processing times to a minimum. Other methods reduce humic contamination during the isolation stage, such as using HTP spin columns where most of the brown contaminants are bound to the column and are not eluted with the nucleic acid. However, it is often necessary to further purify isolated nucleic acid and there are a plethora of methods available to do this.

Berthelet *et al.* (1996) described a simple spin column method using PVPP that has proven very successful in reducing contamination and subsequent inhibition (Donovan *et al.*, 2004), as does the use of PVP in agarose gel electrophoresis (Young *et al.*, 1993). PVPP is an insoluble polymer that needs to be acid washed prior to use (Holben *et al.*, 1988). If PVPP columns are to be used in RNA extractions, then the acid-washing process must be performed with great care with RNase-free solutions.

Salts have been used to precipitate contaminants, leaving the nucleic acid in solution. Both ammonium and potassium acetate have been used for this purpose (Krsek and Wellington, 1999; Yu and Mohn, 1999), but ammonium acetate can also precipitate high molecular weight RNA

(Sambrook *et al.*, 1989). These methods are not always successful, as it appears that loosely bound contaminants can be precipitated with ease while more tightly bound contaminants are more difficult to remove.

Size exclusion or ion-exchange columns have also been used to remove contaminants. Moran *et al.* (1993) described the use of Sephadex G-75 spin columns to purify DNA. Evidence shows that Sepharose 2B is the most effective size-exclusion matrix at removing contaminants (Miller, 2001). However, if small RNA molecules are the target of isolation, then some of the larger pore-sized size-exclusion columns, such as Sepharose 2B, could actually trap the RNA in their pores. Thus, a wide margin of error should be observed when utilizing these methods. However, they are effective at removing not only contaminants but also salts that may have accumulated in the extraction process and generally give a high-quality product.

There are also a number of proprietary columns that have been used in nucleic acid purification, with varying success. Elutip-d and Elutip-r columns from Schliecher and Schuell (Dassel, Germany), Qiagen spin columns (Crawley, UK), microcon ultracentrifugation columns from Amersham (Amersham, UK), and Sephadex columns from Edge Biosystems, Gaithersburg, MD) have been used. Finally, specific kits are available for nucleic acid extraction from soils in particular. These include the Soilmaster kit (Epicenter, Madison, WI), the Powersoil kits (MoBio, Carlsbad, CA), and the FastDNA and RNA kits by QBiogene (Carlsbad, CA).

Nucleic Acid Precipitation

The final step in most extraction methods is the precipitation of nucleic acid. As with all these steps, a number of methods exist. The commonest is ethanol precipitation (Sambrook *et al.*, 1989). Alternatively, isopropanol has been used, primarily because smaller volumes can be used (Sambrook *et al.*, 1989), and it has been reported that isopropanol precipitation can significantly reduce contamination by PCR inhibitors in DNA extracts (Hänni *et al.*, 1995). Precipitation of nucleic acid can also be performed using polyethylene glycol (PEG) (Selenska and Klingmüller, 1991), which is also very good at reducing contamination, but recovers only about 50% of the nucleic acid.

All purification methods have limitations. The traditional methods of purification may not meet the exacting standards required of real-time PCR or be amenable to the necessities of processing large numbers of samples in an ecological study. In essence, the choice of purification method is constrained by the sample and by the lysis and isolation method used. Table IV details the purification method used in the five methods

TABLE IV
PURIFICATION METHODS USED IN THE FIVE DIFFERENT METHODS DESCRIBED

Method	Sample	Purification methods
Giovannoni *et al.* (1996)	Filtered water	Phenol/ChCl$_3$, isopycnic centrifugation
Ruff-Roberts *et al.* (1994)[a]	Biofilm	Phenol/ChCl$_3$, ethanol precipitation
Tsai and Olson (1991)	Soil	Phenol/ChCl$_3$, Elutip-d column
Purdy *et al.* (1996)	Sediment	HTP columns, ethanol precipitation
Hahn *et al.* (1990)[b]	Plant roots	Phenol/ChCl$_3$, ethanol precipitation

[a] From Ruff-Roberts *et al.* (1994).
[b] From Hahn *et al.* (1990).

compared earlier. Other factors include how time-consuming and expensive a method is, especially when processing large numbers of samples. In the robotics era, the processing of many samples is now a realistic prospect and the method employed may actually be decided by the availability of proprietary purification plates that can be incorporated easily into a robotic protocol.

Nucleic Acid Extraction from Environmental Samples Using Hydroxyapatite Spin Columns

This section describes details of a specific extraction method. This method was originally developed in order to extract RNA from sediments [see Purdy (1997) for details] and has subsequently been used on varied sediments, soils, pure cultures of fungi, bacteria and archaea, and termite guts. Its strengths lie in the quality of the extracted nucleic acid, the ability to separate DNA from RNA if required, to process samples rapidly, and to perform a number of extractions simultaneously (Purdy *et al.*, 1996). DNA, rRNA, and mRNA have been isolated using this method and been used in PCR, RT-PCR, real-time PCR, and slot-blot hybridization studies. RNA quality is usually high and the RNA can be stored for long periods without degradation if precautions to exclude RNAse contamination have been observed.

The method employs a mechanical lysis step with an alkaline phosphate buffer in the presence of phenol and SDS followed by nucleic acid isolation/purification on HTP spin columns. RNA and DNA can then be differentially eluted from the columns, desalted, and purified further before storage and use. An experienced worker can go from a sample to a cleaned extract in the freezer in under 2 h; multiple samples (four to eight) can be processed in 3–4 h.

Sample Pretreatment

Generally, sediment and soil samples have been washed prior to storage to remove extracellular DNA using the method of Tsai and Olson (1991).

1. Place up to 20 g of sample in a 50-ml Oak Ridge tube.
2. Add an equal amount (w/v) of 120 mM sodium phosphate, pH 8.0.
3. Shake at 150 rpm for 15 min at site temperature if possible.
4. Spin at 6000g for 10 min at site temperature and discard supernatant.
5. Wash again in 120 mM sodium phosphate, pH 8.0, and spin as described earlier; discard supernatant.
6. Store washed sediment in 4-g aliquots at $-70°$. Aliquoting samples avoids unnecessary freeze/thawing when samples are required for extraction, which results in cell lysis and degradation of RNA within the samples.

Method Preparation

This method, while simple and rapid, involves many short spins so being prepared helps facilitate the process. Table V indicates what needs to be prepared for each sample for the different types of extraction.

TABLE V

REQUIRED COLUMNS AND EPPENDORF TUBES FOR DIFFERENT TYPES OF HTP SPIN COLUMN EXTRACTIONS[a]

Nucleic acid	HTP column	Sephadex columns	Wash buffer (NaPO₃, pH 7.2)	Elution buffer (K₂HPO₃, pH 7.2)
Total	1	1	120 mM	300 mM
DNA only	1	1	150 mM K₂HPO₃	300 mM
RNA only	2	1	120 mM	150 mM
DNA and RNA	2	2	120 mM	150 mM/ 300 mM

[a] Assumes 2 × 1-g samples (2 × 2-ml screw-cap tubes) in extraction.

Preparation of HTP and Sephadex Spin Columns

Spin column blanks can be purchased, but ensure that they are certified RNase free; alternatively, spin columns can be made in-house using 1-ml syringes for HTP columns and 2-ml syringes for Sephadex columns. Syringes made by Becton Dickinson (Franklin Lakes, NJ) are exactly the right size for the preparation of spin columns. Spin columns are spun in reusable, adapted 50-ml Falcon tubes (spin-column holders, Fig. 1) in a bench centrifuge, preferably in a swing-out rotor. Holes that are just large enough to fit either a 1- or a 2-ml syringe are punched in the Falcon tube lid. It is important that the 1-ml syringes in particular are held solidly in the adapted lid, as the syringe tip may touch the side of the holder. Figure 1

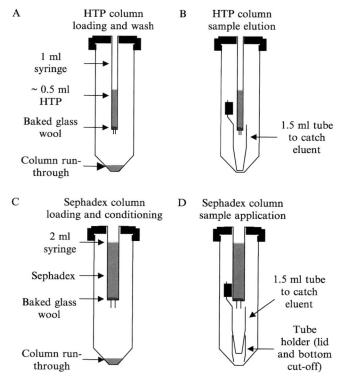

Fig. 1. Diagram of both HTP and Sephadex spin columns in their spin column holders for (A) HTP column loading and washing; (B) HTP column sample elution; (C) Sephadex column preparation and conditioning; and (D) Sephadex column sample application.

shows exactly how the columns and holders are fitted together for sample loading and elution from HTP columns (Figs. 1A and B) and for column washing and desalting for the sephadex columns (Figs. 1C and D). Syringes are packed with a plug of RNase-free (baked) glass-wool, cut prior to baking into lengths of about 1 cm (care must be taken doing this, as glass-wool fibers can be a severe respiratory irritant; use a mask or cut in a fume hood).

1. Push the glass-wool into the syringe with two pairs of RNase-free sharp-pointed tweezers and then tamp down with the syringe plunger. Once plugged, place the syringe into a spin-column holder. Plugged syringes can be preprepared if stored safely (either in their packaging or in a spin-column holder) and the plunger is left in place until required.

2. HTP should be rehydrated as described by the manufacturer (Bio-Rad, Hercules, CA) and then autoclaved (for RNA extracts the HTP is hydrated with sterile RNase-free water and then autoclaved for 1 h). Only standard HTP should be used in spin columns; DNA-grade HTP cannot be used, as the smaller particle size results in columns that load extremely slowly or not at all.

3. Just prior to making the column, resuspend the HTP slurry by gentle swirling; *do not shake*. While it is still in suspension, fill the spin column with HTP using a 1-ml pipette and wide-bore tips. Avoid forming air locks by tilting the column relative to the pipette tip and allow the HTP to run down the side of the syringe slowly. If air locks do form, these will disappear when the column is spun.

4. Once the syringe is full, spin at 100g for 1 min (never spin faster than 100g, as the column will compact and become unusable). This should produce a column containing about 0.5 ml of HTP (Fig. 1A); if not, then add a little extra slurried HTP and respin.

5. Sephadex G-75 or G-50 should be rehydrated as described by the manufacturers and then autoclaved.

6. Sephadex is *not* resuspended but is loaded as a gel. Fill the 2-ml syringe with the gel and then spin at 1600g for 2 min; the final volume of dry Sephadex should be about 2 ml (Fig 1C).

7. Condition the column by adding 400 μl of RNase-free sterile H$_2$O and spinning at 1600g for 8 min: repeat two more times (Moran *et al.*, 1993).

Both HTP and Sephadex columns should be prepared on the day of use but can be stored for a few hours as long as they are completely covered.

Sample Lysis

Multiple tubes can be used for a single sample, but a maximum of two tubes for each HTP column is usually best.

1. Weigh 0.5 g of baked 0.1-mm glass beads into a 2-ml screw-cap Eppendorf tube.

2. Weigh 0.5–1 g of sample into tube and add 0.75 ml of 120 mM sodium phosphate, pH 8.0, 1% (w/v) acid-washed PVPP, 0.5 ml Tris-equilibrated phenol, pH 8.0, and 50 μl 20% (w/v) SDS.

3. Bead beat at 2000 rpm for 30 s three times (Microdismembrator-U, Braun Biotech, Germany) with 30 s on ice between each beating.

4. Spin down sample in microcentrifuge (13,000 rpm, 1 min).

5. Carefully collect the aqueous upper layer and load as much as possible onto a labeled HTP column. Any remaining liquid should be stored in a 1.5-ml Eppendorf tube until it can be loaded onto a column. The column should be spun as soon as possible (see later).

6. Rewash the pellet by adding 0.75 ml 120 mM sodium phosphate, pH 8.0, to the sample tube and bead beat again once for 30 s. Ensure that the pellet is loosened from the bottom of the tube prior to bead beating. Two washes recovers >90% of the extractable nucleic acid (Purdy, 1997).

7. Spin as in step 4, collect aqueous phase as in step 5, and bulk with previous sample.

Processing of samples larger than a few grams can be difficult, as large volumes of lysis and isolation solutions are produced. It is possible to scale up the bead-beating process dependent on exactly which type of equipment is being used. Centrifugation of the lysate can be performed in corex centrifuge tubes (baked at 200°). However, HTP columns become more and more compacted during spinning and so cannot be used to process large volumes of liquid (maximum load is about 4 ml). If large samples must be used, then the crude extract can be precipitated (with ethanol or isopropanol) and then resuspended in a smaller volume (< 2 ml) prior to loading onto the HTP column. An alternative is to use the batch purification method detailed later.

Purifying Total Nucleic Acid Using Hydroxyapatite Spin Columns

1. Load sample onto the column in approximately 0.7-ml aliquots, spinning at 100 g for 2 min, and repeat until all the sample is loaded (Fig. 1A). If an aliquot will not run through in 2 min, continue loading but increase the spin time; *do not* increase spin speed.

2. Once all the sample is loaded, wash the column three times with 0.5 ml of 120 mM sodium phosphate, pH 7.2, spinning at 100g for 2–6 min or as long as it takes to pass all the liquid through the column. This wash elutes all the protein that is bound to the column (Purdy, 1997).

3. Elute total nucleic acid from the column into a sterile Eppendorf tube with 0.4 ml of 300 mM or 500 mM dipotassium phosphate, pH 7.2, spinning at 100g for 2–6 min (see Fig. 1B).

4. Load the eluent onto a preconditioned 2-ml Sephadex spin column, spin at 1600 g for 8 min, and collect nucleic acid in a clean Eppendorf (see Fig. 1D).

5. Add 0.1 volumes of 3 M sodium acetate and 2.5 volumes of ice-cold 100% ethanol. Leave to precipitate at -20° for at least 30 min to 16 h.

6. Spin in a microcentrifuge at 13,000g for 10 min and carefully remove the supernatant by aspiration.

7. Add 1 ml of ice-cold 70% ethanol and vortex for 10 s.

8. Spin in a microcentrifuge at 13,000g for 2 min and carefully remove the supernatant by aspiration.

9. Air dry the pellet for 15–30 min and then resuspend in 50 μl TE or 10 mM Tris, pH 7.5, and store at $-20°$.

DNA only can be collected by changing the 120 mM sodium phosphate wash in step 2 to a 150 mM dipotassium phosphate, pH 7.2, wash instead. This will remove all the RNA along with the bound protein.

Separate Extraction of DNA and RNA

Once the sample is loaded onto a HTP spin column and any contaminating protein has been removed by washing (see earlier discussion), RNA can be eluted using a buffer that will not elute DNA.

1. Elute RNA from the column into a sterile Eppendorf by the addition of 0.7 ml 150 mM dipotassium phosphate, pH 7.2, spinning at 100g for 2–6 min (Fig. 1B). Store the eluent on ice. The exact concentration of the specific RNA elution buffer is batch dependent but has always been 140–150 mM.

2. Repeat twice more into fresh Eppendorfs and store the eluents separately.

3. Dilute all three eluents with 0.7 ml of sterile RNase-free ddH$_2$O each and mix well.

4. Load these onto a new HTP column as described previously.

5. Wash the column once with 120 mM sodium phosphate, pH 7.2.

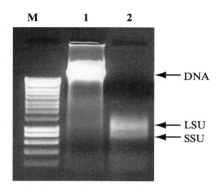

FIG. 2. Separate DNA and RNA extraction using the HTP spin column method from a 1-g (wet weight) sediment sample from the Colne Point saltmarsh in Essex, United Kingdom. Lanes M, Hyperladder 1 (Bioline, UK); 1, DNA extract; and 2, total RNA extract. Used with permission of Dr. Karen Warner.

6. Elute the RNA into a sterile Eppendorf by the addition of 0.4 ml 300 mM K$_2$HPO$_4$, pH 7.2, spinning at 100g for 2–6 min. DNA can be eluted from the first HTP column as described earlier.
7. Load the eluted RNA and DNA onto two *separate* preconditioned 2-ml Sephadex spin columns (Fig 1D), spin at 1600g for 8 min, and collect nucleic acid in clean Eppendorfs.
8. Ethanol precipitate as described earlier.
9. Resuspend RNA in 50-μl sterile, RNase-free ddH$_2$O and store at $-70°$. Resuspend DNA in 50 μl TE or 10 mM Tris, pH 7.5, and store at $-20°$.

Results of a separate DNA and RNA extraction from a sediment sample are shown in Fig. 2.

Additional clean-up steps can be included into this protocol with ease. PVPP spin columns (Berthelet *et al.*, 1996) can be added prior to the Sephadex steps described previously. These reduce humic contamination significantly but can also reduce yields (Donovan *et al.*, 2004). PEG can be used to precipitate clean DNA and RNA in a similar fashion to ethanol (Selenska and Klingmüller, 1991). Parts of both DNAeasy and RNAeasy extraction kits from Qiagen have been used to purify extracts further, but the extract must be in a suitable buffer for the clean-up columns.

Purification of Large Samples

For large samples and large nucleic acid isolation volumes, an alternative method can be used. In this method, spin columns are not used but the nucleic acid is bound to the HTP in batches in 15- or 50-ml Falcon tubes.

1. Isolate nucleic acid as described earlier and collect both washes together. Baked corex centrifuge tubes can be used to process larger volumes of extracts.
2. Place approximately 2 ml of HTP suspension into a 15- or 50-ml Falcon tube and add the sample extracts (DNA-grade HTP can be used for this extraction method).
3. Mix the sample and HTP either in an orbital shaker or on a rotary shaker, preferably at 4° for 15 min.
4. Spin the Falcon tube at maximum speed for 5 min.
5. Discard the supernatant and wash the HTP three times with 3 ml of 120 mM sodium phosphate, pH 7.2, wash buffer and shake and spin again as described earlier each time.
6. Elute total nucleic acid twice with 0.5 ml of 300 mM dipotassium phosphate, pH 7.2, shake, and spin as described previously. Bulk the eluents.
7. This eluent can now be desalted using two Sephadex columns as described earlier and then ethanol precipitated or, if separate DNA and RNA are required, diluted by the addition of 3 ml of RNase-free ddH$_2$O and loaded onto a HTP spin column, following the instructions given earlier.

This method has been adapted for use on specific samples. Li *et al.* (1999) introduced additional pretreatment prior to bead beating that significantly improved the yield of RNA from their lake sediments. Buckley *et al.* (1998) used 2-ml HTP spin columns to purify large amounts of total nucleic acid extracted from soils.

Conclusions

Nucleic acid extraction from environmental samples is the starting point for molecular analysis of microbial communities. This article detailed the steps needed in an effective extraction method and described a very effective method of extracting nucleic acid from a wide variety of samples. With some thought and careful application, all types of samples should be amenable to a particular extraction method and yield an extract that can be manipulated effectively.

References

Bakken, L. R., and Lindahl, V. (1995). Separation and purification of bacteria from soil. *In* "Nucleic Acids in the Environment, Methods and Applications" (J. T. Trevors and J. D. van Elsas, eds.), pp. 8–27. Springer-Verlag, Berlin.
Bernardi, G. (1971). Chromatography of nucleic acids on hydroxyapatite columns. *Methods Enzymol.* **21,** 95–139.

Berthelet, M., Whyte, L. G., and Greer, C. W. (1996). Rapid, direct extraction of DNA from soils for PCR analysis using polyvinylpolypyrrolidone spin columns. *FEMS Microbiol. Lett.* **138,** 17–22.

Blumberg, D. D. (1987). Creating a ribonuclease-free environment. *Methods Enzymol.* **152,** 20–24.

Braid, M. D., Daniels, L. M., and Kitts, C. L. (2003). Removal of PCR inhibitors from soil DNA by chemical flocculation. *J. Microbiol. Methods* **52,** 389–393.

Buckley, D. H., Graber, J. R., and Schmidt, T. M. (1998). Phylogenetic analysis of nonthermophilic members of the kingdom Crenarchaeota and their diversity and abundance in soils. *Appl. Environ. Microbiol.* **64,** 4333–4339.

Courtois, S., Frostegard, A., Goransson, P., Depret, G., Jeannin, P., and Simonet, P. (2001). Quantification of bacterial subgroups in soil: Comparison of DNA extracted directly from soil or from cells previously released by density gradient centrifugation. *Environ. Microbiol.* **3,** 431–439.

de Lipthay, J. R., Enzinger, C., Johnsen, K., Aamand, J., and Sørensen, S. J. (2004). Impact of DNA extraction method on bacterial community composition measured by denaturing gradient gel electrophoresis. *Soil Biol. Biochem.* **36,** 1607–1614.

Dell'Anno, A., and Corinaldesi, C. (2004). Degradation and turnover of extracellular DNA in marine sediments: Ecological and methodological considerations. *Appl. Environ. Microbiol.* **70,** 4384–4386.

DeLong, E. F. (2004). Microbial population genomics and ecology: The road ahead. *Environ. Microbiol.* **6,** 875–878.

Donovan, S. E., Purdy, K. J., Kane, M. D., and Eggleton, P. (2004). Comparison of the euryarchaeal microbial community in guts and food-soil of the soil-feeding termite *Cubitermes fungifaber* across different soil types. *Appl. Environ. Microbiol.* **70,** 3884–3892.

Forney, L. J., Zhou, X., and Brown, C. J. (2004). Molecular microbial ecology: Land of the one-eyed king. *Curr. Opin. Microbiol.* **7,** 210–220.

Fortin, N., Beaumier, D., Lee, K., and Greer, C. W. (2004). Soil washing improves the recovery of total community DNA from polluted and high organic content sediments. *J. Microbiol. Methods* **56,** 181–191.

Gabor, E. M., de Vries, E. J., and Janssen, D. B. (2003). Efficient recovery of environmental DNA for expression cloning by indirect extraction methods. *FEMS Microbiol. Ecol.* **44,** 153–163.

Giovannoni, S. J., De Long, E. F., Schmidt, T. M., and Pace, N. R. (1990). Tangential flow filtration and preliminary phylogenetic analysis of marine picoplankton. *Appl. Environ. Microbiol.* **56,** 2572–2575.

Giovannoni, S. J., Rappé, M. S., Vergin, K. L., and Adair, N. L. (1996). 16S rRNA genes reveal stratified open ocean bacterioplankton populations related to the green non-sulfur bacteria. *Proc. Natl. Acad. Sci. USA* **93,** 7979–7984.

Hahn, D., Kester, R., Starrenburg, M. J. C., and Akkermans, A. D. L. (1990). Extraction of ribosomal RNA from soil for detection of *Frankia* with oligonucleotide probes. *Arch. Microbiol.* **154,** 329–335.

Hänni, C., Brousseau, T., Laudet, V., and Stehelin, D. (1995). Isopropanol precipitation removes PCR inhibitors from ancient bone extracts. *Nucleic Acids Res.* **23,** 881–882.

Holben, W. E., Jansson, J. K., Chelm, B. K., and Tiedje, J. M. (1988). DNA probe method for the detection of specific microorganisms in the soil bacterial community. *Appl. Environ. Microbiol.* **54,** 703–711.

Krsek, M., and Wellington, E. M. H. (1999). Comparison of different methods for the isolation and purification of total community DNA from soil. *J. Microbiol. Methods* **39,** 1–16.

Li, J.-H., Purdy, K. J., Takii, S., and Hayashi, H. (1999). Seasonal changes in ribosomal RNA of sulfate-reducing bacteria and sulfate-reducing activity in a freshwater lake sediment. *FEMS Microbiol. Ecol.* **28,** 31–39.

Lilley, A. K., Fry, J. C., Day, M. J., and Bailey, M. J. (1994). *In situ* transfer of an exogenously isolated plasmid between *Pseudomonas* spp. in sugar beet rhizosphere. *Microbiology* **140,** 27–33.

Marmur, J. (1961). A procedure for the isolation of deoxyribonucleic acid from microorganisms. *J. Mol. Biol.* **3,** 208–218.

Miller, D. N. (2001). Evaluation of gel filtration resins for the removal of PCR-inhibitory substances from soils and sediments. *J. Microbiol. Methods* **44,** 49–58.

Miskin, I. P., Farrimond, P., and Head, I. M. (1999). Identification of novel bacterial lineages as active members of microbial populations in a freshwater sediment using a rapid RNA extraction procedure and RT-PCR. *Microbiology* **145,** 1977–1987.

Moran, M. A., Torsvik, V. L., Torsvik, T., and Hodson, R. E. (1993). Direct extraction and purification of rRNA for ecological studies. *Appl. Environ. Microbiol.* **59,** 915–918.

Moré, M. I., Herrick, J. B., Silva, M. C., Ghiorse, W. C., and Madsen, E. L. (1994). Quantitative cell lysis of indigenous microorganisms and rapid extraction of DNA from sediment. *Appl. Environ. Microbiol.* **60,** 1572–1580.

Ogram, A. V., Sayler, G. S., and Barkay, T. (1987). The extraction and purification of microbial DNA from sediments. *J. Microbiol. Methods* **7,** 57–66.

Pace, N. R., Stahl, D. A., Lane, D. J., and Olsen, G. J. (1986). The analysis of natural microbial populations by ribosomal RNA sequences. *Adv. Micro. Ecol.* **9,** 1–55.

Picard, C., Ponsonnet, C., Paget, E., Nesme, X., and Simonet, P. (1992). Detection and enumeration of bacteria in soils by direct DNA extraction and polymerase chain reaction. *Appl. Environ. Microbiol.* **58,** 2717–2722.

Purdy, K. J. (1997). "The Use of 16S rRNA-Targeted Oligonucleotide Probes to Study the Ecology of Sulphate-Reducing Bacteria." Ph.D. thesis, University of Essex, Colchester, Essex, UK.

Purdy, K. J., Embley, T. M., Takii, S., and Nedwell, D. B. (1996). Rapid extraction of DNA and rRNA from sediments using a novel hydroxyapatite spin-column method. *Appl. Environ. Microbiol.* **62,** 3905–3907.

Purohit, H. J., Kapley, A., Moharikar, A. A., and Narde, G. (2003). A novel approach for extraction of PCR-compatible DNA from activated sludge samples collected from different biological effluent treatment plants. *J. Microbiol. Methods* **52,** 315–323.

Rochelle, P. A., Cragg, B. A., Fry, J. A., Parkes, R. J., and Weightman, A. J. (1995). Effect of sample handling on estimation of bacterial diversity in marine sediments by 16S rRNA gene sequence analysis. *FEMS Microbiol. Ecol.* **15,** 215–226.

Ruff-Roberts, A. L., Kuenen, J. G., and Ward, D. M. (1994). Distribution of cultivated and uncultivated cyanobacteria and *Chloroflexus*-like bacteria in hot spring microbial mats. *Appl. Environ. Microbiol.* **60,** 697–704.

Sambrook, J., Fritsch, E. F., and Maniatis, T. (1989). "Molecular Cloning: A Laboratory Manual," 2nd Ed. Cold Springs Harbor Laboratory Press, Cold Springs Harbor, NY.

Selenska, S., and Klingmüller, W. (1991). DNA recovery and direct detection of *Tn5* sequences from soil. *Lett. Appl. Microbiol.* **13,** 21–24.

Tebbe, C. C., and Vahjen, W. (1993). Interference of humic acids and DNA extracted directly from soil in detection and transformation of recombinant DNA from bacteria and a yeast. *Appl. Environ. Microbiol.* **59,** 2657–2665.

Tsai, Y.-L., and Olson, B. H. (1991). Rapid method for direct extraction of DNA from soil and sediments. *Appl. Environ. Microbiol.* **57,** 1070–1074.

Tsushima, S., Hasebe, A., Komoto, Y., Carter, J. P., Miyashita, K., Yokoyama, K., and Pickup, R. W. (1995). Detection of genetically-engineered microorganisms in paddy soil using a simple and rapid nested polymerase chain-reaction method. *Soil Biol. Biochem.* **27,** 219–227.

Watson, R. J., and Blackwell, B. (2000). Purification and characterization of a common soil component which inhibits the polymerase chain reaction. *Can. J. Microbiol.* **46,** 633–642.

Young, C. C., Burghoff, R. L., Keim, L. G., Minak-Bernero, V., Lute, J. R., and Hintom, S. M. (1993). Polyvinylpyrrolidone-agarose gel electrophoresis purification of polymerase chain-amplifiable DNA from soils. *Appl. Environ. Microbiol.* **59,** 1972–1974.

Yu, Z. T., and Mohn, W. W. (1999). Killing two birds with one stone: Simultaneous extraction of DNA and RNA from activated sludge biomass. *Can. J. Microbiol.* **45,** 269–272.

[17] The Application of Rarefaction Techniques to Molecular Inventories of Microbial Diversity

By Jennifer B. Hughes and Jessica J. Hellmann

Abstract

With the growing capacity to inventory microbial community diversity, the need for statistical methods to compare community inventories is also growing. Several approaches have been proposed for comparing the diversity of microbial communities: some adapted from traditional ecology and others designed specifically for molecular inventories of microbes. Rarefaction is one statistical method that is commonly applied in microbial studies, and this chapter discusses the procedure and its advantages and disadvantages. Rarefaction compares observed taxon richness at a standardized sampling effort using confidence intervals. Special emphasis is placed here on the need for precise, rather than unbiased, estimation methods in microbial ecology, but precision can be judged only with a very large sample or with multiple samples drawn from a single community. With low sample sizes, rarefaction curves also have the potential to lead to incorrect rankings of relative species richness, but this chapter discusses a new method with the potential to address this problem. Finally, this chapter shows how rarefaction can be applied to the comparison of the taxonomic similarity of microbial communities.

Introduction

The increasing ease of inventorying microbial diversity bestows exciting opportunities for microbial ecologists, yet the growing size of molecular inventories challenges researchers to interpret very large datasets in

METHODS IN ENZYMOLOGY, VOL. 397   0076-6879/05 $35.00

biologically informative ways. Microbial ecologists, like other ecologists, seek to understand the distribution of biodiversity. To identify these patterns and the biotic and abiotic factors that drive them, methods are needed to compare microbial communities across time, space, and experimental treatments. As a result, a number of papers address the topic of statistical approaches for microbial community comparisons (Curtis *et al.*, 2002; Dunbar *et al.*, 2001; Hughes *et al.*, 2001; Martin, 2002).

Currently, most molecular inventories use polymerase chain reaction (PCR) amplification of a gene, such as the 16S ribosomal gene, to assess the diversity of a microbial community from a sample of environmental DNA. The molecular methodologies have numerous pitfalls, among them gene duplications, PCR biases, and primer biases. Many authors have noted these biases and have discussed how to minimize these problems (Thompson *et al.*, 2002; von Wintzingerode *et al.*, 1997). Still others have suggested correction factors (Acinas *et al.*, 2004) or new molecular sampling approaches to skip PCR methods all together (Tyson *et al.*, 2004; Venter *et al.*, 2004). These advances are already yielding invaluable information about the extent and consequences of sampling biases for diversity comparisons.

This chapter concentrates on the problem of undersampling of microbial communities, a problem that seems less likely to be alleviated in the near future than the problem of PCR-related biases. For instance, Sargasso Sea data collected by Venter and colleagues (2004) used shotgun sequencing to assess the molecular diversity of seawater microbes. This technique removes PCR and primer biases; however, even with sequencing 1 billion bp and 1164 16S genes, the study still undersampled the microbial community. More than 70% of the "species" of six protein-coding phylogenetic markers in the database were singletons, i.e., they were seen only once. Thus, for microbial ecologists who cannot generate nearly such large datasets, undersampling will certainly present a problem. In contrast, the statistical approaches discussed in this chapter can be applied to samples that knowingly contain methodological biases; as long as these biases are similar (or random) across samples within a study, one can statistically compare community diversity and composition.

This article focuses on one approach, rarefaction analyses, for comparing diversity among communities. Rarefaction is by no means the single best diversity measurement; however, it is probably the most commonly used statistical method in recent microbial diversity studies. This use is for good reason, as it is usually a very good place to begin analysis of a new dataset. We review other diversity statistics used commonly in microbial ecology elsewhere (Bohannan and Hughes, 2003; Hughes and Bohannan, 2004; Hughes *et al.*, 2001). Furthermore, statistics targeted specifically for molecular inventories of microbes are quickly being proposed (e.g., Curtis *et al.*, 2002; Dunbar *et al.*, 2001; Martin, 2002; Singleton *et al.*, 2001).

What Is Rarefaction?

Background

Rarefaction accounts for the fact that large samples have more species (or any taxonomic unit) than small samples even if they are drawn from the same community. Hurlbert (1971) and Sanders (1968) first introduced the idea of scaling down samples of community diversity to the same number of individuals so that richness could be compared across samples. These authors proposed using $E(S_n)$ as a measure of community diversity, i.e., the expected number of species in a sample of n individuals, from a larger collection of N individuals containing S species.

Since then, community ecologists have broadened the idea of rarefaction as a statistical procedure to standardize for sampling effort (Gotelli and Colwell, 2001). Sampling effort can be represented by individuals sampled, as first suggested, or other units such as number of samples or sampling time. Because of sampling constraints, analyses of microbial diversity so far use individual-based rarefaction, thus we concentrate on this approach here. However, it is important to note that the unit of sampling effort used has large consequences for the interpretation of rarefaction analyses (see Gotelli and Colwell, 2001); as microbial diversity inventories begin to include many samples within one study, this issue will become more relevant.

Procedure

An accumulation curve is a plot of the cumulative number of species observed as each individual is sampled and recorded. The curve could be drawn from the notes of a birder walking through a forest and writing down in order the identity of every bird she detects (Fig. 1). Molecular inventories of microbial diversity from one sample usually give a mass capture of individuals, and thus an accumulation curve means very little. For instance, a researcher makes a clone library of PCR products and then randomly picks colonies, or "individuals," to sequence. In contrast to data a birder collects along a transect, the sampling order of the clones from within a clone library does not relate any information about the natural community.

There is, however, useful information for a microbial ecologist in a rarefaction curve: a smoothed, or randomized, accumulation curve.

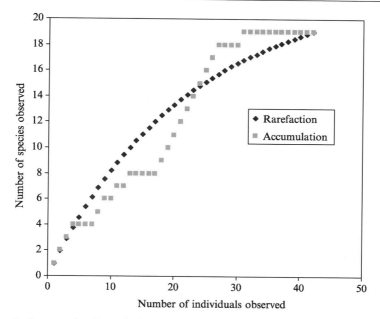

FIG. 1. An example of hypothetical individual-based accumulation and rarefaction curves. The accumulation curve is one possible order of observing the 42 clones. The rarefaction curve was created with the EstimateS program (Colwell, 2004). (See color insert.)

The curve represents the average number of species observed when n individuals are drawn with replacement from the same sample over and over (Fig. 1). In other words, it is the average of all possible accumulation curves. In the case of a clone library where an accumulation curve is an arbitrary ordering of clones, a rarefaction curve is the best way to represent data rather than a random choice of the possible accumulation curves. A rarefaction curve can be estimated by a randomization method or, in the case of an individuals-based sample, by analytic means (Coleman, 1981; Heck *et al.*, 1975).

A key feature of the rarefaction curve is the error bars around the curve. Error bars are so crucial to rarefaction analyses and, at the same time, so often misunderstood. Error bars on a rarefaction curve give a measure of variance around the average accumulation curve; specifically, they represent the variability of the number of species observed [i.e., the variability of $E(S_n)$] when n individuals are drawn from the entire sample. When one individual is drawn, the variance is zero because one species is always observed. Similarly, when all N individuals are drawn, the variance is zero because all S species in the sample are observed. Error bars can be given as variance or standard deviation, but are most useful when reported as 95% confidence intervals (CIs). The 95% confidence limits for a given sample size are $S_{obs} \pm (1.96)*$(the standard deviation of S_{obs}), where S_{obs} is the average number of species observed at that sample size. The 95% CIs represent the range in which 95% of all possible accumulation curves of this particular sample fall. (See later for further information about interpretation of rarefaction error bars.)

A number of software programs will perform rarefaction randomizations and/or calculate analytical formulas for rarefaction curves. In particular, EstimateS randomizes and calculates Coleman formulas and is freely available on the Internet (Colwell, 2004). Although EstimateS can be used for a number of different dataset types, the most common for microbial studies are those for which each clone represents the sampling of one individual. In this case, we find it easiest to load input data as a "Format 3" input file (Fig. 2).

Interpretation of Rarefaction Curves

General Considerations about Diversity Comparisons

Rarefaction analysis of species richness is just one way among many others to compare community diversity between samples. Given the variety available, one must evaluate the utility of different diversity statistics in light of the question of interest and data at hand. As mentioned earlier, even for large-scale molecular inventories of microbial diversity, data at hand are always a minute fraction of the entire community. This fact limits our ability to estimate the true richness of microbial communities with any large degree of confidence. Moreover, without knowing the correct answer, it is impossible to evaluate thoroughly the success of diversity statistics for any microbial community.

Rarefaction differs from other approaches to diversity measurement. Diversity estimators attempt to extrapolate from a sample to the true diversity of a community. Examples of estimators include Chao1 (Chao, 1984) and the Curtis *et al.* (2002) statistic based on a log-normal assumption. Estimators provide two functions: (1) to estimate true richness and (2) to compare these estimates of true diversity among samples. Rarefaction, in contrast, is performed solely for this second purpose, to compare diversity among samples. To discuss the relative benefits of rarefaction compared to the use of diversity estimators, we first discuss some general ideas behind diversity estimation.

D1	▼	ƒx				D1	▼	ƒx		
	A	B	C				A	B	C	D
1	figure 1 data					22	11	20	1	
2	19	42				23	19	21	1	
3	1	1	1			24	18	22	1	
4	2	2	1			25	17	23	1	
5	3	3	1			26	16	24	1	
6	4	4	1			27	15	25	1	
7	1	5	1			28	14	26	1	
8	3	6	1			29	13	27	1	
9	1	7	1			30	12	28	1	
10	6	8	1			31	11	29	1	
11	7	9	1			32	10	30	1	
12	3	10	1			33	9	31	1	
13	5	11	1			34	8	32	1	
14	4	12	1			35	2	33	1	
15	8	13	1			36	2	34	1	
16	1	14	1			37	2	35	1	
17	4	15	1			38	12	36	1	
18	3	16	1			39	12	37	1	
19	6	17	1			40	15	38	1	
20	12	18	1			41	16	39	1	
21	10	19	1			42	17	40	1	
22	11	20	1			43	15	41	1	
23	19	21	1			44	16	42	1	
24	18	22	1			45	-1	-1	-1	

FIG. 2. The EstimateS "Format 3" data file for example data in Fig. 1 Data are entered into three columns in a spreadsheet: an index of the operational taxonomic units (OTU), a nonrepeating index of the number of individuals, and a "1" to indicate a sample size of 1 for each colony. The title of the file must be in the first line, the number of OTU types and the number of clones in the second line, and the final line after data must be "-1,-1,-1." The file is then saved as a tab-delimited text file and is imported into EstimateS. The right-hand screen shows continuation of the columns on the left-hand side.

Any sample comparisons, whether of biomass or species richness of microbial or macrobial communities, must come to terms with three statistical parameters: bias, precision, and accuracy. Ideally, one would like an accurate estimate of species richness, a measurement that yields a very small difference between estimated richness and the true, unknown richness and a consistent estimate of that truth with every sample taken of the community (Hellmann and Fowler, 1999). Often we can only achieve a component of accuracy, either bias or precision. Bias is the difference between the expected value of the estimator (the mean of the estimates from all possible samples of the community) and the true unknown richness of the community being sampled. This difference reveals whether the estimator consistently under- or overestimates the true richness. Precision is the variation of the estimates from all possible samples of the community. This variation represents the repeatability of the richness estimate; i.e., how similar a richness estimate is from one sample to another of the same community. Figure 3 illustrates bias and precision in terms of a dart game. A good dart thrower is accurate because she is both unbiased and precise and therefore always hits the bulls-eye. Less-talented dart throwers are biased, imprecise, or some combination thereof.

Unlike scoring a dart game between two players, evaluating diversity statistics for microbial communities is difficult because we do not know where the bulls-eye is. To test for bias, one needs to know the true richness to compare against the sample estimates. In contrast, precision is relatively easy to assess. With multiple samples, the variance of richness estimates can be calculated and compared. In some cases, an

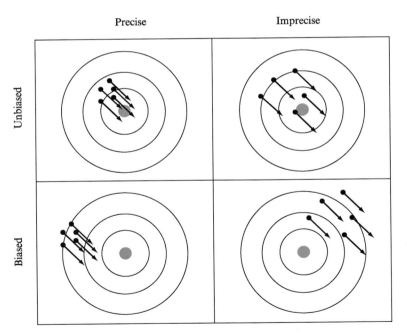

FIG. 3. An illustration of precision and bias. The bulls-eye represents a true value that is trying to be determined. Arrows are individual estimates of the true value. An ideal estimator statistic is precise and unbiased (i.e., accurate), as in the upper left-hand corner. (See color insert.)

estimator may have a closed-form variance that evaluates the precision of the estimate.

What Rarefaction Curves Do and Do Not Tell You

By comparing observed species richness, rarefaction is designed expressly to ignore the issue of bias with respect to true community diversity. Observed richness is always negatively biased, thus rarefaction curves do not say much about the true richness of a community. However, most ecological questions require comparisons of relative diversity (e.g., whether richness is higher or lower in one community or another) rather than an exact number of true richness. For these questions, a diversity statistic that is consistent with repeated sampling (is precise) but biased can be more useful than one that, on average, correctly predicts true richness but is very imprecise (Fig. 4). For relative comparisons, bias is not necessarily a problem as long as the measure is consistently biased. Thus, in theory, rarefaction might be a good approach for relative richness questions if the precision issue is addressed.

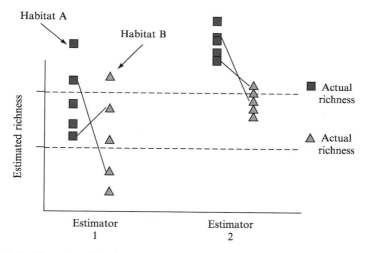

FIG. 4. An illustration of the importance of the precision of richness estimators. Estimator 1 is unbiased, but imprecise. On average, estimator 1 correctly estimates the actual richness of habitat A and B, but the variance of the different estimates is large. Thus, if one compares one estimate of richness from one sample of each habitat, it is easy to incorrectly rank the relative richness of the two habitats. Two example comparisons are shown with solid lines; one pair of estimates correctly orders the relative richness of the habitats, the other pair orders it incorrectly. In contrast, estimator 2 is positively biased, but precise. On average, the estimator overestimates the actual richness; however, because the variance in the estimates is small, any pairwise comparison of estimates correctly ranks the richness of the habitats. (See color insert.)

The most common misconception about rarefaction curves is that the confidence intervals around the curves are a measurement of the precision of $E(S_n)$ (the observed species richness for a given effort). In fact, the CIs do not say anything about the value of $E(S_n)$ if one resamples the community. Error bars only describe the variation of the accumulation curves as one reorders subsamples within the original sample. Specifically, the 95% confidence intervals represent the range in which 95% of all reordered accumulation curves will fall.

This detail, that rarefaction compares *samples* instead of communities, is crucial. A comparison of a rarefaction curve from a grassland and a rarefaction curve from a forest tells you whether the richness of the two samples (standardized for sampling effort) is significantly different (Fig. 5), not whether the richness of grassland and forest communities is significantly different. This difference is particularly important when samples represent a very small fraction of total diversity (as is the case for microbial inventories) so that different samples from the same community can have very different taxonomic representation.

Yet all is not lost for rarefaction. The precision of rarefaction curves can be addressed by sampling from multiple sites or treatment replicates (or see Colwell *et al.*, 2004). The key is to sample multiple times so that many different rarefaction curves are produced. Variation in the curves from multiple samples of a community then yield an estimate of the

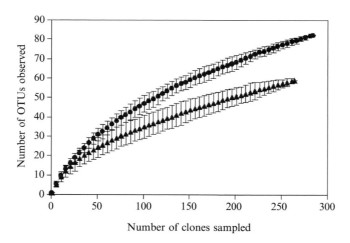

FIG. 5. A comparison of two hypothetical rarefaction curves from a sample of forest soil (●) and a sample of grassland soil (▲). Error bars are 95% confidence intervals. Curves reveal that standardizing for a common sampling effort (at 260 clones), the grassland sample has fewer operational taxonomic units (OTUs) than the forest sample.

precision of $E(S_n)$ for n individuals sampled. For instance, one could sample many grassland and forest sites and draw rarefaction curves for every sample. If the rarefaction curves of the two habitats differ consistently in observed richness for a common sampling effort n, then one could say with some statistical level of confidence whether the richness of grasslands and forest sites differs in richness at that sampling effort. (A technical note: instead of taking multiple samples from the same community, a researcher may prefer to take one very large sample and divide it up. When a large sample is divided into subsamples, the estimated precision of observed richness applies to the number of individuals of the subsamples, not the number of individuals of the total sample. In general, however, it is usually preferable to take independently replicated samples; for instance, independent samples can capture spatial heterogeneity and moderate PCR biases and errors.)

The final nuance of rarefaction is that even with good estimates of precision for rarefaction curves, the precision measure still applies only to observed richness at a particular level of sampling effort (of n individuals). This omission is not a concern if most of the richness of a community has been observed and if the rarefaction curves are asymptotically approaching the true diversity. For microbial inventories, however, sampling effort is low relative to true richness, and the curves are often still steep at the level of sampling effort. As a result, it is possible that the communities differ in their species-abundance patterns so that the rarefaction curves may cross if the sampling effort was increased (Fig. 6). Thus, even if the rarefaction curves are representative of the communities from which they are sampled, at low sample sizes they may suggest an incorrect ordering of relative diversity among communities.

The curve-crossing problem is the primary reason that diversity estimators remain useful even for relative diversity comparisons. Diversity estimators incorporate information about species' abundances in the sample in order to extrapolate true richness. In theory, they predict what happens to the rarefaction curves as one approaches sampling all individuals in the community.

The next section discusses a new technique that might help alleviate the problem of crossing rarefaction curves. However, very little is known overall about the variation in species- or taxa-abundance curves of microbial communities. Therefore, the sampling effort needed to ensure that rarefaction curves do not cross with further sampling is also unknown. Even in the absence of exhaustive surveys of real microbial communities, a few thorough simulation studies could contribute a great deal to this question.

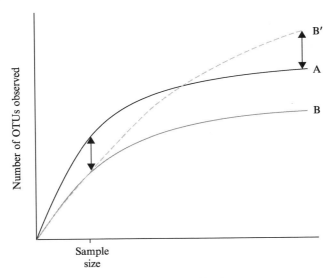

FIG. 6. Hypothetical accumulation curves for two communities: A and B. Rarified curves of the sample size indicated would suggest that OTU richness in A is greater than B. Rarefaction at this sample size cannot distinguish between curves B′ or B″, however. If B′ is the true curve, then richness in B is greater than A.

Future Directions

Rarefaction by Coverage

Cao *et al.* (2002) proposed a modification of rarefaction analysis that attempts to alleviate the problem of crossing rarefaction curves. They suggested that one should standardize by an estimate of the coverage of the sample (the proportion of true richness observed in the sample) rather than sample size.

In any comparison of two samples, samples will vary in their coverage of the communities from which they were drawn. This variation may be due in part to underlying differences between the communities' species-abundance distributions (Brose *et al.*, 2003). Take an example of two jars of colored marbles. The jars contain the same marble richness (i.e., the number of different colors of marbles), but one jar has an even distribution of colors and the other contains 90% blue marbles. If 10 marbles are drawn from each jar, one will almost certainly observe a greater marble richness from the evenly distributed jar than the blue-dominated jar. In this case, standardizing observed richness at 10 marbles falsely suggests that one jar is richer than the other. This is because the samples vary in their coverage. A greater fraction of total marble richness is observed in a 10-marble sample from the evenly distributed jar than the blue-dominated jar.

To account for differences in sample coverage, one can estimate the coverage of the sample and compare observed richness at similar estimated coverage levels (but different sample efforts). Cao and colleagues (2002) estimated coverage by calculating the "autosimilarity" of a sample. [Another commonly used coverage estimator is Good's measure (Good, 1953).] Specifically, they randomly divide the sample in half and estimate the Jaccard coefficient (a similarity index) between the two sample halves. This procedure is repeated at different sampling sizes (for our purposes, the number of individuals) so that one produces a plot of autosimilarity versus individuals sampled (Fig. 7A). In other words, the procedure esti- mates coverage by asking how well one-half of the data reflects the other half of the data at a variety of sample sizes. High coverage will lead to high similarity values. Poor coverage will lead to low similarity values.

The autosimilarity curve can then be used to standardize for coverage on a traditional rarefaction curve. Figure 7B illustrates a case where two rarefaction curves are likely to cross with further sampling. The bottom dashed curve is almost linear, and the top solid curve is leveling off. Under standard rarefaction assumptions, these curves would lead to the conclu- sion that the solid-line community is more diverse than the dashed- line community. Using the Cao method to produce autosimilarity curves (Fig. 7A), one estimates that the highest common coverage of the samples is 30% (a similarity value of 0.3). The autosimilarity curves reveal the sampling effort (in terms of number of individuals) needed to standardize for this coverage value; lines dropping down the x axis in Fig. 7A estimate the number of individuals sampled so that the observed richness represents 30% of the true richness. Reading the observed richness at that sample size on the traditional rarefaction curve in Fig. 7B yields two observed richness values, standardized by coverage. By the coverage-rarefaction method, the dashed curve is now estimated to have a higher observed richness than the solid curve.

How well this method works in general for correctly predicting relative richness remains to be seen. Combined with strategies to sample multiple sites or treatment replicates, it has the potential to improve predictions of relative richness using rarefaction curves.

Rarefaction for Other Community Analyses

The problem of standardizing for sample size rears its head in diversity comparisons other than richness, particularly for comparisons of commu- nity similarity. Community similarity can be calculated by a variety of

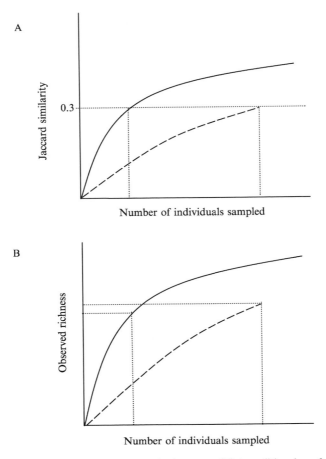

Fig. 7. (A) An autosimilarity versus sample size curve. (B) A traditional rarefaction curve. See text for explanation.

indices, including the Jaccard index, which considers only the presence or absence of taxa, and the Bray–Curtis index, which also considers the abundance of each taxon (Magurran, 1988). This section discusses how similarity indices are biased by sample size and presents a type of rarefaction analysis to account for this bias.

While most ecologists studying large organisms usually ignore the problem of effort standardization for composition comparisons among samples, the problem is especially critical in microbial studies for two reasons. First, estimates of community similarity are highly negatively biased because microbial inventories represent such small fractions of the total community. Second, molecular inventory techniques often produce uneven number of individuals, usually sequences, between samples. Differences in sampling effort among samples will bias the estimates of true community similarity.

This issue is well demonstrated by considering two samples drawn from the same microbial community. For example, imagine two clone libraries made from the same PCR products from a community that has over 500 bacterial "species." Because the samples are taken from the same community, the true similarity is 1. However, if only 10 clones are analyzed from each library, the similarity value is likely to be low. The low similarity value is not because the communities are actually different, but simply because the overall diversity of the microbial community is so high that the chance of drawing the same composition of 10 clones is very small. If sampling effort is increased and 200 clones are drawn, then the estimate of community similarity will increase toward 1.

As with richness, microbial ecologists are usually more interested in measures of relative community similarity than true community similarity. For instance, one may want to know whether the microbial community of a forest gap is more or less similar to the community of the surrounding forest or a nearby grassland. As long as the samples are standardized for sampling effort, then with a rigorous sampling design, biologically relevant comparisons such as these can be made even without knowing the true community similarities.

As an alternative to randomly throwing out data from samples that are overrepresented, one can perform a randomization procedure like that for typical rarefaction curves. This randomization procedure would estimate $E(C_{ij,n})$, i.e., an expected value of similarity (C) between samples i and j given a sampling effort of n individuals. In brief, at the highest common number of individuals sampled, one draws n individuals at random without replacement from each sample. Then, one calculates all the C_{ij} values (the similarity values between all sample pairs). These randomization and calculation steps are repeated over and over to calculate the $E(C_{ij,n})$ values.

As of yet, we know of no software that offers this type of analysis, but this is a relatively simple programming problem. Horner-Devine et al. (2004) used this procedure to compare 26 salt marsh samples of sediment bacteria. A completely unaddressed problem is how likely further sampling might alter relative similarity values among a group of samples; i.e., whether there is an analogous curve-crossing problem as in species richness measurement with rarefaction.

Conclusions

Rarefaction has promise as a reliable method for comparing molecular inventories of microbial communities. The method is easy to perform with freely available software and it is a quick, first-cut approach to surmise potential differences between microbial communities. As with all diversity statistics, however, users must consider carefully the limitations of rarefaction when offering interpretation from its results. First and foremost, rarefaction, like other diversity statistics, performs better as one samples a larger and larger fraction of the diversity of a community. When rarefaction curves are steep and linear, then any diversity statistic calculated from a sample is unlikely to reflect the true community. All diversity analyses are highly suspect when the community is so undersampled that the rarefaction curves are linear and steep (close to a slope of 1).

Specifically with regards to rarefaction, we offer five summary guidelines for its use and interpretation.

- Rarefaction compares observed richness among samples for a given level of sampling effort. It does not attempt to estimate true richness of a community.
- A rarefaction curve must be drawn with confidence intervals to make comparisons against other rarefaction curves. These confidence limits are essential to assess whether variation in the random order of the sampling of individuals may account for apparent differences among the curves.
- Comparisons between two rarefaction curves address whether the observed richness of the *samples* differs, not whether the richness of the *communities* from which the samples were drawn differs. This limitation holds true because the confidence intervals around a rarefaction curve do not give a measure of precision of the observed richness.
- Repeated samples from the same community can be used to estimate the precision of rarefaction and thus compare *communities* from which the samples are taken.
- Finally, rarefaction analyses on small samples do not necessarily yield the correct order of the true richness of the sample, as rarefaction curves may cross with further sampling.

Because of the last point, we recommend that rarefaction be used in concert with other diversity estimators. Diversity measures vary in their performance with respect to bias, precision, and accuracy, and each captures different qualities of a community and has unique benefits and failings (Hellmann and Fowler, 1999; Palmer, 1990; Walther and Morand, 1998). The most robust assessment of a microbial community is one that decides what it aims to accomplish (i.e., comparison of relative diversity or prediction of true diversity) and uses several indices to accomplish that goal. For the most part, the goal of microbial ecologists is to distinguish between relative diversity measures, such as the differentiation and ordering of richness among communities or treatments. If the chosen indices give different assessments, then more sampling is necessary before strong conclusions can be made. If all measures point to the same ordering of relative diversity among samples, then the statistical interpretation is at least robust under a variety of assumptions.

References

Acinas, S. G., Klepac-Ceraj, V., Hunt, D. E., Pharino, C., Ceraj, I., Distel, D. L., and Polz, M. F. (2004). Fine-scale phylogenetic architecture of a complex bacterial community. *Nature* **430,** 551–554.

Bohannan, B. J. M., and Hughes, J. (2003). New approaches to analyzing microbial biodiversity data. *Curr. Opin. Microbiol.* **6,** 282–287.

Brose, U., Martinez, N. D., and Williams, R. J. (2003). Estimating species richness: Sensitivity to sample coverage and insensitivity to spatial patterns. *Ecology* **84,** 2364–2377.

Cao, Y., Williams, D. D., and Larsen, D. P. (2002). Comparison of ecological communities: The problem of sample representativeness. *Ecol. Monogr.* **72,** 41–56.

Chao, A. (1984). Nonparametric estimation of the number of classes in a population. *Scand. J. Stat.* **11,** 265–270.

Coleman, B. D. (1981). On random replacement and species-area relations. *Math. Biosci.* **54,** 191–215.

Colwell, R. K. (2004). EstimateS: Statistical estimation of species richness and shared species from samples. Version 7. User's Guide and application published at http://purl.oclc.org/estimates.

Colwell, R. K., Mao, C. X., and Chang, J. (2004). Interpolating, extrapolating, and comparing incidenced-based species accumulation curves. *Ecology* **85,** 2717–2727.

Curtis, T. P., Sloan, W. T., and Scannell, J. W. (2002). Estimating prokaryotic diversity and its limits. *Proc. Natl. Acad. Sci. USA* **99,** 10494–10499.

Dunbar, J., Ticknor, L. O., and Kuske, C. R. (2001). Phylogenetic specificity and reproducibility and new method for analysis of terminal restriction fragment profiles of 16S rRNA genes from bacterial communities. *Appl. Environ. Microbiol.* **67,** 190–197.

Good, I. J. (1953). The population frequencies of species and the estimation of population parameters. *Biometrika* **40,** 337–364.

Gotelli, N. J., and Colwell, R. K. (2001). Quantifying biodiversity: Procedures and pitfalls in the measurement and comparison of species richness. *Ecol. Lett.* **4,** 379–391.

Heck, K. L., Belle, G. v., and Simberloff, D. (1975). Explicit calculation of the rarefaction diversity measurement and the determination of sufficient sample size. *Ecology* **56,** 1459–1461.

Hellmann, J. J., and Fowler, G. W. (1999). Bias, precision, and accuracy of four measures of species richness. *Ecol. Appl.* **9,** 824–834.

Horner-Devine, M. C., Lage, M., Hughes, J. B., and Bohannan, B. J. M. (2004). A taxa-area relationship for bacteria. *Nature* **432,** 750–753.

Hughes, J. B., and Bohannan, B. J. M. (2004). Application of ecological diversity statistics in microbial ecology. *In* "Molecular Microbial Ecology Manual" (G. A. Kowalchuk, F. J. de Bruijn, I. M. Head, A. D. Akkermans, and J. D. van Elsas, eds.), pp. 1321–1344. Springer, Berlin.

Hughes, J. B., Hellmann, J. J., Ricketts, T. H., and Bohannan, B. J. M. (2001). Counting the uncountable: Statistical approaches to estimating microbial diversity. *Appl. Environ. Microbiol.* **67,** 4399–4406.

Hurlbert, S. H. (1971). The nonconcept of species diversity: A critique and alternative parameters. *Ecology* **52,** 577–585.

Magurran, A. E. (1988). "Ecological Diversity and Its Measurement." Princeton University, Princeton.

Martin, A. P. (2002). Phylogenetic approaches for describing and comparing the diversity of microbial communities. *Appl. Environ. Microbiol.* **68,** 3673–3682.

Palmer, M. W. (1990). The estimation of species richness by extrapolation. *Ecology* **71,** 1195–1198.

Sanders, H. (1968). Marine benthic diversity: A comparative study. *Am. Nat.* **102,** 243–282.

Singleton, D., Furlong, M., Rathbun, S., and Whitman, W. (2001). Quantitative comparisons of 16S rRNA gene sequence libraries from environmental samples. *Appl. Environ. Microbiol.* **67,** 4374–4376.

Thompson, J. R., Marcelino, L. A., and Polz, M. F. (2002). Heteroduplexes in mixed-template amplifications: Formation, consequence and elimination by 'reconditioning PCR.' *Nucleic Acids Res.* **30,** 2083–2088.

Tyson, G. W., Chapman, J., Hugenholtz, P., Allen, E. E., Ram, R. J., Richardson, P. M., Solovyev, V. V., Rubin, E. M., Rokhsar, D. S., and Banfield, J. F. (2004). Community structure and metabolism through reconstruction of microbial genomes from the environment. *Nature* **428,** 37–43.

Venter, J. C., Remington, K., Heidelberg, J. F., Halpern, A. L., Rusch, D., Eisen, J. A., Wu, D. Y., Paulsen, I., Nelson, K. E., Nelson, W., Fouts, D. E., Levy, S., Knap, A. H., Lomas, M. W., Nealson, K., White, O., Peterson, J., Hoffman, J., Parsons, R., Baden-Tillson, H., Pfannkoch, C., Rogers, Y. H., and Smith, H. O. (2004). Environmental genome shotgun sequencing of the Sargasso Sea. *Science* **304,** 66–74.

von Wintzingerode, F., Gobel, U. B., and Stackebrandt, E. (1997). Determination of microbial diversity in environmental samples: Pitfalls of PCR-based rRNA analysis. *FEMS Microbiol. Rev.* **21,** 213–229.

Walther, B. A., and Morand, S. (1998). Comparative performance of species richness estimation methods. *Parasitology* **116,** 395–405.

[18] Culture-Independent Microbial Community Analysis with Terminal Restriction Fragment Length Polymorphism

By Terence L. Marsh

Abstract

Terminal restriction fragment length polymorphism is a polymerase chain reaction (PCR)-based technique that has been used to effectively interrogate microbial communities to determine the diversity of both phylogenetic and functional markers. It requires the isolation of community DNA and knowledge of the target sequence. PCR amplification, performed with fluorescently labeled primers, is followed with restriction digestion and

METHODS IN ENZYMOLOGY, VOL. 397   0076-6879/05 $35.00
          DOI: 10.1016/S0076-6879(05)97018-3

size selection on automated sequencing systems. The fluorescent tag identifies the terminal fragment, and the length polymorphism of the terminal fragments reveals a fraction of the phylogenetic diversity within the target sequence. Because the technique has high-throughput capabilities, it performs well in surveys where a large number of samples must be interrogated to ascertain spatial or temporal changes in community structure.

Introduction

There are many highly complex and seemingly intractable problems remaining in the biological sciences. Among them is the totality of microbial ecology, for in defining this discipline one has staked out the biological collective (bacteria, archaea, and lower eukaryotes) with the greatest phylogenetic, genomic, and metabolic diversity on the earth and placed it on our largest stage, that same earth. The estimated number of microbial species is in the millions if not billions (Cohan, 2001, 2002; Dykhuizen, 1998; Oren, 2004; Rossello-Mora and Amann, 2001; Torsvik *et al.*, 1990, 2002), which is understandable given the number of potential habitats that the abiotic world presents. Add to that all multicellular species to which scores of prokaryotes have coadapted, be it in a commensal, symbiotic, or parasitic association, and the task of defining simply phylogenetic diversity within the microbes becomes staggering. Defining genomic diversity becomes similarly challenging given the apparent propensity for lateral gene transfer in the prokaryotes (Kurland *et al.*, 2003; Lawrence and Hendrickson, 2003; Ochman, 2001). Hence the discipline of microbial ecology is daunting in terms of the enormous number of species and habitats and the correspondingly enormous potential complexity of metabolic webs. We are, however, beginning to detect the order and logic in the ecology of microbes through the establishment of a molecular phylogeny (Woese, 1987), the application of molecular tools (Hugenholtz and Pace, 1996; Lane *et al.*, 1985; Tiedje *et al.*, 1999), and, more recently, genomics as applied to the environment (Eyers *et al.*, 2004; Rodriguez-Valera, 2004; Venter *et al.*, 2004).

Terminal restriction fragment length polymorphism (T-RFLP) is one of those molecular techniques that have been used successfully to answer questions in microbial ecology. Specifically it allows the investigator to interrogate a microbial community in a high-throughput format that estimates the diversity of a gene or genome fragment that has been PCR amplified from community DNA. Because the technique is not dependent on cultivation, it views PCR-amplifiable targets from all populations, even those that have yet to be cultivated. The technique is dependent on automated sequencing and gene mapping systems that make use of fluorescent dye chemistry for fragment detection. By using a 5' fluorescently tagged primer in the PCR amplification step, followed by restriction with an endonuclease and size selection on automated systems, only the restriction fragment that includes the primer, i.e., the terminal fragment, is detected by the fluorescent sensor. Initially the approach was reported as a means of strain comparison (Cancilla *et al.*, 1992) using randomly primed PCR reactions. Brunk's laboratory refined the technique by using it to target a phylogenetically robust marker, 16S rRNA (Avaniss-Aghajani *et al.*, 1994, 1996), and not long after this, the utility of the approach as applied to complex microbial communities was recognized (Bruce, 1997; Clement *et al.*, 1998; Liu *et al.*, 1997). Since these initial descriptions of the technique as applied to communities, over 150 publications report its use to examine microbial communities in a wide diversity of ecosystems. Moreover, a number of papers have provided insight into technical aspects of the method (Blackwood *et al.*, 2003; Dunbar *et al.*, 2001; Kaplan and Kitts, 2003; Kitts, 2001; Marsh, 1999; Marsh *et al.*, 2000; Osborn *et al.*, 2000).

Technically there are several advantages to the method. First, degenerate primers can be employed, thus expanding the potential range of markers targeted by a single primer set. Second, only the terminal fragment is monitored, yielding a single labeled product from an amplified sequence and a rough estimate of target diversity. Third, both phylogenetic and functional markers can be targeted. Finally, the method has high-throughput potential and can be multiplexed using three different fluors. Because of the cost effectiveness of the method, it has become a frequently utilized approach when ecological questions requiring extensive sampling across large temporal or spatial ranges are addressed. The power of the technique is in its ability to compare complex microbial communities. Figure 1 shows an electropherogram of terminal restriction fragments derived from 16S rRNA genes amplified from soil. The fluorescent intensity of the PCR products is plotted on the ordinate and the fragment size, in bases, on the abscissa. Figure 1 shows profiles derived from sampling sites that are adjacent and geochemically similar. As can be seen, the patterns of terminal fragments reveal bacterial communities that appear quite similar. Because reproducibility across lanes and capillaries (depending on the system) is excellent (see later), comparisons between communities are relatively robust.

General Protocol for Community Analysis with T-RFLP

The experimental flow in community analysis using T-RFLP is shown in Fig. 2 with the requisites and potential pitfalls associated with each step shown to the right. The descriptions provided in this section are numbered according to Fig. 2 and are general in nature, serving to provide an overview of the technique. The following protocol section provides details for critical steps in the technique.

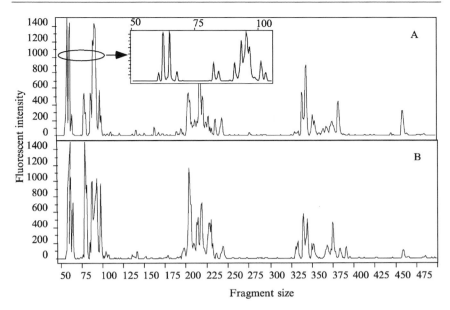

Fᴵɢ. 1. T-RFLP profiles. Electropherogram of terminal fragments derived from *Rsa*I digestion of amplified 16S rRNA genes from soil. (A and B) Two different soil samples taken from geographically proximal sites. The insert in A shows an expanded size range (50–100 bases).

Experimental Design (1.)

As mentioned earlier, T-RFLP is a cost-effective, high-throughput technique that can be applied to ecological questions requiring extensive sampling. In order to reach well-founded conclusions, the quality and quantity of the sampling must be considered carefully (Quinn and Keough, 2002; Scheiner and Gurevitch, 2001; Underwood 1996). Once the experimental design is established and the investigator begins to take samples, sample storage is a concern as community demographics can shift and populations may increase or decrease as a result of inappropriate storage conditions. As a general rule of thumb, samples should be frozen at −80° as soon as possible and stored at the same temperature until processed.

Extraction of Community DNA/RNA (2.)

Extraction of nucleic acids from a variety of habitats and substrata has been described in considerable detail over the past two decades. A detailed discussion of these protocols is beyond the scope of this review. However, commercially available kits have provided rapid and relatively reliable procedures. Purified community DNA can be stored for extended periods as a precipitate in 70% ethanol at −20°.

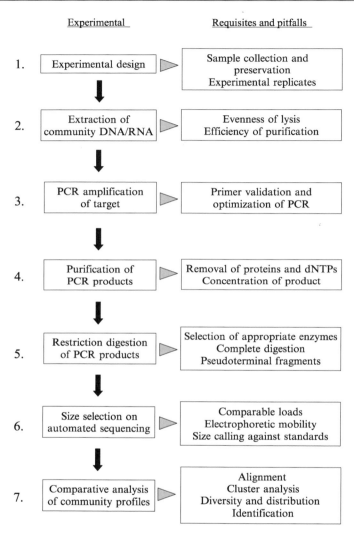

FIG. 2. Experimental protocol of T-RFLP. The series of experimental steps for T-RFLP analysis is given in the left column (solid black arrows), whereas requisites and potential pitfalls for each step are indicated in the right column.

PCR Amplification of Target (3.)

a. Extracted DNA is tested for PCR amplification with validated unlabeled primers. The reason for this pilot experiment is to confirm that the expected product is obtained under the selected PCR conditions with no evidence of other products. Small volume (15 μl) pilot reactions can be run with the resulting products screened on agarose gels (Sambrook and Russell, 2001).

b. To reduce the possibility of nonspecific priming, hot starts or touchdown PCR protocols are recommended (Don *et al.*, 1991).

c. As with any PCR amplification, there is an optimum balance of components and cycling conditions that will, in the best circumstances, lead to efficient and specific product formation. In general, the template and magnesium concentrations, as well as the annealing temperature, are the variables manipulated first to achieve optimization (Ekman, 1999). More elaborate schemes have been proposed that attempt to take all possible variables into consideration (Cao *et al.*, 2004).

d. The temperatures of cycling are usually established through optimization, the specific requirements for the *Taq* employed, and the predicted stability of primer–template interactions. The number of cycles should be kept to 20–25 rather than 30–40 so that PCR bias (Suzuki and Giovannoni, 1996) is minimized.

e. After optimization has been confirmed with unlabeled primers, a fluorescently tagged primer is substituted and the reaction volume is increased to 100 μl. The fluors used most frequently are HEX, FAM, TAMRA, and ROX. Frequently, two to three 100-μl PCR reactions will be required to provide enough PCR product for subsequent digestion and sample replications. Duplicate reactions are combined and purified as described later.

f. Note that the maximum length of read on a sequencing gel or capillary is approximately 600 or 1000 bases, respectively. Thus, large fragments (>1000 bases) cannot be sized accurately and are, in fact, not sized at all if they are outside of the range of the size standards. If the PCR-amplified target is greater than 1000 bp, using both forward- and reverse-labeled primers may be required for complete coverage. For example, in the case of amplified 16S/18S rDNAs, labeling both forward and reverse primers will provide coverage of the entire amplified gene. Different fluors can be used so that the individual forward- and reverse-labeled reactions can be combined prior to digestion and loaded onto the same lane/capillary.

Purification of PCR Products (4.)

Labeled PCR products can be purified using several commercially available kits (e.g., Qiagen or Promega). This is an important step because it strips away any constituents of the amplification that may adversely influence the restriction digestion and permits a more accurate estimate of product concentration.

Restriction Endonuclease Digestion (5.)

Digestion of the purified PCR products with a restriction endonuclease reveals the evolutionarily based polymorphism within the amplified target sequence. For this reason the digestion reaction should follow the recommendations of the manufacturer to the letter. Digestions should be checked for completeness by employing a PCR product from a pure culture or clone, run in parallel as well as mixed with sample PCR product. Under no circumstances should digestions be allowed to incubate for extended periods of time (>4 h) so as to avoid star or other illegitimate activities. Digestions are stopped by 65° incubations (10 min) or as recommended by the vendor. If electrophoresis is to be on a gel, then the samples are ready immediately after digestion. If capillaries are to be used, the digestion products may require desalting to ensure efficient injection onto the capillary. This is achieved easily and economically by ethanol precipitation of the products.

a. The selection of restriction endonucleases is critical because the conservation of restriction site positions within the target sequence varies considerably. Thus, some endonucleases will reveal diversity considerably better than others, some may reveal specific populations against a community background but not reveal total diversity, and others may reveal little because the site is highly conserved. The selection of endonucleases is made based primarily on the need to detect as much diversity as possible or to track a specific population.

b. Egert and Friedrich (2003) pointed out the possibility of pseudo-terminal fragments in the T–RFLP procedure. Such a fragment is produced when a restriction endonuclease recognizes a target sequence that has been formed within heterologous duplexes or even single-stranded products that have folded intramolecularly forming restriction sites. The rapid thermocycling of the PCR amplification can lead to these nonequilibrium structures. Treatment with single-strand-specific nucleases after PCR amplification has been shown to remove these structures (Egert and Friedrich, 2003).

Sequencing Gels and Capillaries (6.)

Size selection of the digested PCR products occurs on an automated sequencing system capable of detecting the fluorescent label. Both slab gels and capillaries from several manufacturers have been used successfully. Several aspects of the technique are relevant to this step, including (a) the fluorescent quantity loaded, (b) estimation of fragment size based on internal standards, and (c) the electrophoretic mobility. These are discussed below.

a. As mentioned earlier, the strength of the technique lies in comparative community analysis. Hence establishing comparable fluorescent loads across all of the communities in the comparison is imperative. Significant differences in the amount of fluorescently labeled product from one lane/capillary to another can lead to substantial differences in the number of detectable fragments, which in turn could lead to false conclusions. To ensure that loads are comparable, the PCR product should be quantitated prior to digestion using spectrophotometry, fluorometry, or other reliable techniques. Following gel/capillary electrophoresis, the total fluorescent load of each community should be calculated by summing either the peak heights or areas within the size range limitations of the system and dropping community profiles that differ by more than 25%. Note that this is an arbitrary cutoff point that can influenced by, for example, disparity in diversity between communities. Hence, the investigator must determine a reasonable cut-off for each particular experiment. This is done effectively by a serial dilution of a test sample to determine the point at which two samples that differ in load begin to diverge significantly.

b. Estimates of fragment size are derived from comparisons made with known standards that are incorporated into every lane or capillary. These size standards are labeled with an alternate fluor and are used to construct a standard curve that is, in turn, used to size each fragment based on a user-selected algorithm implemented in the software of the analyzer. For example, ABI GeneScan (Applied Biosystems Instruments, Foster City, CA) software permits the user to select second- or third-order least-squares analysis, cubic spline interpolation, and local or global Southern methods. These estimates are performed after the known values for the size standards have been entered and the electrophoretic mobility for each detectable fragment has been calculated. Because each lane/capillary has an internal size standard, the sizing calls made from one lane/capillary to the next are quite reproducible. Figure 3 shows the reproducibility of a collection of fragment sizes ranging from 75 to 1000 bases run on 16 different capillaries. Figure 3A shows the range in bases by which the sizing varied across the capillaries, and Fig. 3B shows the standard deviation for each fragment size. In 16 different capillaries, identical fragment lengths ranged from 0.12 to 0.35 base differences in length with greater differences seen at larger fragment sizes. Hence, across lanes/capillaries we expect to see less than a 0.5 base difference in the called sizes of identical fragments. This has been our experience with both gels and capillaries.

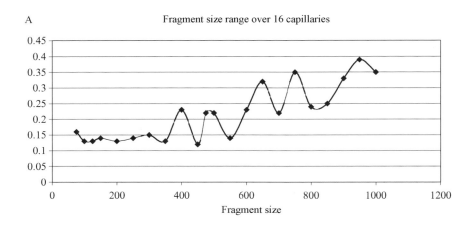

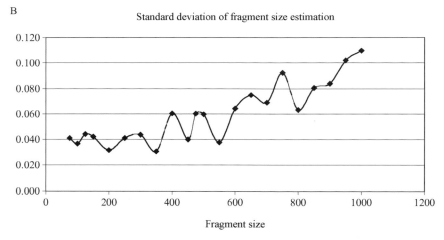

FIG. 3. Reproducibility of fragment sizing across 16 separate capillaries and 22 size standards ranging from 75-1000 bases in length. Standards were obtained from Bioventures Inc. (MapMarker-1000). Two identical standards (same sequence for each oligomer) labeled with either ROX or FAM were loaded and run on an ABI 3100 capillary machine. FAM-labeled oligomers were sized with ABI GeneScan™ software using the ROX-labeled standards. (A) Range of estimated sizes detected in 16 capillaries. (B) Standard deviation of estimated fragment sizes in 16 capillaries.

c. It has been pointed out by Kaplan and Kitts (2003) that there can be variation between the observed and the true terminal fragment length. The mobility of an oligonucleotide is a function of its mass, net charge, and specific sequence. Differences in sequence composition can translate into different local secondary structures that are influenced by temperature and denaturant concentration. Ultimately these differences may translate into electrophoretic mobility differences. Kaplan and Kitts (2003) measured mobility differences within phylogenetic groups, presumably due to sequence differences, and also detected striking mobility variations caused by differences in ambient temperature between runs. Yet another sizing variable is imposed by the fluor. In a typical T-RFLP run, the size standards have fluors that are chemically distinct from the fluor used to label the PCR amplification products. This permits, as mentioned previously, the inclusion of size markers and unknowns in the same lane/capillary. Unfortunately, different fluors impart different electrophoretic mobilities to oligomers. Figure 4A shows an electropherogram of two oligomer collections of identical sequence labeled with different fluors and run on an ABI 3100 Prism genetic analyzer. The filled peaks are labeled with 6-FAM (carboxyfluorescein) and the unfilled peaks with ROX (carboxy-X-rhodamine). Note the structural differences in these two fluors shown in Fig. 5. Figure 4B shows the size of the miscalls caused by fluor-dependent electrophoretic mobilities. The effect is the greatest in the oligomer size range of 50–125 where the miscall can exceed 5 bases. The miscall size drops to 2.5 bases from 200 to 800 bases and then increases again, probably more a function of diminishing resolution at greater fragment size rather than fluor chemistry. The net result of sequence and fluor differences is that the accuracy of absolute sizing has a relatively large uncertainty factor that could be as great as ± 7 bases. This has obvious consequences for the identification of phylotypes as discussed later.

Comparative Analysis of Community Profiles (7.)

When the detection of changes in a complex community as a function of time, chemical gradient, or experimental perturbation is desired, T-RFLP offers a powerful and cost-effective approach. The number of terminal fragments used in any comparison of communities depends on the number of targets interrogated and the level of restriction polymorphism within the targets. Thus, complementing T-RFLP analysis of 16S/18S rDNA with other targets could increase the number of potentially discriminating terminal fragments to hundreds if not thousands. In comparative analyses where demographic shifts are large, i.e., involving many populations, detection of the change may be relatively easy. However, in complex communities where a single invading species changes the demographics only to the extent of addition of the invader, detecting change may be difficult without *a priori* knowledge of the target (invader).

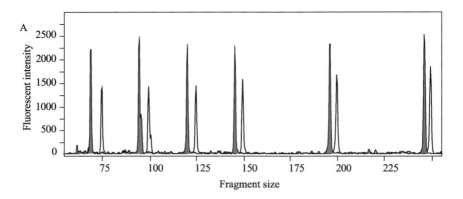

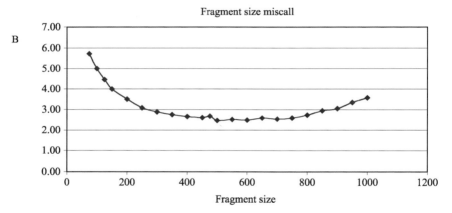

FIG. 4. Size of fragment miscalls derived from fluor-imparted differences in electrophoretic mobility. Size standards labeled with ROX or FAM, were loaded and run as described in Fig. 3. (A) Electropherogram of ROX and FAM (solid) labeled fragments in the size range 50–250 bases. (B) The size of the fragment size miscall in bases is reported for each of 22 FAM-labeled oligomers ranging in size from 75 to 1000 bases based on comparison with ROX-labeled standards using the local Southern algorithm as implemented in GeneScan.

Specific Protocols

Isolation of Community DNAs

As mentioned earlier, many protocols have been designed to provide efficient lysis/rupture of cells and recovery of DNA from problematic contaminating compounds such as humics (e.g., Burgmann *et al.*, 2001; Zhou *et al.*, 1996). Many investigators have resorted to commercially

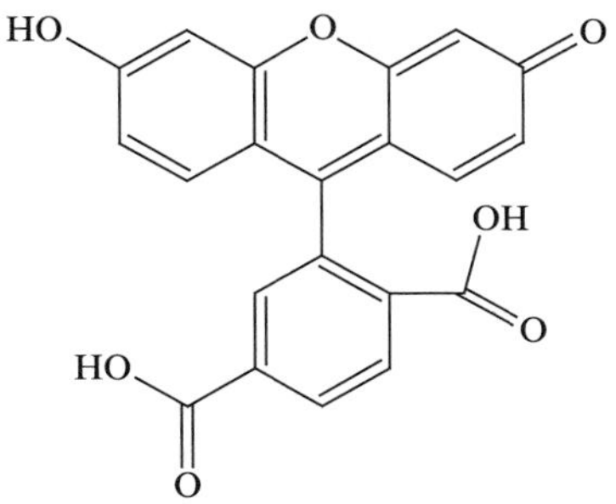

6-FAM
MW, 273.39; Abs, 496; Em, 516

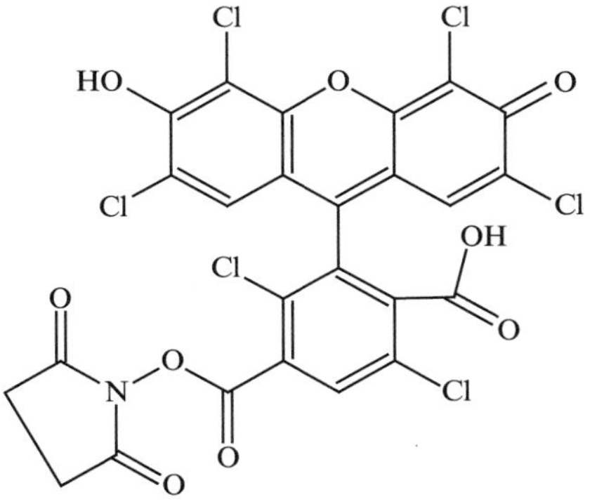

6-HEX
MW, 680.07; Abs, 533; Em, 550

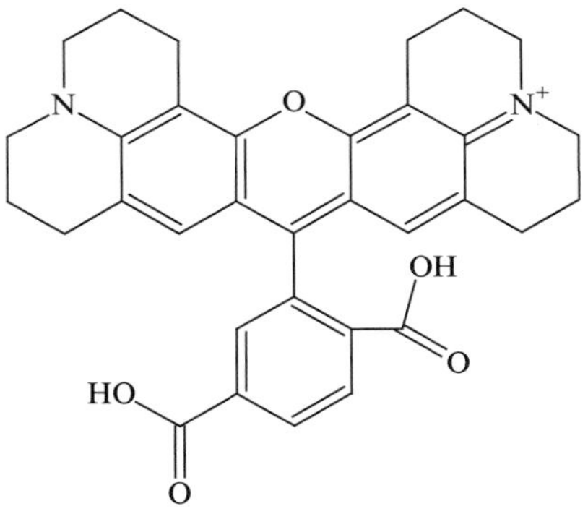

6-ROX
MW, 534.61; Abs, 570; Em, 590

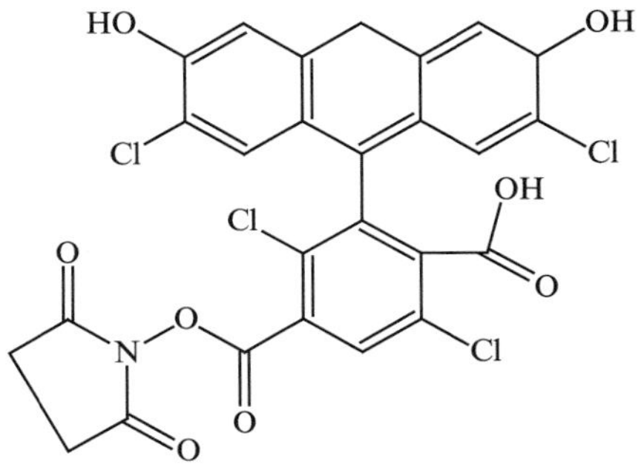

6-TET
MW, 611.18; Abs, 521; Em, 536

FIG. 5. Structure of fluors used frequently for T-RFLP. Data for structure, molecular weight (MW), absorbance maximum (Abs), and emission maximum (Em) are from Molecular Probes (http://probes.invitrogen.com).

available kits that provide an efficient lysis system (bead beating) coupled with strong chemical denaturation and DNA purification systems (Bio101 and MoBio). The physical lysis achieved with aggressive bead beating ensures that even tough spores are broken. The effectiveness of lysis should be checked microscopically.

PCR Amplification of Target

Selection of Primers. As mentioned earlier, any target with a known sequence can be profiled. Amplification products that are at least 1000 bp long will take advantage of the average frequency of four base restriction targets as well as the size range capabilities of sequencing gels and capillaries. For 16S/18S rRNA genes, the primers listed in Table I have been used successfully to give a near full-length product (~1500 bp). Note that degeneracy within primers has no appreciable effect on electrophoretic mobility, thus primers can be designed to target both broad and narrow sequence diversity. Many 16S/18S phylogenetically general and specific primers and probes have been designed that can be used effectively (Amann *et al.*, 1995; Lipski *et al.*, 2001; Marchesi *et al.*, 1998; Raskin *et al.*, 1994) and a probe database is maintained at the University of Vienna (http://www.microbial-ecology.de/probebase/).

Validation of Primers. Primers, especially primers with a sequenced-imposed phylogenetic specificity, must be validated in amplification pilot reactions. The easiest approach for this is to run replicate PCR amplifications in a thermocycler capable of establishing a thermal gradient across the block. The annealing temperature at which the cognate target is

TABLE I
COMMONLY USED PRIMERS FOR AMPLIFICATION OF 16S rRNA GENES

Name	Target on rRNA	Specificity	Sequence
1511R[a]	1526–1511	Universal	5' YGC AGG TTC ACC TAC
1492R[b]	1506–1492	Universal	5' ACC TTG TTA CGA CTT
27F[b]	27-Aug	Bacteria 16S	5' AGA GTT TGA TCM TGG CTC AG
63F[c]	42–63	Bacteria 16S	5' CAG GCC TAA YAC ATG CAA GTC
1387R[c]	1404–1387	Bacteria 16S	5'-GGG CGG WGT GTA CAA GGC
21F[b]	21-Feb	Archaea 16S	5' TTC CGG TTG ATC CYG CCG GA
25F[a]	25-Sep	Eukarya 18S	5' CTG GTT GAT CCT GCC AG

[a] From Medlin *et al.* (1988).
[b] From Amann *et al.* (1995).
[c] From Marchesi *et al.* (1998).

amplified and noncognate targets are not is the annealing temperature with suitable specificity.

Pilot Reactions. In order to conserve fluorescently labeled primers, pilot level reactions are run to establish optimum conditions. In many cases, DNA extracted from environmental sources carry salts and metals that vary from habitat to habitat. Thus, each sample may need to be optimized. The stability of the primer–template association is the main driver of PCR conditions. The primers in general use for amplification of 16S/18S rDNA targets do not require unusual protocols. The ionic conditions that provide a reasonable starting place for optimization are presented in Table II. Initial reactions are set up using a short range of template concentrations (0.01, 0.1, 1.0, 2.0 ng/μl) under fairly general thermocycling parameters, i.e., a hot start (94°) followed by 25 cycles (1 min at 94°, 45 s at 56°, 1 min at 72°). The touchdown PCR protocol may provide additional specificity (Don *et al.*, 1991). As with any PCR amplification, both positive and negative controls should be included in each experiment. PCR products are visualized on ethidium bromide (or equivalent)-stained 1% agarose gels run in TAE or TBE buffer (Sambrook and Russell, 2001). A successful amplification is indicated by evidence of ample product of the expected size and no evidence of illegitimate products. In samples where secondary products are detected on agarose gels, the reactions need to be optimized. As a first approach to optimization, the annealing temperature and/ or [Mg^{2+}] should be manipulated. Increasing the annealing temperature increases the stringency under which the primer–template complex forms while decreasing the [Mg^{2+}] destabilizes duplexes, especially any nonspecific primer–template complexes with base mismatches. Effective

TABLE II
REACTION COMPONENTS FOR T-RFLP PCR AMPLIFICATION

	Total reaction volume			
Stock solution	15 μl	25 μl	50 μl	100 μl
Sterile deionized H_2O	9.15 μl	15.25 μl	30.5 μl	61 μl
10× PCR buffer	1.5 μl	2.5 μl	5.0 μl	10 μl
dNTP mix (2 mM each)	1.5 μl	2.5 μl	5.0 μl	10 μl
$MgCl_2$ (50 mM)	0.45 μl	0.75 μl	1.5 μl	3.0 μl
Bovine serum albumin (10 mg/ml)	0.15 μl	0.25 μl	0.5 μl	1.0 μl
27F primer (10 μM)	0.3 μl	0.5 μl	1.0 μl	2.0 μl
1492R primer (10 μM)	0.3 μl	0.5 μl	1.0 μl	2.0 μl
Taq polymerase (10 U/μl)	0.15 μl	0.25 μl	0.5 μl	1.0 μl
Template DNA	1.5 μl	2.5 μl	5.0 μl	10 μl
Total volume	15 μl	25 μl	50 μl	100 μl

magnesium concentrations are usually in the range of 1.5–3.0 mM. Identifying a reasonable template concentration to use throughout the analysis is also an important consideration. Detectably different T-RFLP profiles can be obtained from the same template when different template concentrations are used in the PCR amplification. Optimal template concentrations are usually in the range of 0.5–2.0 ng/μl of PCR reaction volume. Minor populations in the community may be amplified at low undetectable levels if the total template concentration is too low.

Reactions with Labeled Primers. Because the fluors are photoreactive, it is recommended that all handling of the primers and resulting PCR products be conducted in subdued or zero light. Primer stocks should be kept at $-80°$ in nontranslucent or foil-wrapped tubes. PCR reactions to obtain labeled product can be run by simply substituting labeled primer for unlabeled, mole for mole (although HEX labeled primers are frequently less efficient and may require increasing primer concentration). These reactions should be checked on an agarose gel in the same manner as the unlabeled pilot reactions.

Purification and Quantitation of PCR Products

After PCR amplification with labeled primers, 100-μl reactions are combined (if duplicates were run) and purified using a commercially available PCR cleanup kit (e.g., Qiagen or Promega). After elution from the DNA-binding matrices, the product should be quantitated spectrophotometrically ($A_{260\ nm}$) or fluorometrically (picogreen, Molecular Probes).

Restriction Digestion

Selection of Enzymes. There is no a priori way of knowing which enzyme will be most revealing for a particular community. However, if the goal is an assessment of diversity or detection of a demographic shift, then an enzyme with good phylogenetic resolving power will be the most efficacious. In this regard there is a clear hierarchy of preferred enzymes revealed by *in silico* digestion of rRNA sequences. Judging by the number of unique terminal restriction fragments generated from a labeled 5′ terminal primer (63F), the descending order of preferred enzymes is *Nci*I > *Ava*II > *Taq*I > *Dpn*I > *Mae*I > *Hha*I > *Rsa*I > *Mae*II > *Mse*I > *Bst*NI > *Tsp*E1 > *Msp*I > *Hae*III > *Alu*I > *Aci*I > *Acc*I > *Sau*96I (see Table III). Enzymes that have been used with success are *Rsa*I, *Hha*I, *Msp*I, and *Hae*III.

TABLE III
NUMBER OF 5' PROXIMAL RESTRICTION TARGET SEQUENCES ON 16S rDNA[a]

Restriction enzyme		No. sequences primed (63F) and digested	No. of unique terminal fragments
NciI	CCCGG	18,478	917
AvaII	GGACC	15,365	911
TaqI	TCGA	18,865	874
DpnI	GATC	18,795	818
MaeI	CTAG	18,888	812
HhaI	GCGC	18,913	755
RsaI	GTAC	18,927	730
MaeII	ACGT	18,923	723
MseI	TTAA	18,928	697
BstNI	CCWGG	18,913	690
TspE1	AATT	18,926	663
MspI	CCGG	18,914	524
HaeIII	GGCC	18,917	503
AluI	AGCT	18,928	497
AciI	CCGC	18,929	428
AccI	CGCG	18,929	397
Sau96I	GGNCC	18,923	385

[a] Based on 27,060 sequences downloaded from the Ribosomal Database Project (http://rdp.cme.msu.edu/index.jsp), each with a minimum length of 1200 bases. The database was screened with PatScan (Dsouza et al., 1997) for primer and restriction sites.

Digestions. Digestion reactions should be conducted with deference to the protocol supplied by the endonuclease vendor as mentioned earlier. The recipe for a 15-μl restriction digestion is given below. Note that the range of PCR product input is large. Each automated sequencing or gene-mapping system has different sensitivities and the appropriate fluorescent load may need to be calibrated empirically. For the ABI gel systems, we routinely digested 600 ng and then loaded 60–200 ng/lane. The increased sensitivity of the capillary systems allows one to reduce the load by approximately a factor of three.

- 200–600 ng labeled PCR product.
- 5–10 units of restriction endonuclease.
- 1.5 μl 10× buffer.
- H_2O to 15 μl.
- Incubate at recommended temperature for no more than 4 h.
- Stop reaction as recommended by the vendor.

Desalting. If the digestion products are to be loaded onto a gel system, then no further steps are required. If products are to be loaded onto a capillary system, then an ethanol precipitation immediately after digestion will reduce the ionic concentration of the digestion and ensure that the products are injected onto the capillaries efficiently.

Electrophoresis of Digests

In addition to digested samples, it is recommended that uncut PCR amplifications be run on the sequencing system as well. The sensitivity of the fluorescent detection system is greater than ethidium-stained agarose gels, hence the uncut samples will serve as final confirmation of amplification legitimacy.

Load Mixtures. For gels, 1–2 μl of restriction digests is mixed with 1 μl of size standards (e.g., ABI Tamara 2500) and 6 μl deionized formamide. The mixture is denatured at 95° for 2 min and chilled (4°) until loaded. For capillaries, the ethanol precipitate is pelleted and resuspended in loading buffer (deionized formamide) and 1 μl of size standards (e.g., MM1000, Bioventures Inc.), denatured at 95° for 2 min, chilled, and injected for 10–30 s.

ABI Model 373A Polyacrylamide Gel. Thirty-six-centimeter, 6% denaturing polyacrylamide gels are run for 16–20 h in GeneScan mode with limits of 2500 V and 40 mA.

ABI 3100 Prism Genetic Analyzer (Capillary). The 36-cm capillaries containing POP4 polymer are loaded by injection and run for 2 h at 50 kV.

Sizing of Fragments. Software provided with the sequencing/mapping units includes algorithms for sizing unknowns by comparing electrophoretic mobilities with size standards. Most investigators have used the local Southern method for determining the size of sample fragments.

Comparative Analysis of T-RFLP Profiles

A detailed discussion of gene mapper software or cluster analysis algorithms is beyond the scope of this chapter. Such details are idiosyncratic and will change with each updated software version. Hence general broad issues are discussed with the hope that the user can apply these regardless of the specific software available. In general, T-RFLP data are used to monitor change or differences in community structure and to identify those phylotypes that are constant or changing.

Downloading Data. Depending on the automated analyzer, data may be downloaded in several formats. For ABI prism analyzers running GeneScan, data can be viewed within the software in the form of an electropherographic trace of fluorescent detection data or as a list of numbered fragments detected along with associated information (dye, peak height, peak area, and two columns of information related to the scan number during electrophoresis). This information can be downloaded as asci text and then opened in a spreadsheet.

Aligned Profiles. For a comparative analysis of T-RFLP profiles, proper alignment is essential. In this regard, aligning T-RFLP data is analogous to aligning sequence data for phylogenetic analysis. In the ABI system, Genotyper software allows the user to import GeneScan profiles and align, or bin, the collected data with user-selected parameters that can specify threshold values for peak height and bin width for identifying common fragments among a collection of profiles. The processed data can then be exported in tabular form, suitable for importing into a spreadsheet. ABI has introduced a new piece of software (GeneMapper v3.5) designed to supplant GenoScan and Genotyper. Note that aligned data provide the investigator with a first comparative look inasmuch as the tabulated data set reveals the phylotypes that are common to all samples as well as unique to certain environmental attributes. Once T-RFLP data and sample associated information (geochemistry, for example) are combined, data can be sorted to reveal T-RFLP patterns that correlate with sample-associated information.

Comparative Analyses. Aligned data in tabular format can be reformatted for any number of programs capable of classification analysis, for example, cluster analyses such as UPGMA and neighbor joining (Sokal and Sneath, 1963), or ordination techniques such as principal components analysis and correspondence analysis (Quinn and Keough, 2002; Underwood, 1996; Pielou, 1969) and genetic algorithms (Dollhopf *et al.*, 2001). The specific format of the data file will depend on the program.

Phylogenetic Analysis. Because T-RFLP produces a product whose length can be determined with some accuracy, there is the temptation to use profiles and specific fragment lengths diagnostically, with reference to a database. In the case of 16S/18S rRNAs, several databases (Marsh *et al.*, 2000; Kent *et al.*, 2003) are available on the Web that will inform the user what archived sequences have the same (or close) terminal fragment size for a given restriction endonuclease. The caveat for this was pointed out in Marsh *et al.* (2000), one cannot identify a species based on one or two restriction sites. One can argue that the detection of a specific fragment is consistent with the presence of a particular species, but the presence of a fragment of a particular size cannot be construed as confirmatory. This is especially true when one considers variability in sizing because of sequence (Kaplan and Kitts, 2003) or fluor-dependent sizing miscalls (see earlier discussion). A plus or minus seven base range would, in many cases, include far too many species to make identification even a remote possibility.

Final Caveats

T-RFLP is a robust technique for microbial community analysis. The attributes of high-throughput and excellent resolution and reproducibility make it an ideal technique for microbial community analysis. However, as a PCR-based technique, it is subject to all of the potential pitfalls of PCR (von Wintzingerode *et al.*, 1997). Moreover, in communities where the microbial diversity is high (e.g., soil with over 1000 species/2–3 g), even T-RFLP will resolve only a fraction of the total number of populations present using 16S rRNA and a single primer set. Whether or not the population(s) critical to the investigator's question will be detected and tracked is dependent on many factors, including primer–template matches, marker abundance, and restriction site distributions to name a few. Clearly if one is viewing only 10% of the populations within a community using T-RFLP, detecting important change may be a consequence of the fates rather than clever experimental strategies. Nonetheless, the technique has been used successfully to probe many different habitats. Used in conjunction with clone libraries of 16S rRNA genes from selected habitats, T-RFLP is a powerful technique. As our knowledge of genomics evolves, so too will our ability to target multiple markers, thus expanding the utility and sensitivity of the technique.

References

Amann, R. I., Ludwig, W., and Schleifer, K.-H. (1995). Phylogenetic identification and *in situ* detection of individual microbial cells without cultivation. *Microbiol. Rev.* **59**, 143–169.

Avaniss-Aghajani, E., Jones, K. A. H., Aronson, T., Glover, N., Boian, M., Froman, S., and Brunk, C. F. (1996). Molecular technique for rapid indentification of Mycobacteria. *J. Clin. Microbiol.* **34**, 98–102.

Avaniss-Aghajani, E., Jones, K., Chapman, D., and Brunk, C. (1994). A molecular technique for identification of bacteria using small subunit ribosomal RNA sequences. *Biotechniques* **17**, 144–149.

Blackwood, C. B., Marsh, T., Kim, S. H., and Paul, E. A. (2003). Terminal restriction fragment length polymorphism data analysis for quantitative comparison of microbial communities. *Appl. Environ. Microbiol.* **69**, 926–932.

Bruce, K. D. (1997). Analysis of mer gene subclasses within bacterial communities in soils and sediments resolved by fluorescent-PCR-restriction fragment length polymorphism profiling. *Appl. Environ. Microbiol.* **63**, 4914–4919.

Burgmann, H., Pesaro, M., Widmer, F., and Zeyer, J. (2001). A strategy for optimizing quality and quantity of DNA extracted from soil. *J. Microbiol. Methods* **45**, 7–20.

Cancilla, M. R., Powell, I. B., Hillier, A. J., and Davidson, B. E. (1992). Rapid genomic fingerprinting of *Lactococcus lactis* strains by arbitrarily primed polymerase chain reaction with 32P and fluorescent labels. *Appl. Environ. Microbiol.* **58**, 1772–1775.

Cao, Y., Zheng, Y., and Fang, B. (2004). Optimization of polymerase chain reaction-amplified conditions using the uniform design method. *J. Chem.Tech. Biotech.* **79**, 910–913.

Clement, B. G., Kehl, L. E., De Bord, K. L., and Kitts, C. L. (1998). Terminal restriction fragment patterns [TRFPs], a rapid, PCR-based method for the comparison of complex bacterial communities. *J. Microbiol. Methods* **31**, 135–142.

Cohan, F. M. (2001). Bacterial species and speciation. *Syst. Biol.* **50**, 513–524.

Cohan, F. M. (2002). What are bacterial species? *Annu. Rev. Microbiol.* **56**, 457–487.

Dollhopf, S. L., Hashsham, S. A., and Tiedje, J. M. (2001). Interpreting 16S rDNA T-RFLP data: Application of self-organizing maps and principal component analysis to describe community dynamics and convergence. *Microb. Ecol.* **42**, 495–505.

Don, R. H., Cox, P. T., Wainwright, B. J., Baker, K., and Mattick, J. S. (1991). 'Touchdown' PCR to circumvent spurious priming during gene amplification. *Nucleic Acids Res.* **19**, 4008.

Dsouza, M., Larsen, N., and Overbeek, R. (1997). Searching for patterns in genomic data. *Trends Genet.* **13**, 497–498.

Dunbar, J., Ticknor, L. O., and Kuske, C. R. (2001). Phylogenetic specificity and reproducibility and new method for analysis of terminal restriction fragment profiles of 16S rRNA genes from bacterial communities. *Appl. Environ. Microbiol.* **67**, 190–197.

Dykhuizen, D. E. (1998). Santa Rosalia revisited: Why are there so many species of bacteria? *Antonie Van Leeuwenhoek* **73**, 25–33.

Egert, M., and Friedrich, M. W. (2003). Formation of pseudo-terminal restriction fragments, a PCR-related bias affecting terminal restriction fragment length polymorphism analysis of microbial community structure. *Appl. Environ. Microbiol.* **69**, 2555–2562.

Ekman, S. (1999). PCR optimization and troubleshooting, with special reference to the amplification of ribosomal DNA in lichenized fungi. *Lichenologist* **31**, 517–531.

Eyers, L., George, I., Schuler, L., Stenuit, B., Agathos, S. N., and El Fantroussi, S. (2004). Environmental genomics: Exploring the unmined richness of microbes to degrade xenobiotics. *Appl. Microbiol. Biotechnol.* **66**(2), 123–130.

Hugenholtz, P., and Pace, N. R. (1996). Identifying microbial diversity in the natural environment: A molecular phylogenetic approach. *Trends Biotechnol.* **14**, 190–197.

Kaplan, C. W., and Kitts, C. L. (2003). Variation between observed and true terminal restriction fragment length is dependent on true TRF length and purine content. *J. Microbiol. Methods* **54**, 121–125.

Kent, A. D., Smith, D. J., Benson, B. J., and Triplett, E. W. (2003). Web-based phylogenetic assignment tool for analysis of terminal restriction fragment length polymorphism profiles of microbial communities. *Appl. Environ. Microbiol.* **69**, 6768–6776.

Kitts, C. L. (2001). Terminal restriction fragment patterns: A tool for comparing microbial communities and assessing community dynamics. *Curr. Issues Intest. Microbiol.* **2**, 17–25.

Kurland, C. G., Canback, B., and Berg, O. G. (2003). Horizontal gene transfer: A critical view. *Proc. Natl. Acad. Sci. USA* **100**, 658–662.

Lane, D. J., Pace, B., Olsen, G. J., Stahl, D. A., Sogin, M. L., and Pace, N. R. (1985). Rapid determination of 16S ribosomal RNA sequences for phylogenetic analyses. *Proc. Natl. Acad. Sci. USA* **82**, 6955–6959.

Lawrence, J. G., and Hendrickson, H. (2003). Lateral gene transfer: When will adolescence end? *Mol. Microbiol.* **50,** 739–749.

Lipski, A., Friedrich, U., and Altendorf, K. (2001). Application of rRNA-targeted oligonucleotide probes in biotechnology. *Appl. Microbiol. Biotechnol.* **56,** 40–57.

Liu, W. T., Marsh, T. L., Cheng, H., and Forney, L. J. (1997). Characterization of microbial diversity by determining terminal restriction fragment length polymorphisms of genes encoding 16S rRNA. *Appl. Environ. Microbiol.* **63,** 4516–4522.

Marchesi, J., Sato, T., Weightman, A., Martin, T., Fry, J., Hiom, S., and Wade, W. (1998). Design and evaluation of useful bacterium-specific PCR primers that amplify genes coding for bacterial 16S rRNA. *Appl. Environ. Microbiol.* **64,** 795–799.

Marsh, T. L. (1999). Terminal restriction fragment length polymorphism (T-RFLP): An emerging method for characterizing diversity among homologous populations of amplification products. *Curr. Opin. Microbiol.* **2,** 323–327.

Marsh, T. L., Saxman, P., Cole, J., and Tiedje, J. (2000). Terminal restriction fragment length polymorphism analysis program, a web-based research tool for microbial community analysis. *Appl. Environ. Microbiol.* **66,** 3616–3620.

Medlin, L., Elwood, H. J., Stickel, S., and Sogin, M. L. (1988). The characterization of enzymatically amplified eukaryotic 16S-like rRNA-coding regions. *Gene* **71,** 491–499.

Ochman, H. (2001). Lateral and oblique gene transfer. *Curr. Opin. Genet. Dev.* **11,** 616–619.

Oren, A. (2004). Prokaryote diversity and taxonomy: Current status and future challenges. *Phil. Trans. R. Soc. Lond. B Biol. Sci.* **359,** 623–638.

Osborn, A. M., Moore, E. R. B., and Timmis, K. N. (2000). An evaluation of terminal-restriction fragment length polymorphism (T-RFLP) analysis for the study of microbial community structure and dynamics. *Environ. Microbiol.* **2,** 39–50.

Pielou, E. C. (1969). "An Introduction to Mathematical Ecology." Wiley-Interscience, New York.

Quinn, G. P., and Keough, M. J. (2002). "Experimental Design and Data Analysis for Biologists." Cambridge University Press, Cambridge.

Raskin, L., Stromley, J. M., Rittmann, B. E., and Stahl, D. A. (1994). Group-specific 16S rRNA hybridization probes to describe natural communities of methanogens. *Appl. Environ. Microbiol.* **60,** 1232–1240.

Rodriguez-Valera, F. (2004). Environmental genomics, the big picture? *FEMS Microbiol. Lett.* **231,** 153–158.

Rossello-Mora, R., and Amann, R. (2001). The species concept for prokaryotes. *FEMS Microbiol. Rev.* **25,** 39–67.

Sambrook, J., and Russell, D. W. (2001). "Molecular Cloning: A Laboratory Manual," 3rd Ed. Cold Spring Harbor Laboratory Press, Cold Spring Harbor, NY.

Scheiner, S. M., and Gurevitch, J. (2001). "Design and Analysis of Ecological Experiments." Oxford University Press, Oxford.

Sokal, R., and Sneath, P. H. A. (1963). "Principles of Numerical Taxonomy." Freeman, San Francisco, CA.

Suzuki, M. T., and Giovannoni, S. J. (1996). Bias caused by template annealing in the amplification of mixtures of 16S rRNA genes by PCR. *Appl. Environ. Microbiol.* **62,** 625–630.

Tiedje, J. M., Asuming-Brempong, S., Nusslein, K., Marsh, T. L., and Flynn, S. J. (1999). Opening the black box of soil microbial diversity. *Appl. Soil Ecol.* **13,** 109–122.

Torsvik, V., Goksoyr, J., and Daae, F. L. (1990). High diversity in DNA of soil bacteria. *Appl. Environ. Microbiol.* **56,** 782–787.

Torsvik, V., Overeas, L., and Thingstad, T. F. (2002). Prokaryotic diversity: Magnitude, dynamics, and controlling factors. *Science* **296,** 1064–1066.

Underwood, A. J. (1996). "Experiments in Ecology: Their Logical Design and Interpretation Using Analysis of Variance." Cambridge University Press, Cambridge.

Venter, J. C., Remington, K., Heidelberg, J. F., Halpern, A. L., Rusch, D., Eisen, J. A., Wu, D., Paulsen, I., Nelson, K. E., Nelson, W., Fouts, D. E., Levy, S., Knap, A. H., Lomas, M. W., Nealson, K., White, O., Peterson, J., Hoffman, J., Parsons, R., Baden-Tillson, H., Pfannkoch, C., Rogers, Y. H., and Smith, H. O. (2004). Environmental genome shotgun sequencing of the Sargasso Sea. *Science* **304,** 66–74.

von Wintzingerode, F., Gobel, U. B., and Stackebrandt, E. (1997). Determination of microbial diversity in environmental samples: Pitfalls of PCR-based rRNA analysis. *FEMS Microbiol. Rev.* **21,** 213–229.

Woese, C. R. (1987). Bacterial evolution. *Microbiol. Rev.* **51,** 221–271.

Zhou, J., Bruns, M. A., and Tiedje, J. M. (1996). DNA recovery from soils of diverse composition. *Appl. Environ. Microbiol.* **62,** 316–322.

[19] Quantitative Community Analysis: Capillary Electrophoresis Techniques

By Jeremy D. Semrau and Jong-In Han

Abstract

This chapter presents methodologies for RNA extraction from soils coupled with competitive reverse transcription–polymerase chain reaction and capillary electrophoresis techniques. Combined, these approaches provide new capabilities to quantify gene expression in different environments and can aid our understanding of not only community composition, but also community activity. Such information will prove important for enhancing our knowledge of how microbial communities respond to changing geochemical parameters (e.g., temperature, pH, redox conditions, substrate levels) *in situ*.

Introduction

DNA-based methods, such as the analysis of 16S ribosomal DNA and functional genes, have been used extensively for determining microbial diversity and for identifying particular microbial abilities. Such information, however, especially when partially determined from classical cultivation-dependent studies, can be incomplete due to the bias inherent in laboratory enrichments of environmental samples. Furthermore, although these techniques are very important, it must be noted that DNA techniques are inherently limited in the information they can provide, particularly as DNA is known to be stable in dead as well as living cells (Lindahl, 1993).

METHODS IN ENZYMOLOGY, VOL. 397   0076-6879/05 $35.00
          DOI: 10.1016/S0076-6879(05)97019-5

Compared with DNA methods, analysis of specific gene expression by monitoring mRNA levels can provide one with information as to the activity of cells as a result of the short half-life of mRNA ($\sim$ minutes), provided regulation is primarily at the level of transcription (Kusher, 1996; Sheridan *et al.*, 1998; Spring *et al.*, 2000). This chapter describes a new methodology to monitor mRNA transcript levels in pure cultures using new mRNA extraction techniques coupled with competitive reverse transcription (RT)–polymerase chain reaction (PCR) and capillary electrophoresis for product separation and data analyses.

Quantitative RT-PCR Techniques: Promises and Pitfalls

One can quantify gene expression a number of ways, including using competitive RT-PCR (cRT-PCR). In this procedure, a fixed amount of target mRNA is reverse transcribed and amplified together with known various amounts of an internal RNA standard of a different size (typically smaller) using the same primer set in the same reactions. The quantity of target RNA is then determined by measuring the relative amounts of target and standard products resulting from RT-PCR with the initial target concentration calculated as the value of the standard that gives the same amount of amplification. cRT-PCR is especially useful in quantifying mRNA levels from environmental samples, as tube-to-tube variations in either the RT or the PCR step can be accounted for during the procedure (Arnal *et al.*, 1999; Freeman *et al.*, 1999).

An alternative to this approach, real-time RT-PCR, is often used with external standards for quantification (Corbella and Puyet, 2003; Pfaffl and Hageleit, 2001) but tube–tube variations in enzyme efficiency are not possible to quantify or to remove using external standards. By referencing measured amounts of mRNA to those of a gene with invariant expression as an internal control (typically a housekeeping gene), real-time RT-PCR can overcome this problem. Accurate quantification, however, can be difficult, as the expression of housekeeping genes can also fluctuate, especially during varying growth phases and between samples (Bustin, 2002). The use of cRT-PCR, however, removes the uncertainty of using a housekeeping gene for reference, as quantification of gene expression with cRT-PCR is not affected by growth phase (Han and Semrau, 2004).

cRT-PCR, however, has two limitations that must be addressed. One, an internally modified RNA standard is necessary, whose construction can be a time- and cost-consuming process. Two, the value of the target mRNA concentration is typically unknown, thus a large range of competitive standard concentrations must be used to interpolate the target value successfully. To overcome these limitations, we have developed a simple and rapid method based on reverse-inner reverse primer sets to generate internal RNA standards. This technique allows for in-tube synthesis of the RNA standards, does not require digestion and cloning, and does not utilize restriction enzyme sites for the deletion of the internal fragment. Second, capillary electrophoresis (CE) can overcome the linear range limitation by allowing for complete automation from sample injection to data analysis. This provides fast separation of standard and target products, requires small samples, and enables on-column detection with negligible buffer waste while providing high precision and accuracy. Thus, more samples can be analyzed in the same amount of time, allowing one to extend the range of standard concentrations performed.

Although this RNA quantification technique has significant potential, its successful application for environmental samples can be accomplished only with an efficient and accurate RNA extraction method from soils. Poor cell lysis, RNA degradation by ubiquitous RNases, sorption onto soil surfaces, and coextraction of humic materials that affect enzymatic activity can all pose significant problems when trying to extract RNA from soils (Alm and Stahl, 2000; Lorenz and Wackernagel, 1987; Moran et al., 1993). Thus, this chapter describes a method for RNA extraction from soils, as well as competitive RT-PCR methodology coupled with CE to accurately monitor in situ microbial function and activity.

Construction of Internal RNA Standards

To ensure consistent amplification efficiency of both target and standard, homologous internal RNA standards must be designed with similar base sequences and of similar size that are amplified with the same primers used for target mRNA. A schematic outline of the procedure to create the RNA standards is shown in Fig. 1. In this procedure, chromosomal DNA is used as a template for standard construction using PCR site-directed mutagenesis by overlap extension. Here the T7 RNA polymerase promoter sequence is appended to the forward primer and a reverse-inner reverse primer set is used to create a deletion in the standard RNA, which can then be reverse transcribed and amplified using the same forward–reverse primer set for the target mRNA. It should be stressed that the choice of reverse-inner reverse primer set should yield standard RNA as close as possible in size to the target RNA to avoid any subsequent PCR amplification biases, yet the size difference must be large enough to be distinguishable via electrophoresis.

In vitro transcription is then performed using the riboprobe T7 in vitro transcription system and manufacturer's instructions (Promega) followed by DNase treatment. The PCR products can then be visualized using slab

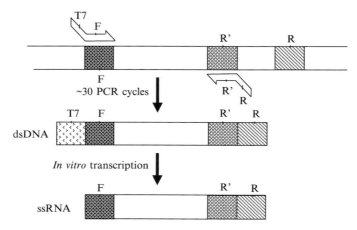

FIG. 1. Schematic diagram for construction of internal competitive standards using fused T7-RNA polymerase promoter region and forward primer sequence(T7-F) and reverse-inner reverse primer sequences (R-R') followed by *in vitro* transcription.

gel electrophoresis and the corresponding bands excised and purified using Qiagen QIAspin columns according to the manufacturer's guide (Qiagen, Valencia, CA). The quantity and quality of the transcribed RNA standards are determined by measuring the absorbance at 260 and 280 nm. To ensure that the RNA standards exclude DNA contaminants, which could later be amplified in RT-PCR assays, PCR reactions should be performed on the standards using the forward–reverse primer set with and without reverse transcription.

Extraction of mRNA

Extraction from Pure Cultures

RNase-free disposable plasticware should be used as well as glassware oven baked at 240° for at least 4 h to prevent RNA degradation. Furthermore, all solutions should be made with RNase-free compounds and diethyl pyrocarbonate (DEPC, Sigma)-treated water. For pure cultures, 2 ml of a culture grown to an $OD_{600} = \sim 0.3$ (equivalent to approximately 0.16 mg protein) pelleted by centrifugation at 12,000 rpm for 5 min at 4° yields sufficient RNA for RT-PCR experiments. After removing the supernatant, the cell pellet can be used directly in the total RNA isolation procedure or frozen and stored at −80° for later analysis.

A variety of techniques have been examined with the most simple, successful RNA extractions found using the Qiagen Total RNeasy kit (Qiagen). Briefly, cells are first resuspended in 0.7 ml extraction buffer containing guanidinium isothiocyanate and β-mercaptoethanol to inactivate RNases. The cell suspension is then disrupted mechanically, essential for quantitative RNA extraction from bacteria. In our laboratory, cell lysis is found to be most efficient using bead beating. Cell disruption is best if 1.6 g of 0.1-mm-diameter zirconia/silica beads is placed in the cell suspension followed by disruption in a minibead beater (BioSpec Products, Bartlesville, OK) six times for 30 s, with 1 min on ice between each cycle. After cell debris is removed by centrifugation at 12,000 rpm for 2 min at room temperature, the supernatant should be transferred to a new tube with an equal volume of 70% ethanol then added. The sample is then applied to an RNeasy minispin column where total RNA is bound to a silica gel membrane. Total RNA can then be eluted using 50 μl DEPC-treated H_2O. Finally, the RNA should be treated with RNase-free DNase I at 37° for 30 min and total RNA extracted using the Qiagen Total RNeasy kit and eluted using 50 μl DEPC-treated H_2O.

Extraction of mRNA from Soils

Quantitative RNA extractions from soils are challenging not only as a result of poor cell lysis and RNA degradation by ubiquitous RNases, but also due to the sorption of RNA to soil surfaces. This adsorption becomes even more problematic if conditions are extreme, such as basic or acidic pH and high salt concentrations (Lorenz and Wackernagel, 1987; Melzak *et al.*, 1996; Ogram *et al.*, 1994). Unfortunately, RNA extraction buffers typically utilize high salt conditions to inactivate RNases and acidic conditions to selectively isolate RNA from DNA (Chomczynski and Sacchi, 1987). In our laboratory, we have found that a silica-based spin column method in which RNA samples are bound onto a silica column in an extraction buffer, washed, and finally eluted with RNase-free water works well for the isolation of mRNA from soils. This approach does utilize high salt concentrations, but it has an obvious advantage in that RNA sorbed on soil particles can be recovered, as the soil itself acts as another binding matrix. RNA recovery from soil is made possible by rinsing this soil matrix with DEPC-treated water, enabling one to collect nucleic acids sorbed onto soils (Lorenz and Wackernagel, 1987; Melzak *et al.*, 1996).

For extraction of mRNA from soil samples, approximately 0.5 g of soil/ soil slurry with 0.7 ml of extraction buffer in a 2-ml microcentrifuge tube should be used, either for immediate total RNA isolation or frozen and stored at −80° for later analysis. Cells should be lysed as described earlier for pure cultures, and after bead beating, the soil/bead suspension can be sampled by piercing the bottom of the micro-centrifuge tube with a small (no larger than 22- gauge) needle. This tube is then placed within a second "collection" microcentrifuge tube. The two are thereafter placed in a 15-ml screw-cap polypropylene tube. After centrifugation for 5 min at 1500 g, the extraction buffer containing the disrupted cells will be recovered in the collection tube along with small soil particles, while zirconia/silica beads and large soil particles remain in the sample tube. Based on preliminary experiments, beads and large soil particles were found not to adsorb RNA significantly. One volume of 70% ethanol can then be added to the suspension. This suspension is then passed through a Qiagen column via relatively slow centrifugation (4000 rpm) to prevent suspended solids from passing through the column. Impurities such as humics, which can inhibit both reverse transcriptase and DNA polymerase (Alm and Stahl, 2000; Lorenz and Wackernagel, 1987; Moran et al., 1993), are removed by washing the columns using the Qiagen washing buffers twice or more until filtrates are no longer brown colored. The total RNA is then collected using a small volume of DEPC-treated water ($\sim$100 μl). This simple elution step with RNase-free water allows the recovery of RNA sorbed onto small soil particles, as well as the column matrix, and in essence treats the soil as an additional binding matrix. Because the final volume is sufficiently small, a concentration step is not necessary and the eluant can be treated with RNase-free DNase I directly and then RNA extracted using the Qiagen Total RNeasy kit.

Competitive RT-PCR Assay

RT-PCR assays are performed with the Qiagen OneStep RT-PCR kit (Qiagen). The RT-PCR reaction is carried out in 50 μl consisting of 5 μl of serially diluted standard RNA and 45 μl of a master mix containing 5 μl total RNA, 10 μl 5$\times$ RT-PCR buffer, 10 μl 5$\times$ Q solution, 400 μM each of dNTPs, 0.6 μm of each primer, and 2 μl of Qiagen OneStep RT-PCR enzyme. Standard RNAs can be diluted in 2- or 3.3-fold series. Following RT incubation at 50° for 30 min and then heating to 95° for 15 min, PCR is conducted at 94° for 1 min; 55° for 1 min; and 72° for 1 min. All samples are finally extended at 72° for 10 min. The number of PCR cycles should be one that produces exponential amplification, which can be determined by withdrawing PCR samples at every other cycle and running either on an agarose gel or via capillary electrophoresis. For soil samples, bovine serum albumin at a concentration of 400 ng $\cdot$ μl^{-1} should be added to reduce inhibition of both reverse transcriptase and DNA polymerase (Kreader, 1996).

Capillary Electrophoresis Analysis of RT-PCR Products

Capillary electrophoresis can be used for the quantification of RT-PCR products using uncoated capillary columns and UV detection. The analysis of PCR products by CE is performed in the reverse polarity mode with 6 kV applied voltage. Uncoated silica capillaries with a 21-cm effective length and a 31-cm total length (75 μm i.d) are adequate for effective separation of RT-PCR products. The separation/flush buffer consists of 50 mM HEPES sodium salt (N-2-hydroxyethylpiperazine-N'-2-ethanesulfonic acid), 65 mM boric acid, 0.5% hydroxypropylmethylcellulose (HPMC), 6% mannitol, and 1 μg · ml^{-1} ethidium bromide. The buffer must be degassed by either sonication or evacuation (or both) to obtain stable baselines. Before a run, the capillary column should be rinsed sequentially with 0.2 N NaOH, 0.2 N HCl, and finally the separation buffer for at least 3 min each. Theoretically, rinsing with NaOH and HCl can be skipped, as this step is used to regenerate surface charges on an uncoated silica column, which actually should be suppressed for DNA separation. Sometimes, however, this rinsing helps produce sharper separation due to unknown reasons. Samples are applied by pressure injection for 50 s, at 1 psi. The amount of samples can be increased by raising injection time, pressure, or both if peaks are too small. The ultraviolet absorbance is monitored at 254 nm. To quantify RT-PCR products, peak areas on the resulting electropherogram can be calculated using standard laboratory software, e.g., P/ACE Station software (V.1.0, Beckman Instruments, Palo Alto, CA). A sample electropherogram of separation of DNA standards ranging from 110 to 1114 bp as little as 14 bp different is shown in Fig. 2. It should be noted that effective DNA separation is possible with this technique, as HPMC acts as a sieving agent and the HEPES-borate buffer shields DNA against capillary wall interaction while mannitol interacts with HPMC in the presence of boric acid to form a dynamic flexible network in the sieving matrix (Han et al., 1998). With these additions, DNA separation in uncoated capillaries is comparable to commercial coating columns, but the cost is reduced greatly. The addition of ethidium bromide helps enhance UV detection, as well as separate DNA (Shihabi, 1999).

Summary

This chapter described the development of a methodology to accurately monitor in situ microbial function and activity. This methodology consists of using competitive RT-PCR to reliably amplify mRNA transcripts from both pure cultures and soil samples in combination with CE to analyze the RT-PCR products accurately and conveniently. This approach can allow

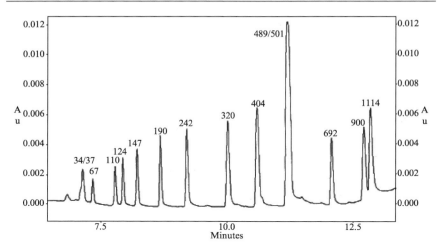

FIG. 2. Separation of DNA molecular weight standards in an uncoated capillary column filled with HEPES-boric acid buffer system containing 50 mM HEPES, 0.5% HPMC, 6% mannitol, 65 mM boric acid, and 1 μg $\cdot$ ml^{-1} ethidium bromide with 6 kV applied voltage. Numbers next to peaks indicate size in base pairs.

one to monitor even growth phase-dependent expression of specific gene (s). Quantification of expression of multiple genes is possible, provided the sizes of all target and standard fragments are of sufficiently different size to allow separation. This methodology can thus help measure *in situ* gene expression and determine what geochemical parameters affect such expression.

Acknowledgment

Support from the United States National Science Foundation (MCB-9708557) to JDS for this research is gratefully acknowledged.

References

Alm, E. W., and Stahl, D. A. (2000). Critical factors influencing the recovery and integrity of rRNA extracted from environmental samples: Use of an optimized protocol to measure depth-related biomass distribution in freshwater sediments. *J. Microbiol. Methods* **40,** 153–162.

Arnal, C., Ferre-Aubinmeau, V., Mignotte, B., Imbert-Marcille, B.-M., and Billaudel, S. (1999). Quantification of hepatitis A virus in shellfish by competitive reverse transcription-PCR with coextraction of standard RNA. *Appl. Environ. Microbiol.* **65,** 322–326.

Bustin, S. A. (2002). Quantification of mRNA using real-time reverse transcription PCR (RT-PCR): Trends and problems. *J. Mol. Endocrinol.* **29,** 23–39.

Chomczynski, P., and Sacchi, N. (1987). Single-step method of RNA isolation by acid guanidinium thiocyanate-phenol-chloroform extraction. *Anal. Biochem.* **162,** 156–159.

Corbella, M. E., and Puyet, A. (2003). Real-time reverse transcription-PCR analysis of expression of halobenzoate and salicylate catabolism-associated operons in two strains of *Pseudomonas aeruginosa. Appl. Environ. Microbiol.* **69,** 2269–2275.

Freeman, W. M., Walker, S. J., and Vrana, K. E. (1999). Quantitative RT-PCR: Pitfalls and potential. *Biotechniques* **26,** 112–125.

Han, F., Xue, J., and Lin, B. (1998). Mannitol influence on the separation of DNA fragments by capillary electrophoresis in entangled polymer solutions. *Talanta* **46,** 735–742.

Han, J.-I, and Semrau, J. D. (2004). Quantification of sMMO and pMMO gene expression by competitive RT-PCR and capillary electrophoresis. *Environ. Microbiol.* **6,** 388–399.

Kreader, C. A. (1996). Relief of amplification inhibition in PCR with bovine serum albumin or T4 gene 32 protein. *Appl. Environ. Microbiol.* **62,** 1102–1106.

Kusher, S. R. (1996). mRNA decay. *In* "*Escherichia coli* and *Salmonella*: Cellular and Molecular Biology" (F. C. Neidhardt, R. Curtiss, J. L. Ingraham, E. C. C. Lin, K. B. Low, B. Magasanik, W. S. Reznikoff, M. Riley, M. Schaechter, and H. E. Umbarger, eds.), Vol. 1, pp. 849–860. American Society of Microbiology, Washington, DC.

Lindahl, T. (1993). Instability and decay of the primary structure of DNA. *Nature* **362,** 709–715.

Lorenz, M. G., and Wackernagel, W. (1987). Adsorption of DNA to sand and variable degradation rates of adsorbed DNA. *Appl. Environ. Microbiol.* **53,** 2948–2952.

Melzak, K. A., Sherwood, C. S., Turner, R. F. B., and Haynes, C. A. (1996). Driving forces for DNA adsorption to silica in perchlorate solutions. *J. Colloid Interface Sci.* **181,** 635–644.

Moran, M. A., Torsvik, V. L., Torsvik, T., and Hodson, R. E. (1993). Direct extraction and purification of rRNA for ecological studies. *Appl. Environ. Microbiol.* **59,** 915–918.

Pfaffl, M. W., and Hageleit, M. (2001). Validities of mRNA quantification using recombinant RNA and recombinant DNA external calibration curves in real-time RT-PCR. *Biotechnol. Lett.* **23,** 275–282.

Sheridan, G. E. C., Masters, C. I., Shallcross, J. A., and Mackey, B. M. (1998). Detection of mRNA by reverse transcription-PCR as an indicator of viability in *Escherichia coli* cells. *Appl. Environ. Microbiol.* **64,** 1313–1318.

Shihabi, Z. K. (1999). Capillary electrophoresis of double-stranded DNA in an untreated capillary. *J. Chromatogr. A* **853,** 349–354.

Spring, S., Schulze, R., Overmann, J., and Schleifer, K.-H. (2000). Identification and characterization of ecologically significant prokaryotes in the sediment of freshwater lakes: Molecular and cultivation studies. *FEMS Microbiol. Rev.* **24,** 573–590.

[20] *In Situ* Functional Gene Analysis: Recognition
of Individual Genes by Fluorescence
In Situ Hybridization

By Katrin Zwirglmaier, Katrin Fichtl, and Wolfgang Ludwig

Abstract

Fluorescence *in situ* hybridization (FISH) using specific probes certainly is one of the most commonly applied molecular techniques in microbial ecology. Monitoring of community composition and dynamics can be combined with localization and identification of individual cells *in situ*. However, the resolution power of the method is limited by the need for high target numbers per cell. Apart from standard targets (ribosomal RNAs), mRNAs could be used successfully for *in situ* visualization in some cases. A new promising variant of *in situ* hybridization could be established that should provide access to any low copy number nucleic acid targets, such as chromosomal genes. The recognition of individual genes by FISH (RING-FISH) technology is based on target visualization mediated by polynucleotide probe network formation. The specificity of the approach is provided by intracellular probe-target hybridization. This initial hybridization apparently acts as a focal point for inter-probe hybridization within and mainly in the periphery of the cells. Probe-conferred fluorescence typically appears halo-like around the cells. RING-FISH can be used in combination with conventional oligonucleotide FISH. Thus, genetic potential can be assigned to *in situ* identified cells.

Introduction

In microbial ecology, in addition to monitoring the community composition and dynamics, assignment of physiological potential to the individual taxa is certainly among the major tasks. Modern molecular biological techniques, such as the so-called full rRNA cycle (Amann *et al.*, 1995), provide access not only to qualitative and quantitative analysis of taxon composition and change, but also *in situ* localization of the respective microbial cells. The development of culture-independent techniques certainly marks a milestone in the history of microbial ecology. This is documented impressively by the rapid growth of specialized rRNA sequence databases (Ludwig *et al.*, 2004; Maidak *et al.*, 2004; Wuyts *et al.*, 2004). Meanwhile the number of "environmental" small subunit rRNA (rDNA)

METHODS IN ENZYMOLOGY, VOL. 397   0076-6879/05 $35.00
Copyright 2005, Elsevier Inc. All rights reserved.   DOI: 10.1016/S0076-6879(05)97020-1

sequences indicating thus far uncultured organisms exceeds that of culture-derived data. Only part of those clone sequences can be assigned to known taxa, whereas a substantial fraction indicates the existence of a variety of unknown taxa in environmental samples. The establishment of polymerase chain reaction (PCR) and cloning-based sequence retrieval directly from original sample material in combination with taxon-specific probe and diagnostic PCR technologies allowed only indirect monitoring of community dynamics and function even if quantitative techniques were applied. The correct assignment of measured data to cell numbers was not possible. The break-through with respect to these limitations was marked by the development of fluorescence *in situ* rRNA probing methods (FISH) (Amann *et al.*, 1995). Now, rRNA-targeted taxon-specific probes not only allowed quantitative analyses, but also *in situ* localization of the probe-identified cells. There are, however, two major limitations of the FISH technology, given that the method relies on the natural amplification of the probe target, the rRNA molecules that occur in high copy numbers in active cells. First, the resolution power for taxon identification is limited to the species and higher taxonomic ranks. As evolutionary conserved molecular fossils, the rRNA sequences usually do not contain differentiating information for closely related unities such as strains (Ludwig and Klenk, 2001). Second, identification based on rRNA targets may appear to indicate certain physiological properties known for close cultured relatives. However, this relationship does not prove physiological potential or activity. Thus, rRNA-based *in situ* methods cannot be used to study and prove microbial function. Several approaches were described to circumvent these limitations. *In situ* PCR (Hodson *et al.*, 1995), which is a common method for the detection of genes in eukaryotic tissues, was applied with limited success for the detection of microbial genetic information *in situ*. The benefit of multiple target molecules was used in combination with signal amplification techniques for the visualization of mRNA, indicating transcription of the respective non-rRNA genes (Wagner *et al.*, 1998). The combination of microautoradiography and conventional fluorescence *in situ* hybridization (FISH-MAR) (Lee *et al.*, 1999; Ouverney and Fuhrmann, 1999) allows assigning the capacity for accumulation or uptake of radioactively labeled compounds to FISH-identified cells. An alternative promising polynucleotide *in situ* hybridization technique has been published, which in the future might allow detection of any genetic information at the cellular level (Zwirglmaier *et al.*, 2004). Thus far, the potential of RING-FISH has been demonstrated for a selection of chromosome and plasmid-encoded genes, as well as organisms representing different cell envelope types.

History and Background

In the preoligonucleotide era of taxon-specific probing, polynucleotide probes were used for identification by hybridization to purified nucleic acids (Ludwig *et al.*, 1994). During the early years of FISH developments, attempts were made to test the applicability of *in vitro* transcripts of well-approved rRNA/rDNA-targeted polynucleotide probes for fluorescence *in situ* cell hybridization (Trebesius *et al.*, 1994). After adaptation of the techniques, general applicability, as well as specificity of polynucleotide transcript probes, could be shown. Surprisingly, the probe-mediated fluorescence was concentrated as a halo in the cell periphery, resembling surface-directed antibody-conferred immune fluorescence. These findings indicated that part of the probe molecules remained outside the cells and induced the development of techniques for taxon-specific cell enrichment or depletion (Stoffels *et al.*, 1999). This approach is based on extracellular probe-mediated immobilization of specifically halo-labeled cells on solid supports. Typically, this is achieved by a second hybridization to surfaces coated with probe complementary DNA. Later, as a result of experimental trials as well as theoretical analyses, the network hypothesis was formulated (Zwirglmaier *et al.*, 2003). By successively shortening the transcript probe molecules and including high copy number plasmids as targets, it became evident that the specificity of the approach is mediated by intracellular target hybridization as a focal point for probe network formation. The latter most probably results from interprobe molecule hybridization. *In silico* secondary structure prediction of all probe variants used successfully for halo formation indicated a potential for intramolecular helix formation. Thus, identical probe molecules are capable of network formation via base pairing of complementary helix halves. Subsequently, the number of target molecules per cell was reduced designing probes for low copy number plasmids and finally chromosomal genes. Finally, the RING-FISH (*r*ecognition of *in*dividual *g*enes by FISH) method was established, allowing *in situ* visualization of single genes at the cellular level (Zwirglmaier *et al.*, 2004).

Method

RING-FISH methodology is composed of five major steps: probe design, probe generation, cell permeabilization, *in situ* hybridization, and hybrid detection.

Probe Design

The size range of polynucleotide probes applied successfully corresponds to 60–840 nucleotides. In contrast to standard oligonucleotide probes for conventional *in situ* hybridization to rRNA targets, appropriate powerful software tools for *in silico* probe design and evaluation are not yet available. In case the respective target sequences are known, *in silico* secondary structure analysis might be helpful to avoid the construction of ineffective probes. The primary structure of the probes should exhibit a certain degree of self-complementarity to enable network formation (Zwirglmaier *et al.*, 2003). Thus far, however, no explicit rules on number, location, length, and quality of potential network formation stretches can be given. The amount of experimental data is still too limited to recognize and formulate rules or minimum standards.

Concerning specificity or discriminating power formulas for polynucleotide hybrid stabilities are available. Thus, in case of available reference sequences, the specificity of the new probes can be roughly checked *in silico*. It has to be taken into account that commonly RNA probes generated by *in vitro* transcription are used.

Probe Generation

According to the current standard protocol, polynucleotide RNA probes are generated via *in vitro* transcription of polymerase chain reaction (PCR)-amplified DNA (Stoffels *et al.*, 1999; Trebesius *et al.*, 1999; Zwirglmaier *et al.*, 2003, 2004). The target region of the DNA is amplified via PCR using appropriate primer pairs. Examples are given in Table I. One of the primers flanking the target-specific region comprises the promotor for the T3 RNA polymerase, which is necessary for the subsequent *in vitro* transcription. During transcription, the emerging probe is labeled with biotin, digoxigenin, or fluorescein using labeled UTP in a ratio of 0.65/0.35 labeled/unlabeled UTP. Higher amounts of labeled UTP decrease the yield, whereas lower amounts result in insufficient labeling (Stoffels *et al.*, 1999). Up to 20 μg RNA transcript probe can be expected per microgram of DNA template.

All components used for *in vitro* transcription according to the following recipe are from Roche (Mannheim, Germany).

Nucleotide triphosphate (NTP) mix: 3.9 μl ATP (100 mM), 3.9 μl CTP (100 mM), 3.9 μl GTP (100 mM), 1.4 μl UTP (100 mM), BIO-16-UTP or DIG-11-UTP or 25 μl FLUOS-12-UTP (10 mM)

Transcription mix (per reaction): 3 μl NTP mix, 3 μl 10$\times$ transcription buffer, 3 μl T3-RNA polymerase, 1.5 μl RNase inhibitor, 1–2 μg PCR product, and up to 30 μl with ultrapure H_2O

Further reagents: 10 M ammonium acetate, 0.2 M EDTA, pH 8.0, absolute EtOH, and 70% EtOH

Enzymes: DNase I (RNase free) and RNase inhibitor

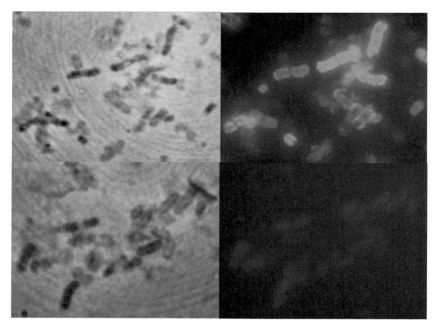

FIG. 1. RING-FISH hybridization of *Synechococcus* WH7803 (top) and *Synechococcus* PCC7942 (bottom) using a RNA polymerase gene-specific polynucleotide transcript probe. The probe targets a 690-bp fragment of the RNA polymerase gene *rpoC1* of *Synechococcus* WH7803. Probe-conferred halos (fluorescein) appear only in the case of *Synechococcus* WH7803 target cells. (Left) Phase contrast and (right) epifluoresence image. From K. Zwirglmaier and D. Scanlan, unpublished results. (See color insert.)

Experimental Procedure

1. Incubate transcription mix for 3 h at 37°.
2. Add 3 μl DNase I (RNase free) to degrade template DNA and incubate for 15 min at 37°
3. Add 3 μl EDTA.
4. Precipitate RNA at least 1 h at $-20°$ by adding 16 μl ammonium acetate and 156 μl absolute EtOH.
5. Centrifuge 15 min at 14,000 rpm.
6. Wash pellet with 70% EtOH.
7. Suspend pellet in 30 μl H_2OMQ + 1 μl RNase inhibitor.
8. Run standard agarose gel (2%) to check integrity of transcript probe (50% formamide stabilizes the transcript probe).
9. For storage, freeze transcript probe solution and store at $-20°$

TABLE I
EXAMPLES OF PRIMER PAIRS USED FOR GENERATION OF RNA POLYNUCLEOTIDE PROBES[a]

Transcript probe		Target[c]	Hybridization		
Primer pair sequence[b]	Designation		Size[d]	Time[e]	% FA[f]
GAGTATTCAACATTTCCG *ATAGGTATTAAACCCTCACTAAAGGG* ACCAATGCTTAATCAGTG	betaLact-V betaLact-RT3	β-Lactamase (sense)	840	24	15
ATAGGTATTAAACCCTCACTAAAGGG GAGTATTCAACATTTCCG ACCAATGCTTAATCAGTG	betaLact-VT3 betaLact-R	β-Lactamase (antisense)	840	24	15
CTAACAAATAGCTGGTGG *ATAGGTATTAAACCCTCACTAAAGGG* CAGTTTCGTCAGTCAGGA	GAP-F-V GAP-F-RT3	Glyceraldehyde-3-phosphate dehydrogenase	338	30	10
ACAACCTCTACATGCCAG *ATAGGTATTAAACCCTCACTAAAGGG* CGCAGCCAGCAACTTGAA	Xantho6.8-V Xantho6.8-RT3	Prepilin leader peptidase	140	30	10

[a] From Zwirglmaier et al. (2004).
[b] Sequences are given in 5'-3' orientation; T3 RNA polymerase promotor sequence indicated by italics.
[c] Gene comprising the target region of the probe.
[d] Number of monomers of the probe construct.
[e] Recommended incubation time(hr) for in situ hybridization according to the standard protocol.
[f] Recommended formamide concentration for in situ hybridization according to the standard protocol.

Cell Permeabilization

As with any *in situ* cell hybridization procedure, the envelope of target cells has to be permeabilized for the probe molecules (Amann *et al.*, 1995). The major task for such a procedure is to enable the probe to enter the cell and to access its target while keeping the cell morphology as native as possible. Given the variety of cell envelope structures and composition, no universal fixation method can be proposed. The following protocol was applied successfully for polynucleotide FISH of various organisms with gram-negative cell walls, a selection of gram-positive rods, and cocci such as bacilli, strepto-, and enterococci. However, alternative protocols, as well as additional cell wall lytic compounds or procedures, might be needed, especially in the case of gram-positive cocci with rigid cell walls.

The cyanobacterial cells shown in Fig. 1 were incubated with lysozyme (5 mg/ml, 30 min at 37°) and proteinase K (1 mg/ml, 10 min at room temperature) following the standard paraformaldehyde fixation.

Reagents

Paraformaldehyde (PFA): 4% (w/v) in PBS (130 mM NaCl, 1.5 mM K$_2$HPO$_4$, 8.0 mM Na$_2$HPO$_4$, 2.7 mM KCl, pH 7.0). Heat PBS to 60° and dissolve PFA by using additively NaOH. Titrate to pH 7.0
Absolute EtOH

Standard Protocol for Cultured Cells

1. Harvest cells by centrifugation at 5000 rpm for 5–10 min
2. Suspend cell pellet in PBS (volume of PBS dependent on cell pellet volume) and add 3 volumes of 4% PFA.
3. Incubate for 5–12 h at 4°.
4. Centrifuge at 5000 rpm for 5–10 min.
5. Wash with appropriate amount of PBS (dependent on cell pellet volume) and centrifuge at 5000 rpm for 5–10 min.
6. Suspend in appropriate volume of PBS:EtOH absolute in a 1:1 ratio (dependent on cell pellet volume).
7. Store at −20° up to 1 year.

In Situ *Hybridization*

In principle, the hybridization procedure for RING-FISH is essentially the same as described for rRNA targeted transcript probes (Zwirglmaier *et al.*, 2003). However, given the lower thermal stability of RNA-DNA-hybrids as compared to that of RNA-RNA-hybrids and the reduced number of target molecules (1–10 copies of chromosomal genes versus up to 10^5

rRNA molecules per cell), the composition (formamide concentration) of the hybridization buffer and the hybridization time have to be adjusted. The optimal formamide concentration, hybridization temperature, and incubation time for probes targeting genomic DNA depend on the characteristics of the respective probe–target and probe–nontarget pairs such as size, nucleotide composition, and sequence. The formamide concentration typically ranges between 5 and 60%, the hybridization temperature between 37 and 53°, and the hybridization time between 24 and 30 h. The following protocols are used routinely for RING-FISH. However, the hybridization parameters have to be optimized experimentally for each particular probe–target and probe–nontarget reference (defining specificity).

> Hybridization buffer: 75 mM NaCl, 20 mM Tris-HCl, pH 8.0, 0.01% SDS, and 5–60% formamide (depends on probe–target characteristics)
>
> Washing buffer (hybridization in solution): 5 mM NaCl, 20 mM Tris-HCl, pH 8.0, and 0.01% SDS

Two variants of a hybridization protocol are given to comply with the conditions of *in situ* hybridization of cells in suspension or immobilized on microscope slides. Hybridization on slides is performed for subsequent microscopic visualization and analysis.

1. Apply an aliquot of 2–5 μl of a suspension of PFA-fixed cells per field of slide.
2. Dry slides at 60° for 5 min.
3. Dehydrate cells subsequently in 50, 80, and 100% ethanol, 2 min each.
4. Apply 12 μl hybridization buffer onto each field of slide.
5. Add 4 μl transcript probe solution (~3–4 μg; see earlier discussion).
6. Put tissue paper in a 50-ml tube and moisten with 1–2 ml hybridization buffer.
7. Insert slide into 50-ml tube and close tube.
8. Denature for 20 min at 80°.
9. Incubate for 24–30 h at 53° (or appropriate temperature).
10. Rinse slide with ultrapure H$_2$O.
11. Dry slide under air flow and store at −20°.

Hybridization of suspended cells is preferred if the probe-conferred network is used for cell enrichment or depletion by probe-mediated immobilization. Nevertheless, cells hybridized in solution can be subsequently immobilized upon slides.

The following protocol was designed for 0.5-ml reaction tubes.

1. Use 5–10 μl of a suspension of PFA-fixed cells per tube.
2. Add 200 μl PBS and centrifuge for 3 min at 12,000 rpm to remove residual ethanol.
3. Suspend cell pellet in 30 μl hybridization buffer.
4. Add 5 μl transcript probe solution (~4–5 μg).
5. Denature 20 min at 80°.
6. Incubate for 24–30 h at 53° (or appropriate temperature).
7. Centrifuge for 5 min at 14,000 rpm.
8. Suspend pellet in 100 μl washing buffer.
9. Incubate for 30 min at hybridization temperature.
10. Centrifuge for 5 min at 14,000 rpm.
11. Suspend pellet in 15 μl PBS.
12. Apply on slides and proceed with detection/storage or use cell suspension for depletion/enrichment procedures.

Hybrid Detection

Direct (fluorescent dye-linked nucleotides) or indirect (digoxigenin or biotin) labeling of polynucleotide transcript probes was successfully applied for hybridization monitoring. Probe networks of directly fluorescence-labeled molecules can be visualized by epifluorescence or laser-scanning microscopy (Figs. 2 and 3). For the detection of biotin and digoxigenin labels, fluorescent dye-conjugated streptavidin and anitidigoxigenin antibody variants can be used, respectively. Commercial detection kits for both indirect approaches and different fluorochromes are available.

Designing a New RING-FISH System

Specificity and sensitivity are crucial tasks whenever a new visualization, detection, or identification system based on nucleic acid techniques should be designed. Usually, probe size influences both. A larger size often helps increase sensitivity, whereas at the same time commonly reduces specificity. Recent experiments, however, showed that size is no longer a major criterion for sensitivity in RING-FISH approaches. Probes comprising only 60 nucleotides were successfully applied for specific halo formation. However, further reducing the size may diminish network formation capacity. By selecting appropriate target regions, smaller probes usually benefit the discriminating power of the assay. Even if *in silico* estimation of specificity by comparative sequence analysis (localization and evaluation

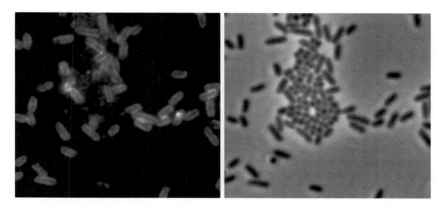

FIG. 2. RING-FISH hybridization of a cell mixture of *Escherichia coli* and *Neisseria canis* using a glyceraldehyde-3-phosphate dehydrogenase gene-specific polynucleotide transcript probe (Table I; Zwirglmaier *et al.*, 2004). The probe was constructed as specific for the respective gene of *E. coli*. RING-FISH (left) and phase-contrast (right) micrographs of the same microscopic field. (See color insert.)

of diagnostic differences between target and similar but nontarget sequence stretches) is possible, *in vitro* evaluation of probe specificity, as well as adjustment of appropriate hybridization parameters, is highly recommended. This can be done easily if reference organisms are available as negative standards exhibiting diagnostic morphologies, which can be used for microscopic differentiation. Alternatively, or in addition, conventional rRNA target diagnostic oligonucleotide FISH can be combined with RING-FISH to verify the identity of the respective cells (Zwirglmaier *et al.*, 2004; Fig. 3). Given the differences in experimental parameters, as well as hybrid stabilities of oligonucleotide FISH and RING-FISH, polynucleotide hybridization has to be performed first.

Obviously, availability, adaptation, or development of an appropriate permeabilization approach is essential when developing a new RING-FISH system. Often, trying published standard procedures in combination with conventional rRNA targeting oligonucleotide FISH (as positive standard) is helpful for first tests. In this case potential problems resulting from low target amount or failing network formation are not of relevance. If rRNA targeting oligonucleotide FISH fails after cell fixation, polynucleotide FISH most probably will not work. If the proposed standard fixation procedure does not provide satisfying results, stepwise changes of incubation time and temperatures might be helpful. Furthermore, pretreatment with cell wall lytic enzymes such as lysozyme or mutanolysin might be advantageous (K. Fichtl, unpublished result). It has to be mentioned that sometimes it is

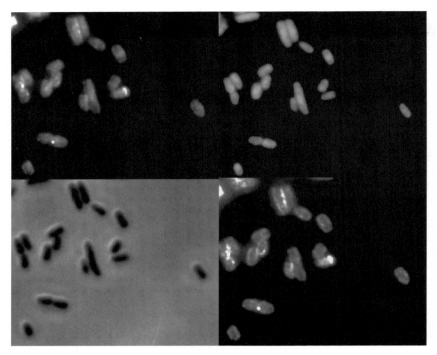

Fig. 3. Combination of RING-FISH and conventional rRNA FISH of *Escherichia coli* cells carrying plasmid pCCR2.1. A Cy3-labeled polynucleotide transcript probe specific for plasmid-encoded β-lactamase (Table I; Zwirglmaier *et al.*, 2004) was used in combination with a fluorescein-labeled oligonucleotide probe targeting 16S rRNA positions 444–462. Micrographs of the same microscopic field are shown: phase contrast (lower left), polynucleotide probe signal (lower right), oligonucleotide probe signal (upper right), and overlay of both signals (upper left). (See color insert.)

difficult to trim the parameters so that all cells of a given taxon behave the same way. It was observed that in cases of generally good RING-FISH results, a certain fraction of the target cells showed cell swelling and bright intracellular, albeit specific, fluorescence. This might be influenced by the growth (phase) status of the individual cells (Zwirglmaier *et al.*, 2004). If target visualization is the primary task, this behavior is not a problem; however, such cells will escape if specific cell immobilization is desired.

Current Limitations

RING-FISH, as any probe-based *in situ* hybridization technique, essentially depends on appropriate cell fixation procedures. Thus, accessibility problems already known from conventional oligonucleotide FISH studies also persist when trying to adapt RING-FISH technology. Although the difficulties concerning certain gram-positive cocci seem to be almost resolved (unpublished studies), there are many taxa of slime, capsule, wax-producing organisms that still remain resistant to *in situ* probing. Furthermore, it might never be possible to generate a common permeabilization procedure for all members of complex (with respect to taxon and hence cell envelope diversity) samples.

Another system-inherited drawback concerns the range of specificity that currently can be addressed by RING-FISH technology. Given the demand of polynucleotide probes, the potential with respect to the number of diagnostic sequence positions needed for clear differentiation of target and nontarget cells is clearly reduced in comparison to oligonucleotide FISH. While the presence or absence of target genes can be visualized clearly, the differentiation of gene variants depends on the number, character, and position of diagnostic nucleotide differences within the target region.

It was shown that very high amounts of probe molecules are needed to enforce intensive network formation allowing microscopic probe–target hybrid visualization (Zwirglmaier *et al.*, 2004). Thus, RING-FISH currently is still much more expensive in material and costs than conventional poly- or oligonucleotide FISH methods. RING-FISH in its current version can be used to study and visualize the presence or absence of target genes or noncoding DNA, as well as for differentiation of sufficiently diverged gene or sequence variants. However, it cannot be used for the selective detection of mRNA, i.e., expression studies. This was documented by studies using sense and antisense transcript probes (Zwirglmaier *et al.*, 2004).

Current Developments

Current studies (unpublished) focus on optimization of the RING-FISH technology toward oligonucleotide specificity. First, successful constructs give rise to for hope that the problem of reduced specificity or differentiation power could be circumvented by designing oligo-oligonucleotide probes. Such probes comprise multiple oligonucleotide sequences and can be designed as monospecific (repeats of a certain oligonucleotide sequence) or polyspecific (composed of different oligonucleotides). Thus, the minimum size requirement would be fulfilled without losing oligonucleotide specificity. Also, probes comprising specific oligonucleotide sequences and nonsense (should not hybridize to any target) tails or spacers are being tested. The major problem still concerns a balanced distribution of self-complementary stretches within the probe constructs needed for network formation.

Concluding Remarks

RING-FISH technology certainly has to be modified, adapted, and optimized further to develop into a standard method as generally applied as the conventional rRNA-targeted oligonucleotide FISH approach. Nevertheless, it opens the door to the visualization of any genetic information at the cellular level. Thus, studies in microbial ecology should benefit from the new technique with respect to the assignment of functional potential to community structure. Furthermore, combining RING-FISH and probe-mediated cell immobilization now provides access to enrichment or depletion of microbial cells according to any genetic similarity or peculiarity. This approach could especially be helpful in environmental genomics of complex habitats.

Acknowledgments

The underlying work was supported by the German Ministry for Education and Research. The authors thank David Scanlan from the University of Warwick, United Kingdom, for sharing unpublished data.

References

Amann, R., Ludwig, W., and Schleifer, K. H. (1995). Phylogenetic identification and *in situ* detection of individual microbial cells without cultivation. *Microbiol. Rev.* **59**, 143–169.

Hodson, R. E., Dustman, W. A., Garg, R. P., and Moran, M. A. (1995). *In situ* PCR for visualization of microscale distribution of specific genes and gene products in prokaryotic communities. *Appl. Environ. Microbiol.* **61**, 4074–4082.

Lee, N., Nielsen, P. H., Andreasen, K. H., Juretschko, S., Nielsen, J. L., Schleifer, K. H., and Wagner, M. (1999). Combination of fluorescence in situ hybridisation and microautoradiography: A new tool for structure-function analyses in microbial ecology. *Appl. Environ. Microbiol.* **65**, 1289–1297.

Ludwig, W., Dorn, S., Springer, N., Kirchhof, G., and Schleifer, K. H. (1994). PCR-based preparation of 23S rRNAtargeted group-specific polynucleotide probes. *Appl. Environ. Microbiol.* **60**, 3236–3244.

Ludwig, W., and Klenk, H. P. (2001). Overview: A phylogenetic backbone and taxonomic framework for prokaryotic systematics. *In* "Bergey's Manual of Systematic Bacteriology" (G. Garrity, ed.), 2nd Ed. pp. 49–65. Springer, New York.

Ludwig, W., Strunk, O., Westram, R., Richter, L., Meier, H., Yadhukumar, Buchner, A., Lai, T., Steppi, S., Jobb, G., Förster, W., Brettske, I., Gerber, S., Ginhart, A. W., Gross, O., Grumann, S., Hermann, S., Jost, R., König, A., Liss, T., Lüßmann, R., May, M., Nonhoff, B., Reichel, B., Strehlow, R., Stamatakis, A., Stuckmann, N., Vilbig, A., Lenke, M., Ludwig, T., Bode, A., and Schleifer, K. H. (2004). ARB: A software environment for sequence data. *Nucleic Acids Res.* **32**, 1363–1371.

Ouverney, C. C., and Fuhrman, J. A. (1999). Combined microautoradiography–16S rRNA probe technique for determination of radioisotope uptake by specific microbial cell types in situ. *Appl. Environ. Microbiol.* **65**, 1746–1752.

Stoffels, M., Ludwig, W., and Schleifer, K. H. (1999). rRNA probe-based cell fishing of bacteria. *Environ. Microbiol.* **1,** 259–271.

Trebesius, K. H., Amann, R., Ludwig, W., Muhlegger, K., and Schleifer, K. H. (1994). Identification of whole fixed bacterial cells with nonradioactive 23S rRNA-targeted polynucleotide probes. *Appl. Environ. Microbiol.* **60,** 3228–3235.

Wagner, M., Schmid, M., Juretschko, S., Trebesius, K. H., Bubert, A., Goebel, W., and Schleifer, K. H. (1998). *In situ* detection of a virulence factor mRNA and 16S rRNA in *Listeria monocytogenes. FEMS Microbiol. Lett.* **160,** 159–168.

Wuyts, J., PerrieÁre, G., and Van de Peer, Y. (2004). The European ribosomal RNA database. *Nucleic Acids Res.* **32,** D101–D103.

Zwirglmaier, K., Ludwig, W., and Schleifer, K. H. (2003). Improved fluorescence in situ hybridization of individual microbial cells using polynucleotide probes: The network hypothesis. *Syst. Appl. Microbiol.* **26,** 327–337.

Zwirglmaier, K., Ludwig, W., and Schleifer, K. H. (2004). Recognition of individual genes in a single bacterial cell by fluorescence in situ hybridization–RING-FISH. *Mol. Microbiol.* **51,** 89–96.

Further Reading

Maidak, B. L., Cole, J. R., Lilburn, T. G., Parker, C. T., Jr., Saxman, P. R., Farris, R. J., Garrity, G. M., Olsen, G. J., Schmidt, T. M., and Tiedje, J. M. (2001). The RDP-II (Ribosomal Database Project). *Nucleic Acids Res.* **29,** 173–174.

[21] Simultaneous Fluorescence *In Situ* Hybridization of mRNA and rRNA for the Detection of Gene Expression in Environmental Microbes

By ANNELIE PERNTHALER and JAKOB PERNTHALER

Abstract

A protocol is presented for the detection of gene expression in environmental microorganisms by means of fluorescence *in situ* hybridization (FISH). Messenger RNA (mRNA) is hybridized with digoxigenin (DIG)- or fluorescein (FLUOS)-labeled ribonucleotide probes. Subsequently the hybrid is detected immunochemically with a horseradish peroxidase (HRP)-labeled antibody and tyramide signal amplification (catalyzed reporter deposition, CARD). After mRNA FISH, microorganisms can be identified by rRNA FISH with oligonucleotide probes labeled either with a fluorochrome or with HRP. Sample preparation and cell permeabilization strategies for various microbial cell types are discussed. The synthesis of DIG- and FLUOS-labeled probes, as well as custom labeling of tyramides with different fluorochromes, is described. As a case study, we describe in detail mRNA FISH of the particulate methane-monooxygenase, subunit A (*pmoA*) in endosymbiotic bacteria from tissue sections of a marine mollusc. *PmoA* is used as a marker gene for methanotrophy.

Introduction

One of the long-standing goals of environmental microbiology is to assign specific activities to individual microbial species *in situ* and thus assess their biogeochemical impact. For the identification of microbes at the single-cell level, fluorescence *in situ* hybridization using rRNA-targeted probes is a routinely used tool. To link a specific metabolic activity to the identity of a microorganism *in situ*, without the need for incubation with tracers or substrates, it is necessary to detect either the key enzyme of a particular process or its corresponding mRNA in combination with rRNA FISH. *In situ* hybridization (ISH) of mRNA sequences is a popular technique to study gene expression in eukaryotic cells and tissues. Presently, microbiology is lacking stable protocols for *in situ* detection and quantification of gene expression in environmental samples. Since the first ISH experiments (John *et al.*, 1969), mainly radioactive nucleotides have been used to synthesize labeled probes (Gerfen, 1989). The advantage of

METHODS IN ENZYMOLOGY, VOL. 397
Copyright 2005, Elsevier Inc. All rights reserved.

radiolabeled probes is their ability to detect very low levels of transcripts. The major limitations are poor spatial resolution and the long exposure times of microautoradiography, depending on the radioisotope and the amount of target molecules in the cell. More recently the application of nonradioactively labeled nucleotides (e.g., biotin-, digoxigenin-UTP) has considerably improved ISH (Farquharson *et al.*, 1990; Hahn *et al.*, 1993; Morris *et al.*, 1990; Singer and Ward, 1982). Of all nonradioactive labeling methods developed, digoxigenin (DIG)-based detection has proven to be the most appropriate for rare transcripts. Because DIG is synthesized only in plants of the genus *Digitalis*, background problems due to unspecific antibody binding in cells of other organisms are avoided (Farquharson *et al.*, 1990). In many mRNA ISH protocols, precipitating substrates are used for the chromogenic detection of these probes (Braissant and Wahli, 1998; Hahn *et al.*, 1993). Fluorescently labeled tyramides are increasingly recognized as a more sensitive alternative for immunochemical detection systems (Van Gijlswijk *et al.*, 1997; Van heusden *et al.*, 1997). Catalyzed reporter deposition has been applied to increase the signal intensities in various immunochemical and FISH applications (Pernthaler *et al.*, 2002; Speel *et al.*, 1997; van de Corput *et al.*, 1998; Yang *et al.*, 1999). By the activity of the horseradish peroxidase, numerous tyramide molecules, pre-conjugated with either haptens or fluorescent reporters, are deposited in close vicinity to the HRP-binding site, resulting in superior spatial resolution and high signal intensity. In combination with HRP-labeled antibodies, the CARD-FISH method has the potential to detect low abundance mRNAs, potentially even single copies (Speel, 1999, 1997; van de Corput *et al.*, 1998; Yang *et al.*, 1999).

This chapter describes an improved protocol for the simultaneous detection of mRNA and rRNA in environmental microorganisms (Pernthaler and Amann, 2004). As a case study we detect the expression of *pmoA*, the gene encoding subunit A of the particulate methane monooxygenase (pMMO) in endosymbionts of the hydrothermal vent mussel *Bathymodiolus puteoserpentis* (Cosel and Métivier, 1994). Methane monooxygenases (MMO) catalyze the first step in the aerobic methane oxidation pathway, the oxidation of methane into methanol. There are two distinct types of MMO enzymes: a soluble, cytoplasmic enzyme complex (sMMO) and a membrane-bound pMMO. Virtually all methanotrophic *Bacteria* possess pMMO, many also sMMO. The *pmoA* is often used as a phylogenetic marker and also as an indicator for methanotrophy (Murrel *et al.*, 2000; Stolyar *et al.*, 2001; Tchawa *et al.*, 2003). Copper ions have been shown to play a key role in regulating the MMO expression. When the copper-to-biomass ratio is high, pMMO is expressed and the expression of sMMO is inhibited (Murrel *et al.*, 2000). In the case of inducible genes, like

MMO genes, the presence of the respective mRNA is an indicator for an ongoing metabolic process, as transcription in prokaryotes is tightly linked to translation. Thus, by combining mRNA and rRNA FISH, key players of specific biogeochemical processes can be identified.

Messenger RNA *In Situ* Hybridization

General Precautions

1. Wear powder-free gloves during all procedures.
2. Plasticware, such as pipette tips and tubes, must be autoclaved for 45 min.
3. Glassware must be baked for 6 h at 160°.
4. Laboratory bench and pipettes must be cleaned with with RNaseZAP (Ambion, Huntington, UK) or equivalent solutions.
5. To inactivate RNases, water and buffers must be treated with 0.1% (v/v) diethylpyrocarbonate (DEPC) overnight at room temperature and then autoclaved for 20 min. For buffers and solutions that cannot be treated with DEPC, e.g., Tris-HCl, autoclaving for 2 h is sufficient.
6. Steps involving the use of xylene, formalin, DEPC, and formamide should be performed in a fume hood.

Synthesis of mRNA-Targeted Polyribonucleotide Probes by
  In Vitro Transcription

In vitro transcription requires a purified linear DNA template containing a promoter, ribonucleotide triphosphates, a buffer system that includes dithiothreitol (DTT) and magnesium ions, and an appropriate phage RNA polymerase (e.g., SP6, T7, T3). Transcription templates include plasmid constructs engineered by cloning and linear templates generated by polymerase chain reaction (PCR). Many commercially available plasmid cloning vectors include phage polymerase promoters. They often contain two distinct promoters, one on each side of the multiple cloning site, allowing the transcription of either strand of an inserted sequence. Depending on the orientation of cDNA sequence relative to the promoter, the template may be designed to produce sense strand or antisense strand RNA. When using PCR products as templates for transcription, a promoter can be added to the PCR product by including the promoter sequence at the 5′ end of either the forward or the reverse PCR primer.

The following example is for *PmoA* primers (Costello and Lidstrom, 1999) containing T7 promoter sequences (bold):

A189-Fwd 5′-**TTA ATA CGA CTC ACT ATA GGG** GGN GAC TGG GAC TTC TGG-3′

Mb661-Rev 5′-**TTA ATA CGA CTC ACT ATA GGG** CCG GMG CAA CGT CYT TAC C-3′

These bases become double-stranded promoter sequences during the PCR reaction. When designing a transcription template, it must be decided whether sense or antisense transcripts are needed. If the RNA is to be used as a probe for hybridization to messenger RNA (e.g., Northern blots, ISH), complementary antisense transcripts are required. In contrast, sense strand transcripts are used as control probes for mRNA ISH (Fig. 1).

The exact conditions used in the transcription reaction depend on the amount of RNA needed for a specific application.

Reagents

10× Wilkinson's transcription buffer (400 mM Tris-HCl, pH 8.0, 60 mM MgCl$_2$, 20 mM spermidine), store at −20°

10× DTT (100 mM), store at −20°, avoid repeated freeze–thawing

10× nucleotide mix (10 mM ATP, 10 mM GTP, 10 mM CTP, 8 mM UTP, 2 mM DIG-UTP or FLUOS-UTP), store at −20°

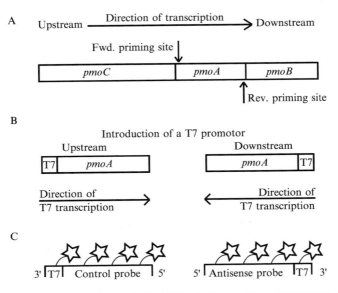

Fig. 1. Flow diagram of probe synthesis. (A) Particulate methane monooxygenase operon. (B) Subunit A is PCR amplified with a T7 polymerase promoter on either the forward (left) or the reverse primer (right). PCR amplicons are then used as templates for the synthesis of fluorescein-labeled transcript probes (C).

Deionized water (MilliQ water, Millipore, Eschborn, Germany)
RNA polymerase (25 U μl^{-1}, Ambion) (keep in freezer as long as possible; when in use, keep on ice)

Transcription Reaction

1. Mix the following reagents at room temperature in a 200-μl tube: add MilliQ water to a final volume of 20 μl, 2 μl Wilkinson's transcription buffer, 2 μl DTT, 2 μl nucleotide mix, 0.5–1 μg template DNA, and 2 μl RNA polymerase.

2. Mix gently and incubate at 37° for 2 h.

3. Add 1 μl of RNase-free DNase (1 U μl^{-1}) and incubate for another 15 min at 37°. Run 2 μl of probe on a 1.5% agarose gel to check the length of the probe. In the meantime, precipitate probe in a 0.5-ml reaction tube by adding 80 μl MilliQ water, 10 μl 3 M sodium acetate (pH 5.2), and 300 μl absolute ethanol. Incubate at $-80°$ for 2 h or at $-20°$ overnight. Centrifuge (e.g., in an Eppendorf centrifuge) at 4° at full speed for 20 min. Wash pellet with chilled 70% (v/v) ethanol and leave to air dry for 10 min. Dissolve the probe with 50 μl MilliQ water and measure the concentration photometrically (the concentration of probe after a successful transcription reaction should range between 300 and 500 μg ml^{-1}).

4. Dilute probe with hybridization buffer (see later) to a concentration of 50 ng μl^{-1}, denature probe at 80° for 5 min, and immediately store the probe at $-20°$. This mix will not freeze due to the formamide in the hybridization buffer and is stable for at least 1 year.

Fixation and Preparation of Tissue Samples

The success of ISH on mRNAs depends strongly on the integrity of the target mRNAs in the cell. Another requirement is a powerful reporter system capable of revealing low numbers of probe-mRNA-hybrids while keeping background staining as low as possible. Variables that influence the sensitivity and reproducibility of the ISH technique include (a) the effects of cell fixation on target mRNA preservation and accessibility to probes, (b) the type and quality of the probe, (c) the efficiency of hybrid formation, (d) the stability of *in situ* formed hybrids during posthybridization treatments, (e) the method of detection of the hybrid, and, finally, (f) background noise masking the hybridization signal. The following section describes and discusses variables that need to be optimized for FISH of *pmoA* mRNA in endosymbiotic bacteria in animal tissue sections (Table I).

TABLE I

Summary Steps for *pmoA* mRNA and rRNA FISH in Symbionts of *B. Puteoserpentis*
(As Shown in Fig. 3)

Sample fixation
  2% (para)formaldehyde in 1× PBS, 10–15 h at 4°
Sample preparation and immobilization
  Dehydrate, infiltrate with polyester wax, let blocks harden, cut sections, and mount onto
  slides. Store at −20°
Pretreatment and permeabilization
  Dewax in absolute ethanol (3 × 5 min), let slides dry, encircle sections with rubber
  pen (if needed), rehydrate in 70% ethanol for 5 min. Incubate for 12 min in 0.1% active
  DEPC in 1× PBS at room temperature, and then wash in 1× PBS, and MilliQ water.
  Place slide in sodium citrate and heat with microwave twice for 4 min. Place
  immediately in MilliQ water at room temperature
Hybridization
  Prehybridize in hybridization buffer without probe for 1 h at 58°. Add probe to tissue
  sections (final concentration 250 pg μl^{-1}) and hybridize overnight at 58°
Posthybridization washes
  Wash in 1× SSC, 50% formamide for 1 h at 58°. Wash in 0.2× SSC, 0.01% SDS for
  30 min at 58°
Immunocytochemistry
  Block for 30 min in 1× PBS, 0.5% blocking reagent at room temperature. Incubate for
  1 h with antibody in 1× PBS, 1% blocking reagent, 1% BSA at room temperature. Wash at
  room temperature once in 1× PBS, 0.5% blocking reagent for 20 min, and once in 1× PBS
  for 10 min
CARD
  Incubate in amplification buffer containing fluorescein (for simple mRNA FISH)- or
  Alexa$_{488}$ (for multiple staining)-labeled tyramide (0.25–0.5 μg ml^{-1}) for 5 min at 37°.
  Wash in PBS and MilliQ water
rRNA FISH
  Mix 100 μl of hybridization buffer (35% formamide) with 5 μl of fluorescently
  labeled oligonucleotide probe.[a] Pipette onto sections and incubate for 2 h at 46° in a
  humidified chamber. Wash with washing buffer at 48° for 5 min. Rinse with deionized tap
  water, let air dry, and embed in mountant. Slides are now ready for microscopy.

[a] Two rRNA probes (Pernthaler and Amann, 2004) were used for hybridization shown in
  Fig. 3, one for each of the two symbionts in *B. puteoserpentis*.

Fixation

Preparation of 20% (w/v) Paraformaldehyde (PFA). Commercially available 35–37% (v/v) formalin solution is often stabilized with methanol, which decreases FISH signal intensity and tends to precipitate upon longer storage. For optimal results, a buffered paraformaldehyde fixative should be prepared as follows.

1. Add 20 g of PFA to 70 ml of MilliQ water (use mouth protection for handling of PFA powder—irritant if inhaled).
2. Heat under fume hood to approximately 60° while stirring (must not boil!) and add solid NaOH until suspension is clear (takes approximately 30 min).
3. Add 5 ml of 20× phosphate-buffered saline (PBS), pH 7.6 (2.74 M NaCl, 54 mM KCl, 0.2 M Na$_2$HPO$_4$, 40 mM KH$_2$PO$_4$).
4. Adjust pH to 7.6 with fuming HCl and add dH$_2$O to 100 ml final volume.
5. Filter through a 0.2-μm filter.
6. Bubble with N$_2$ or helium for 3 to 5 min to remove oxygen and close bottle airtight with a rubber stopper.
7. If kept under N$_2$ in the dark at room temperature, the PFA fixative can be stored for more than 1 year. PFA fixative not treated with N$_2$ should be utilized within days.

After dissection of the fresh mussel, tissue samples are placed into chilled 2% (w/v) PFA in 1× PBS for 10 to 15 h (overnight) at 4°. Tissue is then washed three times in 1× PBS, dehydrated by successive baths in ice-cold 50% (v/v), 70% (v/v), and absolute ethanol for 1 h each. The tissue can then be embedded into wax for sectioning or can be stored at −80° in absolute ethanol for several weeks until further processing.

Embedding in Steedman's Wax and Sectioning

Polyester (PE) wax, developed by Steedman (1957), is a ribboning, low melting point wax that reduces heat-induced artifacts such as tissue shrinkage and loss of accessible target RNA and DNA. It is recommended for heat-labile tissues, to minimize heat-induced hardening in difficult tissues, and is an ideal medium for combined light and scanning electron microscopy of animal tissues (Norenburg and Barrett, 1987). The properties of the wax also facilitate immunohistochemical investigations as antigenic determinants are well preserved. The main advantages of this medium are a low melting point and infiltration directly from absolute ethanol, permitting a near isothermic processing schedule for mammalian tissues. In our laboratory we observed a dramatic increase in mRNA and rRNA FISH signals in tissues embedded in PE wax when compared to tissues embedded in paraffin. PE wax is no longer available commercially and must be prepared from the basic ingredients.

Preparation of Steedman's Wax. Melt 90 g of 400 polyethylene glycol distearate at 60°, add 10 g of 1-hexadecanol (cetyl alcohol), and shake until cetyl alcohol is dissolved.

Embedding Procedure. After the ethanol dehydration (as described earlier), tissue is further dehydrated and impregnated with wax as follows:

    1 h in absolute ethanol at room temperature
    1 h in mix of absolute ethanol (three parts) and PE wax (one part) at 37°
    1 h in mix of absolute ethanol (two parts) and PE wax (one part) at 37°
    1 h in mix of absolute ethanol (one part) and PE wax (one part) at 37°
    Three times PE wax, 1 h each, at 37°

Fill fresh PE wax into a plastic peel-away mold, place tissue into mold, and let wax harden at room temperature for approximately 3 to 4 h. Then store block at −20°.

The wax has good water tolerance and is soluble in most histological dehydrants. PE wax adheres to metal embedding molds and therefore paper boat or plastic peel-a-way molds are recommended. Normally, blocks are cut at ambient room temperatures. For sections thinner than 7 μm, Steedman's wax is cut more conveniently on a cryotome at −5° to 22° or on a microtome with a chilled knife. Sections are affixed to aminoalkylsilane pretreated glass slides. Sections are stretched directly onto the slide at room temperature with a few drops of sterile dH$_2$O. The water is then soaked off with a soft paper tissue and slides are left to air dry overnight. Blocks and sections are stored at −20°.

The steps from fixation to embedding and sectioning of the tissue cannot be performed RNase free. Some reagents might not be available in RNase-free quality and can also not be treated with DEPC. The highest contamination comes from the tissue itself, as all cells contain endogenous RNases. This is not a problem as long as the tissue is in a fixative (PFA or ethanol), frozen at −80°, or embedded in wax. However, one should work as cleanly and as fast as possible.

Dewaxing and Pretreatment of Tissue Sections

PE wax is removed from the slides in absolute ethanol (the use of inexpensive ethanol denatured with methylethylketone is also suitable). Tissue sections are then rehydrated and decarbethoxylated (treatment with DEPC) to inactivate RNases. For better penetration of detection molecules (probes, antibodies), the tissue can then be permeabilized in different ways. We discuss three possibilities: proteinase K digestion, autoclaving, and microwaving. The latter method can increase FISH signal intensity (Olivier *et al.*, 1997). Heat treatment of sections also allows for subsequent immunodetection of proteins (e.g., simultaneous mRNA FISH and immunocytochemical staining of the respective protein) (Fig. 2).

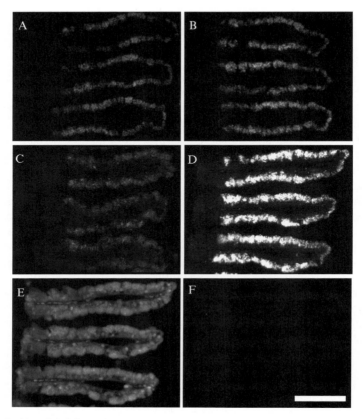

Fig. 2. (A to D) Effect of sample pretreatment on hybridization signals of an antisense probe against *pmoA* mRNA. Cross sections of gill filaments of a juvenile *B. puteoserpentis* are shown. (A) Only DEPC-treated tissue, (B) proteinase K (2 μg ml⁻¹) treatment, (C) autoclaved tissue, and (D) microwaved tissue with (E) corresponding to DAPI staining and (F) background signals with a control (sense) probe. Bar: 100 μm. Exposure times of all FISH images: 0.5 s.

Dewaxing (at Room Temperature)

1. Three times absolute ethanol for 5 min each (dewaxing).
2. Let slides air dry, and encircle section with a rubber pen (PAP Pen, Kisker Biotechnology, Steinfurt, Germany).
3. Five minutes in 70% (v/v) ethanol.
4. Twelve minutes in 0.1% (v/v) active DEPC in 1× PBS (prepare when needed, as DEPC has a half-life of 5 min in aqueous solution).
5. Rinse with PBS, and then rinse with MilliQ water.

Permeabilization

(A). Immerse slides in 1× PBS (also Tris-HCl, pH 8.0, or other buffers are suitable) containing various concentrations of proteinase K (must be tested for each tissue type; for *B. puteoserpentis* sections, 2 μg ml^{-1} is optimal) for 30 min at 37°.

(B). Immerse slides in 200 ml of sodium citrate (10 m*M*), pH 6.0, in a glass container and autoclave at 121° for 3 min.

(C). Immerse slides in 200 ml of sodium citrate (10 m*M*), pH 6.0, in a glass container, place the glass container into another glass container with 350 ml of cold tap water (water level should equal the sodium citrate level), and place both into a kitchen microwave, in the middle of a rotating platform. Heat at full power (800 W) for 4 min. Replace the hot tap water with 350 ml of cold tap water and heat again at full power for 4 min. The sodium citrate will boil during the last minute.

The duration of microwaving and the heat development, as well as the pH and molarity of the citrate buffer, influence the permeability of the tissue and the accessibility of the target mRNA (and therefore the mRNA FISH signal intensity).

For permeabilization approach (A), the proteinase K has to be washed off with three to four rinses in MilliQ water. For options (B) and (C), the hot slides have to be placed into MilliQ water very quickly to avoid drying of the tissue.

Pretreated slides have to be processed immediately for mRNA FISH, without drying of the tissue. The time between rehydration of the tissue and hybridization is the most critical time for mRNA. The tissue is now in buffer or water. If RNases are present, they will be active under these conditions, and mRNA will be degraded even faster at higher temperatures. As soon as the tissue is covered with hybridization buffer, RNases will be deactivated by the formamide.

Messenger RNA FISH

Hybridization buffers contain reagents to maximize nucleic acid duplex formation and to inhibit nonspecific binding of probe. Formamide is used to lower the optimal hybridization temperature in order to minimize cell damage and to enhance the specificity of probe binding. Formamide is also a nuclease inhibitor and allows the salt concentration to be adjusted to physiological osmolarity. Detergents such as SDS and heterologous nucleic acids inhibit background due to charge or nonspecific interaction with nucleic acids. Prehybridization without probe in the hybridization buffer proved to be crucial in avoiding high background fluorescence

(Braissant and Wahli, 1998; Speel *et al.*, 1998; Yang *et al.*, 1999) in most tissues.

Several strategies can increase the sensitivity of the hybridization. The probe concentration should be in excess of target mRNA, resulting in a "probe-driver" condition. However, too high a probe concentration can result in unacceptably high levels of background signal. Dextran sulfate is included in the buffer as a high molecular weight "volume excluding" polymer, in effect concentrating the probe into a smaller physical space, therefore increasing hybridization rates (Hrabovszky and Petersen, 2002; Wahl *et al.*, 1979).

Preparation of Hybridization Buffer. In a 10-ml tube, pipette 5 ml of formamide, 1 ml of 20× SSC (300 mM sodium citrate, 3 M sodium chloride), 1 ml of 10% [w/v] blocking reagent (Roche, Mannheim, Germany), 0.2 ml of 50× Denhard's solution (Sigma), 0.2 ml of yeast RNA [(1 mg ml^{-1}), Ambion], 0.2 ml sheared salmon sperm DNA [(1 mg ml^{-1}), Ambion], 2.4 ml MilliQ water, and 5 μl of 20% (w/v) SDS. Add 1 g of dextran sulfate and dissolve by heating (up to 60°) and shaking. The hybridization buffer can be stored at −20° for up to 1 year.

In Situ Hybridization. To block tissue sections against unspecific probe binding, prehybridization without a probe in the hybridization buffer is performed. Pipette 20 to 100 μl (depending on the size of the section) of hybridization buffer onto the wet tissue section. Place slides into a chamber, e.g., a 50-ml plastic tube, humidified with 50% formamide and incubate at 58° for 60 to 120 min. Mix hybridization buffer with probe stock to obtain a concentration of 500 pg μl^{-1}. Vortex for a few seconds. Then take the slides out of the hybridization chamber and mix the prehybridization buffer that is already on the slide with the buffer–probe mix in a 1:1 ratio by gentle pumping up and down with the pipette (final concentration of probe 250 pg μl^{-1}, do not touch section with pipette tip). Close the humidity chamber again and hybridize overnight at 58°.

Stringent Washing. Prepare 45 ml of washing buffer 1 [50% (v/v) formamide in 1× SSC] and 45 ml of washing buffer 2 [0.2× SSC, 0.01% (w/v) SDS] in 50-ml tubes. Preheat washing buffers at 58°. Then place slides (maximum two slides per tube) into washing buffer 1 and incubate at 58° for 1 h, and subsequently put the slides into washing buffer 2 for 30 min.

Some researchers recommend an RNase A digestion step after hybridization to remove the unhybridized probe, thus decreasing background staining. In our protocol, this step is omitted, as background staining is minimal.

Immunochemical Detection of Hybrids. The hybridized probe is detected immunochemically by an anti-DIG- or anti-FLUOS-antibody, labeled with a HRP. Antibodies can contain RNases. However, this does not seem to be critical, as double-stranded RNA (i.e., the hybrid) is much more stable than single-stranded RNA (i.e., unhybridized probe).

To block against unspecific binding of the antibody, the slides have to be incubated in a blocking buffer.

1. Mix 40 ml of 1× PBS with 2 ml of 10% (w/v) blocking reagent in 50-ml tubes.

2. Incubate slides for 30 min at room temperature.

3. In the meantime, centrifuge the antibody at 10,000 rpm at 4° for 10 min to remove precipitates that may form during storage and that can increase background staining.

4. In a 2-ml tube, pipette 1510 μl of MilliQ water, 200 μl of 10% (w/v) blocking reagent, 200 μl of 10% (w/v) bovine serum albumin (BSA, initial heat shock fractionation, Sigma) in 1× PBS, 90 μl of 20× PBS, and 4 μl of anti-DIG- or anti-FLUOS-antibody (150 U ml^{-1}, Fab fragment, HRP-labeled, Roche).

5. Mix gently by inverting the tube.

6. Place slides into a humidified chamber, spread antibody mix onto the sections, and incubate the slides for 60 to 90 min at room temperature.

7. Wash off unbound antibody by placing the slides back into the 1× PBS/blocking reagent mix for 20 min at room temperature. Then wash in 50 ml of 1× PBS for another 10 to 20 min.

The antibody-delivered HRP is then detected by catalyzed reporter deposition.

Catalyzed Reporter Deposition: Synthesis of Tyramide Conjugates

The tyramide-labeling procedure is modified from Hopman *et al.* (1998).

Reagents

Dimethylformamide
Triethylamine
Tyramine HCl
Succinimidyl esters of 5- (and 6-) carboxyfluorescein, Alexa$_{546}$, Alexa$_{488}$, Alexa$_{633}$, or Alexa$_{350}$ (Molecular Probes, Leiden, The Netherlands)

Succinimidyl esters can hydrolyze rapidly, therefore all reagents have to be water free. The active dye stock as well as the tyramine HCl stock must be prepared a few minutes before use.

Solutions

Tyramine HCl stock: 10 μl triethylamine, 1 ml dimethylformamide, and 10 mg tyramine HCl

Active dye stock: 1 mg succinimidyl ester (Alexa$_{546}$, Alexa$_{488}$, Alexa$_{633}$) and

100 μl dimethylformamide; 5 mg succinimidyl ester (Alexa$_{350}$) and 500 μl dimethylformamide; or 100 mg succinimidyl ester [5- (and 6-) carboxyfluorescein] and 10 ml dimethylformamide

Reaction

1. Add active dye ester in 1.1-fold molar excess to tyramine HCl stock solution:
   100 μl Alexa$_{488}$ stock + 25.2 μl tyramine HCl stock
   100 μl Alexa$_{546}$ stock + 14.7 μl tyramine HCl stock
   100 μl Alexa$_{633}$ stock + 13.1 μl tyramine HCl stock
   500 μl Alexa$_{350}$ stock + 193 μl tyramine HCl stock
   10 ml [5- (and 6-) carboxyfluorescein] stock + 3.3 ml tyramine HCl stock
2. Incubate for 6 to 12 h at room temperature in the dark. Then, dilute reaction mixture with absolute ethanol to a final concentration of 1 mg active dye per milliliter.
3. Dispense portions of 20 μl and desiccate them in a freeze dryer or under vacuum at room temperature in the dark. Desiccated tyramides are stable for years if stored at $-20°$. For use, reconstitute tyramides in 20 μl of MilliQ water or in 20 μl of dimethylformamide containing 20 mg ml^{-1} p-iodophenylboronic acid (IPBA).

In our laboratory we dissolve tyramides labeled with Alexa$_{546}$, Alexa$_{488}$, Alexa$_{633}$, or fluorescein in dimethylformamide (final concentration 1 mg ml^{-1}) containing 20 mg ml^{-1} IPBA. Water-free dimethylformamide prevents rapid hydrolysis of the IPBA, which enhances the deposition of several fluorescently labeled tyramides (Bobrow *et al.*, 2002). Alexa$_{350}$-labeled tyramide should be dissolved in MilliQ water. Tyramides in dimethylformamide can be stored in the freezer; tyramides in aqueous solution should be stored in the refrigerator.

The tyramide signal amplification can be enhanced by the addition of salts (Bobrow *et al.*, 2002). The deposition of tyramides labeled with Cy3, fluorescein, Alexa$_{350}$, Alexa$_{488}$, and Alexa$_{546}$ is enhanced by the presence of NaCl. Preferably, concentrations of NaCl in the amplification buffer range from 2 M to saturation. IPBA (20 mg IPBA per 1 mg of tyramide) will also enhance the CARD-FISH signal. This works well for tyramides labeled with fluorescein, Alexa$_{488}$, Alexa$_{633}$, and Alexa$_{546}$, but not for tyramides labeled with Cy3 and Alexa$_{350}$. For other fluorescent labels, one will need to test either of the salts as well as the combination of both.

Preparation of Amplification Buffer. In a 50-ml tube, pipette 2 ml of 20× PBS, 0.4 ml of 10% (w/v) blocking reagent, 16 ml of 5 M NaCl, and add sterile MilliQ water to a final volume of 40 ml. Add 4 g of dextran sulfate. Heat (40 to 60°) and shake until the dextran sulfate has dissolved completely. The amplification buffer can be stored in the refrigerator for several weeks.

1. Prepare fresh 100× H_2O_2 stock by mixing 1000 μl of 1× PBS with 5 μl of 30% (v/v) H_2O_2.

2. Mix 1000 μl of amplification buffer with 10 μl of the 100× H_2O_2 stock and 0.5 to 2 μl of fluorescently labeled tyramide.

3. Pipette the amplification buffer with the tyramide onto the sections. Incubate for 5 to 10 min in the dark at 37°.

4. Wash slides in 50 ml of 1× PBS for 5 to 15 min at room temperature in the dark.

5. Wash sections three times in dH$_2$O. If you plan to subsequently hybridize rRNA, proceed with section on "Hybridization of rRNA after mRNA FISH." For mRNA FISH alone, proceed to step 6.

6. Sections can now be counterstained with 4′,6′-diamidino-2′-phenyl-indole (DAPI, 1 μg ml^{-1} of deionized water, stain for 5 to 10 min at room temperature, wash with deionized water, and let air dry). Microscopy is performed after embedding in a low fluorescence glycerol mountant containing an antibleaching agent (e.g., Citifluor AF1, Citifluor Ltd., London). The embedding medium can also be amended directly with DAPI (final concentration, 1 μg ml^{-1}), but such a mix should be freshly prepared each month. Stained preparations can also be stored at −20° until further processing.

Controls

For the detection of gene expression in environmental microorganisms, at least two controls are needed.

A. The specificity of the probe has to be shown on the induced and noninduced target organism. This option might often be unavailable for microorganisms in environmental samples. Instead, the probes should be tested on induced and noninduced expression clones. A similar strategy has been described for the optimization of hybridization temperatures of rRNA-targeted oligonucleotide probes for uncultured bacteria (Clone-FISH) (Schramm *et al.*, 2002).

B. A control probe, which is the reverse complement to the antisense probe, has to be used in parallel hybridizations of the environmental samples. This is to ensure that the signal obtained by the antisense probe is due to hybridization and not due to "sticking" (unspecific binding) of the probe to any kind of material in the sample. If no hybridization signal is obtained at all, expression clones are also suitable to test the quality of the antisense probe.

For Clone-FISH, *pmoA* is cloned and expressed in pBAD vectors (Invitrogen, Karlsruhe, Germany) according to the manufacturer's instructions. Overnight cultures of Top10 *Escherichia coli* are diluted 1:100 and grown for 2 h at 37°. Cells are then induced with 0.2% (w/v) L-arabinose for 3 h and subsequently fixed with 2% (w/v) PFA for 30 min at room temperature. Cells are then centrifuged and washed once with 1 ml of 1× PBS, and twice with 1 ml of 50% ethanol in PBS. After resuspension in absolute ethanol, cells can be and stored at −80° until further processing.

Two to 10 μl of a suspension of *pmoA* expression clones is spotted onto aminoalkylsilane slides in fields encircled with a rubber pen (PAP Pen) and then air dried. Slides are then carbethoxylated by incubation in 50 ml of freshly prepared 0.1% (v/v) DEPC in 1× PBS for 12 min at room temperature, which is subsequently removed by washing in 50 ml of 1× PBS and 50 ml of MilliQ water for 1 min each. Next, cells are permeabilized with highly purified lysozyme (5 mg ml^{-1}) (from chicken egg white, nuclease free, Sigma) in 0.1 *M* Tris-HCl, 0.01 *M* EDTA, pH 8.0, for 30 min at room temperature by pipetting 1 ml of lysozyme solution onto the slide. The lysozyme is washed off with 50 ml of MilliQ water, and slides are then processed immediately for mRNA FISH as described earlier.

When using probes that are significantly longer or shorter than 450 nucleotides (which is the approximate length of the *pmoA* probe described here), one has to perform test hybridizations on expression clones at different temperatures or with different formamide concentrations. For example, for a probe that is 1000 nucleotides in length, we would suggest temperatures between 58 and 68°, or formamide concentrations in the hybridization buffer and washing buffer 1 between 50 and 70% (v/v) respectively. At the optimum stringency, the noninduced clone should show no hybridization signal and the induced clone should be stained very brightly with mRNA FISH.

At temperatures above 60° the pH of the hybridization and the washing buffers might change considerably. Therefore one has to check the pH at the given temperature.

Hybridization of rRNA after mRNA FISH

Preparation of Hybridization Buffer. In a 50-ml tube, mix 3.6 ml of 5 *M* NaCl, 0.4 ml of 1 *M* Tris-HCl, 20 μl of 20% (w/v) SDS, x ml of MilliQ water (see Table II), x ml of formamide (see Table II), and 2.0 ml 10% (w/v) blocking reagent. Add 2.0 g of dextran sulfate. Heat (40 to 60°) and shake until the dextran sulfate has dissolved completely. Small portions of the buffer can then be stored at –20° for several months.

Preparation of Washing Buffer (Produce Freshly When Needed). In a 50-ml tube, mix 0.5 ml of 0.5 *M* EDTA, pH 8.0, 1.0 ml of 1 *M* Tris-HCl, pH 8.0, x μl of 5 *M* NaCl (see Table III), add Milli-Q water to a final volume of 50 ml, and 25 μl of 20% (w/v) SDS.

The NaCl concentration in the washing buffer, as well as the formamide concentration of the hybridization buffer, determines the stringency of the hybridization at the selected temperature.

Prior to 16S rRNA hybridization with HRP-labeled oligonucleotide probes, the anti-DIG-HRP used for mRNA FISH has to be inactivated: Incubate slides in 50 ml of 0.01 *M* HCl for 10 min at room temperature. Rinse slides with 50 ml of PBS and 50 ml of MilliQ water.

Mix 300 μl of hybridization buffer with 1 μl of HRP-labeled oligonucleotide probe (50 ng of DNA μl^{-1}). Pipette onto sections, place slides into a humidified chamber (e.g., a 50-ml plastic tube), and incubate at 35° for 2 h.

For stringent washing, prepare washing buffer and preheat at 37°. Wash slides after hybridization for 10 min in 50 ml of washing buffer.

Do not let the sections run dry after hybridization! This will reduce the activity of the HRP.

TABLE II
VOLUMES OF FORMAMIDE AND WATER FOR 20 ML OF HYBRIDIZATION
BUFFER FOR rRNA FISH

% formamide in hybridization buffer	ml formamide	ml water
20	4	10
25	5	9
30	6	8
35	7	7
40	8	6
45	9	5
50	10	4
55	11	3
60	12	2
65	13	1
70	14	0

TABLE III
VOLUMES OF 5 M NaCl IN 50 ML OF WASHING BUFFER FOR rRNA FISH
WITH CORRESPONDING FORMAMIDE CONCENTRATION IN THE
HYBRIDIZATION BUFFER[a]

% formamide in hybridization buffer	μl of 5 M NaCl
20	1350
25	950
30	640
35	420
40	270
45	160
50	90
55	30
60	0
65	0
70	0

[a] The Na^+ concentration is calculated for stringent washing at 37°
after hybridization at 35°.

Perform CARD as described earlier, but using higher tyramide concentrations (1:200 to 1:1000 dilutions) and a longer incubation (15 min) at 37°. For double or triple hybridization, we use the following tyramides in combination: Alexa$_{488}$, Alexa$_{564}$, and Alexa$_{633}$. Wash and counterstain slides with DAPI as described earlier.

Troubleshooting

Possible Causes for High Background Fluorescence (mRNA FISH)

1. Too high probe and/or antibody concentration. Check the concentration of both stock solutions. Test lower concentrations of probe and antibody.
2. Too short washing after the hybridization, antibody reaction or CARD.

Possible Causes for Low mRNA FISH Signal Intensity or No FISH Signal at All

1. No target mRNA is present (either not expressed or lost during sampling and pretreatment) or the mRNA is masked by cross-link formation due to overfixation.

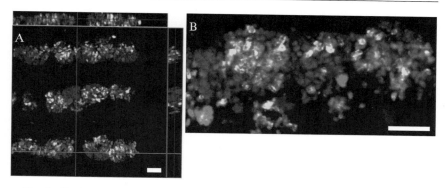

FIG. 3. Confocal images of gill cross section of a juvenile *B. puteoserpentis*. (A) Optical sectioning; red line indicates *z* section in *x* direction (upper frame), green line shows *z* section in *y* direction (right frame). (B) Detailed image of bacteriocytes. Composite of three images: Red: Methanotrophic symbionts hybridized with 16S rRNA-targeted, cy5- labeled probe Baz_meth_845I. Blue: Thiotrophic symbionts hybridized with 16S rRNA-targeted, cy3-labeled probe Baz_thio_193. Green: *PmoA* mRNA antisense probe, visualized with Alexa$_{488}$-labeled tyramide. Bars: 10 μm. (See color insert.)

2. The probe is not labeled sufficiently, hydrolyzed, or is not matching the target mRNA. Because the probe is diluted with hybridization buffer, the length cannot be checked anymore by running it on an agarose gel. Therefore, one should always perform a positive and a negative control, e.g., with *E. coli* clones expressing the respective mRNA (see earlier discussion). If this does not work, consider repeating the probe synthesis.

3. The antibody-delivered HRP has too low or no activity. The antibody solution should be thawed only once and should not be stored in the refrigerator for more than 6 months.

4. Tyramide signal amplification failed. Check the pH of the PBS, it should be 7.6. Check the H_2O_2 concentration and its age (should be prepared freshly from original stock prior to incubation). Test CARD with increased tyramide concentration. Check the reactivity of the tyramide (e.g., by performing rRNA FISH with pure cultures).

5. The probe and/or the antibody cannot penetrate the tissue. Modify the permeabilization protocol.

A protocol for the mRNA hybridization in bacteria in sediment is given in Pernthaler and Amann (2004). A more detailed protocol for rRNA CARD-FISH is given in Pernthaler *et al.* (2004).

Acknowledgments

We are grateful to N. Dubilier for support. F. Zielinski and C. Borowski are acknowledged for collecting *B. puteoserpentis* specimens, as well as the Crew of the R/V METEOR. R. Amann and J. Wulf are acknowledged for critical reading of the manuscript. This work was funded by the Max Planck Society, the German Research Foundation (DFG) within the framework of the DFG Priority Program 1144: "From Mantle to Ocean: Energy-, Material-, and Life-cycles at Spreading Axes," as well as the European Union (EVK3-CT-2002–00078 BASICS).

References

Bobrow, M. N., Adler, K. E., and Roth, A. (2002). "Enhanced Catalyzed Reporter Deposition." Bobrow, MN.

Braissant, O., and Wahli, W. (1998). A simplified *in situ* hybridization protocol using non-radioactively labeled probes to detect abundant and rare mRNAs on tissue sections. *Biochemica* **1**, 10–16.

Cosel, R. V., and Métivier, B. (1994). Three new species of *Bathymodiolus* (Bivalvia: Mytilidae) from hydrothermal vents in the Lau Basin and the North Fiji Basin, Western Pacific, and the Snake Pit Area, Mit-Atlantic Ridge. *Veliger* **37**, 374–392.

Costello, A. M., and Lidstrom, M. E. (1999). Molecular characterization of functional and phylogenetic genes from natural populations of methanotrophs in lake sediments. *Appl. Environ. Microbiol.* **65**, 5066–5074.

Farquharson, M., Harvie, R., and McNcol, A. M. (1990). Detection of messenger RNA using a digoxigenin end-labeled oligonucleotide probe. *J. Clin. Pathol.* **43**, 423–428.

Gerfen, C. R. (1989). Quantification of *in situ* hybridization histochemistry for analysis of brain function. *In* "Methods in Neuroscience" (P. M. Conn, ed.), pp. 79–97. Academic Press, San Diego.

Hahn, D., Amann, R. I., and Zeyer, J. (1993). Detection of mRNA in Streptomyces cells by whole-cell hybridization with digoxigenin-labeled probes. *Appl. Environ. Microbiol.* **59**, 2753–2757.

Hopman, A. H. N., Ramaekers, F. C. S., and Speel, E. J. M. (1998). Rapid synthesis of biotin-, digoxigenin-, trinitrophenyl-, and fluorochrome-labeled tyramides and their application for *in situ* hybridization using CARD amplification. *J. Histochem. Cytochem.* **46**, 771–777.

Hrabovszky, E., and Petersen, S. L. (2002). Increased concentrations of radioisotopically-labeled complementary ribonucleic acid probe, dextran sulfate, and dithiothreitol in the hybridization buffer can improve results of *in situ* hybridization histochemistry. *J. Histochem.Cytochem.* **50**, 1389–1400.

John, H. A., Birnstiel, M. L., and Jones, K. W. (1969). RNA-DNA hybrids at the cytological level. *Nature* **223**, 582–587.

Morris, R. G., Arends, M. J., Bishop, P. E., Sizer, K., Duvall, E., and Bird, C. C. (1990). Sensitivity of digoxigenin and biotin-labeled probes for detection of human papillomavirus by *in situ* hybridization. *J. Clin. Pathol.* **43**, 800–805.

Murrel, C. J., Gilbert, B., and McDonald, I. R. (2000). Molecular biology and regulation of methane monooxygenase. *Arch. Microbiol.* **173**, 325–332.

Norenburg, J. L., and Barrett, M. J. (1987). Steedman's polyester wax embedment and de-embedment for combined light and scanning electron microscopy. *J. Electron Microsc. Techniques* **6**, 35–41.

Olivier, K. R., Heavens, R. P., and Sirinathsinghji, D. J. S. (1997). Quantitative comparison of pretreatment regimes used to sensitize *in situ* hybridization using oligonucleotide probes on paraffin-embedded brain tissue. *J. Histochem. Cytochem.* **45**, 1707–1713.

Pernthaler, A., and Amann, R. (2004). Simultaneous fluorescence *in situ* hybridization of mRNA and rRNA in environmental bacteria. *Appl. Environ. Microbiol.* **70,** 5426–5433.

Pernthaler, A., Pernthaler, J., and Amann, R. (2002). Fluorescence *in situ* hybridization and catalyzed reporter deposition (CARD) for the identification of marine bacteria. *Appl. Environ. Microbiol.* **68,** 3094–3101.

Pernthaler, A., Pernthaler, J., and Amann, R. (2004). Sensitive multi-color fluorescence *in situ* hybridization for the identification of environmental microorganisms. *In* "Molecular Microbial Ecology Manual" (G. Kowalchuk, F. J. de Bruijn, I. M. Head, A. D. L. Akkermans, and J. D. van Elsas, eds.), pp. 711–726. Kluwer Academic Press, Dordrecht.

Schramm, A., Fuchs, B. M., Nielsen, J. L., Tonolla, M., and Stahl, D. A. (2002). Fluorescence *in situ* hybridization of 16S rRNA gene clones (Clone-FISH) for probe validation and screening of clone libraries. *Environ. Microbiol.* **4,** 713–720.

Singer, R. H., and Ward, D. C. (1982). Actin gene expression visualized in chicken muscle tissue culture by using *in situ* hybridization with a biotinated nucleotide analog. *Proc. Natl. Acad. Sci. USA* **79,** 7331–7335.

Speel, E. J. M. (1999). Detection and amplification systems for sensitive, multiple- target DNA and RNA *in situ* hybridization: Looking inside cells with a spectrum of colors. *Histochem. Cell Biol.* **112,** 89–113.

Speel, E. J. M., Saremaslani, P., Komminoth, P., and Hopman, A. H. N. (1997). Card signal amplification: An efficient method to increase the sensitivity of DNA and mRNA *in situ* hybridization. *Am. J. Pathol.* **151,** 1499.

Speel, E. J. M., Saremaslani, P., Roth, J., Hopman, A. H. N., and Komminoth, P. (1998). Improved mRNA *in situ* hybridization on formaldehyde-fixed and paraffin-embedded tissue using signal amplification with different haptenized tyramides. *Histochem. Cell Biol.* **110,** 571–577.

Steedman, H. F. (1957). Polyester wax: A new ribboning embedding medium for histology. *Nature* **4574,** 1345.

Stolyar, S., Franke, M., and Lidstrom, M. E. (2001). Expression of individual copies of *Methylococcus capsulatus* bath particulate methane monooxygenase genes. *J. Bacteriol.* **183,** 1810–1812.

Tchawa, Y. M., Dunfield, P. F., Ricke, P., Heyer, J., and Liesack, W. (2003). Wide distribution of a novel *pmoA*-like gene copy among type II methanotrophs, and its expression in *Methylocystis* strain SC2. *Appl. Environ. Microbiol.* **69,** 5595–5602.

van de Corput, M. P. C., Dirks, R. W., van Gijlswijk, R. P. M., van de Rijke, F. M., and Raap, A. K. (1998). Fluorescence *in situ* hybridization using horseradish peroxidase-labeled oligodeoxynucleotides and tyramide signal amplification for sensitive DNA and mRNA detection. *Histochem. Cell Biol.* **110,** 431–437.

Van Gijlswijk, R. P. M., Zijlmans, H. J. M. A. A., Wiegant, J., Bobrow, M. N., Erickson, T. J., Adler, K. E., Tanke, H. J., and Raap, A. K. (1997). Fluorochrome-labeled tyramides: Use in immuno-cytochemistry and fluorescence *in situ* hybridization. *J. Histochem. Cytochem.* **45,** 375–382.

Van heusden, J., de Jong, P., Ramaekers, F., Bruwiere, H., Borgers, M., and Smets, G. (1997). Fluorescein-labeled tyramide strongly enhances the detection of low bromodeoxyuridine incorporation levels. *J. Histochem. Cytochem.* **45,** 315–320.

Wahl, G. M., Stern, M., and Starck, G. R. (1979). Efficient transfer of large DNA fragments from agarose gels to diazobenzyloxymethly-paper and rapid hybridization by using dextran sulfate. *Proc. Natl. Acad. Sci. USA* **76,** 3683–3687.

Yang, H., Wanner, I. B., Roper, S. D., and Chaudhari, N. (1999). An optimized method for *in situ* hybridization with signal amplification that allows the detection of rare mRNAs. *J. Histochem. Cytochem.* **47,** 431–445.

[22] Community-Level Analysis of Phototrophy: psbA Gene Diversity

By GIL ZEIDNER and ODED BÉJÀ

Abstract

Photosynthetic organisms play a crucial role in the marine environment. In vast areas of the oceans, most of this marine production is performed by cells smaller than 2–3 μm (picoplankton). This chapter describes molecular analyses of the conserved photosynthetic *psbA* gene (protein D1 of photosystem II reaction center) as a diversity indicator of naturally occurring marine oxygenic picophytoplankton and of marine cyanophages carrying photosynthesis genes.

Introduction

Cultivation-independent surveys of microbial rRNA genes have greatly expanded the known phylogenetic diversity of microbial species on Earth (Pace, 1997). It is now well established that many new microbial groups discovered via cultivation-independent rRNA surveys represent major components of natural microbial assemblages in marine systems (Karl, 2002). Community analysis of bacteria using molecular methods such as polymerase chain reaction (PCR) amplification of 16S rRNA gene in combination with temperature or denaturing gradient gel electrophoresis (DGGE) is commonly performed in microbial ecology (Muyzer and Smalla, 1998). Apart from these rRNA-based surveys, studies have focused on functional protein-coding genes such as [NiFe] hydrogenase (Wawer and Muyzer, 1995), *rpoB* (Dahllöf *et al.*, 2000), *nifH* (Lovell *et al.*, 2001; Rosado *et al.*, 1998), or *amoA* (Oved *et al.*, 2001) in combination with DGGE for microbial community structure analyses.

Both cyanobacteria and eukaryotes containing chlorophyll pigments utilize light as their main energy source. Microbial oxygenic ("green plant") photosynthesis has been a key metabolic process on earth and a major driver in the evolution of plant and metazoan species. Their photosynthetic membrane contains two photochemically active photosystems, of which photosystem II (PSII) mediates the transfer of electrons and protons from water, the terminal electron donor, to the plastoquinone pool. The D1 and D2 proteins form the reaction center dimer of PSII that binds the primary electron donors and acceptors. Zeidner *et al.* (2003) surveyed

METHODS IN ENZYMOLOGY, VOL. 397
0076-6879/05 $35.00
DOI: 10.1016/S0076-6879(05)97022-5

picophytoplankton diversity by designing general degenerate DNA primers based on well-conserved amino acid regions of protein D1. Its gene, *psbA*, is found in both cyanobacteria and chloroplast genomes of algae and higher plants and is widely used as a phylogenetic marker (Morden and Golden, 1989; Morden *et al.*, 1992; Scherer *et al.*, 1991; Zhang *et al.*, 2000).

These primers were designed to target all known oxygenic photosynthetic lineages, both prokaryotic (cyanobacteria) and eukaryotic (algae and higher plants). Use of *psbA* primers provided a marker for photosynthetic picoplankton groups such as picoeukaryotic green algae (prasinophytes) and cyanobacteria (Zeidner *et al.*, 2003). Using *psbA* as a marker, Zeidner *et al.* (2003) suggested that the marine prasinophyte *Ostreococcus tauri* (Courties *et al.*, 1994) is similar to the previously reported "uncultured Prasinophyceae" group (Béjà *et al.*, 2000; Rappé *et al.*, 1997, 1998) and is probably abundant in the marine environment. Furthermore, a new uncultured cyanobacterial-like group related to marine *Synechococcus* was identified. Further work combining BAC libraries and denaturing gradient gel electrophoresis data analyses (Zeidner and Béjà, 2004) has linked *psbA* genes belonging to this group to marine bacteriophages.

Previous reports of genes related to oxygenic photosynthesis in the genomes of phages from two viral families (Myoviridae and Podoviridae) have highlighted just how little is known of the dynamics of the cyanobacterial photosynthesis gene pool (Lindell *et al.*, 2004; Mann *et al.*, 2003; Millard *et al.*, 2004). Our present knowledge is based on two cases: the genome of a phage isolated from the cultured marine cyanobacterium *Synechococcus* WH7803 and another isolated from both high light- and low light- adapted *Prochlorococcus* strains (Lindell *et al.*, 2004; Mann *et al.*, 2003; Millard *et al.*, 2004). Nevertheless, a variety of viral photosynthesis genes were identified: *psbA*, encoding the D1 protein; *psbD*, encoding the D2 protein; *hli* genes, encoding the HLIPs (high light-inducible proteins); and the *petE* and *petF* genes encoding the photosynthetic electron transport proteins plastocyanin and ferredoxin, respectively. In addition, Zeidner *et al.* (2005) described a potential *psb*A gene swapping between *Prochlorococcus* and *Synechococcus* populations via viral intermediates, which again indicates the great potential for ecosystem regulation by phages.

The following protocols describe methods that are used to detect *psbA* genes directly in the environment and to examine their potential influence on the community structure. These include methods for constructing environmental *psbA* genomic libraries directly from seawater or from viral fractionated waters, DGGE analyses, and screening of large insert BAC libraries. These methods are not limited to photosynthesis genes and could be applied to any chosen genes.

Construction of Environmental *psbA* Libraries

Environmental DNA Collection and Extraction

In the last years, several protocols for extraction of DNA from seawater and sediments have been developed and improved. At the same time, much effort has also been devoted to improving DNA extraction efficiency and to minimizing biases due to DNA contamination. The most commonly utilized technique for the extraction of DNA from seawater is based on direct cell lysis by physical procedures (e.g., bead mill homogenization, ultrasonication, and freeze–thawing) and/or chemical procedures [e.g., the use of sodium dodecyl sulfate (SDS) or Sarkosyl]. Although this lysis method has the potential to circumvent problems caused by biased representation of the microbial community (i.e., by ensuring that cells from all groups of microorganisms are lysed in equal proportions), in this procedure extracellular DNA is coextracted with nucleic acids released from the lysed cells, which could lead to misinterpretation of the composition of the target community derived from molecular analysis.

For our analysis, 5 liter of seawater is prefiltered through a GF/A ($\sim$1.6-μm-pore-size) fiberglass filter, and the picoplankton remaining in the filtrate is collected on a 0.2-μm Sterivex filter (Milipore) using a peristaltic pump. The Sterivex filters are filled with 1.8 ml of lysis buffer and stored at $-20°$. The pore size of the prefilter is optional. and anyone interested in picoeukaryotic phytoplankton should use a prefiltration mesh 3–5 μm.

DNA extraction begins with the addition of 20 μl of 50 mg/ml lysozyme to the Sterivex filter unit and incubation at $37°$ for 30 min with rotation. Then, 50 μl of 10 mg ml^{-1} proteinase K and 20 μl of 20% SDS are added, and the filter is incubated at $55°$ for 1 h. The lysate is recovered from the filter and extracted with phenol. The aqueous phase is extracted with phenol-chloroform-isoamyl alcohol (25:24:1; pH 8.0) and with chloroform-isoamyl alcohol (41:1). A 1/10 volume of 3 M NaOAc and 2 volume absolute ethanol are added to the aqueous extract and kept at $-20°$ overnight. Following 15 min centrifugation at maximal speed at $4°$, the pellet is washed with 1 ml ice-cold 80% ethanol, incubated for 2 min on ice, and centrifuged for 5 min at maximal speed at $4°$. The pellet should be air dried and resuspended in TE buffer.

Environmental psbA PCR Amplification

The enzymatic amplification of specific segments of DNA (target sequence) by PCR has been used extensively for phylogenetic analysis (using the 16S rDNA as a molecular marker). In order to focus our analysis on

TABLE I

ENVIRONMENTAL *psbA* PRIMERS

Primer name	Sequence(5'-3')	Length	Temperature
*psbA*F/GC[a]	GTNGAYATHGAYGGNATHMGNGARCC	26	53°
*psbA*F			
*psbA*R	GRAARTTRTGNGCRTTNCKYTCRTGCAT	29	57°
*psbA*R-short	CATNARDATRTTRTGYTCNGCYTG	24	48°

[a] Denotes that a GC-rich sequence (GC clamp) is attached to the 5' end of the primer.

marine oxygenic photosynthetic populations, we use a "functional gene" (*psbA*) approach to determine the diversity.

psbA genes are usually amplified by PCR directly from environmental samples using degenerate primers. The general degenerate primers used to amplify the *psbA* gene from environmental samples were designed based on the conserved regions: 58-VDIDGIREP-66 and 331-MHERNAHNFP-340 (see Table I). PCRs are carried out using proofreading BIO-X-ACT DNA polymerase (BIOLINE). Amplification of DNA fragments is achieved by adding 2 μl of template DNA from environmental samples, 10 μl of OptiBuffer, 4 μl of 50 mM MgCl$_2$, 10 μl of dNTPs (25 mM each), 1 μl of 100 μM each primer, and 1 μl of polymerase (4 u μl^{-1}) in a final volume of 100 μl. The PCR program include 3 min of denaturation in 94°, 35 cycles of denaturation (1 min, 92°), annealing (1 min, gradient 53°), and extension (1 min, 68°); after the last cycle we add an additional 7 min of extension and cooling to 4°. The products of the reaction are analyzed on an agarose gel containing 1% Seaken agarose (BMA), dissolved in TAE buffer and 0.5 μg ml^{-1} EtBr.

RFLPs Analysis for Environmental PCR Products

Restriction site variation has been used previously to address questions in systematic and evolutionary biology. This method has several characteristics that make it particularly appropriate for phylogenetic analysis. Restriction fragment patterns can be characterized either as a simple list of fragment lengths or as mapped restriction sites. For our analysis and usually when comparing closely related taxa with a low level of sequence divergence (in our case), it may be possible to infer restriction site differences by inspecting fragment profiles. The choice of restriction enzymes for the analysis is usually based on number of fragments generated. One should also keep in mind that because many restriction enzymes recognize multiple sites, they should be avoided (or considered carefully) in the case of use of multiple restriction enzymes.

Our environmental cloned *psbA* plasmids are digested with restriction endonucleases *Eco*RI, *Alu*I, and *Rsa*I (New England Bio Labs) at 37° for 3 h according to the manufacturer's instructions. The *Eco*RI used for extraction from the multiple cloning site in the vector and *Rsa*I (four cutter enzyme) combined with *Alu*I for gene restriction. Restricted DNA fragments are analyzed by electrophoresis in agarose gel containing 1% Seaken agarose (BMA) and dissolved in TAE buffer and 0.5 μg ml^{-1} EtBr. The RFLP screen is used for rapidly assessing the quality of the clone library rapidly and for measuring the distribution of different RFLP genotypes. Usually, at least two clones from each RFLP group are sent for sequencing.

Phylogenetic Analysis

Sequence data of restriction patterns survey can be summarized in a phylogeny tree. Sequences are first aligned based on protein sequences. This alignment can then be used to align their corresponding DNA, in the analysis of GC content. Neighbor joining (NJ) and maximum parsimony (MP) analyses are conducted on protein data sets using version 4.0b10 of PAUP (Swofford, 2002). Default parameters are used in all analyses. Bootstrap resampling of NJ (1000 replicates) and MP (1000 replicates) trees are performed in all analyses to provide confidence estimates for the inferred topologies. It is important to note here that these phylogenetic analyses are not mainly used to assess the exact phylogenetic relationships of different sequences but rather to monitor related environmental sequences and to look for their closest cultured organisms.

Preparation of Viral-Enriched Water Fractions

It has been reported that *psb*A genes were found in cyanophages both infecting *Synechococcus* sp. and *Prochlorococcus* sp. This phage–host interaction had raised questions regarding the physiology ecology and evolution of the most important light-harvesting mechanism in the ocean. For viral fractionation analysis, 25 liters of waters is collected; the sample volume can vary depending on the initial concentration of viruses (a typical seawater sample viral concentration is about 10^7 ml). Seawater is prefiltered through GF/A Whatman filters (1.6 μm). As mentioned before, the option for a different size prefiltration filter should be planned ahead according to the different planktonic group investigated. Viral fractionation is essentially prepared as described by Steward (2001). Samples are filtered through a 0.22-μm pore size Sterivex GS (Millipore) filter to remove bacteria and phytoplankton. The filtrate is concentrated to 50 ml using a Pellicon XL (Millipore) filtration system and is concentrated further by ultracentrifugation at 225,000*g* at 10° for 1.5 h. The final viral pellet is washed twice with 5× TE buffer in a Microcon-100 to a final volume of 50 μl. Samples are scanned using SEM for a visual confirmation of nonbacterial contamination. The viral-concentrated lysate is treated with DNase and RNase to prevent bacterial genomic contamination, and viral DNA is extracted using the standard phenol/chloroform extraction protocol.

DGGE Analyses of *psbA* Clones

Materials, Reagents, and Equipment

> D-Gene system (Bio-Rad, Hercules)
> UV transillumination table (302 nm)-Herolab UVT-28M
> Kodak digital camera EDAS-290

Methods

Environmental psbA PCR Amplification. DGGE is a PCR-based technique for the physical separation of DNA sequences that have only minor differences. For our purposes, fragments (~750 bp) of the *psbA* gene are amplified with the general degenerate primers used to amplify the *psbA* gene from environmental samples, with the addition of a GC clamp (5'-CCGCCGCGCGGCGGGCGGGGCGGGGGCACGGGG-3') attached to the forward primer in order to increase DGGE gel separation (Muyzer *et al.*, 1993). In addition, because the recommended optimal PCR product fragment length for DGGE is up to 500 bp, a short degenerate primer was designed based on the conserved region: FQAEHNILM 5'-CATNAR-DATRTTRTGYTCNGCYTG-3'. Amplification with psbAF and psbAR-SHORT results in a PCR fragment of (~450 bp).

PCR amplification is carried out in a total volume of 25 μl containing 10 ng of template DNA, 200 μM dNTPs, 1.5 mM MgCl$_2$, 0.2 μM primers, and 2.5 U of BIO-X-ACT DNA polymerase (Bioline Ltd., London, UK). The amplification conditions comprise steps at 92° for 4 min and 35 cycles at 92° for 1 min, 50° or 57° for 1 min (for short vs long psbA reverse primer, respectively), and 68° for 1 min. PCR products are verified and quantified by agarose gel electrophoresis using the 2-log ladder standard (New England Biolabs, MA).

Gel Casting and Separation Procedure. DGGE is caste using the following ingredients and conditions: 1× TAE [40 mM Tris, 20 mM acetic acid, 1 mM EDTA (pH 8.3)] and 1-mm-thick gels. For the ~750-bp PCR product, we cast a double denaturant gradient of 15 to 45% urea-formamide (100% denaturant agent is 7 M urea and 40% formamide) and 6–8% acrylamide; the ~450-bp PCR products are separated on a 40–60% gel. One microgram of PCR product is loaded for each sample, and the separation is conducted at 60° ; 250 V for 3 h; alternatively, for psbAF-psbAR fragments, the running time is 16 h at 90 V or 3.5 h at 250 V. The combination of double denaturant gradient, lower voltage, and longer running time seems to give better separation, especially with the long DNA fragments. Ethidium bromide-stained DGGE gels are photographed on a UV transillumination table (302 nm) with a Kodak digital camera, and inverse images are prepared with Photoshop version 4.0. DGGE bands are excised from the gel and kept in 20 μl of sterilized water overnight. One microliter of the supernatant is used for reamplification with the original primer set. PCR products are purified with the DNA gel extraction kit (Millipore, France) cloned into pDrive vector and sequenced. Two main objectives instruct us in designing the DGGE experiments: the ability to study spatial and temporal variability of phytoplankton populations in combination with the ability to get phylogenetic information from these populations. The combination of both long and short PCR products achieves both objectives. The use of short PCR products gives better band separation, which allows us to observe inter-species differences, whereas the use of long PCR products allows better phylogeny analyses, as the sequence information is doubled.

Screening of Large Insert BAC Libraries

The screening of BAC libraries using different target genes enables us to look at the genomic environment of these genes and not only in the narrow context of direct PCR amplification. This has already proved worthy in identifying a psbA gene on a phage genomic fragment (Zeidner and Béjà, 2004).

Minipreparation of pooled plates from environmental BAC libraries is performed using multiplex protocols (Stein et al., 1996). Plate pools are screened for psbA containing clones as described earlier using forward and reverse original primers. All positive plates are screened further using rows and column pools for the specific positive clone position in the plate. For inserts, size estimation positive clones are grown overnight at 37° in 2 ml of LB and 12 μg ml^{-1} chloramphenicol. DNA is extracted as described (Béjà et al., 2000), cut using NotI, and loaded onto a PFGE. The gel is run on a PFGE machine at 14° , 6 V cm^{-1} for 13 h with 5- to 15-s pulses in 0.5× TBE buffer. For reference, we use the yeast chromosome PFG marker (New

England Bio-Labs). Chosen BAC clones are usually first analyzed using a 1× sequencing coverage; based on these results, a decision on the amount of further sequencing coverage is taken.

References

Béjà, O., Suzuki, M. T., Koonin, E. V., Aravind, L., Hadd, A., Nguyen, L. P., Villacorta, R., Amjadi, M., Garrigues, C., Jovanovich, S. B., Feldman, R. A., and DeLong, E. F. (2000). Construction and analysis of bacterial artificial chromosome libraries from a marine microbial assemblage. *Environ. Microbiol.* **2**, 516–529.

Courties, C., Vaquer, A., Troussellier, M., Lautier, J., Chrétiennot-Dinet, M. J., Neveux, J., Machado, M. C., and Claustre, H. (1994). Smallest eukaryotic organism. *Nature* **370**, 255.

Dahllöf, I., Baillie, H., and Kjelleberg, S. (2000). *rpoB*-based microbial community analysis avoids limitations inherent in 16S rRNA gene intraspecies heterogeneity. *Appl. Environ. Microbiol.* **66**, 3376–3380.

Karl, D. M. (2002). Hidden in a sea of microbes. *Nature* **415**, 590–591.

Lindell, D., Sullivan, M. B., Johnson, Z. I., Tolonen, A. C., Rohwer, F., and Chisholm, S. W. (2004). Transfer of photosynthesis genes to and from *Prochlorococcus* viruses. *Proc. Natl. Acad. Sci. USA* **101**, 11013–11018.

Lovell, C. R., Friez, M. J., Longshore, J. W., and Bagwell, C. E. (2001). Recovery and phylogenetic analysis of *nifH* sequences from diazotrophic bacteria associated with dead aboveground biomass of *Spartina alterniflora*. *Appl. Environ. Microbiol.* **67**, 5308–5314.

Mann, N. H., Cook, A., Millard, A., Bailey, S., and Clokie, M. (2003). Bacterial photosynthesis genes in a virus. *Nature* **424**, 741.

Millard, A., Clokie, M. R. J., Shub, D. A., and Mann, N. H. (2004). Genetic organization of the *psbAD* region in phages infecting marine *Synechococcus* strains. *Proc. Natl. Acad. Sci. USA* **101**, 11007–11012.

Morden, C. W., Delwiche, C. F., Kuhsel, M., and Palmer, J. D. (1992). Gene phylogenies and the endosymbiotic origin of plastids. *Biosystems* **28**, 75–90.

Morden, C. W., and Golden, S. S. (1989). *psbA* genes indicate common ancestry of prochlorophytes and chloroplasts. *Nature* **337**, 382–385.

Muyzer, G., de Waal, E. C., and Uitterlinden, A. G. (1993). Profiling of complex microbial populations by denaturing gradient gel electrophoresis analysis of polymerase chain reaction-amplified genes coding for 16S rRNA. *Appl. Environ. Microbiol.* **59**, 695–700.

Muyzer, G., and Smalla, K. (1998). Application of denaturing gradient gel electrophoresis (DGGE) and temperature gradient gel electrophoresis (TGGE) in microbial ecology. *Antonie Van Leeuwenhoek* **73**, 127–141.

Oved, T., Shaviv, A., Goldrath, T., Mandelbaum, R. T., and Minz, D. (2001). Influence of effluent irrigation on community composition and function of ammonia-oxidizing bacteria in soil. *Appl. Environ. Microbiol.* **67**, 3426–3433.

Pace, N. R. (1997). A molecular view of microbial diversity and the biosphere. *Science* **276**, 734–740.

Rappé, M. S., Kemp, P. F., and Giovannoni, S. J. (1997). Phylogenetic diversity of marine coastal picoplankton 16S rRNA genes cloned from the continental shelf off Cape Hatteras, North Carolina. *Limnol. Oceanogr.* **42**, 811–826.

Rappé, M. S., Suzuki, M. T., Vergin, K. L., and Giovannoni, S. J. (1998). Phylogenetic diversity of ultraplankton plastid small-subunit rRNA genes recovered in environmental nucleic acid samples from the Pacific and Atlantic coasts of the United States. *Appl. Environ. Microbiol.* **64**, 294–303.

Rosado, A. S., Duarte, G. F., Seldin, L., and Van Elsas, J. D. (1998). Genetic diversity of *nifH* gene sequences in *Paenibacillus azotofixans* strains and soil samples analyzed by denaturing gradient gel electrophoresis of PCR-amplified gene fragments. *Appl. Environ. Microbiol.* **64,** 2770–2779.

Scherer, S., Herrmann, G., Hirschberg, J., and Boger, P. (1991). Evidence for multiple xenogenous origins of plastids: Comparison of *psbA*-genes with a xanthophyte sequence. *Curr. Genet.* **19,** 503–507.

Stein, J. L., Marsh, T. L., Wu, K. Y., Shizuya, H., and DeLong, E. F. (1996). Characterization of uncultivated prokaryotes: Isolation and analysis of a 40-kilobase-pair genome fragment from a planktonic marine archaeon. *J. Bacteriol.* **178,** 591–599.

Steward, G. F. (2001). Fingerprinting viral assemblages by pulsed field gel electrophoresis (PFGE). *In* "Methods in Microbiology Series" (J. H. Paul, ed.), Vol. 30, pp. 85–103. Academic Press, London.

Swofford, D. L. (2002). "PAUP*. Phylogenetic Analysis Using Parsimony (*and Other Methods)." Sinauer Associates, Sunderland, MA.

Wawer, C., and Muyzer, G. (1995). Genetic diversity of *Desulfovibrio* spp. in environmental samples analyzed by denaturing gradient gel electrophoresis of [NiFe] hydrogenase gene fragments. *Appl. Environ. Microbiol.* **61,** 2203–2210.

Zeidner, G., and Béjà, O. (2004). The use of DGGE analyses to explore eastern Mediterranean and Red Sea marine picophytoplankton assemblages. *Environ. Microbiol.* **6,** 528–534.

Zeidner, G., Bielawski, J. P., Shmoish, M., Scanlan, D. J., Sabehi, G., and Béjà, O. (2005). Potential photosynthesis gene swapping between Prochlorococcus & Synechococcus via viral intermediates. *Environ Microbiol.* in press.

Zeidner, G., Preston, C. M., Delong, E. F., Massana, R., Post, A. F., Scanlan, D. J., and Béjà, O. (2003). Molecular diversity among marine picophytoplankton as revealed by *psbA* analyses. *Environ. Microbiol.* **5,** 212–216.

Zhang, Z., Green, B. R., and Cavalier-Smith, T. (2000). Phylogeny of ultra-rapidly evolving dinoflagellate chloroplast genes: A possible common origin for sporozoan and dinoflagellate plastids. *J. Mol. Evol.* **51,** 26–40.

[23] Quantitative Analysis of *nifH* Genes and Transcripts from Aquatic Environments

By STEVEN M. SHORT and JONATHAN P. ZEHR

Abstract

The availability of fixed inorganic nitrogen often plays a fundamental role in regulating primary production in both aquatic and terrestrial ecosystems. Because biological nitrogen fixation is an important source of nitrogen in marine environments, the study of N_2-fixing microorganisms is of fundamental importance to our understanding of global nitrogen and carbon cycles. Quantitative molecular tools have made it possible to examine uncultivated N_2-fixing microorganisms directly in the environment. Currently, we are using quantitative polymerase chain reaction (PCR;

METHODS IN ENZYMOLOGY, VOL. 397
Copyright 2005, Elsevier Inc. All rights reserved.

0076-6879/05 $35.00
DOI: 10.1016/S0076-6879(05)97023-7

Q-PCR) and quantitative reverse transcriptase PCR (Q-RT-PCR) to study the ecology and gene expression of N_2-fixing bacteria in aquatic environments. Using these methods, we discovered that specific estuarine diazotrophs have distinct nonrandom distributions and that some diazotrophs in the open ocean have different diel patterns of *nifH* gene expression. This chapter describes briefly our 5' nuclease assay protocols for Q-PCR and Q-RT-PCR of *nifH* gene fragments in environmental samples and discusses some important methodological considerations for the quantitative molecular examination of microbes in aquatic environments.

Introduction

Because nitrogen is an essential element in proteins and nucleic acids, the availability of fixed inorganic nitrogen often plays a fundamental role in regulating primary production in aquatic and terrestrial ecosystems. In the ocean, the fluxes of nitrogen are intimately linked to carbon cycling and its export from surface waters. Thus, the availability of nitrogen indirectly affects other biogeochemical processes and global climate regimes.

Although nitrogen is abundant in the ocean, most of it is in the form of nitrogen gas (N_2) and therefore is not available to most marine life. The transformation of N_2 to biologically available nitrogen, biological nitrogen fixation (BNF), is mediated only by prokaryotes (both autotrophic and heterotrophic). Because BNF is an important source of nitrogen in the marine environment, the study of N_2-fixing microorganisms in aquatic environments is of fundamental importance to our understanding of global nitrogen and carbon cycles.

Zehr and McReynolds (1989) used a molecular approach to identify N_2-fixing microorganisms in the marine environment using degenerate PCR primers to amplify the *nifH* gene. Molecular investigations have led to the discovery of numerous heterotrophic N_2-fixing microbes in microbial mats (Zehr *et al.*, 1995), oligotrophic oceans (Zehr *et al.*, 1998), and deep sea and hydrothermal vents (Mehta *et al.*, 2003). This approach has proved particularly useful in the discovery that unicellular cyanobacteria in the subtropical North Pacific Ocean have the genes necessary for nitrogen fixation and also express them (Zehr *et al.*, 2001). Molecular tools have revolutionized the way oceanic nitrogen cycling and fixation are studied (Zehr and Ward, 2002; Zehr *et al.*, 2000).

Advances in molecular techniques are facilitating studies of *nifH* diversity and gene expression in aquatic microbial communities. Quantitative PCR was employed in an oceanographic context by Suzuki, who examined the abundance of different prokaryote taxa in marine samples (Suzuki *et al.*, 2000, 2001). Wawrick *et al.* (2002) used quantitative reverse transcriptase

PCR to study the expression of Rubisco genes (*rbcL*) in eukaryotic marine primary producers. Other applications of this powerful tool in the marine sciences include studies of the abundance of both uncultured (Labrenz *et al.*, 2004) and cultured (Skovhus *et al.*, 2004) bacteria and toxic phytoplankton species in the coastal waters of the United States (Popels *et al.*, 2003) and Spain (Galluzzi *et al.*, 2004). Currently, we are using Q-PCR and Q-RT-PCR to study the autecology and the expression of *nifH* phylotypes in different marine environments (Short *et al.*, 2004). The following is a brief description of our 5' nuclease assay protocols for Q-PCR and Q-RT-PCR and some important methodological considerations.

Materials

Sample Collection

    25-mm 0.2-μm pore size Supor filters (Pall Corporation)
    25-mm in-line filter holder (Pall Corporation)
    25-mm polysulfone filter funnels (Pall Corporation)
    2-ml bead beater vials (BioSpec Products)
    0.1-mm glass beads (BioSpec Products)
    RLT buffer (Qiagen)
    β-Mercaptoethanol (Sigma)

DNA Extraction from Environmental Samples

    Potassium ethyl xanthogenate (Sigma)
    Tris(hydroxymethyl)aminomethane (Trizma base)(Sigma)
    EDTA (Sigma)
    Sodium dodecyl sulfate (SDS)(Sigma)
    Isopropanol (Sigma)
    Ethanol (Sigma)
    UltraClean microbial DNA kit (Mo Bio)
    PicoGreen dsDNA quantitation kit (Molecular Probes)

RNA Extraction from Environmental Samples

    RNeasy mini kit (Qiagen)
    RNase-free DNase set (Qiagen)
    Ethanol (Sigma)
    RiboGreen RNA quantitation kit (Molecular Probes)

Q-PCR and Q-RT-PCR

    TaqMan PCR core reagents kit (Applied Biosystems)

TaqMan TAMRA probe (Applied Biosystems)
Sequence detection primer (Applied Biosystems)
SuperScript III first-strand synthesis system (Invitrogen)
ABI PRISM optical tubes (Applied Biosystems)
ABI PRISM optical caps (Applied Biosystems)

Description of Methods

Sample Collection

Surface water samples are obtained using a bucket or by simply immersing well-rinsed bottles into the water to be sampled. Samples from deep water or open ocean sites are usually collected using GO-FLO bottles (General Oceanics) mounted on a rosette water sampler (e.g., General Oceanics Model 1015) deployed from a ship. Samples are vacuum filtered gently (pressure should not exceed 100 mm Hg) onto 25-mm-diameter, 0.2-μm pore-size filters using a vacuum pump and filtration funnels mounted in a vacuum manifold or flask. Alternatively, samples can be pressure filtered using a peristaltic pump and in-line filter holders. We routinely filter from 100 to 1000 ml depending on the amount of particulate material in the sample. The samples can be size fractionated using different pore size filters to assay different size fractions or to eliminate large particles or aggregates, if desired. It is important to note the volume filtered in order to calculate the nucleic acid yield for each sample. For DNA extraction, samples should be filtered until air is drawn through the filter. Using clean forceps, filters should be placed into 2-ml bead beater vials, frozen in liquid nitrogen, and stored at $-80°$ until nucleic acids are extracted. The same procedure should be followed for samples collected for RNA extraction except that the filters are placed into 2-ml vials containing 600 μl RLT buffer, 1% β-mercaptoethanol, and ca. 0.2 g of 0.1-mm glass beads.

DNA Extraction and Purification

We routinely use a modified xanthogenate protocol (Tillett and Neilan, 2000) to recover DNA from samples collected on 0.2-μm pore-size filters. Briefly, 1 ml of XS buffer (1% potassium ethyl xanthogenate; 100 mM Tris-HCl, pH 7.4; 1 mM EDTA, pH 8.0; 1% sodium dodecylsulfate; 800 mM ammonium acetate) and ca. 0.2 g of 0.1-mm glass beads are added to the vials containing the filters. The vials are tightly capped, placed in a bead beater (e.g., Qbiogene fastPrep) and agitated three times at medium speed for 45 s. Care must be taken to let the samples sit for 5 min between agitation steps and at the end because the tubes heat up significantly. After the final agitation, samples are incubated at 70° for 60 min. Following the incubation, the XS buffer is transferred to a 2-ml centrifuge tube, vortexed for 10 s, and placed on ice for 30 min. Cell debris is removed by centrifugation at room temperature at 10,000g for 12 min in a bench-top centrifuge and the supernatants decanted into 2-ml centrifuge tubes containing 1 ml of isopropanol at room temperature. Samples are incubated at room temperature for 10 min, and the precipitated DNA is pelleted by centrifugation at 10,000g for 16 min at room temperature. Isopropanol is then decanted, and the DNA pellets are washed with 70% ethanol, air dried, and resuspended in 100 μl of 10 mM Tris-HCl (pH 8.5).

Alternatively, if DNA of higher purity is required, a Mo Bio UltraClean microbial DNA kit can be used. The first step of our protocol differs slightly from that of the manufacturer. The 25-mm filter is placed directly in the MicroBead tubes provided. To the tubes containing the filters, 300 μl of the MicroBead and 50 μl of the MD1 solutions from the kit are added. The tubes are then placed in a bead beater and agitated as described in the preceding paragraph. The tubes are centrifuged at 10,000g for 30 s, supernatants transferred to clean microcentrifuge tubes, and the remaining steps in the manufacturer's protocol followed.

Samples are stored at −20° until further use. We quantify the sample DNA concentration and extraction yield using a Cary Eclipse fluorescence spectrophotometer (Varian) and the PicoGreen dsDNA quantification kit following the manufacturer's protocols.

RNA Extraction and Purification

Our extraction procedure is based on the protocol outlined for the Qiagen RNeasy kit. Tubes containing the filters, RLT with 1% β-mercaptoethanol, and glass beads are thawed briefly at room temperature. After the tubes have thawed, they are placed in a bead beater and agitated as described earlier. After agitation, the tubes are spun at 10,000g to pellet the glass beads, and the supernatant is transferred to new centrifuge tubes. An equal volume of 70% ethanol is added, the tubes are inverted to mix thoroughly, 0.7 ml is pipetted directly onto the center of an RNeasy spin column, the column is centrifuged at 10,000g for 30 s (all subsequent spins are carried out at the same speed for the same amount of time), and the effluent is discarded. The remaining supernatant is added to the RNeasy spin column, the column is centrifuged a second time, and the effluent is discarded again. Three hundred fifty microliters of RW1 buffer (provided in the RNeasy kit) is added to column, and the column is centrifuged and the effluent discarded. After this step we perform optional on-column

DNase digestion according to the manufacturer's protocol. The remaining steps described in the kit protocol should be followed and the purified RNA eluted using 50 μl of the RNase-free H_2O provided in the kit. We have found that an additional DNase treatment is necessary for some samples to eliminate all contaminating DNA. Samples are stored at -80° until further use. We quantify sample RNA concentration and yield using a Cary Eclipse fluorescence spectrophotometer (Varian) and the RiboGreen dsDNA quantification kit following the manufacturer's recommendations.

Q-PCR of nifH Fragments from Environmental Samples

Q-PCR methods include the 5′ nuclease assay, which uses PCR primers and a specific fluorogenic TaqMan probe, and the SYBR Green assay, which uses only PCR primers and a nonspecific fluorescent reporter. Because of its increased specificity, we use the 5′ nuclease assay for Q-PCR and the following sections focus exclusively on this method. After environmental samples have been collected and DNA extracted and quantified, the steps to quantify the number of copies of a particular gene fragment are relatively straightforward. The first step is amplification, cloning, and sequencing genes of interest (Zehr and Turner, 2001). This step is not described here as it has become relatively common and many different genes have been recovered and sequenced from a wide range of aquatic environments (Affourtit *et al.*, 2001; Allen *et al.*, 2001; Auman *et al.*, 2001; Brussaard *et al.*, 2004; Diez *et al.*, 2001; Giri *et al.*, 2004; Huber *et al.*, 2002). Once sequences of interest have been identified, TaqMan primers and probes should be designed to target these sequences. The sequences of several primers and probes used in our laboratory are listed in Table I. We design primers using the ABI Primer Express software, but then it is critical to check the primer design against sequence databases. We use ARB (Ludwig *et al.*, 2004) for this purpose. Standard curves should be constructed from serial dilutions of quantified, linearized plasmids containing cloned fragments of the target sequences.

Sample DNA is amplified in a Q-PCR thermal cycler and the resulting Ct (cycle threshold) values are compared to a standard curve to provide an estimate of the number of gene fragments in a particular sample. Knowing the initial volume of sample filtered, the total amount of DNA recovered from the extraction, and the amount of DNA added to each reaction allow a simple calculation to estimate the number of gene copies in any sample.

nifH *Fragment Quantification Using the 5′ Nuclease Assay.* To reduce assay cost, we routinely reduce our reaction volume to 25 μl from the 50-μl volume recommended by the TaqMan PCR Core Reagents kit protocol. The 25-μl reactions contain 1× TaqMan buffer (Applied Biosystems),

TABLE I
SEQUENCES OF THE OLIGONUCLEOTIDE PRIMERS AND PROBES USED IN THIS STUDY[a]

Oligonucleotide/target	Name	Sequence (5'-3')
Upstream primer/ nifH clone 912h4	912h4-F	ggttatggacaaggtccgtgaa
Downstream primer/ nifH clone 912h4	912h4-R	agccgcgtttacagacatcttc
TaqMan probe / nifH clone 912h4	912h4-P	FAM-agttccagatcctcaacggtgccga-TAMRA
Upstream primer/ nifH clone 907h22	907h22-F	acggcggaacttggtgtgt
Downstream primer/ nifH clone 907h22	907h22-R	aataccgcgacctgcacaac
TaqMan probe / nifH clone 907h22	907h22-P	FAM-cggtggtcctgagccgggagtt-TAMRA
Upstream primer / human β-actin	ACTB-F	aggcatcctcaccctgaagtac
Downstream primer/ human β-actin	ACTB-R	tctccatgtcgtcccagttg
TaqMan probe / human β-actin	ACTB-P	FAM-ccatcgagcacggcatcgtca-TAMRA

[a] The 912h4 sequence was amplified and cloned from the Chesapeake Bay.

2.0 mM MgCl2, 200 μM each dATP, dGTP, dCTP, and 400 μM dUTP, 400 nM each forward and reverse primer, 200 nM of fluorogenic probe (primer and probe design is discussed below), 0.25 U of AmErase uracyl N-glycosylase (UNG, Applied Biosystems), 0.625 U AmpliTaq Gold DNA polymerase (Applied Biosystems), and no more than 2 μl of template DNA. For the no-template controls, 2 μl of sterile nuclease-free water should be added to each reaction. For every sample, 5' nuclease reactions should be replicated at least three times and for every reaction the no-template controls should be run in triplicate. We have used a GeneAmp 5700 sequence detection system (PE Applied Biosystems) to carry out our thermal cycling reactions with the following parameters: 50° for 2 min, 95° for 10 min, and 45 cycles of 95° for 15 s followed by 60° for 1 min. It should be noted that this instrument is no longer available and has been replaced with an updated model. Further, many different Q-PCR instruments are now available from several manufacturers.

Methodological Considerations. As stated earlier, we designed PCR primers and TaqMan probes using Primer Express software (Applied Biosystems) and check the specificity of the primers and probes against sequences in our database. The Primer Express software provides multiple sets of primers and probes and we discard those that are homologous to nontarget sequences. We often use probes 5' labeled with the fluorescent reporter FAM (6-carboxyfluorescein) and 3' labeled with the quenching dye TAMRA (6-carboxytetramethylrhodamine). Other fluorescent reporters and quenching dyes are also available from most manufacturers that synthesize custom oligonucleotides. To determine the specificity of the TaqMan primers and probes, we compare results of reactions with plasmids containing the fragment from which the primers and probes were designed (positive control) to results of reactions with plasmids containing closely related inserts (Table II). Results from our specificity experiments indicate that reactions with fragments with less than 84% nucleotide identity to the primers and probes would not contribute to the estimated number of gene copies in a sample, but fragments more than 89% identical could. It is likely that these results are specific to the fragments amplified and, therefore, we recommend determining the specificity of TaqMan primers and probes empirically whenever closely related gene fragments are available.

For each primer and probe set, we make standard curves using triplicate serial dilutions of quantified, linearized plasmids containing the positive control insert. The number of molecules of a plasmid is estimated from the amount of DNA according to Eq. (1). It is worth noting that we use these standard curves to quantify the abundance of genes from uncultured organisms. If the goal of a study is to quantify the abundance of a particular organism, serial dilutions of nucleic acids extracted from quantified cells should be used to generate the standard curves. Standard curves generated from cells rather than cloned gene fragments will help avoid errors associated with variable extraction efficiency from cells or variable numbers of gene copies per cell. For examples, see Galluzzi et al. (2004) and Popels et al. (2003). We have found that our quantitative thermal cycler can reproducibly detect a range of templates from 1 to approximately 10^7 gene copies per reaction with a number of different primer and TaqMan probe sets. Results have demonstrated that the assay is very precise even when two of the triplicate dilution series were analyzed during the same experiment while the third replicate dilution series was analyzed separately several days later (Fig. 1). In our laboratory, plasmid dilutions have been stored for up to 1 month with no significant change in Ct. However, we do not recommend storing dilutions for long term, as DNA degradation could affect the precision of the standard curve. A linear regression of Ct vs $\log_{10}$ number of gene copies should be used to generate the equation for the slope of the relationship. This equation can then be used to estimate the number of gene copies in the sample by the determination of sample Ct.

$$N = (Q/660) \times (1/L) \times (6.022 \times 10^{23}) \tag{1}$$

TABLE II

COMPARISON OF SEQUENCE IDENTITY, OLIGONUCLEOTIDE MISMATCHES, Ct OF REACTIONS, AND PREDICTED NUMBER OF COPIES OF POSITIVE CONTROL AND RELATED *nifH* GENE FRAGMENTS

nifH clone	Percentage identity of fragment compared to positive control[a]	Oligonucleotide	No. of mismatched bases	No. of target copies added to reaction	Ct	No. of copies predicted[b]
Positive control	100	Forward primer	0	1.6×10^5	22	1.53×10^5
		Reverse primer	0			
		Probe	0			
Clone A	81	Forward primer	3	1.4×10^5	45	0
		Reverse primer	6			
		Probe	3			
Clone B	84	Forward primer	0	1.6×10^5	45	0
		Reverse primer	8			
		Probe	5			
Clone C	89	Forward primer	1	1.8×10^5	26	1.03×10^4
		Reverse primer	0			
		Probe	2			

[a] Identity comparisons are based on entire cloned *nifH* fragments (ca. 330 bp).
[b] The number of copies predicted was calculated from the equation of the standard curve from dilutions of the positive control fragment. See text for further details on standards.

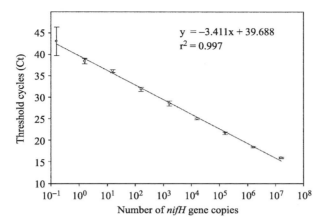

FIG. 1. Regression of Q-PCR threshold cycles (Ct) versus the base-10 logarithm of the number of *nifH* gene copies. The equation of the slope and the coefficient of determination are given. Error bars show the standard deviation from triplicate reactions.

where N is number of molecules, Q is quantity of DNA in grams, 660 is average molecular weight of a nucleotide in grams per mole, L is length of the molecule in base pairs, and 6.022×10^{23} is Avogadro's number (molecules/mole).

To check the environmental extracts for compounds that inhibit Q-PCR reactions, a common problem when amplifying DNA extracted from environmental samples, we estimate the efficiency of each Q-PCR reaction by comparing the Ct from amplification of a positive control alone to the Ct in reactions with the positive control plus an environmental template. For a conservative estimate of inhibition, we normally add an amount of positive control (e.g., 10^4 gene copies) that will produce a strong amplification signal (low Ct) relative to the sample. Amplification efficiency can be estimated with a simple calculation according to Eq. (2). Assuming an efficiency of 1, the Ct value from the amplification of plasmid alone can be used to calculate X_n. To estimate sample amplification efficiency, the calculated X_n and the Ct value from the amplification of plasmid plus sample can be used to solve Eq. (2) for E_x. E_x can be converted to a percentage to provide an estimate of the efficiency of the reaction.

$$X_n = X_0 \times (1 + E_x)^n \tag{2}$$

where X_n is number of target molecules at cycle n, X_0 is initial number of target molecules, E_x is efficiency of target amplification, and n is number of cycles (Ct).

As a final consideration, before Q-PCR reactions on environmental DNA extracts are performed, the amount of template added to the reaction mixtures should be optimized. Using the method and equation described earlier, we calculate the reaction efficiency for reactions containing plasmid plus 10, 5, 2, 1, or 0.5 ng of DNA from different extracts; environmental DNA samples are diluted so that the volumes added to each reaction are constant (2 μl). For several different primer and TaqMan probe sets, and numerous environmental extracts, 2 ng of template DNA consistently gave us the highest signal with the least evidence of inhibition. Because we have observed differences in amplification efficiency depending on the amount of template added, we recommend empirically determining the optimum amount of template to add to each reaction for different reaction conditions and different primer and TaqMan probe sets.

Q-RT-PCR of nifH Fragments from Environmental Samples

Like our Q-PCR protocol given earlier, the first step in Q-RT-PCR is to amplify, clone, and sequence genes of interest. For the sake of brevity, this step is not described here. An example is the recently published study of diversity of expressed nitrogenase genes in microbial mats (Omoregie et al., 2004). After sequences have been obtained from expressed genes of interest, primers and TaqMan probes are designed as described previously. Unlike the protocol for Q-PCR, Q-RT-PCR includes a first step to synthesize cDNA from environmental RNA extracts. After this, quantitative amplification methods are performed exactly as described for Q-PCR.

Although most vendors sell one-step quantitative RT-PCR kits, we commonly reverse transcribe our extracted RNA in a separate step before Q-PCR reactions. Q-RT-PCR reactions are performed in this way for two specific reasons. First, this allows us to check the cDNA synthesis step for inhibition to ensure that cDNA synthesis performed with equal efficiency for each sample. This is necessary because many environmental extracts contain compounds that can inhibit enzymatic reactions. Second, large Q-RT-PCR and Q-PCR experiments are simplified because the same reaction components and the same set of standards can be used for both assays. Therefore, the following sections describe cDNA synthesis and methodological considerations only. The quantitative step is performed as described in the section detailing Q-PCR. It is worth mentioning that we have never observed Q-PCR inhibition from cDNA and, therefore, no longer check each cDNA synthesis for PCR inhibition.

cDNA Synthesis. To produce cDNA templates for Q-RT-PCR, we reverse transcribe extracted RNA using the SuperScript III first-strand synthesis system following the manufacturer's protocol. We use from 20 to 80 ng environmental RNA per cDNA synthesis reaction. For each sample to be compared, we use the same amount of RNA for each cDNA synthesis. As the manufacturer recommends, each synthesis reaction should contain 0.2 μM each antisense (downstream) gene-specific primers, 1 mM dNTP mix, $1\times$ RT buffer, 5 mM MgCl$_2$, 10 mM dithiothreitol, 40 U RNaseOUT, and 200 U SuperScript III RT. After cDNA synthesis is completed, residual RNA is digested by incubating the reactions with RNase H, and the cDNA is stored at $-20°$ until utilized in Q-RT-PCR assays.

Methodological Considerations. Following cDNA synthesis, our Q-RT-PCR protocol is the same as the protocol described earlier for Q-PCR. The standard curves are produced in a similar way and the same considerations apply. Nonetheless, because Q-RT-PCR includes a step to synthesize cDNA, methodological issues surrounding this technique must be considered. To ensure that environmental RNA extracts do not contain compounds that inhibit cDNA synthesis, we often run a control cDNA synthesis experiment. For each sample to be checked, the control experiment involves two cDNA syntheses and Q-PCR reactions. The first cDNA synthesis reaction should be carried out with the control reagents (total HeLa RNA, and antisense control primer) included in the SuperScript III first-strand synthesis kit and is used to synthesize cDNA of human β-actin transcripts. The second parallel cDNA synthesis should be carried out with control reagents plus the same amount of environmental RNA as used in cDNA synthesis. This reaction is used to evaluate the efficiency of human β-actin cDNA synthesis in the presence of environmental RNA to determine if compounds are present in the environmental RNA samples that inhibit reverse transcriptase activity. To provide a conservative estimate of the potential inhibition, we synthesize human β-actin cDNA using the same amount of HeLa RNA and environmental RNA. After cDNA synthesis, the number of β-actin cDNA copies in the two parallel syntheses should be quantified and compared using Q-PCR. The sequences of the primers and TaqMan probe we designed to quantify β-actin cDNA molecules are listed in Table II. Because the parallel reactions can be compared directly to estimate cDNA synthesis efficiency, there is no need to produce a standard curve for human β-actin cDNA.

As a minimal normalization, the number of transcripts can be estimated for a given amount of environmental RNA. By knowing the volume of water sampled, the total amount of RNA recovered, and the proportion of RNA included in each Q-RT-PCR assay, the number of transcripts per volume sampled can be estimated. This provides an estimate of relative transcription for any set of samples. As a more rigorous normalization, the absolute number of transcripts per gene copy can be estimated by comparing the total number of transcripts to the number of gene copies. This provides an estimate of cellular activity by taking into account the number of transcripts produced per gene copy. If it is assumed that the gene studied is a single-copy gene, this results in an estimate of gene expression per cell. For example, diel patterns of *nifH* gene expression for several phylotypes have been observed in the Pacific Ocean near Hawaii (M. Church and J. Zehr, unpublished data). It is worth noting that this normalization is not always possible, as the number of gene copies may not be detectable, even though the transcripts are. Nonetheless, even if this normalization is not possible, quantification of the transcription of functional genes can still provide answers to questions about the number of transcripts of a specific gene that are present in a given volume of water or in a sample of total RNA.

Concluding Remarks

The ability to quantify specific genes or the expression of specific genes is a relatively new addition to the toolbox of molecular techniques for environmental studies. Using the Q-PCR methods described in here, we have initiated studies on the autecology of diazotrophic prokaryotes in aquatic environments. These methods have opened new avenues of investigation and will allow the examination of the distribution, abundance, and physiology of marine nitrogen-fixing microorganisms.

Our initial study using Q-PCR focused on the abundance of specific diazotrophs in the Chesapeake Bay. In this study, we learned that the distribution of two different N_2-fixing bacteria differed spatially and temporally, demonstrating that diazotrophs in the Chesapeake Bay have distinct nonrandom distributions (Short *et al.*, 2004). A study of the distribution and abundance of N_2-fixing prokaryotes in the subtropical North Pacific (Church *et al.*, 2005) demonstrated that diazotrophs were not randomly distributed vertically through the water column. More importantly, this study demonstrated for the first time that unicellular cyanobacteria were one of the most abundant open ocean diazotrophs.

Currently, we are studying the expression patterns of diazotrophs in both estuarine and open ocean systems. From our preliminary results it appears that not all diazotrophs have the same diel patterns of expression. Some diazotrophs have maximum expression of *nifH* during the night and others during the day. These quantitative molecular tools have made it possible to examine uncultivated N_2-fixing microorganisms directly in the environment. Ultimately, these studies will add to our understanding of nitrogen cycling in aquatic environments and how N cycling influences the biogeochemistry and food webs in aquatic environments.

Acknowledgments

We thank M. J. Church and C. M. Short for their valuable contributions to this manuscript and to B. Ward and all collaborators on the Chesapeake Bay Biocomplexity Project. This research was supported by grants from the National Science Foundation (OCE-9981437, OCE-0131762 and OCE-9977460 to JPZ) and the Gordon and Betty Moore Foundation.

References

Affourtit, J., Zehr, J. P., and Paerl, H. W. (2001). Distribution of nitrogen fixing microorganisms along the Neuse River Estuary, North Carolina. *Microb. Ecol.* **41,** 114–123.

Allen, A. E., Booth, M. G., Frischer, M. E., Verity, P. G., Zehr, J. P., and Zani, S. (2001). Diversity and detection of nitrate assimilation genes in marine bacteria. *Appl. Environ. Microbiol.* **67,** 5343–5348.

Auman, A. J., Speake, C. C., and Lidstrom, M. E. (2001). *nifH* sequences and nitrogen fixation in type I and type II methanotrophs. *Appl. Environ. Microbiol.* **67,** 4009–4016.

Brussaard, C. P. D., Short, S. M., Frederickson, C. M., and Suttle, C. A. (2004). Isolation and phylogenetic analysis of novel viruses infecting the phytoplankton *Phaeocystis globosa* (Prymnesiophyceae). *Appl. Environ. Microbiol.* **70,** 3700–3705.

Church, M. V., Jenkins, B. D., Karl, D. M., and Zehr, J. P. (2005). Vertical distributions of nitrogen-fixing phylotypes at Station ALOHA in the oligotrophic North Pacific Ocean. *Aquat. Microb. Ecol.* **38,** 3–14.

Diez, B., Pedros-Alio, C., and Massana, R. (2001). Study of genetic diversity of eukaryotic picoplankton in different oceanic regions by small-subunit rRNA gene cloning and sequencing. *Appl. Environ. Microbiol.* **67,** 2932–2941.

Galluzzi, L., Penna, A., Bertozzini, E., Vila, M., Garces, E., and Magnani, M. (2004). Development of a real-time PCR assay for rapid detection and quantification of *Alexandrium minutum* (a dinoflagellate). *Appl. Environ. Microbiol.* **70,** 1199–1206.

Giri, B. J., Bano, N., and Hollibaugh, J. T. (2004). Distribution of RuBisCO genotypes along a redox gradient in Mono Lake, California. *Appl. Environ. Microbiol.* **70,** 3443–3448.

Huber, J. A., Butterfield, D. A., and Baross, J. A. (2002). Temporal changes in archaeal diversity and chemistry in a mid-ocean ridge subseafloor habitat. *Appl. Environ. Microbiol.* **68,** 1585–1594.

Labrenz, M., Brettar, I., Christen, R., Flavier, S., Botel, J., and Hofle, M. G. (2004). Development and application of a real-time PCR approach for quantification of uncultured bacteria in the central Baltic Sea. *Appl. Environ. Microbiol.* **70,** 4971–4979.

Ludwig, W., Strunk, O., Westram, R., Richter, L., Meier, H., Yadhukumar, Buchner, A., Lai, T., Steppi, S., Jobb, G., Forster, W., Brettske, I., Gerber, S., Ginhart, A. W., Gross, O., Grumann, S., Hermann, S., Jost, R., Konig, A., Liss, T., Lussmann, R., May, M., Nonhoff, B., Reichel, B., Strehlow, R., Stamatakis, A., Stuckmann, N., Vilbig, A., Lenke, M., Ludwig, T., Bode, A., and Schleifer, K. H. (2004). ARB: A software environment for sequence data. *Nucleic Acids Res.* **32,** 1363–1371.

Mehta, M. P., Butterfield, D. A., and Baross, J. A. (2003). Phylogenetic diversity of nitrogenase (*nifH*) genes in deep-sea and hydrothermal vent environments of the Juan de Fuca ridge. *Appl. Environ. Microbiol.* **69,** 960–970.

Omoregie, E. O., Crumbliss, L. L., Bebout, B. M., and Zehr, J. P. (2004). Determination of nitrogen-fixing phylotypes in *Lyngbya* sp. and *Microcoleus chthonoplastes* cyanobacterial mats from Guerrero Negro, Baja California, Mexico. *Appl. Environ. Microbiol.* **70,** 2119–2128.

Popels, L. C., Cary, S. C., Hutchins, D. A., Forbes, R., Pustizzi, F., Gobler, C. J., and Coyne, K. J. (2003). The use of quantitative polymerase chain reaction for the detection and enumeration of the harmful alga *Aureococcus anophagefferens* in environmental samples along the United States East Coast. *Limnol. Oceanogr. Methods* **1,** 92–102.

Short, S. M., Jenkins, B. D., and Zehr, J. P. (2004). Spatial and temporal distribution of two diazotrophic bacteria in the Chesapeake Bay. *Appl. Environ. Microbiol.* **70,** 2186–2192.

Skovhus, T. L., Ramsing, N. B., Holmstrom, C., Kjelleberg, S., and Dahllof, I. (2004). Real-time quantitative PCR for assessment of abundance of *Pseudoalteromonas* species in marine samples. *Appl. Environ. Microbiol.* **70,** 2373–2382.

Suzuki, M. T., Preston, C. M., Chavez, F. P., and Delong, E. F. (2001). Quantitative mapping of bacterioplankton populations in seawater: Field tests across an upwelling plume in Monterey Bay. *Aquat. Microbial Ecol.* **24,** 117–127.

Suzuki, M. T., Taylor, L. T., and Delong, E. F. (2000). Quantitative analysis of small-subunit rRNA genes in mixed microbial populations via 5′-nuclease assays. *Appl. Environ. Microbiol.* **66,** 4605–4614.

Tillett, D., and Neilan, B. A. (2000). Xanthogenate nucleic acid isolation from cultured and environmental cyanobacteria. *J. Phycol.* **36,** 251–258.

Zehr, J. P., Carpenter, E. J., and Villareal, T. A. (2000). New perspectives on nitrogen-fixing microorganisms in tropical and subtropical oceans. *Trends Microbiol.* **8,** 68–73.

Zehr, J. P., and McReynolds, L. A. (1989). Use of degenerate oligonucleotides for amplification of the *nifH* gene from the marine cyanobacterium *Trichodesmium thiebautii*. *Appl. Environ. Microbiol.* **55,** 2522–2526.

Zehr, J. P., Mellon, M., Braun, S., Litaker, W., Steppe, T., and Paerl, H. W. (1995). Diversity of heterotrophic nitrogen-fixation genes in a marine cyanobacterial mat. *Appl. Environ. Microbiol.* **61,** 2527–2532.

Zehr, J. P., Mellon, M. T., and Zani, S. (1998). New nitrogen-fixing microorganisms detected in oligotrophic oceans by amplification of nitrogenase (*nifH*) genes. *Appl. Environ. Microbiol.* **64,** 3444–3450.

Zehr, J. P., and Turner, P. J. (2001). Nitrogen fixation: Nitrogenase genes and gene expression. *Methods Microbiol.* **30,** 271–286.

Zehr, J. P., and Ward, B. B. (2002). Nitrogen cycling in the ocean: New perspectives on processes and paradigms. *Appl. Environ. Microbiol.* **68,** 1015–1024.

Zehr, J. P., Waterbury, J. B., Turner, P. J., Montoya, J. P., Omoregie, E., Steward, G. F., Hansen, A., and Karl, D. M. (2001). Unicellular cyanobacteria fix N_2 in the subtropical North Pacific Ocean. *Nature* **412,** 635–638.

Further Reading

Wawrik, B., Paul, J. H., and Tabita, F. R. (2002). Real-time PCR quantification of rbcL (ribulose-1,5-bisphosphate carboxylase/oxygenase) mRNA in diatoms and pelagophytes. *Appl. Environ. Microbiol.* **68,** 3771–3779.

[24]   Community Level Analysis: Genetic and Biogeochemical Approaches to Investigate Community Composition and Function in Aerobic Ammonia Oxidation

By Bess B. Ward and Gregory D. O'Mullan

Abstract

Aerobic ammonia oxidation is the process that converts ammonium to nitrate and thus links the regeneration of organic nitrogen to fixed nitrogen loss by denitrification. It is performed by a phylogenetically restricted group of *Proteobacteria* (ammonia-oxidizing bacteria, AOB) that are autotrophic and obligately aerobic. This chapter describes methods for the measurement of ammonia oxidation in the environment, with a focus on seawater systems and stable isotopic tracer methods. It also summarizes the current state of molecular ecological approaches for detection of AOB in the environment and characterization of the composition of AOB assemblages.

Introduction

Ammonia oxidation, as the first step in aerobic nitrification, the oxidation of ammonium to nitrate via nitrite, is an essential component of the nitrogen (N) cycle in aquatic and terrestrial environments. While nitrification does not result in a direct change in the fixed N inventory, it is critical in the linkage between organic N and its eventual loss from the system as N_2 via denitrification. Ammonia-oxidizing bacteria produce nitrite from ammonia, via hydroxylamine, and nitrite-oxidizing bacteria perform the final oxidation of nitrite to nitrate. No organism in culture is known to oxidize both ammonia and nitrite. The bacteria that oxidize nitrite or ammonia are not phylogenetically related to each other.

The overall reaction for ammonia oxidation [Eq. (4)] shows that the process consumes molecular oxygen and produces hydrogen ions, in addition to nitrite. A requirement for molecular oxygen occurs in the first step of the oxidation [Eq. (1)], which is catalyzed by a monooxygenase (ammonia monooxygenase, AMO). Uncharged gaseous ammonia is the actual substrate for AMO, as demonstrated by the pH dependence of the reaction rate (Suzuki *et al.*, 1974; Ward, 1987a,b). Ammonia monooxygenase has never been completely purified and assayed in cell-free conditions, although its complete gene sequence has been obtained from both

METHODS IN ENZYMOLOGY, VOL. 397
0076-6879/05 $35.00
DOI: 10.1016/S0076-6879(05)97024-9

Nitrosomonas (betaproteobacteria) and *Nitrosococcus* (gammaproteobacteria) type AOBs (Alzerreca *et al.*, 1999; Klotz and Norton, 1995; Sayavedra-Soto *et al.*, 1998). AMO contains copper and probably also iron in its active form (Zahn *et al.*, 1996).

The immediate product of AMO is hydroxylamine, which is further oxidized by hydroxylamine oxidoreductase (HAO) to nitrite [Eq. (2)]. Oxygen is also consumed by the terminal oxidase [Eq. (3)] as a result of electron transport generating ATP for cellular metabolism.

$$NH_3 + O_2 + 2H^+ + 2e^- \rightarrow NH_2OH + H_2O \tag{1}$$

$$NH_2OH + H_2O \rightarrow NO_2^- + 5H^+ + 4e- \tag{2}$$

$$2H^+ + 0.5\ O_2 + 2e^- \rightarrow H_2O \tag{3}$$

$$NH_3 + 1.5\ O_2 \rightarrow NO_2^- + H_2O + H^+ \tag{4}$$

AOB are typically not abundant, as a percentage of the total bacterial assemblage, but nonetheless can have a large impact on biogeochemical cycling. This results from the relative inefficiency of autotrophy at the expense of ammonia oxidation; a vast amount of inorganic nitrogen must by oxidized to fix a small amount of CO_2 into the biomass. Exceptions to the low abundance of AOB occur in wastewater treatment plants, where biofilms containing clusters of AOB cells can comprise on the order of 20% of the total biomass (Juretschko *et al.*, 1998).

Methods for Measuring Ammonia Oxidation in Seawater and Sediments

Nutrient Inventory Approach

Direct measurement of the rate of nitrification is problematic for several reasons having to do with the sensitivity of the methods and potential artifacts introduced by incubation methods. The most direct experimental design might simply be to incubate samples and measure the concentrations of dissolved inorganic nitrogen (DIN) pools over time. This approach provided some of the earliest evidence for the occurrence of biologically mediated nitrification (Rakestraw, 1936; von Brand *et al.*, 1937). In such an experiment, accumulation of nitrite or nitrate indicates net nitrification. A decrease in the concentration over time could be observed, however, even when nitrification is occurring, if consumption of nitrate or nitrite exceeds production in the incubation bottle. The low concentration relative to high biological demand for fixed nitrogen in the surface waters of the ocean and many lakes means that large fluxes can be obscured by tight coupling between production and consumption terms.

Specific Inhibitors

The addition of specific inhibitors has been used as a modification of the nutrient inventory approach just described. In this approach, chemicals that specifically inhibit ammonia oxidation (e.g., acetylene, allylthiourea, methyl fluoride, N-serve) or nitrite oxidation (chlorate) are added to replicate incubation bottles (Bianchi *et al.*, 1997; Billen, 1976). The method assumes that the nitrite concentration is at steady state in the sample and that nitrification is the only process that produces or consumes nitrite. Clearly, the bottles must be incubated in the dark to prevent DIN consumption by phytoplankton. One need only measure the concentration of nitrite over time in the bottles in which ammonium oxidation was inhibited to estimate the nitrite oxidation rate (equal to the rate of nitrite decrease). The rate of nitrite increase in the bottles to which the nitrite-oxidation inhibitor was added equals the rate of ammonium oxidation. There are some potential problems with this approach: (1) Preventing photosynthesis by phytoplankton probably has cascading effects on the activities of other microbes in the bottle, such that the rate of ammonium mineralization is reduced, therefore changing the source term for the nitrification substrate. (2) Incubating in the dark may release the nitrifying bacteria from light inhibition such that the measured rate exceeds the *in situ* rate. (3) Due to the necessity to detect small changes in DIN concentrations, incubations typically last 48 h, which is sufficient to overcome the lag induced by light inhibition and is also long enough to create quite unnatural conditions.

The search for precisely specific inhibitor compounds has been extensive and has resulted in a plethora of potentially useful compounds. Many are problematic for reasons not directly related to nitrification. For example, acetylene inhibits both nitrifiers and denitrifiers (Balderston *et al.*, 1976; Berg *et al.*, 1982). Thus, its use to measure one process will also inhibit the other, and when one depends on the other (as is the case when denitrification depends on nitrification for nitrate), both rates are affected and the independent measurement of one is not possible. It is reported that the level or length of exposure to acetylene can be optimized to differentiate between its effects on nitrification and denitrification (Kester *et al.*, 1996). N-serve is a commercial preparation that specifically inhibits ammonia-oxidizing bacteria (Goring, 1962) and serves as the basis for the sensitive $^{14}CO_2$ method for the measurement of nitrification rates. Being chemolithoautotrophs, nitrifiers fix CO_2 while oxidizing nitrogen. The amount of CO_2 fixation due to nitrifiers can be computed by differences between incubations with and without addition of an inhibitor that specifically removes the contribution of nitrifiers (Billen, 1976; Dore and Karl, 1996; Somville, 1978). Then a conversion factor is used to translate the CO_2 fixation into ammonium and nitrite oxidation rates. This conversion factor has been shown to vary by a factor of five in pure cultures of AOB (Billen, 1976; Glover, 1985), and thus its use introduces some uncertainty because the factor cannot be directly determined in field samples. In addition, because N-serve is not soluble in water, it must be dissolved in an organic solvent prior to addition to samples. This solvent can affect the other members of the community, and the use of ethanol has been shown to stimulate dark CO_2 incorporation by heterotrophic bacteria (Owens, 1986; Ward, 1986). Thus, especially in systems where nitrifiers are a very small part of the overall assemblage, the N-serve approach may yield artifacts (Priscu and Downes, 1985; Viner, 1990; Ward, 1986).

N Isotope Tracer Methods

In the tracer approach, a "trace" amount (an amount low enough to avoid perturbation of the ambient substrate concentration) of a labeled substrate (a radio or stable isotope) is added to a sample. After incubation, the amount of label in the product is used to compute the transformation rate. Unfortunately, a direct radioisotope tracer method is not very useful for measuring rates of nitrification in the environment. Capone *et al.* (1990) demonstrated the use of ^{13}N to quantify nitrification rates, but the isotope is so short-lived (half life $= 10$ min) that its use is usually impractical.

The main approach to measuring nitrification rates directly is to use the stable isotope, ^{15}N, as a tracer (Olson, 1981; Ward *et al.*, 1984). This approach is not without its problems, mainly because ^{15}N has a significant natural abundance and it must be measured using a mass spectrometer or emission spectrometer, both more expensive and difficult than using a scintillation counter for radioisotopes. Due to the great sensitivity of isotope ratio mass spectrometry, much shorter incubations (compared to the inhibitor and inventory methods) are possible (a few hours to 24 h are commonly used). The signal of transfer of the tracer from substrate to product pool (e.g., $^{15}NH_4$ to $^{15}NO_2$) can be detected regardless of what other processes are occurring in the incubation (so *in situ* light conditions can be used) and no assumptions of steady state need be made. The major drawback of this method is the necessity of adding a tracer, sometimes in excess of the natural concentration of substrate. This problem has been largely overcome by the advent of more sensitive mass spectrometers, however, and estimates obtained under conditions approaching *in situ* are possible.

The ^{15}N approach is most useful in water samples because complete mixing of the tracer is possible. In sediments and soils, rate measurements are constrained by the heterogeneous nature of the sample and the dependence of rates on the structure of the environment. In this situation, fluxes between overlying water and sediment cores can be analyzed to obtain areal rates. In conjunction with ^{15}N tracer addition, estimates of nitrification rates can be obtained from the isotopic dilution of nitrite or nitrate in the overlying water due to the production of nitrate or nitrite from processes in the sediments (Capone *et al.*, 1992). The isotope dilution approach is essentially the opposite of the tracer approach. The product pool is amended with isotopically labeled product. During incubation, product with the natural abundance isotope signature is produced from a naturally occurring substrate in the sample, effectively diluting the label in the product pool. The rate of dilution is used to compute the rate of product formation. The isotope pairing method for the measurement of denitrification (Nielsen, 1992; Rysgaard *et al.*, 1993) is essentially a modification of an isotope dilution approach and provides information on the rates of both denitrification and nitrification simultaneously. Inhibitor approaches similar to those described earlier for water samples have been used in sediments (Henricksen *et al.*, 1981; Miller *et al.*, 1993). The methylfluoride method (Caffrey and Miller, 1995; Miller *et al.*, 1993) seems particularly promising because the gas can diffuse thoroughly into the core with a minimal disturbance of microzones and gradients. This ammonium oxidation inhibitor is added to cores and the accumulation of NH_4^+ over time is assumed to represent the net rate of nitrification. Other processes that consume ammonium would proceed without inhibition or competition for ammonium and therefore lead to an underestimate of the nitrification rate.

To overcome the biasing resulting from uneven dispersal of a tracer or an inhibitor, sediment rate measurements are often made in slurries that destroy the gradient structure of sediments, which is essential to the *in situ* fluxes. Slurries may provide useful information on potential rates, but not *in situ* rates. Even if rate measurements in sediments are made using whole core incubations, e.g., when the inhibitor is a gas, it is still difficult to determine the depth distribution of the rate because an areal rate is usually obtained. A sophisticated measurement and model-based system that avoids direct rate measurements has been used to overcome this problem. Microelectrodes, which have very high vertical resolution, are used to measure the fine scale distribution of oxygen and nitrate in freshwater sediments. By assuming that the observed vertical gradients represent a steady-state condition, reaction-diffusion models can then be used to estimate the rates of nitrification, denitrification, and aerobic respiration and to compute the location of the rate processes in relation to the chemical profiles (e.g., Binnerup *et al.*, 1992; Jensen *et al.*, 1994). Recent advances in biosensor design may overcome interferences that have been problematic for microelectrode measurements of dissolved nitrate in seawater.

Measurement of Ammonium Oxidation Rates in Seawater Using ^{15}N

Sample Collection and Incubation. Samples are collected from the desired depth using Niskin bottles deployed on a rosette or with a peristaltic pump for shallower depths. Because nitrification, especially ammonia oxidation, is oxygen sensitive, precautions should be taken to maintain the *in situ* oxygen conditions. This can be done by overfilling the incubation bottles (preferably polycarbonate, to avoid potential trace metal contamination from glass) and then sealing the caps without introducing bubbles into the bottle. Below the photic zone, incubations can be carried out in gas-impermeable plastic bags (trilaminates produced by Pollution Measurement Corporation, Oak Park, IL, or in equivalent impermeable plastics). These are convenient and versatile incubation containers because they can be purchased in various sizes (we have found 500 ml to 10 liter to be useful for various applications) and can be aseptically subsampled without perturbing the gas concentrations in the remaining sample.

Incubations are performed under simulated *in situ* conditions (using screening to simulate *in situ* light conditions) in controlled temperature incubators or in running seawater incubators (for surface samples). For single end point measurements, the length of the incubation should be kept as short as possible, usually 1–3 h. Even in short incubations, the atom-% of the substrate pool can be seriously diluted by ammonium regeneration occurring in the same bottle and, during long incubations, can make it impossible even to account for the dilution in the rate calculations (see later).

The tracer [$^{15}NH_4$ as $Na^{15}NH_4$ or $(^{15}NH_4)_2SO_4$ at 99 at-%] can be added by a syringe through a silicone seal in the cap or with a pipettor just prior to sealing. In the latter case, it is a good idea to cool the tracer stock below the temperature of the incubation so that the added tracer solution sinks into the bottle and is not lost when the cap is secured. It is considered optimal to add tracer at a level that increases the ambient pool by no more than 10%. This is not always possible because the ambient pool size may not be known in real time and it may be so low as to be impossible to add a large enough signal without overwhelming the *in situ* substrate levels. In that case, an addition of 50 to 100 nM is commonly used. The sensitivity of the isotope tracer method is in the range of nM d^{-1} and can be optimized by adjusting incubation volume, length of incubation, and level of tracer addition.

Prior to incubation, it is also advisable to add about 250 nM natural abundance NO_2^-. Because the ambient nitrite pool is usually quite small, any $^{15}NO_2^-$ that is produced is essentially lost immediately to dilution, uptake by phytoplankton, or oxidation to nitrate. Addition of the carrier nitrite allows the recently produced nitrite to be diluted into a larger pool, which can then be recovered at the end of the incubation. The amount of nitrite added is too small to influence the ammonia oxidizers, although it may stimulate phytoplankton and nitrite oxidizers.

At the end of the incubation, the sample is filtered through precombusted glass fiber or silver filters to remove particulate material (which can be used for the determination of ammonium uptake rates) and the filtrate can be frozen for storage until analysis.

$^{15}NO_2$ *Analysis.* In order to assay its ^{15}N content, the dissolved nitrite must be removed from solution and concentrated for introduction to the mass spectrometer. The most common way to concentrate the nitrite involves converting it to an azo dye that partitions into organic solvents and then concentrating the dye by either solvent or solid phase extraction. The solvent extraction method described here is modified from that described by Olson (1981) applied to filtrate volumes of 4 liter (necessary for sensitivity in open ocean environments), but can be scaled to any volume. The nitrite is first complexed with aniline: add 25 ml of aniline reagent (5 ml aniline in 1 liter of 2 N HCl) at room temperature, shake to mix well, and allow to sit for 15 min. Then add 25 ml of β-naphthol (5 g per liter of 3 N NaOH) to form the dye. Shake well; the solution will turn slightly orange and the base will form a precipitate in seawater. Add 5 ml of concentrated HCl and mix well to dissolve the precipitate. Then add 20–30 ml of trichloroethylene and shake well, venting frequently (conveniently done using a Teflon separatory funnel if one of sufficient volume is available). The orange color will partition into the organic phase. Drain the TCE out of the funnel into a clean glass beaker and add 20–30 ml more TCE to the water and repeat the extraction. Each extraction is about 80% efficient, so two extractions are sufficient to remove 98% of the initial dye; further extractions are usually not useful. Combine the extracts and store at this step or proceed to the washing steps (see later). If extracting very large volumes of filtrate, the procedure can be more cumbersome and it is necessary to pour off the water layer into a holding reservoir and then pour the TCE off into a beaker. The water must then be returned to the extraction vessel and the extraction repeated with a new aliquot of TCE.

Wash the extracts in 250-ml Teflon separatory funnels. This is done conveniently in small batches of samples using a lazy Susan or other custom rack to hold the separatory funnels. Add 100 ml 1 N HCl and mix well with frequent venting. Allow the layers to separate and discard the acid. Repeat with another 100 ml of 1 N HCl. Then wash twice with 100 ml of 2 N NaOH and twice with 100 ml of dH$_2$O. Because TCE is denser than water, it will form the lower layer in the separatory funnel. Therefore, it must be decanted into a beaker at every wash step and then returned to the funnel for subsequent washes. After the last wash, be sure to achieve a clean separation between layers and avoid allowing any water into the cleaned TCE. The cleaned extract can be stored at this step or evaporated in a hood immediately and stored dry. The dye dries as an orange scum in the beaker or small glass vial. To introduce the dye into the mass spec, dissolve it in 200 μl TCE (adjust volume to allow replication) and wick or pipette the solution onto a small fragment of a precombusted glass fiber filter. Encapsulate the filter in foil manually or using a pellet press.

An alternative approach uses essentially the same chemistry with a solid phase extraction (Kator *et al.*, 1992). In this case, the azo dye is removed from solution by pumping it through a C-18 ion-exchange column, followed by wash solutions, and elution in fresh TCE. The solid phase extraction is less demanding in terms of physical labor with large volumes of filtrate and results in elution in a small volume of solvent. In both permutations of the method, the samples can be processed to the extract phase and then the initial extract stored, either in glass vials or on the C-18 columns, for transport back to the laboratory for washing and further processing. This avoids the necessity to freeze, store, and transport large volumes of water, which can be inconvenient if many experiments are carried out on a long research cruise.

Because TCE dissolves many plastics, use glass or Teflon throughout this protocol for separatory funnels, graduated cylinders, and holding and storage vessels. All glass utensils and vessels should be combusted at 450° for 2 h prior to use. Forceps, foil, glass surfaces used for cutting filters, and so on should be cleaned with ethanol between samples. All of the extraction steps should be performed in a hood or very well-ventilated area.

Rate Calculations. Using mass or emission spectrometry, determine the atom-% ^{15}N of the sample (Fiedler and Proksch, 1975). It is very straightforward to compute the rate of ^{15}NO$_2$ production if you can assume that the amount of ammonium oxidized represents an infinitesimal fraction of the ambient ammonium pool. This may be a reasonable assumption if very short incubations are used and the substrate pool was sizable in the beginning. This is often not the case in seawater where ammonium regeneration (ammonification of N-containing organic materials) often proceeds at a rate comparable to the rate of oxidation. In this case, the atom-% of the substrate pool is continually diluted throughout the incubation and failure to account for this change will cause an underestimate of the ammonium oxidation rate. A comparison of the most commonly used equations and a comparison of the effect of accounting for or ignoring isotope dilution of the substrate pool can be found in Ward *et al.* (1989b).

Detection and Evaluation of *amoA* Genes in the Environment

Phylogeny of Nitrifying Bacteria

The number and diversity of bacterial strains that are identified as autotrophic-nitrifying bacteria are rather limited, in comparison, for example, to the number and diversity of organisms that are capable of denitrification or of nitrogen fixation. A description of the species of ammonia oxidizers and nitrite oxidizers recognized on the basis of morphology (cell shape and distribution of internal membranes) and physiology can be found in Bock and Koops (1992), Koops and Moller (1992), and Koops *et al.* (2004), respectively.

The generic and species affiliations of nitrifiers were reassessed on the basis of DNA sequences of the 16S ribosomal RNA genes from the cultured strains (Head *et al.*, 1993; Teske *et al.*, 1994). The phylogeny of nitrifiers (Teske *et al.*, 1994) shows that most of them are descendants of a common ancestor that was photosynthetic rather than descending from a common ancestral nitrifier. The ammonia oxidizers are found in the β and γ subdivisions of the Proteobacteria. The nitrite oxidizers are found in the α, δ, and γ subdivisions. An excellent description of the phylogeny of the AOB was provided by Purkhold *et al.* (2002) as well as in Koops *et al.* (2004); the nitrite oxidizers have received much less attention.

The β proteobacterial AOB species, containing the genera *Nitrosospira* and *Nitrosomonas*, have been the subject of many recent studies. The main tool for investigating them in both culture and field conditions is by amplification using the polymerase chain reaction (PCR) followed by DNA sequencing. On the basis of 16S rRNA sequence analysis, several clusters and a large amount of microdiversity within clusters have been detected in a wide variety of environments, both terrestrial and aquatic.

The γ subdivision ammonia oxidizers are represented by a single genus and two species, *Nitrosococcus oceani* and *N. halophilus*, which have been reported only from marine or saline environments. Several different strains of *N. oceani* exist in culture, but they all appear to be closely related. They have been detected by PCR in various marine environments (O'Mullan and Ward, 2005; Ward and O'Mullan, 2002) and in saline lakes in Antarctica (Voytek *et al.*, 1999) and by immunofluorescence in various marine systems (Ward and Carlucci, 1985; Ward et al., 1989a), including the Mediterranean Sea (Zaccone et al., 1996). N. halophilus was isolated from salt lakes and lagoons (Koops et al., 1990) and its distribution has not been investigated intensely.

The restricted nature of AOB phylogeny makes it feasible to detect and evaluate their community structure on the basis of 16S rRNA genes. This is an advantage of working with AOB because 16S-based analyses are much more fraught with uncertainty and potential errors for other microbial groups that are defined primarily by function rather than phylogeny. In what is now a traditional approach, PCR primers that are designed to be specific for AOB 16S rRNA genes are used to amplify genes from DNA extracted from environmental samples. Several sets of primers with varying levels of specificity are commonly used. This approach was reviewed elsewhere (Kowalchuk and Stephen, 2001) for terrestrial environments. At the time of their review, relatively few reports from marine systems were available; those data are accumulating, however, and they paint a picture of major dominance by a so far uncultivated clade related to Nitrosospira (Bano and Hollibaugh, 2000; O'Mullan and Ward, 2005). The genus Nitrosococcus appears to be restricted to saline environments, but current evidence does not indicate that it is the dominant form.

When the question focuses more on the function of AOB in the environment, rather than more narrowly on diversity and community structure, it is preferable to use the gene amoA, which encodes the signature enzyme in aerobic ammonia oxidation, ammonia monooxygenase. While this enzyme is probably related evolutionarily to the similar monooxygenase that catalyzes the first step in aerobic methane oxidation (Holmes et al., 1995), it is relatively straightforward to apply the traditional PCR approach to investigate AOB on the basis of their amoA gene sequences. amoA sequences offer greater resolution (greater sequence divergence) and hold the promise of molecular regulation information at the level of functional genes that are directly involved in the biogeochemical transformation of interest.

Venter et al. (2004) suggested that ammonia oxidizers may also be found within Archaea, but at present there is insufficient evidence to support this assertion. The presence of an archaeal gene with limited similarity to amoA has been detected in metagenomic libraries from both terrestrial and aquatic environments (Schleper et al., 2005; Venter et al., 2004) but there are currently no data to support Archaeal ammonia oxidation. Regardless, this highlights the continuing need for the cultivation of novel microbes as a complement to the application of molecular ecological techniques, and vice versa, for the study of ammonia oxidizers.

Sample Collection and Storage

Particulate material is collected from water samples by filtration, either onto flat polysulfone filters by vacuum filtration or onto filter capsules using a peristaltic pump. The filtering should be done as rapidly as possible (the capsule filters are generally faster) and all tubing and supplies rinsed thoroughly with 10% HCl and ddH$_2$O between samples. Flat filters can be folded and placed in 2-ml cryovials. It is not necessary to add a buffer to the filters or filter capsules, but it is necessary to drop them into liquid nitrogen immediately. Soil or sediment samples are collected by coring or with clean scoops, placed into 2-ml cryovials, and frozen immediately in liquid N$_2$. Once frozen, filters and soil samples should be stored on dry ice or at $-80°$ prior to extraction. Either RNA or DNA can usually be obtained from samples handled in this manner.

DNA Extraction and Gene Amplification

There are many DNA extraction protocols, most of them essentially permutations on a few basic approaches. This section describes some tested approaches that have worked on a variety of sample formats in our hands. New kits and protocols become available continually, but the need to modify or optimize such methods usually ensures that once a protocol is perfected, it is not always profitable to switch to the next best thing.

If flat filters are used, DNA can be extracted with a phenol/chloroform protocol (Ausubel *et al.*, 1987), including a CTAB step, in which case the filter partially dissolves and is discarded at the first phenol centrifugation step. The chloroform and phenol extractions can be repeated as necessary if the extract is highly colored or if required to obtain a clean interface. Capsule filters (e.g., Sterivex from Millipore) are conveniently extracted using the Gentra Pure-Gene kit (the tissue protocol), which is modified by doubling or tripling the initial reagent volumes and the initial extraction is performed inside the capsule. Obtaining clean DNA extracts from sediments is often much more problematic than from water, and the most reliable kit in our hands is the Bio101 Soil DNA kit. The washes at the bead step can be repeated, and elution in a larger volume than the kit suggests is usually advised.

In order to obtain RNA from an environmental sample, the usual protocols to avoid RNase contamination must be observed. For particulate material collected from water onto capsule filters, we use the RNAqueous 4PCR kit from Ambion (or the RNeasy kit from Qiagen). Obtaining RNA from sediment samples is still very problematic. A new kit designed for extraction of RNA from soil has been produced by Bio101 and, in our hands, yields small quantities of RNA, but shows promise. It is advisable to synthesize cDNA immediately following elution of RNA from the spin columns or membranes used in the extraction protocols, using, for example, the Superscript kit from Invitrogen with random oligonucleotides. The cDNA so produced is much more stable and less prone to degradation than the original RNA, but the RNA can be stored at $-80°$ for future experiments. Check the quality of the cDNA by PCR amplification with EUB ("universal" eubacterial 16S rRNA) primers. Include the RNA as a template in control reactions. If the RNA control amplifies, it could indicate that the extract still contains DNA (even after a DNase treatment as part of the RNA extraction protocol). However, most *Taq* polymerase enzymes contain a very small amount of reverse polymerase activity and can result in a faint EUB amplification from RNA.

Once the DNA extract or cDNA is obtained, the first PCR reaction should be performed with a universal bacterial primer set simply to verify the PCR quality of the extract. As a positive control, include a reaction with DNA extracted from a pure culture of bacteria. It is often necessary to optimize a new primer set or extraction protocol at the PCR level, and optimal conditions for the same primer set may vary depending on the nature of the template. The Taguchi approach (Cobb and Clarkson, 1994) is extremely useful in this regard; it prescribes an initial set of nine PCR reactions that usually suffice to identify optimal conditions, or at the least to indicate how to further modify the conditions in order to maximize yield of the correct product. If the positive control reaction yields the expected product and the environmental DNA extract does not amplify, the problem is at least as likely to be DNA quality as target abundance. Ethanol precipitation is sometimes effective as a cleanup, as are additional phenol extractions. We have found that PCR ability is improved by performing the bead binding and washing steps from the Bio101 kit, regardless of whether the DNA extract was initially obtained with that kit or with another protocol. Simply dilute a small portion of the extract (e.g., 25 μl) with 5–10 volumes of the binding buffer (we use the PB buffer from the Qiaquick spin kits), add 200 μl of the Bio101 binding matrix, mix 2 min, wash as indicated (use larger volume multiple washes), and elute in up to 500 μl of warmed ddH$_2$O. Impurities that might inhibit PCR are either removed or diluted by this procedure. Use up to 10 μl of the cleaned DNA in the PCR reaction, adjusting water volumes accordingly.

Probably because AOB usually comprise a very small component of the total bacterial assemblage in most environments, many workers use a nested PCR protocol to amplify both 16S rRNA and *amoA* genes from environmental DNA extracts. For 16S rRNA, the first step is amplification with universal eubacterial (EUB) primers (e.g., *Escherichia coli* position 9 to 1542 primers of Liesack *et al.*, 1991) and a few alternative comparable primers are now available. The EUB PCR product is then used as the template in the specific AOB amplification [both β and γ proteobacterial

AOB 16S rRNA gene fragments can be obtained in this manner, using appropriate primer pairs (e.g., O'Mullan and Ward, 2005)]. Alternatively, many workers have succeeded in amplifying AOB 16S rRNA genes directly without the nested approach (e.g., Bano and Hollibaugh, 2000) (Table I).

TABLE I
AOB PRIMERS

Primer name	Position in *E. coli*	Sequence 5'-3'	Reference
Target 16S rRBA genes			
EUB1	9–27	GAGTTTGATCCTGGCTCAG	Liesack *et al.* (1991)
EUB2	1525–1542	AGAAAGGAGGTGATCCAGCC	Liesack *et al.* (1991)
βAMOf	142–162	TGGGGRATAACGCAYCGAAAG	McCaig *et al.* (1994)
βAMOr	1302–1321	AGACTCCGATCCGGACTACG	McCaig *et al.* (1994)
NITA	137–160159	CTTAAGTGGGGAATAACGCATCG	Voytek and Ward (1995)
NITB	12143–1234	TTACGTGTGAAGCCCTACCCA	Voytek and Ward (1995)
CTO 189f	189–207	GGAGRAAAGYAGGGGATCG	Kowalchuck *et al.* (1997)
CTO 654r	632–653	CTAGCYTTGTAGTTTCAAACGC	Kowalchuck *et al.* (1997)
NOC1	25–45	CGTGGGAATCTGGCCTCTAGA	Voytek *et al.* (1999)
NOC2[a]	111668–11881686	AGATTAGCTCCGCATCGCGT	Voytek *et al.* (1999)
Target *amoA* gene			
Position in *N. europaea*			
AmoA-1F	332–349	GGGGTTTCTACTGGTGGT	Rotthauwe *et al.* (1997)
AmoA-2R	802–822	CCCCTCKGSAAAGCCTTCTTC	Rotthauwe *et al.* (1997)
AMO189		GGNGACTGGGACTTCTGG	Holmes *et al.* (1995)
AMO682		GAASGCNGAGAAGAASGC	Holmes *et al.* (1995)

[a] The sequence given here for NOC2 is slightly different from that published by Ward *et al.* (2000). It has been modified to make it slightly more broad and to enhance its ability to amplify *Nitrosococcus halophilus*, the only other cultivated member of the marine *Nitrosococcus* genus.

A variety of primer sets have been developed to amplify specifically the 16S rRNA genes from AOB. The three most commonly used primers sets for the amplification of β proteobacterial AOB 16S rRNA genes are βAMOf and βAMOr (McCaig et al., 1994), NITA and NITB (Voytek and Ward, 1995), and CTO 189f and CTO 654r (Kowalchuk et al., 1997). The first two primer sets amplify approximately 1200- and 1100-bp fragments, respectively, while the last set yields an approximately 465-bp fragment that is often used for DGGE studies. The βAMO primers were instrumental in expanding the range of recognized AOB 16S rRNA sequences but were designed with relatively low specificity for AOB in an effort to obtain sequences beyond the known cultured strains. This lack of specificity can be problematic in environmental analyses because non-AOB sequences are often amplified, but they can often be applied successfully to studies on enrichment cultures. In contrast, the NIT primer set has greater specificity for the known AOB, but has been criticized for its bias toward a preferential amplification of some *Nitrosomonas* lineages. The CTO primer set amplifies a short fragment, making it useful for DGGE but less powerful for phylogenetic discrimination. We have found that the NITA,B primer set is useful for most environmental samples, where the amplification of many false negatives would be overly problematic. The NITA,B primers also produce phylogenetic patterns that are consistent with *amoA* results when both genes are amplified from the same set of marine samples (Caffrey et al., 2003; O'Mullan and Ward, 2005), including those samples dominated by *Nitrosospira*-like sequences. Koops et al. (2004) provided a useful overview of the specificity of these primer sets as well as many hybridization probes.

Analysis of γ proteobacterial 16S rRNA gene sequences is much less common, and the NOC1,2 primer set (Voytek et al., 1999) is used exclusively for specific amplification of this group. We have found that the diversity detected in marine environments with this primer set is extremely limited, likely reflecting the limited sequence information available for their design (Ward and O'Mullan, 2002). An altered version of this primer set was designed but did not yield an amplification product from marine sediments (Freitag and Prosser, 2003). The isolation of new AOB strains and the steady increase in the number of identified environmental sequences require that primer design be reevaluated constantly for specificity and altered as appropriate.

Different PCR primer sets are used to amplify the *amoA* gene fragments from β proteobacterial and γ proteobacterial AOB. Primers AmoA-1F and AmoA-2R of Rotthauwe et al. (1997) are specific to the β AOB; we usually decrease the annealing temperature to 55° and otherwise use the protocols of Rotthauwe et al. (1997). Minor alterations have been made to this primer set by Stephen *et al.* (1999), and a suite of nondegenerate, but otherwise similar, primers have been used for DGGE and competitive PCR (Bjerrum *et al.*, 2002; Nicolaisen and Ramsing, 2002). The primers AMO189 and AMO682 of Holmes *et al.* (1996) amplify both β and γ proteobacterial AOB *amoA* gene fragments, and we have found them to be useful in combination with the Rotthauwe primers in a nested arrangement in order to increase sensitivity with difficult templates. A set of *amoA* primers designed specifically to amplify a gene fragment from the γ proteobacterial strains in culture (Purkhold *et al.*, 2000) have apparently not been widely used in environmental applications. It can be problematic to amplify *amoA* from environmental samples; whether this is simply a problem of target abundance (i.e., *amoA* is likely to be much less abundant than 16S rRNA from the same organisms due to the difference in gene copy number), a problem of extract quality, or of sequence divergence in natural assemblages remains to be resolved.

The procedures for cloning, sequencing, and analyzing the resulting gene sequences are standard and are not described here. The same PCR primers used for detection and diversity studies are applicable for Q-PCR studies (Okano *et al.*, 2004). The details of controls and quantification for Q-PCR described in the *nifH* chapter by Zehr and Short should be applicable to analogous experiments with *amoA* and will not be repeated here.

Microbial ecology is likely to make the most progress while utilizing the "simplest" of microbial groups rather than the most inclusive, universal datasets. It has been argued that AOB provide a powerful model system for microbial ecology because of their limited phylogenetic diversity, parallel topology of 16S rRNA and *amoA* genealogies, and clear identification by functional as well as phylogenetic criteria (Kowalchuk and Stephen, 2001). In many environments it appears that even limited sampling of clone libraries (20 to 50 clones) for AOB genes may adequately identify dominant sequences rather than simply revealing stochastic variation in a poorly sampled, complex dataset. The limited diversity of this group lends itself to rarefaction analysis, analyses of molecular variation, and community fingerprinting approaches. When such analyses are integrated with the biogeochemical measures described earlier, the coupling of bacterial dynamics to clear measures of functional variation is a goal within reach.

References

Alzerreca, J. J., Norton, J. M., and Klotz, M. G. (1999). The amo operon in marine, ammonia-oxidizing gamma-proteobacteria. *FEMS Microbiol. Lett.* **180,** 21–29.

Ausubel, F. M., Brent, R., Kingston, R. E., Moore, E. E., A, S. J., Sideman, J. G., and Struhl, K. (1987). "Current Protocols in Molecular Biology." Wiley, New York.

Balderston, W. L., Sherr, B., and Payne, W. J. (1976). Blockage by acetylene of nitrous oxide reduction in *Pseudomonas perfectomarinus*. *Appl. Environ. Microbiol.* **31,** 504–508.

Bano, N., and Hollibaugh, J. T. (2000). Diversity and distribution of DNA sequences with affinity to ammonia-oxidizing bacteria of the beta subdivision of the class Proteobacteria in the Arctic Ocean. *Appl. Environ. Microbiol.* **66,** 1960–1969.

Berg, P., Klemedtsson, L., and Roswall, T. (1982). Inhibitory effects of low partial pressure of acetylene on nitrification. *Soil Biol. Biochem.* **14,** 301–303.

Bianchi, M., Feliatra, Treguer, P., Vincendeau, M. A., and Morvan, J. (1997). Nitrification rates, ammonium and nitrate distribution in upper layers of the water column and in sediments of the Indian sector of the Southern Ocean. *Deep-Sea Res.* **44,** 1017–1032.

Billen, G. (1976). Evaluation of nitrifying activity in sediments by dark ^{14}C-bicarbonate incorporation. *Wat. Res.* **10,** 51–57.

Binnerup, S. J., Jensen, K., Revsbech, N. P., Jensen, M. H., and Sorensen, J. (1992). Denitrification, dissimilatory reduction of nitrate to ammonium, and nitrification in a bioturbated estuarine sediment as measured with N-15 and microsensor techniques. *Appl. Environ. Microbiol.* **58,** 303–313.

Bjerrum, L., Kjaer, T., and Ramsing, N. B. (2002). Enumerating ammonia-oxidizing bacteria using competitive PCR. *J. Microbiol. Methods* **51,** 227–239.

Bock, E., and Koops, H.-P. (1992). The genus *Nitrobacter* and related genera. *In* "The Prokaryotes" (A. Ballows, H. G. Trüper, W. Dworkin, W. Harder, and K. H. Schleifer, eds.), pp. 2302–2637. Springer-Verlag, New York.

Caffrey, J. M., Harrington, N., Solem, I., and Ward, B. B. (2003). Biogoechemical processes in a small California estuary. 2. Nitrification activity, community structure and role in nitrogen budgets. *Marine Ecol. Progr. Ser.* **248,** 27–40.

Caffrey, J. M., and Miller, L. G. (1995). A comparison of two nitrification inhibitors used to measure nitrification rates in estuarine sediments. *FEMS Microbiol. Ecol.* **17,** 213–219.

Capone, D. G., Dunham, S. E., Horrigan, S. G., and Duguay, L. E. (1992). Microbial nitrogen transformations in unconsolidated coral-reef sediments. *Mar. Ecol. Progr. Ser. Ecol.Progr. Ser.* **80,** 75–88.

Capone, D. G., Horrigan, S. G., Dunham, S. E., and Fowler, J. (1990). Direct determination of nitrification in marine waters by using the short-lived radioisotope of Nitrogen, N-13. *Appl. Environ. Microbiol.* **56,** 1182–1184.

Cobb, B. D., and Clarkson, J. M. (1994). A simple procedure to optimising the polymerase chain reacation (PCR) using modified Taguchi methods. *Nucleic Acids Res.* **22,** 3801–3805.

Dore, J. E., and Karl, D. M. (1996). Nitrification in the euphotic zone as a source for nitrite, nitrate, and nitrous oxide at station ALOHA. *Limnol. Oceanogr.* **41,** 1619–1628.

Fiedler, R., and Proksch, G. (1975). The determination of nitrogen-15 by emission and mass spectrometry in biochemical analysis: A review. *Anal. Chim. Acta* **78,** 1–62.

Freitag, T. E., and Prosser, J. I. (2003). Community structure of ammonia-oxidizing bacteria within anoxic marine sediments. *Appl. Environ. Microbiol.* **69,** 1359–1371.

Glover, H. E. (1985). The relationship between inorganic nitrogen oxidation and organic carbon production in batch and chemostat cultures of marine nitrifying bacteria. *Arch. Microbiol.* **74,** 295–300.

Goring, C. A. I. (1962). Control of nitrification by 2-chloro-6-(trichloromethyl)pyridine. *Soil Sci.* **93,** 211–218.

Head, I. M., Hiorns, W. D., Embley, T. M., McCarthy, A. J., and Saunders, J. R. (1993). The phylogeny of autotrophic ammonia-oxidizing bacteria as determined by analysis of 16S Ribosomal-RNA gene-sequences. *J. Gen. Microbiol.* **139,** 1147–1153.

Henricksen, K., Hansen, J. I., and Blackburn, T. H. (1981). Rates of nitrification, distribution of nitrifying bacteria, and nitrate fluxes in different types of sediment from Danish waters. *Mar. Biol.* **61,** 299–304.

Holmes, A. J., Costello, A., Lidstrom, M. E., and Murrell, J. C. (1995). Evidence that particulate methane monooxygenase and ammonia monooxygenase may be evolutionarily related. *FEMS Microbiol. Lett.* **132,** 203–208.

Jensen, K., Sloth, N. P., Risgaard-Petersen, N., Rysgaard, S., and Revsbech, N. P. (1994). Estimation of nitrification and denitrification from microprofiles of oxygen and nitrate in model sediment systems. *Appl. Environ. Microbiol.* **60,** 2094–2100.

Juretschko, S., Timmermann, G., Schmid, M., Schleifer, K.-H., Pommerening-Roser, A., Koops, H.-P., and Wagner, M. (1998). Combined molecular and conventional anlyses of nitrifying bacterium diversity in activated sludge: *Nitrosococcus mobilis* and *Nitrospira*-like bacteria as dominant populations. *Appl. Environ. Microbiol.* **64,** 3042–3051.

Kator, H., Morris, L. J., Wetzel, R. L., and Koepfler, E. T. (1992). A rapid chromatographic method for recovery of $^{15}NO_2^-$ and $^{15}NO_3^-$ produced by nitrification in aqueous samples. *Limnol. Oceanogr.* **37,** 900–907.

Kester, R. A., de Boer, L., and Laanbroek, H. J. (1996). Short exposure to acetylene to distinguish between nitrifier and denitrifier nitrous oxide production in soil and sediment samples. *FEMS Microbiol. Ecol.* **20,** 111–120.

Klotz, M. G., and Norton, J. M. (1995). Sequence of an ammonia monooxygenase subunit a-encoding gene from *Nitrosospira* sp NpAV. *Gene* **163,** 159–160.

Koops, H.-P., Bottcher, B., Moller, U. C., Pommereningroser, A., and Steher, G. (1990). Descripton of a new species of *Nitrosococcus*. *Arch. Microbiol.* **154,** 244–248.

Koops, H.-P., and Moller, U. C. (1992). The lithotrophic ammonia-oxidizing bacteria. *In* "Prokaryokes" (Balows *et al.*, eds.), pp. 2625–2637. Springer-Verlag, New York.

Koops, H.-P., Purkhold, U., Pommerening-Roser, A., Timmermann, G., and Wagner, M. (2004). The lithoautotrophic ammonia-oxidizing bacteria. *In* "The Prokaryotes: An Evolving Electronic Resource for the Microbiological Community." Release 3.13, http://link.springer-ny.com/link/service/books/10125/. Springer-Verlag, New York.

Kowalchuk, G. A., and Stephen, J. R. (2001). Ammonia-oxidizing bacteria: A model for molecular microbial ecology. *Annu. Rev. Microbiol.* **55,** 485–529.

Kowalchuk, G. A., Stephen, J. R., De Boer, W., Prosser, J. I., Embley, T. M., and Woldendorp, J. W. (1997). Analysis of ammonia-oxidizing bacteria of the beta subdivision of the class proteobacteria in coastal sand dunes by denaturing gradient gel electrophoresis and sequencing of PCR-amplified 16S ribosomal DNA fragments. *Appl. Environ. Microbiol.* **63,** 1489–1497.

Liesack, W., Weyland, H., and Stackebrandt, E. (1991). Potential risks of gene amplification by PCR as determined by 16S rDNA analysis of a mixed culture of strict barophilic bacteria. *Microbial Ecol.* **21,** 191–198.

McCaig, A. E., Embley, T. M., and Prosser, J. I. (1994). Molecular analysis of enrichment cultures of marine ammonia oxidisers. *FEMS Microbiol. Lett.* **120,** 363–367.

Miller, L. G., Coutlakis, M. D., Oremland, R. S., and Ward, B. B. (1993). Selective inhibition of nitrification (ammonium oxidation) by methylfluoride and dimethyl ether. *Appl. Environ. Microbiol.* **59,** 2457–2464.

Nicolaisen, M. H., and Ramsing, N. B. (2002). Denaturing gradient gel electrophoresis (DGGE) approaches to study the diversity of ammonia-oxidizing bacteria. *J. Microbiol. Methods* **50,** 189–203.

Nielsen, L. (1992). Denitrification in sediment determined from nitrogen isotope pairing. *FEMS Microbiol. Ecol.* **86,** 357–362.

Okano, Y., Hristova, K. R., Leutenegger, C. M., Jackson, L. E., Denison, R. F., Gebreyesus, B., Lebauer, D., and Scow, K. M. (2004). Application of real-time PCR to study effects of ammonium on population size of ammonia-oxidizing bacteria in soil. *Appl. Environ. Microbiol.* **70,** 1008–1016.

Olson, R. J. (1981). ^{15}N tracer studies of the primary nitrite maximum. *J. Mar. Res.* **39,** 203–226.

O' Mullan, G. D., and Ward, B. B. (2005). Comparison of temporal and spatial variability of ammonia-oxidizing bacteria to nitrification rates in Monterey Bay, CA. *Appl. Environ. Microbiol* **71,** 697–705.

Owens, N. J. P. (1986). Estuarine nitrification: A naturally-occurring fluidized-bed reaction. *Estuarine Coast. Shelf Sci.* **22,** 31–44.

Priscu, J. C., and Downes, M. T. (1985). Nitrogen uptake, ammonium oxidation and nitrous oxide (N_2O) levels in the coastal waters of Western Cook Strait, New Zealand. *Estuarine Coast. Shelf Sci.* **20,** 529–542.

Purkhold, U., Pommerening-Roser, A., Juretschko, S., Schmid, M. C., Koops, H. P., and Wagner, M. (2000). Phylogeny of all recognized species of ammonia oxidizers based on comparative 16S rRNA and *amoA* sequence analysis: Implications for molecular diversity surveys. *Appl. Environ. Microbiol.* **66,** 5368–5382.

Rakestraw, N. W. (1936). The occurrence and significance of nitrite in the sea. *Biol. Bull.* **71,** 133–167.

Rysgaard, S., Risgaardpetersen, N., Nielsen, L. P., and Revsbech, N. P. (1993). Nitrification and denitrification in lake and estuarine sediments measured by the N-15 dilution technique and isotope pairing. *Appl. Environ. Microbiol.* **59,** 2093–2098.

Sayavedra-Soto, L. A., Hommes, N. G., Alzerreca, J. J., Arp, D. J., Norton, J. M., and Klotz, M. (1998). Transcription of the amoC, amoA and amoB genes in *Nitrosomonas europaea* and *Nitrosospira* sp, NpAV. *FEMS Microbiol. Lett.* **167,** 81–88.

Schleper, C., Jurgens, G., and Jonuscheit, M. (2005). Genomic studies of uncultivated Archaea. *Nat. Rev. Microbiol.* **3,** 479–488.

Somville, M. (1978). A method for the measurement of nitrification rates in water. *Wat. Res.* **12,** 843–848.

Suzuki, I., Dular, U., and Kwok, S. (1974). Ammonia or ammonium ion as substrate for oxidation by *Nitrosomonas europaea* cells and extracts. *J. Bacteriol.* **120,** 556–558.

Teske, A., Alm, E., Regan, J. M., Toze, S., Rittmann, B. E., and Stahl, D. A. (1994). Evolutionary relationships among ammonia- and nitrite-oxidizing bacteria. *J. Bacteriol.* **176,** 6623–6630.

Venter, C. J., Remington, K., Heidelberg, J. G., Halpern, A. L., Rusch, D., Eisen, J. A., Wu, D., Paulsen, I., Nelson, K. E., Nelson, W., Fouts, D. E., Levy, S., Knap, A. H., Lomas, M. W., Nealson, K., White, O., Peterson, J., Hoffman, J., Parsons, R., Baden-Tillson, H., Pfannkoch, C., Rogers, J.-H., and Smith, H. O. (2004). Environmental genome shotgun sequencing of the Sargasso Sea. *Science* **304,** 66–74.

Viner, A. B. (1990). Dark ^{14}C uptake, and its relationships to nitrification and primary production estimates in a New Zealand upwelling region. *N. Z. J. Mar. Freshwat. Res.* **24,** 221–228.

von Brand, T., Rakestraw, N., and Renn, C. (1937). The experimental decomposition and regeneration of nitrogenous organic matter in sea water. *Biol. Bull.* **72,** 165–175.

Voytek, M. A., and Ward, B. B. (1995). Detection of ammonium-oxidizing bacteria of the beta-subclass of the class Proteobacteria in aquatic samples with the PCR. *Appl. Environ. Microbiol.* **61,** 1444–1450.

Voytek, M. A., Priscu, J. C., and Ward, B. B. (1999). The distribution and relative abundance of ammonia-oxidizing bacteria in lakes of the McMurdo Dry Valley, Antarctica. *Hydrobiologia* **401,** 113–130.

Ward, B. B. (1986). Nitrification in marine environments. *In* "Nitrification" (J. I. Prosser, ed.), pp. 157–184. IRL Press, Oxford.

Ward, B. B. (1987a). Kinetic studies on ammonia and methane oxidation by *Nitrosococcus oceanus. Arch. Microbiol.* **147,** 126–133.

Ward, B. B. (1987b). Nitrogen transformations in the Southern California Bight. *Deep-Sea Res.* **34,** 785–805.

Ward, B. B., and Carlucci, A. F. (1985). Marine ammonium- and nitrite-oxidizing bacteria: Serological diversity determined by immunofluorescence in culture and in the environment. *Appl. Environ. Microbiol.* **50,** 194–201.

Ward, B. B., Glover, H. E., and Lipschultz, F. (1989a). Chemoautotrophic activity and nitrification in the oxygen minimum zone off Peru. *Deep-Sea Res.* **36,** 1031–1051.

Ward, B. B., Kilpatrick, K. A., Renger, E., and Eppley, R. W. (1989b). Biological nitrogen cycling in the nitracline. *Limnol. Oceanogr.* **34,** 493–513.

Ward, B. B., and O' Mullan, G. D. (2002). Worldwide distribution of *Nitrosococcus oceani,* a marine ammonia-oxidizing gamma-proteobacterium, detected by PCR and sequencing of 16S rRNA and amoA genes. *Appl. Environ. Microbiol.* **68,** 4135–4157.

Ward, B. B., Talbot, M. C., and Perry, M. J. (1984). Contributions of phytoplankton and nitrifying bacteria to ammonium and nitrite dynamics in coastal water. *Continental Shelf Res.* **3,** 383–398.

Zaccone, R., Caruso, G., and Azzaro, M. (1996). Detection of *Nitrosococcus oceanus* in a Mediterranean lagoon by immunofluorescence. *J. Appl. Bacteriol.* **80,** 611–616.

Zahn, J. A., Arciero, D. M., Hooper, A. B., and Di Spirito, A. A. (1996). Evidence for an iron center in the ammonia monooxygenase from *Nitrosomonas europaea. FEBS Lett.* **397,** 35–38.

[25] Community-Level Analysis: Key Genes of Aerobic Methane Oxidation

By Marc G. Dumont and J. Colin Murrell

Abstract

Aerobic methane-oxidizing bacteria (methanotrophs) are a diverse group of bacteria that are currently represented by 13 recognized genera. They play a major role in the global methane cycle and are widespread in nature with representatives found in soils, freshwater, seawater, freshwater and marine sediments, peat bogs and at extremes of temperature, salinity, and pH. There has been an interest in methanotrophs for their potential in bioremediation processes. Methanotroph diversity and ecology are often studied using the "functional" genes *pmoA, mmoX,* and *mxaF,* encoding subunits of the particulate methane monooxygenase, soluble methane monooxygenase, and the methanol dehydrogenase, respectively. This chapter describes methods used to detect and analyze these functional genes.

Introduction

Aerobic methane-oxidizing bacteria (methanotrophs) can use methane as a sole source of carbon and energy (Hanson and Hanson, 1996). Globally, the bulk of biological methane oxidation activity occurs in the aerobic zones overlaying anoxic sediments where methane is generated by methanogens.

METHODS IN ENZYMOLOGY, VOL. 397
Copyright 2005, Elsevier Inc. All rights reserved.

0076-6879/05 $35.00
DOI: 10.1016/S0076-6879(05)97025-0

In addition to having an important role in the global carbon cycle, methane has a high greenhouse warming potential, and the consumption of methane by methanotrophs impacts atmospheric chemistry. Upland soils also act as a sink for atmospheric methane (Conrad, 1996).

Most methanotrophs are incapable of growth on compounds containing carbon–carbon bonds. The ability of methanotrophs to oxidize methane is conferred by a methane monooxygenase (MMO) enzyme that converts methane to methanol. There are two types of MMO: a cytoplasmic version (sMMO) and a membrane-bound version (pMMO) (Murrell *et al.*, 2000). Nearly all characterized methanotrophs possess either pMMO only or both pMMO and sMMO; members of the *Methylocella* genus are the only methanotrophs in which a pMMO has not been detected. Apart from the oxidation of methane to methanol, the enzymatic and biochemical pathways generating energy and cellular carbon are similar to those found in non-methane utilizing gram-negative methylotrophs (Lidstrom, 2001). After the conversion of methane to methanol by the MMO enzyme, the methanol is oxidized to formaldehyde by a pyrroloquinoline quinone (PQQ)-dependent methanol dehydrogenase (MDH) (Anthony and Dales, 1996). Carbon is assimilated at the level of formaldehyde by either the ribulose monophosphate (RuMP) or the serine pathway depending on the organism (Anthony, 1982). Alternatively, formaldehyde can be oxidized completely to carbon dioxide, which generates reducing equivalents for cellular metabolism.

The methanotroph genes targeted primarily in cultivation-independent studies include the 16S rRNA gene, *pmoA*, *mmoX,* and *mxaF*. The three "functional" genes encode the active site protein of their respective enzyme: *pmoA* corresponds to the pMMO; *mmoX* to the sMMO; and *mxaF* to the MDH. There is scope for targeting other methylotroph functional genes, such as some unique enzymes of the serine pathway for formaldehyde assimilation and enzymes involved in methanopterin biosynthesis or C1-transfer pathway (see Kalyuzhnaya and Chistoserdova, 2005). The aim of this chapter is to provide a summary of several basic techniques often used to detect methanotroph functional genes in environmental samples.

Methanotroph Diversity and Phylogeny

There are currently 13 recognized methanotroph genera and they fall into the α and γ subdivisions of the Proteobacteria: *Methylobacter* (γ), *Methylococcus* (γ), *Methylocaldum* (γ), *Methylomicrobium* (γ), *Methylomonas* (γ), *Methylosarcina* (γ), *Methylosphaera* (γ), *Methylohalobius* (4), *Methylothermus* (γ), *Methylocystis* (α), *Methylosinus* (α), *Methylocella* (α) and *Methylocapsa* (α). Phylogenetic analyses of γ-Proteobacteria methanotrophs indicate that they form a distinct branch and have been assigned the family Methylococceae (Bowman, 2000). Among the α-Proteobacteria methanotrophs, *Methylocystis* and *Methylosinus* form a lineage and are designated Methylocystaceae (Bowman, 2000), whereas *Methylocella* and *Methylocapsa* are more closely related to the non-methane utilizer *Beijerinckia* (Dedysh *et al.*, 2000, 2002).

Methanotrophs exist in many different habitats (Hanson and Hanson, 1996) and at several environmental extremes (Trotsenko and Khmelenina, 2002). The biological oxidation of atmospheric methane also occurs in upland soils by a group of methanotrophs that are as yet uncultivated (Hanson and Hanson, 1996). A distinct *pmoA* sequence type occurs frequently in soils that exhibit atmospheric methane oxidation and is believed to belong to this group of methanotrophs (Henckel *et al.*, 2000; Holmes *et al.*, 1999).

DNA Extraction from Environmental Samples for Analysis of Methanotrophs

Many methanotroph ecology studies have used a bead-beating mechanical lysis protocol for the preparation of DNA from environmental samples. DNA can be isolated from 0.3-g (wet weight) samples using the FastPrep DNA extraction kit (Bio101) in less than 1 h (Radajewski *et al.*, 2002; Yeates and Gillings, 1998). This protocol has also been scaled up and adapted to purify DNA from 3-g samples (Morris *et al.*, 2002). The advantage of the method is that it is rapid and efficient; however, the protocol requires a bead-beating machine. In general, mechanical disruption is not essential to lyse methanotroph cells and isolate DNA. Therefore, if a bead-beating apparatus is not available, it is possible to use a chemical cell lysis protocol and extract the DNA by a method that does not require special equipment (Selenska and Klingmuller, 1991; Zhou *et al.*, 1996) as performed in some studies (Holmes *et al.*, 1999; Lin *et al.*, 2004; McDonald *et al.*, 1995). It has been reported that α-Proteobacteria methanotrophs can be more difficult to lyse than γ-Proteobacteria methanotrophs and therefore the former may be underestimated in clone libraries (Kolb *et al.*, 2003).

Organic compounds that inhibit *Taq* DNA polymerase often contaminate DNA extracted from soil and additional purification steps are often required (see Purdy, 2005). The purity of the DNA sample can be assessed by performing a polymerase chain reaction (PCR) with bacterial 16S rRNA PCR primers (27f/1492r) (Lane, 1991) with approximately 10 ng, 1 ng, 100 pg, and 10 pg of DNA template. The aim is to be able to use approximately 10 ng or more of template with the methanotroph gene primers, but DNA frequently requires further purification to obtain PCR product even with a very dilute template. In some instances it is adequate to gel purify DNA that has been resolved by electrophoresis through a 1% agarose gel. Gel purification is performed using a clean electrophoresis tank and recovering high molecular weight DNA by means of a standard commercial kit such as Geneclean II (Bio 101) or QIAquick (Qiagen); alternatively, a low-melting point agarose can be used and the segment of gel containing the DNA can be melted and used directly in the PCR (Steinkamp *et al.*, 2001). More extensive steps such as CsCl ethidium bromide gradient centrifugation (Sambrook and Russell, 2001) and combinations of methods to purify DNA are sometimes used, but are rarely needed. A concentration of DNA template that generates abundant PCR product with the bacterial 16S rRNA primers (Lane, 1991) can then be used in PCR to target methanotroph genes.

PCR Amplification of Methanotroph Genes

pmoA

The *pmoA* gene is the most frequent target in molecular ecology studies of methanotrophs. It has the advantage of being present in all known methanotrophs with the exception of *Methylocella* (Dedysh *et al.*, 2000; A.R. Theisen and J.C. Murrell, unpublished result). There is a large dataset of *pmoA* sequences available from characterized methanotroph strains, which make it possible to identify a methanotroph based on its *pmoA* gene sequence. Heyer and co-workers (2002) analyzed the *pmoA* genes of Methylocystaceae isolates and found no evidence for recent *pmoA* gene transfer events. Although a similar investigation into the *pmoA* genes of Methylococcaceae representatives has not been performed, to our knowledge there is no evidence of any recent *pmoA* gene transfer in methanotrophs. There are several sets of PCR primers that target *pmoA* (Table I). Methods have also been developed to quantitatively detect *pmoA* genes using real-time PCR (Kolb *et al.*, 2003), and *pmoA* and *mmoX* gene transcripts using competitive rt-PCR (Hanand and Semrau, 2004).

Several characterized methanotrophs possess two copies of the *pmoCAB* operon (which encodes subunits of the pMMO enzyme), but often the copies are nearly identical (Auman *et al.*, 2000; Murrell *et al.*, 2000). A known exception is in some strains of *Methylocystis,* which have an additional *pmoCAB* operon (designated copy 2), which is significantly different from the usual *Methylocystis pmoCAB* (Dunfield *et al.*, 2002; Tchawa Yimga *et al.*, 2003); *pmoA* genes that resemble this copy 2 group frequently appear in clone libraries from environmental samples. A library

TABLE I

Selected PCR Primers That Target Methanotroph Functional Genes

Primer name	Sequence (5′–3′)	Target gene	Annealing position[a]	Paired primer	Product size (bp)[a]	Reference
A189	GGNGACTGGGACTTCTGG	pmoA/amoA	136–153			Holmes et al. (1995)
A682	GAASGCNGAGAAGAASGC	pmoA/amoA	666–649	A189	531	Holmes et al. (1995)
mb661r	CCGGMGCAACGTCYTTACC	pmoA	643–625	A189	508	Costello and Lidstrom (1999)
A650	ACGTCCTTACCGAAGGT	pmoA	635–619	A189	500	Bourne et al. (2001)
pmof1	GGGGGAACTTCTGGGITGGAC	pmoA	319–335			Cheng et al. (1999)
pmor	GGGGGRCIACGTCITTACCGAA	pmoA	638–622	pmof1		Cheng et al. (1999)
pmoA f325	TGGGGYTGGACCTAYTTCC	pmoA	325–343			Fjellbirkeland et al. (2001)
pmoA r643	CCGGCRCRACGTCCTTACC	pmoA	643–625	pmoA f325	319	Fjellbirkeland et al. (2001)
pmoAfor	TTCTGGGGNTGGACNTAY TTYCC	pmoA	322–344			Steinkamp et al. (2001)
pmoArev	TCNACCATNCKDATRTAY TCNGG	pmoA	602–580	pmoAfor	281	Steinkamp et al. (2001)
mmoXA	ACCAAGGARCARITCAAG	mmoX	166–183			Auman et al. (2000)
mmoXD	CCGATCCAGATDCCRCCCCA	mmoX	956–937	mmoXA	791	Auman et al. (2000)
mmoXB	TGGCACTCRTARCGCTC	mmoX	1400–1384	mmoXA	1235	Auman et al. (2000)
mmoX206f	ATCGCBAARGAATAYGCSCG	mmoX	187–206			Hutchens et al. (2004)

(continued)

TABLE I (*continued*)

Primer name	Sequence (5′–3′)	Target gene	Annealing position[a]	Paired primer	Product size (bp)[a]	Reference
mmoX886r	ACCCANGGCTCGACYTTGAA	mmoX	905–886	mmoX206f	719	Hutchens et al. (2004)
mmoX1	CGGTCCGCTGTGGAAGGGCA TGAAGCGCGT	mmoX	531–560			Miguez et al. (1997)
mmoX2	GGCTCGACCTTGAACTTGGA GCCATACTCG	mmoX	899–870	mmoX1	369	Miguez et al. (1997)
mmoXr901	ACCCAGCGGTTCCASGTYTTS ACCCA	mmoX	926–901	mmoX1	396	Shigematsu et al. (1999)
534f	CCGCTGTGGAAGGGCATGAA	mmoX	535–554			Horz et al. (2001)
1393r	CACTCGTAGCGCTCCGGCTC	mmoX	1397–1378	534f	863	Horz et al. (2001)
mmoX536f	CGCTGTGGAAGGGCATG AAGCG	mmoX	536–557			Fuse et al. (1998)
mmoX898r	GCTCGACCTTGAACTTGG AGCC	mmoX	898–877	mmoX536f	363	Fuse et al. (1998)
mmoX882f	GGCTCCAAGTTCAAGGTCGAGC	mmoX	877–898			McDonald et al. (1995)
mmoX1403r	TGGCACTCGTAGCGCTCCGGCTCG	mmoX	1400–1377	882f	524	McDonald et al. (1995)
mxaf1003	GCGGCACCAACTGGGGCTGGT	mxaF	800–820			McDonald et al. (1995)
mxar1561	GGGCAGCATGAAAGGGCTCCC	mxaF	1353–1334	mxaf1003	554	McDonald et al. (1995)

[a] Based on the gene sequences from *Methylococcus capsulatus* (Bath) (accession numbers L40804, M90050, and NC_002977 for *pmoA*, *mmoX*, and *mxaF*, respectively).

with a copy 2 *pmoA* gene will likely have a copy 1 *pmoA* from the same organism. There are other unusual *pmoA* genes that can appear in clone libraries generated with the A189/A682 primers. Examples of these include the RA21 sequence (accession number AF148522) (Holmes *et al.*, 1999) and the MR1 sequence (accession number AF200729) (Henckel *et al.*, 2000), which have not been detected in a cultivated organism. Evidently, caution should be used if interpreting the function of the enzyme, if any, to which these genes correspond.

mmoX

Unlike the pMMO enzyme, the sMMO is only found in a subset of methanotrophs and its presence often varies between species of the same genus (Bowman *et al.*, 1993; Heyer *et al.*, 2002). Therefore, although highly specific for methanotrophs, the *mmoX* is the least inclusive of the methanotroph gene markers. Nevertheless, it is a useful target, as it provides information regarding the catalytic diversity of methanotrophs and it complements *pmoA* diversity studies, which will not detect *Methylocella* representatives that lack a pMMO.

The *mmoX* gene appears to be an accurate phylogenetic marker in methanotrophs that possess an sMMO enzyme (Auman and Lidstrom, 2002; Heyer *et al.*, 2002). Heyer *et al.* (2002) noted that an *mmoX* PCR product could not be amplified from two *Methylocystis* strains in their culture collection that exhibited sMMO activity; this may indicate that the primers used in the study do not detect all *mmoX* genes. There are no known examples of *mmoX* gene transfer, and a single copy of the gene is present in the organisms examined. Several sets of *mmoX* PCR primers have been reported and are shown in Table I.

mxaF

The *mxaF* PCR primers available amplify a 555-bp segment of the methanol dehydrogenase gene from most gram-negative methylotrophs (McDonald and Murrell, 1997). Both methanotrophs and other gram-negative methylotrophs (e.g., methanol utilizers) contain a similar methanol dehydrogenase, and therefore it is difficult to design PCR primers that are specific to the *mxaF* of methanotrophs but exclude the genes from the non-methanotrophs. There are no reports of *mxaF* gene transfer between a non-methane utilizer and a methanotroph, and it is usually possible to distinguish the sequences by phylogenetic analysis. Heyer and co-workers (2002) reported that phylogenetic analysis of *mxaF* sequences from Methylocystaceae representatives were not always congruent with 16S rRNA and *pmoA* phylogeny: some *mxaF* sequences from *Methylosinus trichosporium* grouped closely with *mxaF* from *Methylosinus sporium* and others with *mxaF* from *Methylocystis*. Therefore, phylogenetic analysis of *mxaF* sequences should be interpreted with caution.

Selected Primers, PCR Conditions, and RFLP Analysis

The PCR primers described here are those that are used commonly in our laboratory. In addition, many other primer sets that target these genes have been published and we refer the reader to the original manuscripts for further details (Table I). PCR primers specific to methanotroph groups or sequence types are also present in the literature (Auman and Lidstrom, 2002; Kolb *et al.*, 2003; Tchawa Yimga *et al.*, 2003) and are not described here.

Differences in the equipment and reagents may affect the performance of the PCR and the conditions may need to be optimized. Unless otherwise indicated, the conditions given here are as described in the original manuscripts.

pmoA/amoA *PCR*

A189 5′-GGNGACTGGGACTTCTGG-3′
A682 5′-GAASGCNGAGAAGAASGC-3′

This primer set targets the *pmoA* gene and the *amoA* of autotrophic ammonia oxidizers (Holmes *et al.*, 1995). This is the only primer set that amplifies all known *pmoA* genes, including copy 2 *pmoA* of *Methylocystis* representatives (Dunfield *et al.*, 2002). Clone libraries often contain large numbers of *amoA* sequences that can conceal *pmoA* data (Bourne *et al.*, 2001). Full-length sequencing of *pmoCAB* operons indicates that primer mismatching may be considerable, but the amplification of *pmoA* is effective with all pure cultures of methanotrophs known to possess pMMO.

50 μl total volume containing:
DNA template
2.5 U *Taq* DNA polymerase (Invitrogen); added after initial denaturing step
1× *Taq* DNA polymerase buffer (diluted from 10× stock solution supplied with *Taq* DNA polymerase)
1.5 mM MgCl$_2$ (diluted from stock solution supplied with *Taq* DNA polymerase)
200 μM each dNTP

0.2 μM each primer
Initial denaturing step: 94°, 5 min
30 cycles: 94°, 1 min; 56°, 1 min; 72°, 1 min
Final extension: 72°, 5 min.

These conditions have been optimized slightly since the original publication. See also Henckel *et al.* (1999) for an alternative touchdown program.

pmoA-*Specific PCR*

A189 5′-GGNGACTGGGACTTCTGG-3′
mb661r 5′-CCGGMGCAACGTCYTTACC-3′

This primer set targets *pmoA* genes (Costello and Lidstrom, 1999). The reverse primer is specific to *pmoA* and mostly eliminates the amplification of *amoA* targets obtained with the A189/A682 primers. These primers are poor for targeting the upland soil *pmoA* gene clade potentially associated with novel high-affinity methane oxidizers; refer to Bourne *et al.* (2001) and Kolb *et al.* (2003) for primers more useful in targeting these genes.

30 μl total volume containing:
DNA template
2.5 U *Taq* DNA polymerase; added after initial denaturing step
1× *Taq* DNA polymerase buffer
1.5 mM MgCl$_2$
200 μM each dNTP
0.33 μM each primer
25 cycles: 92°, 1 min; 55°, 1.5 min; 72°, 1 min
Final extension: 72°, 5 min.

mmoX *PCR*

mmoXA 5′-ACCAAGGARCARTTCAAG-3′
mmoXD 5′-CCGATCCAGATDCCRCCCCA-3′

These primers were used to amplify *mmoX* from lake sediment and were designed using the available *mmoX* sequences in the database (Auman and Lidstrom, 2002; Auman *et al.*, 2000).

30 μl total volume containing:
DNA template
2.5 U *Taq* DNA polymerase; added after initial denaturing step
1× *Taq* DNA polymerase buffer
1.5 mM MgCl$_2$
167 μM each dNTP

0.33 μM each primer
Initial denaturing step: 94°, 30 s
30 cycles: 92°, 1 min; 60°, 1 min; 72°, 1 min
Final extension: 72°, 5 min

mmoX206f 5'-ATCGCBAARGAATAYGCSCG-3'
mmoX886r 5'-ACCCANGGCTCGACYTTGAA-3'.

These primers amplify a segment of *mmoX* and were designed based on the available *mmoX* sequences in the database (Hutchens *et al.*, 2004; Lin *et al.*, 2004). They amplify the region encoding the proposed active site of the enzyme (Elango *et al.*, 1997).

50 μl total volume containing:
DNA template
2.5 U *Taq* DNA polymerase; added after initial denaturing step
1× *Taq* DNA polymerase buffer
1.5 mM MgCl$_2$
200 μM each dNTP
0.2 μM each primer
Initial denaturing step: 94°, 5 min
30 cycles: 94°, 1 min; 60°, 1 min; 72°, 1 min
Final extension: 72°, 10 min.

mxaF *PCR*

mxaf1003 5'-GCGGCACCAACTGGGGCTGGT -3'
mxar1561 5'-GGGCAGCATGAAGGGCTCCC-3'

These primers target the *mxaF* from all gram-negative methylotrophs (McDonald and Murrell, 1997; McDonald *et al.*, 1995).

50 μl total volume containing:
DNA template
2.5 U *Taq* DNA polymerase; added after initial denaturing step
1× *Taq* DNA polymerase buffer
1.5 mM MgCl$_2$
200 μM each dNTP
0.5 μM each primer
Initial denaturing step: 94°, 5 min
30 cycles: 94°, 1 min; 55°, 1 min; 72°, 1 min
Final extension: 72°, 5 min.

These conditions have been optimized slightly since the original publication. See also Henckel *et al.* (1999) for an alternative touchdown program.

Analysis of PCR Products

PCR products are first cloned into a cloning vector (e.g., TA cloning kit, Invitrogen). A number of clones are isolated, usually 50–100, to create a clone library. The coverage of gene libraries can be assessed by rarefaction analysis (Hughes and Hellmann, 2005).

Restriction Fragment Length Polymorphism (RFLP). Clones are often screened using RFLP analysis by restriction digestion and agarose gel electrophoresis. The appropriate enzyme to release the insert from the TA cloning vector can be included in the digestion mix (*Eco*RI for pCR2.1, Invitrogen); alternatively, the insert can be reamplified before RFLP analysis using a small amount of the clone biomass as template in PCR; this eliminates the problem of vector DNA bands obscuring the clone insert bands. Restriction enzymes that can distinguish phylogenetic groups of clones can be determined *in silico*; examples for *pmoA*, *mmoX*, and *mxaF* are provided in Table II. A quantity of DNA that can be visualized clearly following electrophoresis and ethidium bromide staining should be used and may need to be optimized. Clones are grouped into operational taxonomic units (OTUs) or sequence types based on the banding pattern. Normally, a minimum of one clone from each OTU containing >1 representative is sequenced.

TABLE II
ENZYMES USED FOR RFLP OF CLONES

Primer	Gene target	Restriction enzyme(s)	Reference
A189/A682	*pmoA/*	*Rsa*I	
	amoA	*Hinc*II/*Pvu*II	Radajewski *et al.* (2002)
A189/mb661	*pmoA*	*Hha*I	
		*Msp*I/*Hae*III	Costello and Lidstrom (1999)
mmoXA/	*mmoX*	*Alu*I/*Hha*I	
mmoXD		*Msp*I /*Hae*III	Auman and Lidstrom (2002)
mmoX206F/	*mmoX*	*Ava*II	
mmoX886R		*Rsa*I	Hutchens *et al.* (2004)
mxaF1003/	*mxaF*	*Hinc*II	Radajewski *et al.* (2002)
mxaF1561			

Example analysis; RFLP of A189/A682 clones:

1. The *pmoA* gene is reamplified from the clone by introducing, with a sterile toothpick, a small amount of a colony by dipping it into the PCR mix (see earlier discussion). The product is reamplified using the original PCR conditions.
2. Add approximately 100 ng of the PCR product to tube 1 containing 1 μl of 10× react 1 buffer (Invitrogen) and 1 μl *Rsa*I enzyme in a final volume of 10 μl.
3. Add approximately 100 ng of the PCR product to tube 2 containing 1 μl of 10× react 4 buffer (Invitrogen), 0.5 μl *Hinc*II, and 0.5 μl *Pvu*II in a final volume of 10 μl.
4. Incubate for 1 h at 37°.
5. Resolve the DNA fragments by electrophoresis through a 2% (w/vl) agarose gel containing a DNA size marker; alternatively, a 3% (w/v) NuSieve GTG agarose (FMC) gel can be used (Auman and Lidstrom, 2002; Costello and Lidstrom, 1999).

Alternative Approaches for PCR Product Analysis. The analysis of a clone library is the simplest method of screening for diversity in a PCR library. Alternative approaches include denaturing gradient gel electrophoresis (DGGE), capillary electrophoresis (Semrau and Han, 2005), and terminal restriction fragment length polymorphism (T-RFLP) (Marsh, 2005). A T-RFLP method for the analysis of *pmoA* libraries was developed by Horz *et al.* (2001, 2002). T-RFLP is rapid, sensitive, and avoids the potentially biased cloning reaction. DGGE analysis is also rapid to perform once the conditions have been optimized. DGGE of *pmoA* and *mxaF* PCR products has been described (Bourne *et al.*, 2001; Fjellbirkeland *et al.*, 2001; Henckel *et al.*, 1999; Steinkamp *et al.*, 2001). Microarray technology is the most sophisticated approach to assessing gene diversity in a sample and has been developed and demonstrated for *pmoA* genes (Bodrossy and Sessitsch, 2004; Bodrossy *et al.*, 2003).

References

Anthony, C. (1982). The bacterial oxidation of methane, methanol, formaldehyde and formate. *In* "The Biochemistry of the Methylotrophs," pp. 152–194. Academic Press, London.

Anthony, C., and Dales, S. L. (1996). The biochemistry of methanol dehydrogenase. *In* "Microbial Growth on C1 Compounds" (M. E. Lidstrom and F. R. Tabita, eds.), pp. 213–219. Kluwer Academic Publishers, The Netherlands.

Auman, A. J., and Lidstrom, M. E. (2002). Analysis of sMMO-containing type I methanotrophs in Lake Washington sediment. *Environ. Microbiol.* **4,** 517–524.

Auman, A. J., Stolyar, S., Costello, A. M., and Lidstrom, M. E. (2000). Molecular characterization of methanotrophic isolates from freshwater lake sediment. *Appl. Environ. Microbiol.* **66,** 5259–5266.

Bodrossy, L., and Sessitsch, A. (2004). Oligonucleotide microarrays in microbial diagnostics. *Curr. Opin. Microbiol.* **7,** 245–254.

Bodrossy, L., Stralis, N., Murrell, J. C., Radajewski, S., Weilharter, A., and Sessitsch, A. (2003). Development and validation of a diagnostic microbial microarray for methanotrophs. *Environ. Microbiol.* **5,** 566–582.

Bourne, D. G., McDonald, I. R., and Murrell, J. C. (2001). Comparison of *pmoA* PCR primer sets as tools for investigating methanotroph diversity in three Danish soils. *Appl. Environ. Microbiol.* **67,** 3802–3809.

Bowman, J. (2000). The Methanotrophs: The families *Methylococcaceae* and *Methylocystaceae*. *In* "The Prokaryotes." Springer Verlag, New York.

Bowman, J. P., Sly, L. I., Nichols, P. D., and Hayward, A. C. (1993). Revised taxonomy of the methanotrophs: Description of *Methylobacter* gen. nov., emendation of *Methylococcus*, validation of *Methylosinus* and *Methylocystis* species, and a proposal that the family *Methylococcaceae* includes only the group I methanotrophs. *Int. J. Syst. Bacteriol.* **43,** 735–753.

Cheng, Y. S., Halsey, J. L., Fode, K. A., Remsen, C. C., and Collins, M. L. P. (1999). Detection of methanotrophs in groundwater by PCR. *Appl. Environ. Microbiol.* **65,** 648–651.

Conrad, R. (1996). Soil microorganisms as controllers of atmospheric trace gases (H_2, CO, CH_4, OCS, N_2O, and NO). *Microbiol. Rev.* **60,** 609–640.

Costello, A. M., and Lidstrom, M. E. (1999). Molecular characterization of functional and phylogenetic genes from natural populations of methanotrophs in lake sediments. *Appl. Environ. Microbiol.* **65,** 5066–5074.

Dedysh, S. N., Khmelenina, V. N., Suzina, N. E., Trotsenko, Y. A., Semrau, J. D., Liesack, W., and Tiedje, J. M. (2002). *Methylocapsa acidiphila* gen. nov., sp. nov., a novel methane-oxidizing and dinitrogen-fixing acidophilic bacterium from Sphagnum bog. *Int. J. Syst. Bacteriol.* **52,** 251–261.

Dedysh, S. N., Liesack, W., Khmelenina, V. N., Suzina, N. E., Trotsenko, Y. A., Semrau, J. D., Bares, A. M., Panikov, N. S., and Tiedje, J. M. (2000). *Methylocella palustris* gen. nov., sp. nov., a new methane-oxidizing acidophilic bacterium from peat bogs, representing a novel subtype of serine-pathway methanotrophs. *Int. J. Syst. Bacteriol.* **50,** 955–969.

Dunfield, P. F., Yimga, M. T., Dedysh, S. N., Berger, U., Liesack, W., and Heyer, J. (2002). Isolation of a *Methylocystis* strain containing a novel *pmoA*-like gene. *FEMS Microbiol. Ecol.* **41,** 17–26.

Elango, N., Radhakrishnan, R., Froland, W. A., Wallar, B. J., Earhart, C. A., Lipscomb, J. D., and Ohlendorf, D. H. (1997). Crystal structure of the hydroxylase component of methane monooxygenase from *Methylosinus trichosporium* OB3b. *Protein Sci.* **6,** 556–568.

Fjellbirkeland, A., Torsvik, V., and Ovreas, L. (2001). Methanotrophic diversity in an agricultural soil as evaluated by denaturing gradient gel electrophoresis profiles of *pmoA*, *mxaF* and 16S rDNA sequences. *Antonie van Leeuwenhoek* **79,** 209–217.

Fuse, H., Ohta, M., Takimura, O., Murakami, K., Inoue, H., Yamaoka, Y., Oclarit, J. M., and Omori, T. (1998). Oxidation of trichloroethylene and dimethyl sulfide by a marine *Methylomicrobium* strain containing soluble methane monooxygenase. *Biosci. Biotechnol. Biochem.* **62,** 1925–1931.

Han, J. I., and Semrau, J. D. (2004). Quantification of gene expression in methanotrophs by competitive reverse transcription-polymerase chain reaction. *Environ. Microbiol.* **6,** 388–399.

Hanson, J. L., and Hanson, T. E. (1996). Methanotrophic bacteria. *Microbiol. Rev.* **60,** 439–471.

Henckel, T., Friedrich, M., and Conrad, R. (1999). Molecular analyses of the methane-oxidizing microbial community in rice field soil by targeting the genes of the 16S rRNA, particulate methane monooxygenase, and methanol dehydrogenase. *Appl. Environ. Microbiol.* **65,** 1980–1990.

Henckel, T., Jackel, U., Schnell, S., and Conrad, R. (2000). Molecular analyses of novel methanotrophic communities in forest soil that oxidize atmospheric methane. *Appl. Environ. Microbiol.* **66,** 1801–1808.

Heyer, J., Galchenko, V. F., and Dunfield, P. F. (2002). Molecular phylogeny of type II methane-oxidizing bacteria isolated from various environments. *Microbiology* **148,** 2831–2846.

Holmes, A. J., Costello, A. M., Lidstrom, M. E., and Murrell, J. C. (1995). Evidence that particulate methane monooxygenase and ammonia monooxygenase may be evolutionarily related. *FEMS Microbiol. Lett.* **132,** 203–208.

Holmes, A. J., Roslev, P., McDonald, I. R., Iversen, N., Henriksen, K., and Murrell, J. C. (1999). Characterization of methanotrophic bacterial populations in soils showing atmospheric methane uptake. *Appl. Environ. Microbiol.* **65,** 3312–3318.

Horz, H. P., Raghubanshi, A. S., Heyer, J., Kammann, C., Conrad, R., and Dunfield, P. F. (2002). Activity and community structure of methane-oxidising bacteria in a wet meadow soil. *FEMS Microbiol. Ecol.* **41,** 247–257.

Horz, H. P., Yimga, M. T., and Liesack, W. (2001). Detection of methanotroph diversity on roots of submerged rice plants by molecular retrieval of *pmoA, mmoX, mxaF,* and 16S rRNA and ribosomal DNA, including *pmoA*-based terminal restriction fragment length polymorphism profiling. *Appl. Environ. Microbiol.* **67,** 4177–4185.

Hughes, J. B., and Hellmann, J. J. (2005). The application of rarefaction techniques to molecular inventories of microbial diversity. *Methods Enzymol.* **397,** 292–308.

Hutchens, E., Radajewski, S., Dumont, M. G., McDonald, I. R., and Murrell, J. C. (2004). Analysis of methanotrophic bacteria in Movile Cave by stable isotope probing. *Environ. Microbiol.* **6,** 111–120.

Kalyuzhnaya, M. G., and Chistoserdova, L. (2005). Community-level analysis: Genes encoding methanopterin dependent activities. *Methods Enzymol.* **397**(27), this volume (2005).

Kolb, S., Knief, C., Stubner, S., and Conrad, R. (2003). Quantitative detection of methanotrophs in soil by novel *pmoA*-targeted real-time PCR assays. *Appl. Environ. Microbiol.* **69,** 2423–2429.

Lane, D. J. (1991). 16S/23S rRNA sequencing. *In* "Nucleic Acid Techniques in Bacterial Systematics" (E. Stackebrandt and M. Goodfellow, eds.), pp. 115–175. Wiley, Chichester, UK.

Lidstrom, M. E. (2001). Aerobic methylotrophic prokaryotes. *In* "The Prokaryotes" (E. Stackebrandt, ed.). Springer Verlag, New York.

Lin, J.-L., Radajewski, S., Eshinimaev, B. T., Trotsenko, Y. A., McDonald, I. R., and Murrell, J. C. (2004). Molecular diversity of methanotrophs in Transbaikal soda lake sediments and identification of potentially active populations by stable isotope probing. *Environ. Microbiol.* **6,** 1049–1060.

Marsh, T. L. (2005). Culture-independent microbial community analysis with terminal restriction fragment length polymorphism. *Methods Enzymol.* **397,** 308–329.

McDonald, I. R., Kenna, E. M., and Murrell, J. C. (1995). Detection of methanotrophic bacteria in environmental samples with the PCR. *Appl. Environ. Microbiol.* **61,** 116–121.

McDonald, I. R., and Murrell, J. C. (1997). The methanol dehydrogenase structural gene *mxaF* and its use as a functional gene probe for methanotrophs and methylotrophs. *Appl. Environ. Microbiol.* **63,** 3218–3224.

Miguez, C. B., Bourque, D., Sealy, J. A., Greer, C. W., and Groleau, D. (1997). Detection and isolation of methanotrophic bacteria possessing soluble methane monooxygenase (sMMO) genes using the polymerase chain reaction (PCR). *Microbial Ecol.* **33**, 21–31.

Morris, S. A., Radajewski, S., Willison, T. W., and Murrell, J. C. (2002). Identification of the functionally active methanotroph population in a peat soil microcosm by stable-isotope probing. *Appl. Environ. Microbiol.* **68**, 1446–1453.

Murrell, J. C., Gilbert, B., and McDonald, I. R. (2000). Molecular biology and regulation of methane monooxygenase. *Arch. Microbiol.* **173**, 325–332.

Purdy, K. (2005). Nucleic acid recovery from complex environmental samples. *Meth. Enzymol.* **397**, 271–292.

Radajewski, S., Webster, G., Reay, D. S., Morris, S. A., Ineson, P., Nedwell, D. B., Prosser, J. I., and Murrell, J. C. (2002). Identification of active methylotroph populations in an acidic forest soil by stable-isotope probing. *Microbiology* **148**, 2331–2342.

Sambrook, J., and Russell, D. W. (2001). "Molecular Cloning: A Laboratory Manual." Cold Spring Harbor Laboratory Press, Cold Spring Harbor, NY.

Selenska, S., and Klingmuller, W. (1991). DNA recovery and direct detection of Tn5 sequences from soil. *Lett. Appl. Microbiol.* **13**, 21–24.

Semrau, J. D., and Han, J.-I. (2005). Quantitative community analysis: Capillary electrophoresis techniques. *Methods Enzymol.* **397**, 329–338.

Shigematsu, T., Hanada, S., Eguchi, M., Kamagata, Y., Kanagawa, T., and Kurane, R. (1999). Soluble methane monooxygenase gene clusters from trichloroethylene-degrading *Methylomonas* sp. strains and detection of methanotrophs during *in situ* bioremediation. *Appl. Environ. Microbiol.* **65**, 5198–5206.

Steinkamp, R., Zimmer, W., and Papen, H. (2001). Improved method for detection of methanotrophic bacteria in forest soils by PCR. *Curr. Microbiol.* **42**, 316–322.

Tchawa Yimga, M., Dunfield, P. F., Ricke, P., Heyer, J., and Liesack, W. (2003). Wide distribution of a novel *pmoA*-like gene copy among type II methanotrophs, and its expression in *Methylocystis* strain SC2. *Appl. Environ. Microbiol.* **69**, 5593–5602.

Trotsenko, Y. A., and Khmelenina, V. N. (2002). Biology of extremophilic and extremotolerant methanotrophs. *Arch. Microbiol.* **177**, 123–131.

Yeates, C., and Gillings, M. R. (1998). Rapid purification of DNA from soil for molecular biodiversity analysis. *Lett. Appl. Microbiol.* **27**, 49–53.

Zhou, J., Bruns, M. A., and Tiedje, J. M. (1996). DNA recovery from soils of diverse composition. *Appl. Environ. Microbiol.* **62**, 316–322.

[26] Methyl-Coenzyme M Reductase Genes: Unique Functional Markers for Methanogenic and Anaerobic Methane-Oxidizing *Archaea*

By Michael W. Friedrich

Abstract

In many anoxic environments, methanogenesis is the predominant terminal electron accepting process involved in the mineralization of organic matter, which is catalyzed by methanogenic *Archaea*. These organisms represent a unique but phylogenetically diverse guild of prokaryotes, which can be conveniently tracked in the environment by targeting the *mcrA* gene as a functional marker. This gene encodes the α subunit of the methyl-coenzyme M reductase (MCR), which catalyzes the last step in methanogenesis and is present in all methanogens. Cultivation-independent analysis of methanogenic communities involves the polymerase chain reaction (PCR) amplification of the *mcrA* gene from extracted community DNA, comparative analysis of *mcrA* clone libraries, or PCR-based fingerprinting analysis by terminal restriction fragment polymorphism analysis (T-RFLP). It has also been suggested that anaerobic methane-oxidizing *Archaea* possess MCR, which facilitates detection of this novel group of "reverse methanogens" as well using the *mcrA* gene as a functional marker.

Introduction

In many anoxic environments, methanogenesis is the key terminal biogeochemical process involved in the mineralization of organic matter, which is catalyzed by methanogenic *Archaea*. These microorganisms are a unique guild that share the capability to form methane from a small range of substrates, such as acetate, CO_2 and H_2, and a few other C1 compounds (e.g., formate, methanol, methylamines, methylthiols) as part of their energy metabolism (Thauer, 1998). Methanogens inhabit a large variety of environments, such as wetlands, sediments, digesters, geothermal springs, and hydrothermal vent sites, as well as the digestive tract of animals (Garcia *et al.*, 2000). The phylogenetic diversity, distribution, and dynamics of methanogenic *Archaea* in these environments have been studied intensively over the last decade, with cultivation-independent molecular methods becoming increasingly more important, as this approach allows

METHODS IN ENZYMOLOGY, VOL. 397
0076-6879/05 $35.00
DOI: 10.1016/S0076-6879(05)97026-2

bypassing the inevitable bias involved in the cultivation of microorganisms (Amann *et al.*, 1995). Many studies have relied on the 16S rRNA or its gene as a molecular marker, which has increased our understanding of the ecology of methanogens in their natural habitats markedly. However, the identification of uncultivated methanogens based on 16S rRNA (or its gene) as a marker is generally limited by the fact that methanogenic *Archaea* are not monophyletically (Barns *et al.*, 1996); rather, methanogens form several different major lines of descent within the kingdom Euryarchaeota, some of which are interspersed by lines of descent harboring nonmethanogenic *Archaea* only (Lueders *et al.*, 2001). Currently, five different orders of methanogens have been recognized: Methanosarcinales, Methanomicrobiales, Methanococcales, Methanobacterales, and Methanopyrales (Boone *et al.*, 1993; Garcia *et al.*, 2000).

Therefore, methanogens may be targeted much more efficiently as a physiological coherent guild in molecular ecological analyses by using a specific functional marker gene. Methyl-coenzyme M reductase (EC 2.8.4.1) is the key enzyme of methanogenesis, which catalyzes the final step in methanogenesis, the reduction of the coenzyme M-bound methyl group to methane (Thauer, 1998). In fact, this enzyme appears to be unique to methanogens, whereas other enzymes involved in methanogenesis (e.g., methylene tetrahydromethanopterin dehydrogenase, methenyl tetrahydromethanopterin cyclohydrolase) occur in another guild of C1-utilizing microorganisms as well, the aerobic methanotrophic bacteria (Chistoserdova *et al.*, 1998). Two forms of MCR exist, MCR-I (encoded by the *mcrBDCGA* operon), which is present in all methanogens, and isoenzyme MCR-II (encoded by the *mrtBDGA* operon), which is additionally present in members of the Methanococcales and the Methanobacteriales only (Lueders *et al.*, 2001).

Genes encoding the α subunit of MCR, *mcrA* and *mrtA*, are evolutionarily highly conserved, probably due to functional constraints on the catalytic activity of MCR (Hallam *et al.*, 2003). Comparative phylogenetic studies have clearly shown that the topologies of 16S rRNA-based and *mcrA/mrtA* (or derived amino acid McrA/MrtA)-based trees are largely consistent (Lueders *et al.*, 2001; Springer *et al.*, 1995) (see Fig. 1 for an McrA/MrtA tree). This is an important prerequisite to identifying methanogens in environmental samples by comparative *mcrA/mrtA* sequence analysis.

Based on the conserved sequence of *mcrA* genes, degenerate primers were developed (Hales *et al.*, 1996; Luton *et al.*, 2002; Springer *et al.*, 1995), which allowed to amplify and retrieve environmental *mcrA* sequences using a cloning/sequencing approach from a variety of environments, such as termite guts (Ohkuma *et al.*, 1995), peat bogs (Edwards *et al.*, 1998;

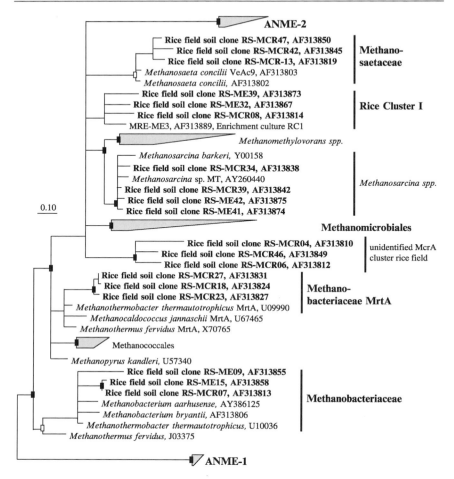

Fig. 1. Phylogenetic consensus tree (based on distance matrix analysis) of deduced partial McrA and MrtA sequences showing the relationships of major lineages of methanogenic Archaea, anaerobic methane-oxidizing *Archaea* (ANME-1 and ANME-2), and selected uncultured methanogens retrieved from rice field soil (Lueders *et al.*, 2001). Rice field soil clones (in bold) retrieved from library RS-MCR and RS-ME were generated using the MCR and ME primer sets, respectively (Table I). Sequences are McrA, unless MrtA is indicated. The tree was reconstructed from distance matrix using neighbor joining. Polytomic nodes indicate branches for which the relative branching order could not be determined unambiguously using neighbor joining, maximum parsimony, and maximum likelihood methods [Tree puzzle, Whelan–Goldman model of amino acid substitution (Whelan and Goldman, 2001); PHYML]. Parsimony bootstrap values for branches are indicated by solid (>90%) or open squares (75 to 90%). Branches without squares had bootstrap values of less than 75%. The scale bar represents 10% sequence divergence.

Lloyd *et al.*, 1998; Nercessian *et al.*, 1999), marine sediments (Bidle *et al.*, 1999), landfill (Luton *et al.*, 2002), rice field soil (Chin *et al.*, 2004; Lueders *et al.*, 2001; Ramakrishnan *et al.*, 2001), and riparian soil (Kemnitz *et al.*, 2004). It was only recognized recently that MCR-targeting primer sets also amplify the *mrtA* gene (Lueders *et al.*, 2001), which has important consequences for the analysis of environmentally derived clone libraries. Because members of the Methanobacterales and Methanococcales contain both *mcrA* and *mrtA* genes (see Fig. 1), they will be represented in clone libraries with both copies, which must be considered; careful analysis of phylogenetic relationships of *mcrA/mrtA* clones detected may help avoid overrepresentation of these organisms when estimating clone frequencies. Moreover, environmental *mrtA* sequences may belong to a member of the Methanobacterales rather than to the closely related methanococcal species (Lueders *et al.*, 2001).

Using *mcrA* as a functional marker in fact has facilitated identifying novel methanogens. Among several novel euryarchaeotal lineages represented only by environmental sequences in rice field soil, we were able to identify the Rice Cluster I *Archaea* (Fig. 1) as novel methanogens (Lueders *et al.*, 2001). *mcrA* gene sequences were obtained from samples containing *Archaea* (i.e., anaerobic methane-oxidizing *Archaea* belonging to sequence clusters ANME-1 and ANME-2; Fig. 1) presumably involved in anaerobic methane oxidation in a variety of marine sediments with high methane concentrations, e.g., methane seeps (Hallam *et al.*, 2003). This finding is remarkable as it demonstrates how molecular ecology can help identify microorganisms involved and unravel hitherto unknown pathways of important biogeochemical processes such as anaerobic methane oxidation. ANME organisms supposedly reverse the reactions of methanogenesis, including that catalyzed by MCR (Hallam *et al.*, 2003); the detection of a conspicuous nickel protein with a modified F_{430} cofactor in anaerobically methane-oxidizing microbial mats provides strong support for the involvement of MCR (Krüger *et al.*, 2003). Currently, ANME-1 microorganisms represent the most deeply branching *Archaea* in McrA/MrtA trees (Fig. 1).

Community analysis by cloning and sequencing of randomly selected clones is time-consuming and may be subject to cloning-inherent bias (von Wintzingerode *et al.*, 1997). Fingerprinting methods, such as terminal restriction fragment length polymorphism (T-RFLP) (Liu *et al.*, 1997), are much more rapid and allow analyzing multiple samples in parallel, which is often necessary in ecological research. For example, following microbial population dynamics or distribution of populations over environmental gradients typically involves sample numbers not manageable by a cloning/sequencing approach. We have developed a T-RFLP method for environmental *mcrA/mrtA* gene fragments (Lueders *et al.*, 2001) and applied this method to the analysis of community structure and dynamics of methanogenic *Archaea* in rice rice field soil (Lueders *et al.*, 2001; Ramakrishnan *et al.*, 2001). Individual terminal restriction fragments (T-RFs) can be identified and assigned to clone sequences if a clone library of *mcrA/mrtA* genes is established and analyzed in parallel from the same sample (see Fig. 2 as an example) (Lueders *et al.*, 2001).

This chapter focuses on analyzing methanogenic communities with the functional marker gene *mcrA/mrtA*, including a discussion of experimental procedures and pitfalls.

Practical Example

Analysis of the methanogenic community in an Italian rice field soil, reported previously (Lueders *et al.*, 2001), is used here as a practical example to illustrate the utility of the *mcrA/mrtA* gene as a functional marker for determining the composition of a methanogenic community. The *mcrA/mrtA* gene was amplified from total community DNA extracts with the MCR and ME primer sets (Table I), and two clone libraries, MCR and ME, were created using the respective PCR products. Fifty clones from each library were selected randomly, some of which were excluded from further analysis as chimers, and a large number of ME clones, because of a

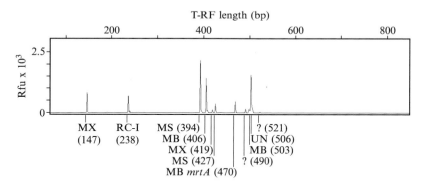

FIG. 2. T-RFLP analysis of *mcrA/mrtA* gene PCR products amplified from a rice field soil community (Lueders *et al.*, 2001). Sequences types were identified by *in silico* analysis of an *mcrA/mrtA* clone library established from the same sample and include distinct T-RFs representing members of the Methanosaetaceae (MX), Rice Cluster I (RC-I), Methanosarcinaceae (MS), Methanobacteriaceae (MB), the *mrtA* gene of Methanobacteriaceae (MB *mrtA*), an unidentified sequence type (UN), and a sequence type not represented in the clone library ("?"). T-RF length of peaks as indicated in parentheses. RFU, relative fluorescence units. From Lueders *et al.* (2001) with permission from Blackwell Science.

TABLE I

PRIMER SETS FOR AMPLIFYING PARTIAL FRAGMENTS OF *mcrA/mrtA* GENES OF
KNOWN METHANOGENIC *ARCHAEA* (PRIMER SET LUTON-*mcrA*, ME, MCR) AND ANAEROBIC
METHANE-OXIDIZING *ARCHAEA* (PRIMER SET AOM)

Primer set	Primer sequence 5′ → 3′	PCR product (bp)	Reference
Luton-mcrA	Forward primer: GGT GGT GTM GGA TTC ACA CAR TAY GCW ACA GC Reverse primer: TTC ATT GCR TAG TTW GGR TAG TT	~414– 438	Luton *et al.* (2002)
ME	Forward primer: ME1 GCM ATG CAR ATH GGW ATG TC Reverse primer: ME2 TC ATK GCR TAG TTD GGR TAG T	~760	Hales *et al.* (1996)
MCR	Forward primer: MCRf TAY GAY CAR ATH TGG YT Reverse primer: MCRr ACR TTC ATN GCR TAR TT	~500	Springer *et al.* (1995)
AOM[a]	Forward primer: AOM39_F[b] GCT GTG TAG CAG GAG AGT CA Reverse primer: AOM40_R GAT TAT CAG GTC ACG CTC AC		Hallam *et al.* (2003)

[a] Reported to be specific for amplifying *mcrA* group b of ANME-1 *Archaea*; the primer set was used for the amplification of *mcrA* gene fragments from fosmid libraries according to Hallam *et al.* (2003); however, primer AOM39_F does not match currently published *mcrA* fosmid sequences of ANME1 *Archaea*.

[b] Primer ANME1b_F 5′-GCT GTG T G CAG GGG AGT CA-3′′ (this report) modified from primer AOM39_F (Hallam *et al.*, 2003) matches ANME1 *mcrA* group b sequences of fosmids with the accession numbers AY327048, AY714819, and AY71482; however, this primer has not been verified experimentally.

wrong insert size. Forty-five MCR clones and 30 ME clones were analyzed phylogenetically based on deduced amino acid sequences (selected clones, representing the major lineages detected, are shown in Fig. 1). Clones from the MCR library fell into six distinct clusters of sequences representing *mcrA* genes of *Methanosarcina* spp., the Methanosaetaceae, the Methano-bacteriaceae, the uncultivated Rice Cluster I (RC-I) *Archaea*, and another cluster of unidentified McrA sequences of rice field *Archaea*, as well as sequences representing *mrtA* genes of the Methanobacteriaceae. Notably, uncultivated RC-I *Archaea* were identified in this study as methanogens by their unique presence in a highly enriched methanogenic culture. In the ME clone library, only clones representing *mcrA* sequences related to *Methanosarcina* spp., Methanosaetaceae, and Methanobacteriaceae were detected. In both libraries, representatives of the Methanomicrobiales were not detected, although these methanogens have been shown in rice field soil before (Großkopf *et al.*, 1998). A T-RFLP fingerprinting assay targeting the *mcrA/mrtA* gene was developed, and a community profile of methanogens in the same rice field soil sample was obtained (Fig. 2). Based on clone sequences of the *mcrA/mrtA* clone libraries, all clones were represented by terminal restriction fragments (T-RFs) as indicated in Fig. 2, including those that represent *mrtA* genes of the Methanobacteriaceae. In addition, two novel T-RFs were detected that were not represented in the clone libraries, however, with a small peak height only suggesting a small size of these populations.

Methodology

Nucleic Acid Extraction

Total community DNA may be extracted from suitable environmental samples following protocols available elsewhere [see Purdy (2005) or Lueders *et al.* (2004)]. Depending on the characteristics of the sample, extraction protocols have to be adjusted to allow for the recovery of sufficient amounts of PCR-amplifiable DNA. For various anoxic soil and sediment samples, the following protocol can be recommended, which involves the direct lysis of cells by bead beating in the presence of 1% sodium dodecyl sulfate (Henckel *et al.*, 1999). Wet soil (or sediment) up to a volume of 500 μl is transferred to a 2-ml screw-capped polyethylene reaction tube, which contains $\sim$1 g of oven-baked (170°, 4 h) zirconia beads (diameter, 0.1 mm; Biospec Products Inc., Bartlesville, OK). Subsequently, 750 μl of sodium phosphate buffer (120 mM; pH 8) and 250 μl of a sodium dodecyl sulfate solution (10% sodium dodecyl sulfate, 0.5 M Tris-HCl, pH 8.0, 0.1 M NaCl) are added to the tube before cells are lysed mechanically by shaking with a cell disintegrator (Model FP120 FastPrep; Savant Instruments Inc., Farmingdale, NY) for 45 s at a setting of 6.5 m s^{-1}; duration and speed setting on the cell disintegrator may have to be adjusted to reduce shear forces, which may allow for a larger size of the genomic DNA. Cell debris, soil, and beads are separated by centrifugation (3 min, 12,000g) from the supernatant, which is transferred to a new tube. Nucleic acids are extracted from the supernatant by consecutive treatments with equal volumes of phenol–chloroform–isoamyl alcohol [PCI, 25:24:1 (v/v/v)] and chloroform-isoamyl alcohol [CI, 24:1 (v:v), Sigma], followed by repeated centrifugations (4 min, 12,000g). Total nucleic acids are precipitated from the aqueous phase layer by adding two volumes of 30% (w/v) polyethelene glycol 6000 (Fluka BioChemika) in 1.6 M NaCl for 2 h at room temperature and centrifugation (30 min, 20,000g, 4°). Pelleted nucleic acids are washed once with ice-cold 70% (v/v) ethanol (−20°), centrifuged again (4 min, 12,000g), and air dried prior to resuspension in 50 to 100 μl of elution buffer (EB; 10 mM Tris-HCl, pH 8.5; Qiagen). RNA present in the total nucleic acid extract typically does not interfere with PCR amplification but may be removed by RNase treatment. To further remove PCR-inhibiting compounds (e.g., humics), the DNA extract may be passed over a spin column (Bio-Rad, Munich, Germany) filled with acid-washed polyvinylpolypyrrolidone (Sigma-Aldrich, Steinhein, Germany) (Berthelet et al., 1996; Egert et al., 2003). The size of the extracted DNA is determined by standard agarose gel electrophoresis of a 5-μl aliquot and documented after ethidium bromide staining. DNA concentrations of extracts are determined photometrically or, more precisely, with the PicoGreen double-stranded DNA quantification kit (Molecular Probes, Leiden, The Netherlands) in 96-well microtiter plates on a Fluorolite 1000 fluorescence microtiter plate reader (Dynatech Laboratories, Chantilly, VA).

PCR Amplification of mcrA/mrtA Gene Fragments

Fragments of the mcrA/mrtA gene can be amplified with primer sets Luton-MCR, MCR, or ME, which are assumed to target all methanogens, with the limitations described later. For amplification with MCR or ME primer sets, the following protocol applies: PCR reactions are set up in 200-μl reaction tubes with a total volume of a 50-μl reaction mixture, which contains 1× Masteramp PCR PreMix B (Epicentre Technologies), 0.5 μM each of one primer pair, 1.25 U of AmpliTaq DNA polymerase (Applera), and 1 μl of environmental DNA extract. The PCR PreMix B contains the PCR enhancer betaine, which has been shown to be effective in improving amplification yield and the specificity to amplify templates with a high G + C content of the DNA (Henke et al., 1997). Reactions should be kept at 4°, e.g., using a precooled aluminum block before placing reaction tubes directly in the preheated Gene Amp 9700 thermal cycler (Applied Biosystems) to avoid unspecific priming and amplification; alternatively, a hot start protocol can be used. The thermal protocol for amplification consists of an initial denaturation step (3 min, 94°) followed by 30–35 cycles of denaturation (45 s, 94°), annealing (45 s, 50°), and extension (90 s, 72°) and a final extension step (5 min, 72°). PCR products are checked for their correct size (see Table I) by analytical gel electrophoresis and staining of gels with ethidium bromide. PCR products are purified using MinElute PCR purification spin columns as described by the manufacturer (Qiagen), and the DNA concentration of PCR products is determined by UV photometry (Biophotometer, Eppendorf). General considerations for the amplification of genes from environmental nucleic acid extracts are available in Wagner *et al.* (2005).

T-RFLP Analysis

For T-RFLP analysis, *mcrA/mrtA* gene fragments are amplified using the protocol described earlier, except for using the MCR primer set with a 5′, 6-carboxyfluorescein (FAM)-labeled forward primer (Table I) (Lueders *et al.*, 2001). Restriction digests are carried out in a total volume of 10 μl in 200-μl reaction tubes. Reaction mixtures contain ~75 ng of purified PCR product, 3U of *Sau*96I (Promega), 1× 4-CORE buffer C (Promega), and 1 μg of bovine serum albumin and are incubated for 2 h at 37°; depending on the underlying diversity of the community, the amount of amplicon can be varied between 40 and 100 ng per digest. Digested amplicons (1.25 μl) are mixed with 1 μl of formamide loading dye (Amersham Pharmacia) and 0.25 μl of GeneScan-1000 (ROX) size standard (Applera) and analyzed on an ABI Prism 377 (Applied Biosystems) in GeneScan mode. A 36-cm-long 6% polyacrylamide gel (w/v) is prepared containing 8.3 M urea and 1× TBE (89 mM Tris-borate, 2 mM EDTA), PCR products are loaded, and fragments are separated electrophoretically for 14 h at 2500 V, 40 mA, and 30 W. Electropherograms are analyzed with Genescan analysis software (version 3.1, Applied Biosystems) using third-order least-squares size calling to calculate the length of terminal fragments (T-RFs) in relation to the internal standard GeneScan-1000 (ROX). The reproducibility of the method is high (Osborn *et al.*, 2000), and the error in size determination is typically <1 bp for the gel-based automated sequencers (ABI 373, ABI377) up to a fragment length of 300 bp.

Peaks <35 bp in size are generally discarded because these peaks can represent primer or primer–dimer artifacts. Signals with a peak height below 100 relative fluorescence units or with a peak area contribution below 1% are regarded as background noise and are excluded from analysis (Osborn *et al.*, 2000).

Typically, 16S rRNA-based T-RFLP analysis of community composition involves determining which T-RFs are present in a given sample and quantifying the relative amount of T-RFs based on peak area or heights. However, in case of the *mcrA/mrtA* gene, quantitative analysis with the current MCR primer system (Table I) cannot be recommended. We found that known *mcrA/mrtA* gene ratios cannot be recovered adequately by T-RFLP analysis of a defined template mixture representing a four-membered community (Lueders and Friedrich, 2003); some populations (*Methanobrevibacter bryanti, Methanococcus janaschii*) were overrepresented, whereas others (*Methanospirillum hungatei*) were underrepresented; however, 16S rRNA gene ratios were reflected adequately by T-RFLP analysis in the same study. Different *mcrA/mrtA* operon copy numbers, differences in the primer binding site of the species tested, and/or the use of the degenerate MCR primer set might cause quantitative PCR selection, which may result in an inadequate reflection of *mcrA/mrtA* gene ratios in PCR-T-RFLP analysis.

Nevertheless, communities of methanogenic *Archaea* can be characterized by comparing the absence and presence of T-RFs in individual T-RFLP profiles by statistical methods similar to those available for other community profiling methods [for an overview on T-RFLP data analysis, see Blackwood *et al.* (2003)]. A very simple approach for sample comparison based on the absence/presence of T-RFs involves calculation of the Sørensen coefficient (Buckley and Schmidt, 2001) or the Jaccard coefficient (Dunbar *et al.*, 2001).

A general limitation of T-RFLP is the lack of direct correlation of individual phylotypes (represented as T-RFs) with phylogenetic information retrieved, for example, by probing or recovery of sequence information. Also, T-RFs may represent many phylogenetically diverse microorganisms, which limits cross sample comparisons, especially when samples originate from different sites and habitats. Therefore, it is highly recommended to combine T-RFLP analysis with the analysis of *mcrA/mrtA* gene clone libraries (see next paragraph) to allow for the assignment of clone sequences to individual T-RFs (see practical example in Fig. 2). Further details on the T-RFLP method are available in Marsh (2005).

Cloning and Sequencing of mcrA/mrtA Gene Fragments

PCR-amplified *mcrA/mrtA* gene fragments from environmental samples are cloned in *Escherichia coli* JM 109 cells using a commercially available TA cloning kit (pGem T vector system II, Promega) following the manufacturer's instructions. Purified PCR products are ligated overnight at 4°, which usually yields a higher number of transformants. Randomly selected clones are checked for their correct insert size by vector-targeted PCR using primers M13F (5′-GAT AAA CGA CGG CCA G-3′) and M13B (5′-CAG GAA ACA GCT ATG AC-3′) according to the manufacturer's instruction and standard analytical agarose gel electrophoresis. Sequencing of M13 PCR products is performed using

BigDye terminator cycle sequencing chemistry on an ABI Prism 377 sequencer (Applied Biosystems).

A sufficient number of clones have to be analyzed to assess community structure, which can be estimated using statistical tools (Dunbar, 2004; Hughes and Bohannan, 2004).

Phylogenetic Analysis

Fragments of *mcrA/mrtA* genes retrieved from clone libraries or from pure cultures are analyzed phylogenetically by using either the gene sequences or the derived amino acid data. For data analysis, sequence alignment, and tree reconstruction, the ARB software package [http://www. arb-home.de (Ludwig *et al.*, 2004)], the Phylip package (Version 3.6; J. Felsenstein, University of Washington, *http://evolution.genetics. washington.edu/phylip.html*), and PHYML (*http://atgc.lirmm.fr/phyml/*; Guindon and Gascuel, 2003) are used. For all details regarding the use of the ARB program, the reader is referred to the documentation of the program.

Steps involved in phylogenetic analysis include the following.

1. Setting up a new ARB database. First, *mcrA/mrtA* reference sequences have to be imported from public databases (e.g., GenBank). As a guideline, a phylogenetic frame work representing the major lineages of methanogens known is required. In addition, BLAST searches can identify close relatives to sequence data to be analyzed, which have to be integrated into the ARB database as well.

2. Alignment. Gene sequences are converted to amino acid sequences using the appropriate reading frame. Subsequently, an alignment of McrA/MrtA sequences is established by using the Clustal W function built into ARB. If necessary, the alignment is corrected manually. Established amino acid alignments may be used to realign gene sequences accordingly using the appropriate function in ARB ("Realign nucleic acid according to aligned protein").

3. Phylogenetic framework. A phylogenetic framework of McrA/MrtA amino acid sequences is established using the longest reference sequences available. Several different algorithms are available in the ARB program, which can be used to generate a core tree (e.g., Fitch, neighbor joining, parsimony, maximum likelihood methods). For tree reconstruction, only alignment positions present in all data sets are used. First, a core tree is generated using distance matrix-based neighbor-joining analysis or Fitch analysis. Distance matrices are calculated using Protdist (Phylip 3.6) with the Dayhoff PAM matrix 001 as the amino acid replacement model. Tree reconstruction from distance matrices by neighbor joining (or Fitch)

(Phylip 3.6) uses randomization of species input order (random number seed 7, jumble seven times) and global rearrangement. Statistical support for tree reconstruction is obtained by bootstrap analysis using neighbor joining. In addition, the topology of the core tree is validated by using parsimony and maximum likelihood methods as implemented in ARB and PHYML.

4. Adding sequence data to a core tree. Deduced amino acid sequences are fitted into the alignment of the McrA database using the Fast aligner tool. Short sequences (<150 amino acids) can be inserted into an existing core tree using the parsimony tool of the ARB program (implemented in the newest version of the ARB program [030822] also for amino acid sequences). Longer sequences (>150 amino acids) may be added by calculating phylogenetic trees as described earlier using reference sequences present in the core tree as well as closest relatives available from public databases. Chimeric clones can be identified by fractional treeing (Ludwig et al., 1998), i.e., by separately adding the N-terminal and the C-terminal stretch of sequence to the core tree.

Interpretation of Community Composition

As with all PCR-based methods, the amplification of *mcrA/mrtA* gene fragments can be subject to molecular bias. Compared to rRNA genes, any PCR assay targeting a functional marker gene is probably more subject to bias due to the typically high degeneracy of the primers used, which results from the degeneracy of the genetic code, especially of the third codon position (see Table I for degeneracy of *mcrA/mrtA* gene-targeting primers). This may result in selective amplification of certain sequence types over others and, consequently, limits inferences on the composition of a microbial community. For example, we found that clone libraries based on the ME primer set (Table I; Hales et al., 1996) had a reduced diversity coverage of methanogens in rice field soil samples compared to a clone library established using the MCR primer set (Lueders et al., 2001; Springer et al., 1995); moreover, even when using the MCR primer set, members of the Methanomicrobiales proven to be present by 16S rRNA analysis were not detectable neither in clone libraries nor in T-RFLP profiles (Lueders et al., 2001). In addition to bias originating from qualitative PCR selection (Wagner et al., 1994) (presence or absence of a certain population), quantitative PCR selection (in the sense of frequency of certain amplicons after PCR relative to the frequency of target templates in the original sample) affects the amplification of *mcrA/mrtA* genes as well (Lueders and Friedrich, 2003) (see previous discussion). A novel *mcrA*-targeting primer set has been developed, but clone libraries established with this primer set also showed an overrepresentation of *mrtA* sequences of

Methanobacteriales over their *mcrA* counterpart (Luton *et al.*, 2002). Thus, any inferences about the community composition of *Archaea* based on *mcrA/mrtA* gene amplification should be made with due caution in view of the PCR- and cloning-inherent limitations of the methods.

Acknowledgments

The author thanks Tillmann Lüders and Bianca Pommerenke for their dedicated work on the development of novel molecular assays. The experimental background for this contribution was carried out in the author's laboratory and was supported by grants from the Deutsche Forschungsgemeinschaft (Bonn, Germany) and the Max Planck Society (Munich, Germany).

References

Amann, R. I., Ludwig, W., and Schleifer, K. H. (1995). Phylogenetic identification and *in situ* detection of individual microbial cells without cultivation. *Microbiol. Rev.* **59,** 143–169.

Barns, S. M., Delwiche, C. F., Palmer, J. D., and Pace, N. R. (1996). Perspectives on archaeal diversity, thermophily and monophyly from environmental rRNA sequences. *Proc. Natl. Acad. Sci. USA* **93,** 9188–9193.

Berthelet, M., Whyte, L. G., and Greer, C. W. (1996). Rapid, direct extraction of DNA from soils for PCR analysis using polyvinylpolypyrrolidone spin columns. *FEMS Microbiol. Lett.* **138,** 17–22.

Bidle, K. A., Kastner, M., and Bartlett, D. H. (1999). A phylogenetic analysis of microbial communities associated with methane hydrate containing marine fluids and sediments in the Cascadia margin (ODP site 892B). *FEMS Microbiol. Lett.* **177,** 101–108.

Blackwood, C. B., Marsh, T., Kim, S. H., and Paul, E. A. (2003). Terminal restriction fragment length polymorphism data analysis for quantitative comparison of microbial communities. *Appl. Environ. Microbiol.* **69,** 926–932.

Boone, D. R., Whitman, W. B., and Rouviere, P. E. (1993). Diversity and taxonomy of methanogens. *In* "Methanogenesis" (J. G. Ferry, ed.), pp. 35–80. Chapman & Hall, New York.

Buckley, D. H., and Schmidt, T. M. (2001). The structure of microbial communities in soil and the lasting impact of cultivation. *Microb. Ecol.* **42,** 11–21.

Chin, K. J., Lueders, T., Friedrich, M. W., Klose, M., and Conrad, R. (2004). Archaeal community structure and pathway of methane formation on rice roots. *Microb. Ecol.* **47,** 59–67.

Chistoserdova, L., Vorholt, J. A., Thauer, R. K., and Lidstrom, M. E. (1998). C-1 transfer enzymes and coenzymes linking methylotrophic bacteria and methanogenic archaea. *Science* **281,** 99–102.

Dunbar, J. (2004). Sampling efficiency and interpretation of diversity in 16S rRNA gene libraries. *In* "Molecular Microbial Ecology Manual" (G. A. Kowalchuk, F. J. De Bruijn, I. M. Head, A. D. Akkermans, and J. D. van Elsas, eds.), pp. 1345–1360. Kluwer, Dordrecht.

Dunbar, J., Ticknor, L. O., and Kuske, C. R. (2001). Phylogenetic specificity and reproducibility and new method for analysis of terminal restriction fragment profiles of 16S rRNA genes from bacterial communities. *Appl. Environ. Microbiol.* **67,** 190–197.

Edwards, C., Hales, B. A., Hall, G. H., McDonald, I. R., Murrell, J. C., Pickup, R., Ritchie, D. A., Saunders, J. R., Simon, B. M., and Upton, M. (1998). Microbiological processes in the terrestrial carbon cycle: Methane cycling in peat. *Atmosph. Environ.* **32,** 3247–3255.

Egert, M., Wagner, B., Lemke, T., Brune, A., and Friedrich, M. W. (2003). Microbial community structure in midgut and hindgut of the humus-feeding larva of *Pachnoda ephippiata* (Coleoptera : Scarabaeidae). *Appl. Environ. Microbiol.* **69,** 6659–6668.

Garcia, J. L., Patel, B. K. C., and Ollivier, B. (2000). Taxonomic phylogenetic and ecological diversity of methanogenic Archaea. *Anaerobe* **6,** 205–226.

Großkopf, R., Janssen, P. H., and Liesack, W. (1998). Diversity and structure of the methanogenic community in anoxic rice paddy soil microcosms as examined by cultivation and direct 16S rRNA gene sequence retrieval. *Appl. Environ. Microbiol.* **64,** 960–969.

Guindon, S., and Gascuel, O. (2003). A simple, fast, and accurate algorithm to estimate large phylogenies by maximum likelihood. *Syst. Biol.* **52,** 696–704.

Hales, B. A., Edwards, C., Ritchie, D. A., Hall, G., Pickup, R. W., and Saunders, J. R. (1996). Isolation and identification of methanogen-specific DNA from blanket bog peat by PCR amplification and sequence analysis. *Appl. Environ. Microbiol.* **62,** 668–675.

Hallam, S. J., Girguis, P. R., Preston, C. M., Richardson, P. M., and De Long, E. F. (2003). Identification of methyl coenzyme M reductase A (mcrA) genes associated with methane-oxidizing Archaea. *Appl. Environ. Microbiol.* **69,** 5483–5491.

Henckel, T., Friedrich, M., and Conrad, R. (1999). Molecular analyses of the methane-oxidizing microbial community in rice field soil by targeting the genes of the 16S rRNA, particulate methane monooxygenase, and methanol dehydrogenase. *Appl. Environ. Microbiol.* **65,** 1980–1990.

Henke, W., Herdel, K., Jung, K., Schnorr, D., and Loening, S. A. (1997). Betaine improves the PCR amplification of GC-rich DNA sequences. *Nucleic Acids Res.* **25,** 3957–3958.

Hughes, J. B., and Bohannan, B. J. M. (2004). Application of ecological diversity statistics in microbial ecology. *In* "Molecular Microbial Ecology Manual" (G. A. Kowalchuk, F. J. De Bruijn, I. M. Head, A. D. Akkermans, and J. D. van Elsas, eds.), pp. 1321–1344. Kluwer, Dordrecht.

Kemnitz, D., Chin, K. J., Bodelier, P., and Conrad, R. (2004). Community analysis of methanogenic archaea within a riparian flooding gradient. *Environ. Microbiol.* **6,** 449–461.

Krüger, M., Meyerdierks, A., Glöckner, F. O., Amann, R., Widdel, F., Kube, M., Reinhardt, R., Kahnt, R., Bocher, R., Thauer, R. K., and Shima, S. (2003). A conspicuous nickel protein in microbial mats that oxidize methane anaerobically. *Nature* **426,** 878–881.

Liu, W. T., Marsh, T. L., Cheng, H., and Forney, L. J. (1997). Characterization of microbial diversity by determining terminal restriction fragment length polymorphisms of genes encoding 16S rRNA. *Appl. Environ. Microbiol.* **63,** 4516–4522.

Lloyd, D., Thomas, K. L., Hayes, A., Hill, B., Hales, B. A., Edwards, C., Saunders, J. R., Ritchie, D. A., and Upton, M. (1998). Micro-ecology of peat: Minimally invasive analysis using confocal laser scanning microscopy, membrane inlet mass spectrometry and PCR amplification of methanogen-specific gene sequences. *FEMS Microbiol. Ecol.* **25,** 179–188.

Ludwig, W., Strunk, O., Klugbauer, S., Klugbauer, N., Weizenegger, M., Neumaier, J., Bachleitner, M., and Schleifer, K. H. (1998). Bacterial phylogeny based on comparative sequence analysis. *Electrophoresis* **19,** 554–568.

Ludwig, W., Strunk, O., Westram, R., Richter, L., Meier, H., Yadhukumar, Buchner, A., Lai, T., Steppi, S., Jobb, G., Forster, W., Brettske, I., Gerber, S., Ginhart, A. W., Gross, O., Grumann, S., Hermann, S., Jost, R., Konig, A., Liss, T., Lussmann, R., May, M., Nonhoff, B., Reichel, B., Strehlow, R., Stamatakis, A., Stuckmann, N., Vilbig, A., Lenke, M., Ludwig, T., Bode, A., and Schleifer, K. H. (2004). ARB: A software environment for sequence data. *Nucleic Acids Res.* **32,** 1363–1371.

Lueders, T., Chin, K. J., Conrad, R., and Friedrich, M. (2001). Molecular analyses of methyl-coenzyme M reductase-subunit (mcrA) genes in rice field soil and enrichment cultures reveal the methanogenic phenotype of a novel archaeal lineage. *Environ. Microbiol.* **3,** 194–204.

Lueders, T., and Friedrich, M. W. (2003). Evaluation of PCR amplification bias by terminal restriction fragment length polymorphism analysis of small-subunit rRNA and *mcrA* genes by using defined template mixtures of methanogenic pure cultures and soil DNA extracts. *Appl. Environ. Microbiol.* **69,** 320–326.

Lueders, T., Manefield, M., and Friedrich, M. W. (2004). Enhanced sensitivity of DNA- and rRNA-based stable isotope probing by fractionation and quantitative analysis of isopycnic centrifugation gradients. *Environ. Microbiol.* **6,** 73–78.

Luton, P. E., Wayne, J. M., Sharp, R. J., and Riley, P. W. (2002). The *mcrA* gene as an alternative to 16S rRNA in the phylogenetic analysis of methanogen populations in landfill. *Microbiology (Reading, England)* **148,** 3521–3530.

Marsh, T. L. (2005). Culture-independent microbial community analysis with terminal restriction fragment length polymorphism. *Methods Enzymol.* **397**[18] this volume (2005).

Nercessian, D., Upton, M., Lloyd, D., and Edwards, C. (1999). Phylogenetic analysis of peat bog methanogen populations. *FEMS Microbiol. Lett.* **173,** 425–429.

Ohkuma, M., Noda, S., Horikoshi, K., and Kudo, T. (1995). Phylogeny of symbiotic methanogens in the gut of the termite *Reticulitermes speratus*. *FEMS Microbiol. Lett.* **134,** 45–50.

Osborn, A. M., Moore, E. R. B., and Timmis, K. N. (2000). An evaluation of terminal-restriction fragment length polymorphism (T-RFLP) analysis for the study of microbial community structure and dynamics. *Environ. Microbiol.* **2,** 39–50.

Purdy, K. (2005). Nucleic acid recovery from complex environmental samples. *Methods Enzymol.* **397**[16] this volume (2005).

Ramakrishnan, B., Lueders, T., Dunfield, P. F., Conrad, R., and Friedrich, M. W. (2001). Archaeal community structures in rice soils from different geographical regions before and after initiation of methane production. *FEMS Microbiol. Ecol.* **37,** 175–186.

Springer, E., Sachs, M. S., Woese, C. R., and Boone, D. R. (1995). Partial gene sequences for the A subunit of methyl-coenzyme M reductase (*mcrI*) as a phylogenetic tool for the family *Methanosarcinaceae*. *Int. J. Syst. Bacteriol.* **45,** 554–559.

Thauer, R. K. (1998). Biochemistry of methanogenesis: A tribute to Marjory Stephenson. *Microbiology* **144,** 2377–2406.

von Wintzingerode, F., Goebel, U. B., and Stackebrandt, E. (1997). Determination of microbial diversity in environmental samples: Pitfalls of PCR-based rRNA analysis. *FEMS Microbiol. Rev.* **21,** 213–229.

Wagner, A., Blackstone, N., Cartwright, P., Dick, M., Misof, B., Snow, P., Wagner, G. P., Bartels, J., Murtha, M., and Pendleton, J. (1994). Surveys of gene families using polymerase chain-reaction; PCR selection and PCR drift. *Syst. Biol.* **43,** 250–261.

Wagner, M., Loy, A., Klein, M., Lee, N., Ramsing, N. B., Stahl, D. A., and Friedrich, M. W. (2005). Functional marker genes for identification of sulfate-reducing prokaryotes. *Methods Enzymol.* **397**[29] this volume (2005).

Whelan, S., and Goldman, N. (2001). A general empirical model of protein evolution derived from multiple protein families using a maximum-likelihood approach. *Mol. Biol. Evol.* **18,** 691–699.

[27] Community-Level Analysis: Genes Encoding Methanopterin-Dependent Enzymes

By Marina G. Kalyuzhnaya and Ludmila Chistoserdova

Abstract

This chapter describes a set of novel tools for the environmental detection of C_1 transfer functions linked to the cofactor methanopterin. These tools include degenerate environmental primers targeting four of the most conserved genes in the methanopterin-linked C_1 transfer pathway in bacteria, *fae*, *mtdB*, *mch*, and *fhcD*, and extensive databases of the respective genes. The tools described are suitable for detecting methanopterin-linked formaldehyde-oxidizing capacity in natural microbial communities and for determining the phylogenetic affiliations of major phyla involved in single-carbon cycling in the environment. The range of detection includes a variety of methano- and methylotrophic groups, other proteobacterial species capable of methanopterin-mediated reactions, and a variety of planctomycetes, as well as groups of microbes with currently unknown phylogenetic affiliations.

Introduction

The pathway for oxidation of formaldehyde involving tetrahydro-methanopterin (H_4MPT)-linked C_1 transfer reactions is one of the central functional metabolic modules in the utilization of single-carbon compounds by bacteria (Chistoserdova *et al.*, 1998; Vorholt *et al.*, 1999). While being a signature of methylotrophy, this pathway is also found in organisms not considered traditional methylotrophs, such as *Burkholderia* and *Rubrivivax* species, and in various groups of Planctomycetes (Chistoserdova *et al.*, 2004; Kalyuzhnaya *et al.*, 2004; Marx *et al.*, 2004; unpublished results). Evidence from Lidstrom and co-workers indicated that these latter bacterial groups may be involved in the co-metabolism or detoxification of C_1 compounds, thus potentially playing a role in local C_1 cycling (Kalyuzhnaya *et al.*, 2005a,b). The H_4MPT-linked formaldehyde oxidation pathway presents an attractive model for broad environmental detection of C_1 utilization functions due to its wide distribution and to the extreme sequence divergence of genes involved in the module, pointing to the long evolutionary history of this pathway (Chistoserdova *et al.*, 2004). A total of 17 genes specifically involved in the pathway are known. Seven of these are

METHODS IN ENZYMOLOGY, VOL. 397
0076-6879/05 $35.00
DOI: 10.1016/S0076-6879(05)97027-4

involved in the catalytic reactions (Fig. 1) and the rest are shown or hypothesized to participate in specific cofactor biosyntheses (Chistoserdova *et al.*, 2003, 2005). We have built databases of each of the 17 genes consisting of sequences representing different groups of methylotrophs, nonmethylotrophic proteobacteria, and Planctomycetes and tested every group of sequences for suitability in environmental primer design. We found that only 4 of the 17 genes, *fae*, *mtdB*, *mch* and *fhcD*, showed enough conservation to enable the design of degenerate polymerase chain reaction (PCR) primers that provide sufficient specificity for environmental detection. Primer design strategy, PCR amplification protocols, and results of testing on both pure culture and environmental DNA are described in this chapter.

Methods

Primer Design and Testing

Overview. The strategy for designing primers targeting highly divergent genes is essentially the same as that used for highly conserved genes, via DNA sequence alignment and detection of the most conserved sequence regions, using Vector NTI Suite 9 AlignX software (InforMax, Frederick, MD). The AlignX program is used to align the sequences, and the BLAST search program is used to test for primer specificity. In the case of highly divergent genes, however, such as H_4MPT-linked C_1 transfer genes, a challenge exists of most genes having so little conservation that even short regions of high sequence conservation could not be found. Of the 17 genes tested, only 4 genes, *fae*, *mtdB*, *mch*, and *fhcD*, showed stretches of sufficient sequence conservation to allow for the design of degenerate environmental primers (Kalyuzhnaya *et al.*, 2004, 2005a). To ensure

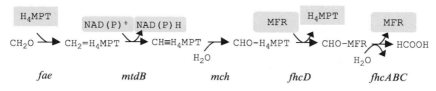

Known/Putative Cofactor Biosynthesis genes
mptG, orf5, orf7, orf9, orf17, orf19
orf20, orf21, orf22, orfY

FIG. 1. The H_4MPT-linked pathway for formaldehyde oxidation. *fae* encodes formaldehyde activating enzyme, *mtdB* encodes methylene-H_4MPT dehydrogenase, *mch* encodes methenyl-H_4MPT cyclohydrolase, and *fhcABCD* encode formyltransferase/hydrolase complex. For more details on enzymes, cofactors, and reactions, see Vorholt (2002) and Chistoserdova *et al.* (2003).

confident environmental detection, two sets of primers are sometimes required for use in sequential (two-step) PCR amplification protocols.

Primers Targeting fae, mch, mtdB, and fhcD. Two sets of primers targeting *fae* were designed based on the alignment of 13 bacterial *fae* sequences that included proteobacterial and planctomycete sequences. Two sets of primers targeting *mtdB* were designed based on the alignment of 15 proteobacterial *mtdB* sequences. One set of primers targeting *mch* has been designed previously based on one bacterial and a number of archaeal sequences (Vorholt *et al.*, 1999). The second set was designed based on the alignment of 27 proteobacterial *mch* sequences (Kaluyzhnaya *et al.*, 2004). A set of primers targeting *fhcD* was designed based on the alignment of 8 bacterial sequences that included proteobacterial and planctomycete sequences (Kalyuzhnaya *et al.*, 2005a). Primer sequences, the respective annealing temperatures, degrees of degeneracy, and amplified fragment sizes are summarized in Table I.

Primer Testing. The specific primer sets are tested on DNA isolated from pure cultures of various methylotrophic strains (*Methylobacterium, Methylosinus, Methylobacter, Methylobacillus,* etc.), nonmethylotrophic species possessing the genes of interest (*Burkholderia xenovorans* LB400; *Rubrivivax gelatinosus),* and representatives of Planctomycetes (*Gemmata, Pirellula, Isosphaera, Planctomyces*; Kalyuzhnaya *et al.*, 2005a). As a negative control, DNA from strains whose genomes are known not to contain genes for H_4MPT-linked reactions, such as *Escherichia coli* or *Deinococcus radiodurans,* is used. To ensure bacterial gene specificity, DNA from strains of methanogenic archaea, such as *Methanococcus maripaludis,* is used as a negative control.

Cycling Protocol. PCR amplifications are carried out in a total volume of 20 μl containing 0.1 μg of template DNA, 200 μM dNTPs, 1.5 mM $MgCl_2$, 5% dimethyl sulfoxide, 0.2 μM primers, and 0.5 U of *Taq*-DNA polymerase (Invitrogen). An initial denaturation of the reaction components at 92° for 4 min is followed by 24 cycles of amplification: 92° for 40 s, the annealing temperature indicated in Table I for 40 s, and 72° for 90 s. This is followed by a 5-min incubation at 72°.

Note. Group-specific, genus-specific, or species-specific primers with a more narrow range of detection may be designed in a similar fashion (Kalyuzhnaya *et al.*, 2004).

Environmental PCR Amplification

DNA Isolation. Total DNA is isolated as described by Kalyuzhnaya *et al.* (2004) or by using the Ultra Clean Soil DNA extraction kit (MOBIO Lab. Inc.). These two methods result in DNA preparations of high quality

TABLE I
PCR PRIMERS USED IN THIS STUDY

Primer name	Primer sequence	Nucleotides[a]	T_m[b]($°C$)	Degeneracy	Fragment size (bp)
fae1f	5′-gtcggcgacggc aaygargtcg-3′	25–46	45	4	380
fae1r	5′-gtagttgwanty ctggatctt-3′	385–405		16	
fae2f	5′-gcacacatcgacc tsatcatsgg-3′	61–83	56	4	305
fae2r	5′-ccagtgratgaav acgccrac-3′	346–366		12	
mtdB1f	5′-ccgttkgaygtg aacatggc-3′	49–68	45	4	744
mtdB1r	5′-ctggtayttgacgtt gccga-3′	774–793		2	
mtdB2f	5′-accggcatcttya tcggcggc-3′	181–201	56	2	509
mtdB2r	5′-ggcggcacggcg ttgacgtc-3′	671–690		0	
mch2a	5′-tgcctcggctckc aatatgcyggbtgg-3′	280–306	45	12	427
mch3	5′-gcgtcgttkgtkck bcccat-3′	688–707		24	
mch1	5′-ctdcgcgmtcgg ctcsggbcc-3′	339–359	56	36	341
mch2	5′-acgaaatcggght ggggcgg-3′	661–680		3	
fhcD-1f	5′- gacaccttygcn gargcsttysc -3′	36–58	58	128	851
fhcD-1r	5′-ccsagnttrccgcc gtagttgcc -3′	865–887		16	

[a] Positions in sequences of *M. extorquens* AM1.
[b] Annealing temperature.

and sufficient concentration (approximately 0.45 and 0.2 mg/g soil or sediment, respectively).

Two-Step PCR Amplification Protocol. The *fae, mtdB,* and *mch* genes are amplified via a two-step PCR amplification protocol. The first step is carried out with primer sets amplifying the larger fragment of each gene, i.e., *fae1f-fae1r, mtdB1f-mtdB1r,* and *mch2a-mch3,* respectively. The cycling protocol described earlier is used. The second amplification step employs primers targeting a smaller fragment of each gene, i.e., *fae2f-fae2r, mtdB2f-mtdB2r,* and *mch1-mch2,* respectively. The second amplification step is carried out in a total volume of 50 μl containing 5 μl of the reaction mixture from the first PCR amplification step and all other reaction components at the concentrations listed for the first step. The amplification conditions are similar to the ones for the first step, except that the temperature for the annealing step is raised to 56°.

Community Analysis

Clone Library Construction. DNA products from PCR amplifications, after visualization in 1.5% agarose gels, are purified using the Qiagen PCR purification kit and are ligated into the pCR2.1 vector (Invitrogen), in accordance with the manufacturer's protocol. One Shot *E. coli* cells (Invitrogen) are transformed as described by the manufacturer. Single colonies are picked for plasmid DNA isolation using the Qiagen Miniprep kit.

Restriction Fragment Length Polymorphism (RFLP) Analysis. The number and the variety of operational taxonomic units (OTU) present in the library are determined via RFLP analysis. For RFLP analysis of libraries containing *fae* gene inserts, plasmid DNA is digested for 2 h at 37° in NEB buffer 4, in the presence of bovine serum albumin (BSA), as recommended by the manufacturer (NEB), by a combination of the following restriction enzymes: *Bcl*I, *Sac*II, *Stu*I, *Hinc*II, and *Nco*I (all from NEB). For the analysis of libraries containing *mtdB* gene inserts, plasmid DNA is digested in NEB buffer 2 with *Alu*I (NEB). For the analysis of libraries containing *mch*, plasmid DNA is digested in NEB buffer 2 in the presence of BSA, by a combination of *Hinc*II, *Nco*I, and *Xho*I (all from NEB). For the analysis of libraries containing *fhcD* gene inserts, plasmid DNA is used as a template to reamplify *fhcD* (see Cycling protocol) and the resulting DNA fragments are digested in NEB buffer 3 in the presence of BSA, by a combination of *Alu*I, *Hinc*II, *Hinf*I, and *Sac*II (all from NEB). Digests are resolved in 1.5 or 2% agarose gels and analyzed manually. OTUs are determined based on different RFLP patterns as exemplified in Fig. 2.

Sequence Analysis. Three to five representatives of each OTU are sequenced using the BigDye 3.1 termination sequencing kit (Applied Biosystems). Reaction analysis is performed using an ABI 3730XL high-throughout capillary DNA analyzer. The resulting sequences are compared to sequences in the nonredundant database (NCBI) using BLAST and are sorted into phylotypes. While a value of 97% cut off at the DNA level is commonly used to discriminate between different phylotypes, or species, based on 16S rDNA sequences (Stackebrandt and Göebel, 1994), a lower value of approximately 94% has been suggested for phylotype

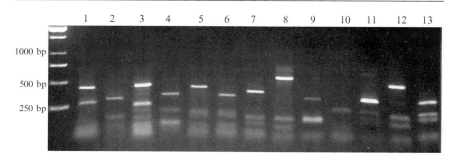

FIG. 2. Electrophoretic separation of RFLP fragments. Shown are 1-kb ladder (Fermentas) and different RFLP patterns produced by restriction digestion (by a combination of *AluI*, *HincII*, *HinfI*, and *SacII*) of *fhcD* fragments amplified, respectively, from (1) *Xanthobacter* sp. LWF 7, (2) *Methylobacterium organophilum* XX, (3) *Methylobacterium nodulans*, (4) *Methylosinus* sp. PW1, (5) *Methylocystis* sp. LW3, (6) *Methylocystis* sp. LW4, (7) *Burkholderia xenovorans* LB400, (8) *Methylomicrobium* sp. AMO, (10) *Methylohalobium crimeensis*, (11) *Methylosarcina lacus* LW14, (12) *Methylomonas* sp. LW13, and (13) *Methylomonas* sp. LW21.

discrimination, based on functional genes (Venter *et al.*, 2004). The sequences then may be subjected to phylogenetic analysis employing one of the commonly used software packages, e.g., the Phylip package (Felsenstein, 2003).

Caveat. Use of some of the primer pairs described in this publication may result in certain group-specific biases in community structure determination. For example, *mch*- and *fae*-specific primer pairs appear to result in underestimating the abundance of sequences belonging to the *Methylobacter/Methylomonas* group, whereas mtdB-specific primer pairs appear to result in underestimating the abundance of sequences belonging to the *Methylosinus/Methylocystis* group (Kalyuzhnaya *et al.*, 2004).

Results and Discussion

Genes for Methanopterin-Dependent Enzymes in Lake Washington

The primer sets described previously were first tested on DNA preparations isolated from a number of methylotrophic and nonmethylotrophic cultures available in our laboratory (Kalyuzhnaya *et al.*, 2004, 2005a). These tests resulted in the amplification of DNA fragments of expected size from strains possessing the H_4MPT-linked functions and in negative amplification from the control strains. We then used these primer sets for environmental amplifications using DNA isolated from Lake Washington sediment (Kalyuzhnaya *et al.*, 2004, 2005a). All of the primer pairs shown in Table I resulted in amplification of products of expected size. However, with environmental DNA, using a single pair of primers for *fae, mtdB,* or *mch,* we occasionally observed amplification of nonspecific DNA fragments. Thus, a two-step PCR amplification protocol is recommended for amplifying these genes. Amplification with the *fhcD*-specific primer pair also occasionally resulted in the amplification of nonspecific fragments, but these were of a different size, while all the fragments of predicted size (approximately 750–830 bp) were specific. Therefore, a single-step amplification protocol is appropriate for *fhcD* detection and analysis. Libraries of 100 to 200 clones were constructed containing, respectively, *fae, mch, mtdB,* and *fhcD* gene inserts. These were subjected to RFLP analysis. Analysis of 200 clones from the *fae* clone library resulted in the identification of 28 RFLP patterns, or OTUs. Analysis of 120 *mtdB*-containing clones revealed the presence of only 4 OTUs. Analysis of 120 *mch*-containing clones resulted in the identification of 8 OTUs. Analysis of 200 *fhcD*-containing clones resulted in 28 OTUs. Representatives of each OTU were sequenced, and the resulting sequences were sorted into phylotypes. We used the cutoff value of 94% (Venter *et al.,* 2004) when comparing DNA or a cutoff value of 95% when comparing protein sequences. Both approaches produced similar phylotype groupings. Using these cutoff values, we were able to identify between 4 and 29 phylotypes, dependent on the specific PCR amplicon library. We attributed the lower numbers of OTUs and, respectively, of phylotype groups for *mtdB* and *mch* genes to possible cloning biases (Kalyuzhnaya *et al.,* 2004). The numbers of phylotypes detected by analyzing *fae-* and *fhcD* PCR-based libraries (28 and 29, respectively) probably represent the lowest estimate of the number of major phylotypes in the site. However, to cover minor phylotypes in environments with great species richness, such as Lake Washington, more exhaustive analyses would be required. We obtained evidence for the existence of additional *fhcD* phylotypes, not covered by the PCR-based library, via analysis of a metagenomic library constructed from Lake Washington DNA (Kalyuzhnaya *et al.,* 2005b). Based on comparisons similar to the ones described earlier, PCR-amplified *fhcD* sequences from the metagenomic library were categorized into 14 unique phylotypes, of which only 4 were identified previously in the PCR-amplified library from total DNA (Kalyuzhnaya *et al.,* 2005a). The combination of the two approaches increased the lowest estimated number of phylotypes in Lake Washington possessing genes for H_4MPT-linked C_1 transfer pathway to 39.

Determining Phylogenetic Affiliations

The ultimate goal of environmental detection is understanding the structure of the community possessing a specific functional capacity, including identification of the phylogenetic affiliation of major species in a given community. For accurate phylogenetic prediction, extensive databases of genes indicative of the function of interest are required, matched to genes traditionally used for phylogenetic analysis (e.g., 16S rRNA genes). We have generated databases of *fae, mtdB, mch,* and *fhcD* genes from a number of cultivated proteobacteria belonging to the α, β, and γ classes (Chistoserdova *et al.,* 2004; Kalyuzhnaya *et al.,* 2004, 2005a,b; all sequences have been deposited with GenBank). In addition, relevant sequences are available for representatives of a deeply branching bacterial division, Planctomycetes (Chistoserdova *et al.,* 2004; Kalyuzhnaya *et al.,* 2004, 2005a). However, at this time, the databases of H_4MPT-linked C_1 transfer genes, including the variety of genes recently identified in Lake Washington, are dominated by sequences not matched to any cultured species. At this time we also cannot discriminate clearly between sequences characteristic of β- and γ—Proteobacteria, as the number of β-proteobacterial sequences is limited in the database. So far the sequences translated from the 17 genes of *Burkholderia xenovorans* LB400, *Rubrivivax gelatinosus,* and *Methylibium petroleophilum* involved in H_4MPT-linked C_1 transfers, with the exception of *fae,* diverge clearly from the sequences of *Methylobacillus flagellatus* and *Methylophilus methylotrophus,* the latter two clustering with the sequences of γ-proteobacterial methylotrophs, possibly reflecting the evolutionary history of these genes in these groups of Proteobacteria. Another complication in phylogenetic analysis comes from the tendency of some of the genes to be present in more than one copy per genome, with low identities between different copies (Kalyuzhnaya *et al.,* 2004). Thus, the accuracy of determination of phylogenetic affiliations for newly uncovered sequences in the future will depend on progress in cultivation and characterization of novel species containing the genes of interest, or environmental genomics approaches. In our model study, using the PCR-amplified libraries of *fae, mtdB, mch,* and *fhcD* genes from Lake Washington, we were able to suggest phylogenetic affiliations for sequences represented by only five phylotypes with high confidence, based on the chosen cutoff values, which were assigned to the genera of *Methylobacterium, Methylocystis, Methylomicrobium, Methylobacter,* and *Burkholderia* (not shown). Association of other phylotypes with α- or β/γ—proteobacteria or Planctomycetes not closely related to any known species was carried out via phylogenetic analyses, as exemplified in Fig. 3, for Fae polypeptides. Using these analyses, up to 7 novel α-proteobacterial phylotypes, up to 11 novel β/γ -proteobacterial phylotypes, and up to 4 novel planctomycete phylotypes were determined in the Lake Washington PCR amplicon databases. In addition, we were able to identify a number of phylotypes that could not be affiliated with any known species, which were tentatively designated as novel phyla (Fig. 3).

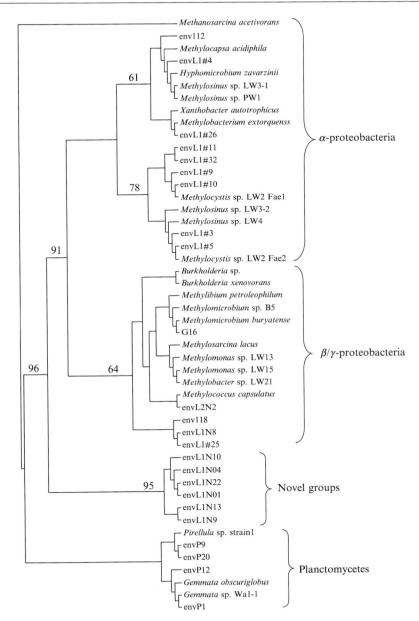

FIG. 3. Neighbor-joining consensus tree including representatives of the 24 phylotypes of Fae recovered from Lake Washington (designated as env) and representatives of the cultivated species. Bootstrap values (%) for major groups are shown.

Concluding Remarks

Which Primer Set Is the Best?

In our experience with the Lake Washington environmental DNA, the *fae-* and the *fhcD*-targeted sets of primers uncovered the widest breadth of sequence divergence. In addition, the *fhcD* primer set resulted in abundant, specific PCR amplicons in one step. Moreover, the fragment amplified with the *fhcD*-targeted primers covered most of the *fhcD* gene, resulting in a more reliable phylogenetic signal. Thus, the *fhcD*-targeted primer set is probably the tool of choice for the initial detection of H_4MPT-linked formaldehyde oxidative capacity. However, it is possible that the true degree of divergence of the H_4MPT-linked functions might still remain undiscovered. Therefore, conclusions on the presence of genes of interest would be more robust if more than one or all four genes were detected.

The Importance of Databases

The validity of employing functional genes for phylogenetic analysis is still a matter of debate (Santos and Ochman, 2004; Wolf *et al.*, 2001). In the case of methanopterin-linked functions in bacteria, we have built extensive databases of functional genes from identified species, specifically Proteobacteria and Planctomycetes, so the newly identified sequences can be matched against these databases. In addition, we have built databases of novel genes not yet affiliated with phyla, indicative of a broader distribution of H_4MPT-linked C_1 transfer capacity in the microbial world than previously known. Work is in progress on the identification and taxonomic characterization of these organisms. The availability of a number of methanogen genomes provides a sufficient database of archaeal gene counterparts. No doubt these databases will continue to grow, providing means for improved phylogenetic classification.

The Future of Environmental Genomics

The approach described here, using a group of genes as a signature of a single functional metabolic module for a specific physiological capability (H_4MPT-linked formaldehyde oxidation) for environmental detection and community structure analysis, is still in early stages of development. However, we already have collected strong evidence for the broad distribution and high variance of this function in the environment. As mentioned earlier, most sequences in the current databases belong to yet unidentified species, and some sequences are sufficiently divergent to imply their affiliation with major, undiscovered phyla. Some of these novel phyla may be cultivated and characterized in the near future; however, it is likely that many of the novel sequences will belong to species resistant to cultivation. Affiliating these sequences with specific phylogenetic groups, including the novel groups, will rely on progress in environmental genomics, especially the recently tested approach of shotgun environmental sequencing to high coverage (Tyson *et al.*, 2004; Venter *et al.*, 2004). As more sequences of the H$_4$MPT-linked formaldehyde oxidation genes are matched to other genes with a strong phylogenetic signal, a better picture of the reliability of these genes for phylogenetic predictions will emerge. At the same time, a better understanding of evolution of these genes in the microbial world will be gained. Conversely, gene databases generated via environmental genomics approaches will likely present opportunities for identifying new sequences indicative of H$_4$MPT-linked formaldehyde oxidation.

Acknowledgment

The authors acknowledge support from the Microbial Observatories program funded by the National Science Foundation (MCB-0131957).

References

Chistoserdova, L., Chen, S.-W., Lapidus, A., and Lidstrom, M. E. (2003). Methylotrophy in *Methylobacterium extorquens* AM1 from a genomic point of view. *J. Bacteriol.* **185,** 2980–2987.

Chistoserdova, L., Jenkins, C., Kalyuzhnaya, M. G., Marx, C. J., Lapidus, A., Vorholt, J. A., Staley, J. T., and Lidstrom, M. E. (2004). The enigmatic planctomycetes may hold a key to the origins of methanogenesis and methylotrophy. *Mol. Biol. Evol.* **21,** 1234–1241.

Chistoserdova, L., Rasche, M. E., and Lidstrom, M. E. (2005). Novel dephosphotetrahy-dromethanopterin biosynthesis genes discovered via mutagenesis in *Methylobacterium extorquens* AM1. *J. Bacteriol* **187,** 2508–2512.

Chistoserdova, L., Vorholt, J. A., Thauer, R. K., and Lidstrom, M. E. (1998). C$_1$ transfer enzymes and coenzymes linking methylotrophic bacteria and methanogenic Archaea. *Science* **281,** 99–102.

Felsenstein, J. (2003). "Inferring Phylogenies." Sinauer, Sunderland, MA.

Kalyuzhnaya, M. G., Lidstrom, M. E., and Chistoserdova, L. (2004). Utility of environmental probes targeting ancient enzymes: Methylotroph detection in Lake Washington. *Microb. Ecol.* Published online on November 9, 2004.

Kalyuzhnaya, M. G., Nercessian, O., Lidstrom, M., and Chistoserdova, L. (2005a). Development and application of polymerase chain reaction primers based on *fhcD* for environmental detection of methonopterin-linked C$_1$ metabolism in bacteria. *Environ. Microbiol.* **7,** 1269–1274.

Kalyuzhnaya, M. G., Bouerman, S., Nercessian, O., Lidstrom, M., and Chistoserdova, L. (2005b). Highly divergent genes for methanopterin-linked C$_1$ transfer reactions in Lake Washington, assessed via metagenomic analysis and mRNA detection. Submitted for publication.

Marx, C. J., Miller, J. A., Chistoserdova, L., and Lidstrom, M. E. (2004). Multiple formaldehyde oxidation/detoxification pathways in *Burkholderia fungorum* LB400. *J. Bacteriol.* **186,** 2173–2178.

Santos, S. R., and Ochman, H. (2004). Identification and phylogenetic sorting of bacterial lineages with universally conserved genes and proteins. *Environ. Microbiol.* **6,** 754–759.

Stackebrandt, E., and Göebel, B. M. (1994). Taxonomic note: A place for DNA-DNA reassociation and 16S sequence rRNA analysis in the present species definition in bacteriology. *Int. J. Syst. Bacteriol.* **44,** 846–849.

Tyson, G. W. J., Chapman, J., Hugenholtz, P., Allen, E. E., Ram, R. J., Richardson, P. M., Solovyev, V. V., Rubin, E. M., Rokhsar, D. S., and Banfield, J. F. (2004). Community structure and metabolism through reconstruction of microbial genomes from the environment. *Nature* **428,** 37–43.

Venter, J. C., Remington, K., Heidelberg, J. F., Halpern, A. L., Rusch, D., Eisen, J. A., Wu, D., Paulson, I., Nelson, K. E., Nelson, W., Fouts, D. E., Levy, S., Knap, A. H., Lomas, M. W., Nealson, K., White, O., Peterson, J., Hoffman, J., Parsons, R., Baden-Tillson, H., Pfannkoch, C., Rogers, Y. H., and Smith, H. O. (2004). Environmental genome shotgun sequencing of the Sargasso Sea. *Science* **304,** 66–74.

Vorholt, J. A. (2002). Cofactor-dependent pathways of formaldehyde oxidation in methylotrophic bacteria. *Arch. Microbiol.* **178,** 239–249.

Vorholt, J. A., Chistoserdova, L., Stolyar, S. M., Thauer, R. K., and Lidstrom, M. E. (1999). Distribution of tetrahydromethanopterin-dependent enzymes in methylotrophic bacteria and phylogeny of methenyl tetrahydromethanopterin cyclohydrolases. *J. Bacteriol.* **181,** 5750–5757.

Wolf, Y. I., Rogozin, I. B., Grishin, N. V., Tatusov, R. L., and Koonin, E. V. (2001). Genome trees constructed using five different approaches suggest new major bacterial clades. *BMC Evol. Biol.* **1,** 8.

Further Reading

Thompson, J. D., Higgins, D. G., and Gibson, T. J. (1994). CLUSTAL W: Improving the sensitivity of progressive multiple sequence alignment through sequence weighting, position-specific gap penalties and weight Matrix choice. *Nucleic Acid. Res.* **22,** 4673–4680.

[28] Community-Level Analysis: Key Genes of CO_2-Reductive Acetogenesis

By Charles R. Lovell and Adam B. Leaphart

Abstract

CO_2-reductive acetogenic bacteria are ubiquitous in anaerobic habitats and are physiologically and phylogenetically diverse. The latter characteristics have rendered their diversity in natural environments, their distributions, and their ecological function(s) difficult to assess. Recently introduced polymerase chain reaction (PCR) primers for specific amplification of the

METHODS IN ENZYMOLOGY, VOL. 397
0076-6879/05 $35.00
DOI: 10.1016/S0076-6879(05)97028-6

structural gene encoding formyltetrahydrofolate synthetase (FTHFS, EC 6.3.4.3), a key enzyme in the acetyl-CoA pathway of acetogenesis, have facilitated studies of acetogen diversity and ecology. These primers amplify an approximately 1100-bp segment of the FTHFS gene. FTHFS sequences have been recovered from authentic acetogens, from sulfate reducing bacteria, and from a variety of other nonacetogenic bacteria. Phylogenetic analyses segregated these sequences into distinct clusters, only one of which contained sequences from known acetogens. This chapter describes the PCR primers, defines conditions for successful amplification of FTHFS sequences, and details the phylogenetic analysis of the FTHFS sequences. Information on the types of sequences that have been recovered from natural acetogen habitats and how they have been interpreted is also included.

Introduction

CO_2-reductive acetogenic bacteria (acetogens) utilize a unique pathway, the acetyl-CoA pathway, for fixation of C_1 compounds (Drake, 1994; Ljungdahl, 1986). This ubiquitous group of anaerobic bacteria can grow autotrophically on H_2 and CO_2 and/or heterotrophically on numerous organic compounds, with mixotrophic growth on H_2 and an organic substrate observed in some species (Breznak and Kane, 1990; Wood and Ljungdahl, 1991). Acetogenesis contributes an estimated 10% of the approximately 10^{13} kg of acetate produced annually in anaerobic environments (Wood and Ljungdahl, 1991), consequently the organisms carrying out this process are very important participants in global carbon cycling. Despite their significance, acetogens remain an understudied functional group of bacteria. This is due in part to the difficulties inherent in detecting, identifying, and characterizing diversity and distributions of acetogenic bacteria. Molecular biological approaches have been very useful for studies of the ecology of many functional groups of microorganisms, but acetogens present some particular challenges.

The acetogens as a group are among the most metabolically versatile anaerobes (Drake et al., 1997, 2002) and are also phylogenetically quite diverse (Tanner and Woese, 1994). In addition to this phylogenetic diversity, acetogen species commonly have sister taxa that are not acetogenic (Drake et al., 2002). This limits the utility of 16S rRNA gene sequence recovery and analysis, the most commonly used set of tools in molecular microbial ecology, for examination of acetogen diversity or distributions in natural environments. As a consequence, function-specific genes have become very important to studies of acetogen ecology.

Some of the enzymes of the acetyl-CoA pathway (Fig. 1) are conserved both structurally and functionally (Lovell *et al.*, 1990; Ragsdale, 1991). In particular, the gene sequence of formyltetrahydrofolate synthetase (FTHFS, EC 6.3.4.3), which catalyzes the ATP-dependent activation of formate ion, is quite conservative (Leaphart and Lovell, 2001; Lovell *et al.*, 1990) and has served as a target sequence for both DNA probe hybridization (Lovell and Hui, 1991) and polymerase chain reaction amplification using specific primers (Leaphart and Lovell, 2001; Leaphart *et al.*, 2003; Salmassi and Leadbetter, 2003). Other apparently conservative enzymes (e.g., Salmassi and Leadbetter, 2003), methenyltetrahydrofolate cyclohydrolase (EC 3.5.4.9) and methylenetetrahydrofolate dehydrogenase (EC 1.5.1.5), are also found in the methyl branch of the acetyl-CoA pathway.

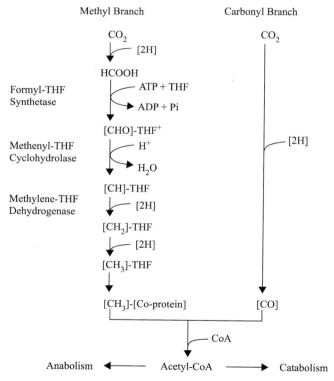

Fig. 1. The acetyl-CoA pathway employed by acetogens for C_1 incorporation into acetate. Reprinted with minor modifications from Drake *et al.* (1997), with permission of IOS Press.

Few sequences from these enzymes, and no probes or PCR primers, are available at present. Consequently, recovery and analysis of FTHFS sequences are the primary focus of this chapter. This functional group approach has yielded some important new insights into the ecology of acetogenic bacteria.

Design of FTHFS PCR Primers

Initial PCR primer design employed the FTHFS sequences from *Moorella thermoacetica* (formerly *Clostridium thermoaceticum*) (Lovell *et al.*, 1990), *Clostridium acidiurici* (Whitehead and Rabinowitz, 1988), and *Clostridium cylindrosporum* (Rankin *et al.*, 1993). The latter two organisms are not actually acetogenic, but ferment purines and amino acids via the glycine synthase-glycine reductase pathway, which also employs FTHFS (Fuchs, 1986). The FTHFS sequences were aligned and stretches of six or more consecutive conserved residues were identified and examined for utility as priming sequences. Initial primers were designed and were used to recover a 1374-bp segment of the approximately 1680-bp FTHFS gene from the authentic acetogens *Acetobacterium woodii*, *Clostridium formicoaceticum*, *Clostridium magnum*, and *Thermoanaerobacter* (formerly *Acetogenium*) *kivui*, but yielded nonspecific products from several other acetogens and were not considered broadly useful (Leaphart and Lovell, 2001).

Examination of the new FTHFS sequences revealed less degenerate PCR primer sequences that amplified an approximately 1100-bp stretch of the FTHFS gene specifically and at high yield (Fig. 2) (Leaphart and Lovell, 2001). The *M. thermoacetica* nucleotide sequence was substantially divergent from other acetogen sequences and was dropped from consideration in the design of these primers (see later). These final primer sequences were

FTHFS forward primer: 5'-TTYACWGGHGAYTTCCATGC-3' (24-fold degenerate)

FTHFS reverse primer: 5'-GTATTGDGTYTTRGCCATACA-3' (12-fold degenerate)

These FTHFS primers are recommended for ecological studies of the acetogens and have been used to recover FTHFS sequences from several acetogen habitats, including horse manure (the source of the original *M. thermoacetica* isolate) (Leaphart and Lovell, 2001), salt marsh plant roots (Leaphart and Lovell, 2001; Leaphart *et al.*, 2003), and termite hindgut (Salmassi and Leadbetter, 2003).

1 2 3 4 5   MW (bp)

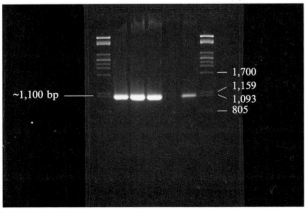

FIG. 2. Agarose gel showing partial FTHFS gene amplicons from (lane 1) *Acetobacterium woodii*, (lane 2) *Clostridium aceticum*, (lane 3) *Eubacterium limosum*, (lane 4) negative control (water blank), and (lane 5) *Clostridium formicoaceticum*. Molecular weight markers are *Pst*I-digested bacteriophage λ DNA.

PCR Reaction Conditions and Testing of FTHFS Primers

The PCR Reaction System

The PCR reaction mixture optimized for the amplification of FTHFS sequences from authentic acetogens and other FTHFS-producing organisms consists of 1× Dynazyme EXT Tbr polymerase buffer (MJ research, Waltham, MA), 1.5 mM MgCl$_2$, 0.2 mM each deoxynucleotide triphosphate, 0.5 μM each FTHFS primer (0.5 pmol μl^{-1}), 0.4 μg μl^{-1} molecular biology grade (nonacetylated) bovine serum albumin (BSA), 25 ng μl^{-1} template DNA, and 1 U μl^{-1} DynaZyme EXT in a reaction volume of 25 μl. BSA substantially improves yields of PCR amplifications from environmental DNA, but may not be required for amplifications from pure culture DNA samples. Salmassi and Leadbetter (2003) reported successful amplification using the listed primers and the Roche *Taq* DNA polymerase.

Thermal Cycling Programs

The thermal cycling program used for FTHFS amplification begins with denaturation at 94° for 2 min. This step is followed by a short touchdown program consisting of 9 cycles of 94° for 30 s, 63° for 30 s (decreasing by

1° per cycle), and 72° for 30 s. A production program consisting of 25 cycles of 94° for 30 s, 55° for 30 s, and 72° for 30 s is used next, followed by a final extension step of 72° for 2 min. This program produces high yields of an amplicon of approximately 1100 bp in length from pure culture and environmental DNA samples without spurious products (Leaphart and Lovell, 2001). An alternative program consisting of 94° for 2 min, followed by 20 cycles of 94° for 30 s, 47° for 30 s, 72° for 30 s, and a final extension step of 72° for 2 min also amplifies FTHFS sequences from some pure culture DNA samples. Lower annealing temperatures, predictably, can produce spurious products, even from pure cultures of authentic acetogens. It should also be noted that the FTHFS primers and the thermal cycling programs listed produce spurious products in addition to the FTHFS amplicon from *M. thermoacetica* and the very closely related *M. thermoautotrophica*.

Primer Testing

PCR amplifications were performed using DNA from a variety of authentic acetogens [including *Acetobacterium psammolithicum*, *A. woodii*, *Clostridium aceticum*, *C. formicoaceticum*, *C. magnum*, *Eubacterium limosum*, *Ruminococcus (formerly Peptostreptococcus) productus*, *Sporomusa ovata*, *Sporomusa termitida*, *T. kivui*, and *Treponema primitia* strains ZAS-1 and ZAS-2], other FTHFS-producing organisms (*Desulfoarculus baarsii*, *Desulfomicrobium bacculatus*, *Desulfovibrio desulfuricans*, *Desulfovibrio piger*, *Desulfovibrio salexigens*, *Desulfovibrio* sp. strain Summer lac-1, Proteus vulgaris, and two uncharacterized strains of sulfate-reducing bacteria, BG8 and BG14), and non-FTHFS-producing organisms (several, see primary sources) (Leaphart and Lovell, 2001; Leaphart et al., 2003). In all cases, amplicons of the correct size were produced from DNA of acetogens and other FTHFS-producing organisms without spurious products (see Fig. 2). An amplicon of the correct size was produced from *M. thermoacetica*, but nonspecific products dominated the amplification. No amplicons were produced from DNA from any non-FTHFS-producing species.

FTHFS Sequence Alignment and Phylogenetic Analysis Methods

FTHFS Amplicon Cloning

DynaZyme EXT and *Taq* DNA polymerase, as well as other DNA polymerases, add a single deoxyadenosine to the 3′ ends of amplicons (Clark, 1988). This facilitates cloning of the amplicons, using pGEM-T

(Promega, Madison, WI) or comparable vectors, for subsequent sequence analysis. Cloning is not required for determination of FTHFS gene nucleotide sequences from pure cultures of most acetogens. However, two different FTHFS gene sequences were recovered from *Treponema primitia* strain ZAS-1 (Salmassi and Leadbetter, 2003), so FTHFS gene duplication apparently occurs in at least one known acetogenic species. Environmental samples often yield large numbers of different sequences, and construction of clonal libraries is currently the most facile method for analysis of acetogen diversity.

FTHFS Sequence Alignment

Nucleotide sequences that encode proteins, such as the partial sequences of FTHFS genes recovered using the methods described earlier, should be aligned maintaining the proper reading frame of the sequences. In addition, the third codon position is very sensitive to genomic GC content, which can account for much of the variation observed in protein-encoding nucleotide sequences (Normand and Bousquet, 1989). A very useful alternative to nucleotide sequence alignment is alignment of translated polypeptide sequences. This approach facilitates comparison of FTHFS sequences from distantly related organisms and those from thermophiles and mesophiles, as it eliminates most of the impact of differences in genomic GC contents. Alignments of translated sequences can be constructed using CLUSTALW (Thompson *et al.*, 1994) or CLUSTALX (Thompson *et al.*, 1997).

Based on the amino acid numbering system for *M. thermoacetica* FTHFS (Lovell *et al.*, 1990), the only complete FTHFS sequence from a true acetogen, the amplicon begins with Val 136 and ends with Val 487 exclusive of primers, and from Phe 129 to Tyr 494 including primers. All sequences should be examined for amino acid residues that are universally conserved among known FTHFS sequences and thought to be important to FTHFS structure or catalysis (Leaphart and Lovell, 2001; Lovell *et al.*, 1990; Radfar *et al.*, 2000). These include a hexapeptide, encompassing residues 197–202, thought to be involved in tetrahydrofolate binding (Lovell *et al.*, 1990; Radfar *et al.*, 2000), and the putative adenosine ring-stabilizing residue, Trp 412 (Radfar *et al.*, 2000). In addition, residues 271–284 are highly conserved in FTHFS sequences. These residues are useful markers in performing accurate sequence alignments and provide additional confidence in the identity of FTHFS sequences recovered from environmental samples. Amino acid residues that are useful in differentiating FTHFS sequences of known acetogens from those from other sources include Lys 187, Lys 256, Thr 266, Leu 314, Tyr 339, Ile 370, Pro

385, Val 406, Ala 407, Val 411, Lys 414, Gly 419, Ile 447, Ala 470, Leu 480, and Lys 484 (Asn 484 in the *M. thermoacetica* sequence).

Phylogenetic Analyses

Construction of a phylogenetic hypothesis (tree) that proposes relationships of the FTHFS sequences to each other can be performed using a variety of methods (Opperdoes, 2003). Distance methods, such as neighbor joining, employ a matrix of pairwise distance values between the aligned sequences. Character-based methods base calculations on individual residues in the sequences and include maximum likelihood and maximum parsimony. Character-based methods are generally slow and computationally intensive for protein sequences. Neighbor joining is a convenient, rapid, and useful tool for assessing relationships among FTHFS sequences. Correction of the distance matrix for multiple hits and back mutations is necessary and numerous methods for making this correction have been employed (Leaphart and Lovell, 2001; Opperdoes, 2003; Salmassi and Leadbetter, 2003). Rooting the tree requires an outgroup, and sequences from *Thermoplasma acidophilum* (NCBI GenBank Accession number AL445067) and *Thermoplasma volcanicum* (AP000991) are ideal for this purpose, as these *Archaea* clearly branched off before all other taxa. Bootstrap or jackknife analyses should always be employed to assess the reliability of the tree topology. It is also advisable to compare tree topologies yielded by different tree-building methods and different orders of sequences in the alignment. Phylogenetic trees provide useful hypothetical models of the evolutionary relationships among FTHFS sequences, but no such model should be considered accurate unless supported by high bootstrap values and validated through a series of independent computations.

Results of Phylogenetic Analyses of FTHFS Sequences

FTHFS sequences from a variety of acetogens, sulfate-reducing bacteria, and other organisms resolved into several distinct sequence clusters (Fig. 3). Cluster A was monophyletic and contained all of the FTHFS sequences from authentic acetogens (Leaphart and Lovell, 2001; Leaphart *et al.*, 2003). In all cases, subdivisions within cluster A were consistent with known relationships of the acetogens. Different FTHFS clones from a single strain (*T. primitia* strain ZAS-1, clones A and B) were identical in peptide sequence and 97.8% similar in nucleotide sequence. FTHFS peptide sequences from different strains of *E. limosum* were 98.9% similar. Greater sequence variability was observed for different species in the same genus, with 88% peptide sequence similarity between *Acetobacterium*

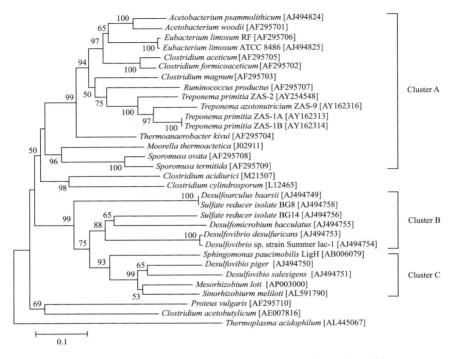

FIG. 3. Phylogenetic analysis of partial FTHFS and FTHFS-like polypeptide sequences. The dendrogram was generated using neighbor-joining and Poisson-corrected distances. *Thermoplasma acidophilum* was used as the outgroup taxon. Values at the nodes are the percentages of 1000 bootstrap replicates supporting the branching order. Bootstrap values below 50% are not shown.

species, 90.4% similarity between *Sporomusa* species, and 92.9% similarity between *C. aceticum* and *C. formicoaceticum*. FTHFS peptide sequences from authentic acetogens averaged 67.6% similar, reflecting the deep bifurcation between the *Moorella/Sporomusa* subcluster and the balance of cluster A. Within cluster A, only *T. primitia* strain ZAS-9 is not known to have the capacity to grow as a true acetogen (Breznak and Leadbetter, 2002). FTHFS sequences from the purinolytic clostridia *C. acidiurici* and *C. cylindrosporum* formed a monophyletic clade placed near but outside cluster A. These sequences averaged 65.2% similarity to their nearest cluster A neighbors, the *Sporomusa* sequences. Sequences from additional known and presumptive FTHFS producing organisms formed additional deeply branching clusters.

FTHFS sequences falling into cluster B (following the cluster designations of Leaphart *et al.*, 2003) were from authentic sulfate-reducing bacteria, including species of *Desulfoarculus*, *Desulfomicrobium*, and *Desulfovibrio*. Numerous sequences allied with this cluster were recovered by Leaphart *et al.* (2003) from salt marsh plant roots, a known hot spot for sulfate reduction activity and high abundance of sulfate-reducing bacteria (Hines *et al.*, 1999; Rooney-Varga *et al.*, 1997; 1998;). Acetate is considered to be quantitatively the most important substrate for sulfate reducers in natural ecosystems (Gibson, 1990; Laanbroek and Pfennig, 1981; Sorensen *et al.*, 1981), and catabolism of acetate by sulfate reducers can employ either the TCA cycle (Brandis *et al.*, 1983; Gebhart *et al.*, 1983) or the acetyl-CoA pathway, functioning in reverse (Fuchs, 1986; Schauder *et al.*, 1986). In addition, *D. baarsii* is capable of autotrophic acetogenesis using FTHFS and the acetyl-CoA pathway (Jansen *et al.*, 1984; 1985). Cluster B, while not as coherent as clusters A or C, reflects a well-defined group of organisms that produce FTHFS.

Cluster C also contained two sequences from known sulfate-reducing bacteria, as well as additional sequences, including a known FTHFS homologue, from other types of organisms. The *Sphingomonas paucimobilis* *lig*H gene encodes an *o*-demethylating enzyme having 60% polypeptide sequence similarity to the FTHFS sequence from *M. thermoacetica* (Nishikawa *et al.*, 1998). Cluster C was monophyletic, with average sequence similarity of 70.5%. While it is tempting to speculate that sequences in this cluster are in fact FTHFS homologues, with no authentic FTHFS sequences included, this has not been demonstrated. Clearly, cluster C was quite coherent and well separated from clusters that seem most likely to contain only FTHFS sequences. In addition, two authentic FTHFS sequences from organisms not known to utilize the acetyl-CoA pathway either in an anabolic or in a catabolic capacity, *Proteus vulgaris* and *Clostridium acetobutylicum*, are highly divergent from all other sequences. The *P. vulgaris* and *C. acetobutylicum* FTHFS peptide sequences averaged only 43.9% similar to the other sequences (excluding the outgroup sequence from *T. acidophilum*) employed in this analysis.

Phylogenetic analyses of FTHFS and FTHFS homologue sequences from several natural environments have also been revealing. While sequences are available from only a handful of acetogen habitats, some patterns are beginning to emerge. DNA purified from fresh horse manure yielded FTHFS sequences strongly allied with cluster A and similar to the *R. productus* sequence (Leaphart and Lovell, 2001). *R. productus* was originally isolated from a sewage digester (Lorowitz and Bryant, 1984) and has been found in the intestinal tracts of a variety of mammals (Wang *et al.*, 1996). DNA from the hindgut of the Pacific dampwood termite,

Zootermopsis angusticollis, yielded numerous sequences allied with cluster A (Salmassi and Leadbetter, 2003). A few of these termite hindgut sequences were also allied with the *R. productus* and the horse manure sequences, but most grouped with sequences from acetogenic spirochaetes in the genus *Treponema*. This result is remarkable in that acetogenesis is the most important hindgut activity impacting the energy metabolism of wood-feeding termites (Breznak, 1994), and spirochaetes represent as much as 50% of prokaryotic cells in virtually all termite guts (Breznak and Leadbetter, 2002; Lilburn *et al.*, 1999; Paster *et al.*, 1996). While it is not clear that spirochaetes are the functional termite gut acetogens, given that FTHFS sequences were also recovered from other acetogen groups or that the FTHFS sequences recovered represent the dominant spirochaetes, the prospect that the functionally important members of the termite gut acetogen assemblage have been identified is certainly exciting.

FTHFS sequences were also recovered from an entirely different acetogen habitat, the roots of salt marsh plants (Leaphart and Lovell, 2001; Leaphart *et al.*, 2003). Salt marsh plants include the keystone salt marsh cordgrass, *Spartina alterniflora*, and strongly impact marsh geomorphology, productivity, and stability (Morris *et al.*, 2002). The roots of these plants are well-established hot spots for activities of oxygen utilizing and anaerobic bacteria and support large populations of a diverse array of organisms (e.g., Rooney-Varga *et al.*, 1997; 1998; Hines *et al.*, 1999; Kostka *et al.*, 2000; Lovell *et al.*, 2000). DNA samples from roots of three salt marsh plant species, *S. alterniflora*, the black needle rush *Juncus roemerianus*, and the common pickleweed, *Salicornia virginica*, all yielded FTHFS sequences and/or FTHFS homologues. While many sequences were recovered, it was noteworthy that only *S. alterniflora* yielded sequences that were allied with the cluster A FTHFS sequences from authentic acetogens (Leaphart and Lovell, 2001; Leaphart *et al.*, 2003). Cluster A *S. alterniflora* root FTHFS sequences included one sequence closely allied with sequences from acetogenic *Sporomusa* species, four sequences similar to those of acetogenic clostridia, and two sequences that were over 99% similar to that of *E. limosum*. This latter finding is of particular interest as members of the *Acetobacterium–Eubacterium* clade have also been detected on and in the roots of a seagrass, *Halodule wrightii* (Küsel *et al.*, 1999). A commonality of these two grass species is their capacity to transport oxygen to their roots, somewhat oxidizing their rhizospheres (Hwang and Morris, 1991; Jackson and Armstrong, 1999; Lovell, 2002; Welsh, 2000). Rhizosphere ventilation serves two important functions for the plants: supply of oxygen for root respiration and detoxification of reduced compounds, such as the sulfides generated by rhizosphere sulfate reduction (Armstrong, 1978; Hemminga, 1998). It is possible that rhizosphere ventilation by some marine plants also protects the acetogens associated with their roots from excessively high levels of reduced toxins. A number of acetogens have been shown to be oxygen tolerant (Karnholz *et al.*, 2002), possibly opening a niche within which acetogens could outcompete other groups of anaerobes in the rhizosphere microenvironment.

Future Prospects

Our ability to use specific PCR primers to recover FTHFS sequences from environmental materials has opened the door to studies of acetogen diversity and distributions. This capacity has already produced some exciting new findings that lend insight into the types of acetogens that succeed in different habitats and it facilitates testing of experimentally motivated, explicit hypotheses concerning acetogen autecology as well as the contributions of these organisms to ecological processes. This is an important advance, but it must be recognized that little is known concerning acetogens and their functions in even the most thoroughly studied natural environments. In order to advance our understanding of these important organisms, we must address some remaining practical issues.

Analysis of FTHFS sequences and exploration of the phylogenetic relationships they can reveal are dependent on the database of sequences available from authentic, formally described acetogens. Only a relative handful of reference sequences are currently available from the more than 100 known acetogenic species and these are predominantly from two phyla, the *Firmicutes* and the *Spirochaetes*. Additional sequences from other acetogen groups, including the *Acidobacteria*, represented by *Holophaga foetida* (Liesack *et al.*, 1994), are particularly needed. Such sequences would allow assessment of the coverage of the FTHFS primers for all major acetogen groups. FTHFS sequences from formally described non-acetogen taxa are also needed. It appears that a cohesive cluster of FTHFS sequences from sulfate reducing bacteria has been identified, but it is presently defined by only a few sequences. Additional sequences from sulfate reducers, including sequences from known acetogenic sulfate reducers, such as *Desulfotignum phosphitoxidans* (Schink *et al.*, 2002) would help resolve this cluster and might aid in understanding why the sequence from *Desulfoarculus baarsii*, the only FTHFS sequence currently available from an authentic sulfate-reducing bacterium that can grow acetogenically, groups with cluster B sequences rather than cluster A sequences. Finally, cluster C remains a puzzle. It is tempting to dismiss these sequences as only FTHFS homologues, but more effort is needed to determine if cluster C organisms utilize, and thus encode, true FTHFS genes. This sequence recovery effort, while unglamorous, is very important to further advances in the molecular exploration of acetogen ecology.

It is also important that we develop an understanding of the extent, if any, of horizontal transfer of the gene encoding FTHFS, as well as other acetyl-CoA pathway functions, among anaerobic bacteria. Salmassi and Leadbetter (2003) found that a nonacetogenic treponeme yielded a cluster A, apparently acetogenic FTHFS sequence. This is the only sequence identified to date representing a nonacetogen that is allied with cluster A FTHFS sequences. Because closely related termite treponemes are acetogenic and also yielded cluster A sequences, a simple explanation for this observation might be horizontal gene transfer, but this supposition remains to be tested. It would be of substantial interest to know if DNA sequences flanking the *Treponema azotonutricium* ZAS-9 FTHFS gene encode other acetyl-CoA pathway enzymes, as is the case for the acetogenic *T. primitia* ZAS-2 (Salmassi and Leadbetter, 2003). If not, horizontal transfer mechanisms might be profitably explored.

Acknowledgments

This research was supported by National Science Foundation Grants MCB-9873606 and MCB-0237854 to C.R.L.

References

Armstrong, W. (1978). Root aeration in the wetland condition. *In* "Plant Life in Anaerobic Environments" (D. D. Hook and R. M. Crawford, eds.), pp. 269–298. Ann Arbor Science Series, Ann Arbor, Michigan.

Brandis, A., Gerhart, N. A., Thauer, R. K., Widdel, F., and Pfennig, N. (1983). Anaerobic oxidation to CO_2 by *Desulfobacter postgatei*. 1. Demonstration of all enzymes required for the operation of the citric acid cycle. *Arch. Microbiol.* **136,** 222–229.

Breznak, J. A. (1994). Acetogenesis from carbon diozide in termite guts. *In* "Acetogenesis" (H. L. Drake, ed.), pp. 303–330. Chapman and Hall, New York.

Breznak, J. A., and Kane, M. D. (1990). Microbial H_2/CO_2 acetogenesis in animal guts: Nature and nutritional significance. *FEMS Microbiol. Rev.* **87,** 309–313.

Breznak, J. A., and Leadbetter, J. R. (2002). Termite gut spirochetes. *In* "The Prokaryotes: An Evolving Electronic Resource for the Microbiological Community" (M. Dworkin, S. Falkow, E. Rosenberg, K. H. Schleifer, and E. Stackbrandt, eds.). Springer, New York.

Clark, J. M. (1988). Novel non-templated nucleotide addition reactions catalyzed by prokaryotic and eucaryotic DNA polymerases. *Nucleic Acids Res.* **16,** 9677–9686.

Drake, H. L. (1994). Acetogenesis, acetogenic bacteria, and the acetyl CoA "Wood/ Ljungdahl" pathway: Past and current perspectives. *In* "Acetogenesis" (H. L. Drake, ed.), pp. 3–60. Chapman and Hall, New York.

Drake, H. L., Daniel, S. L., Küsel, K., Matthies, C., Kuhner, C., and Braus-Stromeyer, S. (1997). Acetogenic bacteria: What are the *in situ* consequences of their diverse metabolic versatilities? *BioFactors* **6,** 13–24.

Drake, H. L., Küsel, K., and Matthies, C. (2002). Ecological consequences of the phylogenetic and physiological diversities of acetogens. *Antonie van Leeuwenhoek* **81,** 203–213.

Fuchs, G. (1986). CO_2 fixation in acetogenic bacteria: Variations on a theme. *FEMS Microbiol. Rev.* **39,** 181–213.

Gebhart, N. A., Linder, D., and Thauer, R. K. (1983). Anaerobic acetate oxidation to CO_2 by *Desulfobacter postgatei*. 2. Evidence from ^{14}C-labeling studies for the operation of the citric acid cycle. *Arch. Microbiol.* **136,** 230–233.

Gibson, G. R. (1990). Physiology and ecology of the sulphate-reducing bacteria. *J. Appl. Bacteriol.* **69,** 769–797.

Hemminga, M. A. (1998). The root/rhizome system of seagrasses: An asset and a burden. *J. Sea Res.* **39,** 183–196.

Hines, M. E., Evans, R. S., Sharak Genthner, B. R., Willis, S. G., Friedman, S., Rooney-Varga, J. N., and Devereux, R. (1999). Molecular phylogenetic and biogeochemical studies of sulfate-reducing bacteria in the rhizosphere of *Spartina alterniflora*. *Appl. Environ. Microbiol.* **65,** 2209–2216.

Hwang, Y.-H., and Morris, J. T. (1991). Evidence for hygrometric pressurization in the internal gas space of *Spartina alterniflora*. *Plant Physiol.* **96,** 166–171.

Jackson, M. B., and Armstrong, W. (1999). Formation of aerenchyma and the processes of plant ventilation in relation to soil flooding and submergence. *Plant Biol.* **1,** 274–287.

Jansen, K., Fuchs, G., and Thauer, R. K. (1985). Autotrophic CO_2 fixation by *Desulfovibrio baarsii*: Demonstration of enzyme activities characteristic for the acetyl-CoA pathway. *FEMS Microbiol. Lett.* **28,** 311–315.

Jansen, K., Thauer, R. K., Widdel, F., and Fuchs, G. (1984). Carbon assimilation pathways in sulfate reducing bacteria: Formate, carbon dioxide, carbon monoxide, and acetate assimilation by *Desulfovibrio baarsii*. *Arch. Microbiol.* **138,** 257–262.

Karnholz, A., Küsel, K., Gossner, A., Schramm, A., and Drake, H. L. (2002). Tolerance and metabolic response of acetogenic bacteria toward oxygen. *Appl. Environ. Microbiol.* **68,** 1005–1009.

Küsel, K., Pinkart, H. C., Drake, H. L., and Devereux, R. (1999). Acetogenic and sulfate-reducing bacteria inhabiting the rhizoplane and deep cortex cells of the sea grass *Halodule wrightii*. *Appl. Environ. Microbiol.* **65,** 5117–5123.

Laanbroek, H. J., and Pfennig, N. (1981). Oxidation of short-chain fatty acids by sulfate-reducing bacteria in freshwater and in marine sediments. *Arch. Microbiol.* **128,** 330–335.

Leaphart, A. B., Friez, M. J., and Lovell, C. R. (2003). Formyltetrahydrofolate synthetase sequences from salt marsh plant roots reveal a diversity of acetogenic bacteria and other bacterial functional groups. *Appl. Environ. Microbiol.* **69,** 693–696.

Leaphart, A. B., and Lovell, C. R. (2001). Recovery and analysis of formyltetrahydrofolate synthetase gene sequences from natural populations of acetogenic bacteria. *Appl. Environ. Microbiol.* **67,** 1392–1395.

Liesack, W., Bak, F., Kreft, J. U., and Stackebrandt, E. (1994). *Holophaga foetida* gen. Nov., sp. nov., a new, homoacetogenic bacterium degrading methoxylated aromatic compounds. *Arch. Microbiol.* **162,** 85–90.

Lilburn, T. G., Schmidt, T. M., and Breznak, J. A. (1999). Phylogenetic diversity of termite gut spirochaetes. *Environ. Microbiol.* **1,** 331–345.

Ljungdahl, L. G. (1986). The autotrophic pathway of acetate synthesis in acetogenic bacteria. *Annu. Rev. Microbiol.* **40,** 415–450.

Lorowitz, W. H., and Bryant, M. P. (1984). *Peptostreptococcus productus* strain that grows rapidly with CO_2 as the energy source. *Appl. Environ. Microbiol.* **47,** 961–964.

Lovell, C. R. (2002). Plant-microbe interactions in the marine environment. *In* "Encyclopedia of Environmental Microbiology" (G. Bitton, ed.), Vol. 5, pp. 2539–2554. Wiley, New York.

Lovell, C. R., and Hui, Y. (1991). Design and testing of a functional group-specific DNA probe for the study of natural populations of acetogenic bacteria. *Appl. Environ. Microbiol.* **57**, 2602–2609.

Lovell, C. R., Piceno, Y. M., Quattro, J. M., and Bagwell, C. E. (2000). Molecular analysis of diazotroph diversity in the rhizosphere of the smooth cordgrass, *Spartina alterniflora. Appl. Environ. Microbiol.* **47**, 961–964.

Lovell, C. R., Przybyla, A., and Ljungdahl, L. G. (1990). Primary structure of the thermostable formyltetrahydrofolate synthetase from *Clostridium thermoaceticum. Biochemistry* **29**, 5687–5694.

Morris, J. T., Sundareshwar, P. V., Nietch, C. T., Kjerfve, B., and Cahoon, D. T. (2002). Responses of coastal wetlands to rising sea level. *Ecology* **83**, 2869–2877.

Nishikawa, S., Sonoki, T., Kasahara, T., Obi, T., Kubota, S., Kawai, S., Morohoshi, N., and Katayama, Y. (1998). Cloning and sequencing of the *Sphingomonas* (*Pseudomonas*) *paucimobilis* gene essential for the O-demethylation of vanillate and syringate. *Appl. Environ. Microbiol.* **64**, 836–842.

Normand, P., and Bousquet, J. (1989). Phylogeny of nitrogenase sequences in *Frankia* and other nitrogen-fixing microorganisms. *J. Mol. Evol.* **29**, 436–447.

Opperdoes, F. R. (2003). Phylogenetic analysis using protein sequences. *In* "The Phylogenetic Handbook: A Practical Approach to DNA and Protein Phylogeny" (M. Salemi and A.-M. Vandamme, eds.), pp. 207–235. Cambridge University Press, Cambridge.

Paster, B. J., Dewhirst, F. E., Cooke, S. M., Fussing, V., Poulsen, L. K., and Breznak, J. A. (1996). Phylogeny of not-yet-cultured spirochetes from termite guts. *Appl. Environ. Microbiol.* **62**, 3905–3907.

Radfar, R., Shin, R., Sheldrick, G. M., Minor, W., Lovell, C. R., Odom, J. D., Dunlap, R. B., and Lebioda, L. (2000). The crystal structure of N^{10}-formyltetrahydrofolate synthetase from *Moorella thermoacetica. Biochemistry* **39**, 3920–3926.

Ragsdale, S. W. (1991). Enzymology of the acetyl-CoA pathway of CO_2 fixation. *Crit. Rev. Biochem. Mol. Biol.* **26**, 261–300.

Rankin, C. A., Haslam, G. C., and Himes, R. H. (1993). Sequence and expression of the gene for N-10-formyltetrahydrofolate synthetase from *Clostridium cylindrosporum. Prot. Sci.* **2**, 197–205.

Rooney-Varga, J. N., Devereux, R., Evans, R. S., and Hines, M. E. (1997). Seasonal changes in the relative abundance of uncultivated sulfate-reducing bacteria in a salt marsh sediment and in the rhizosphere of *Spartina alterniflora. Appl. Environ. Microbiol.* **63**, 3895–3901.

Rooney-Varga, J. N., Genthner, B. R. S., Devereux, R., Willis, S. G., Friedman, S. D., and Hines, M. E. (1998). Phylogenetic and physiological diversity of sulfate-reducing bacteria isolated from a salt marsh sediment. *Syst. Appl. Microbiol.* **21**, 557–568.

Salmassi, T. M., and Leadbetter, J. R. (2003). Analysis of genes of tetrahydrofolate-dependent metabolism from cultivated spirochaetes and the gut community of the termite *Zootermopsis angusticollis. Microbiology* **149**, 2529–2537.

Schauder, R., Eikmanns, B., Thauer, R. K., Widdel, F., and Fuchs, G. (1986). Acetate oxidation to CO_2 in anaerobic bacteria via a novel pathway not involving reactions of the citric acid cycle. *Arch. Microbiol.* **145**, 162–172.

Schink, B., Thiemann, V., Laue, H., and Friedrich, M. W. (2002). *Desulfotignum phosphitoxidans* sp. nov., a new marine sulfate reducer that oxidizes phosphite to phosphate. *Arch. Microbiol.* **177**, 381–391.

Sorensen, J., Christensen, D., and Jorgensen, B. B. (1981). Volatile fatty acids and hydrogen as substrates for sulfate-reducing bacteria in anaerobic marine sediment. *Appl. Environ. Microbiol.* **42,** 5–11.

Tanner, R. S., and Woese, C. R. (1994). A phylogenetic assessment of the acetogens. *In* "Acetogenesis" (H. L. Drake, ed.), pp. 254–269. Chapman and Hall, New York.

Thompson, J. D., Gibson, T. J., Plewniak, F., Jeanmougin, F., and Higgins, D. G. (1997). The CLISTAL_X windows interface: Flexible strategies for multiple-sequence alignment aided by quality analysis tools. *Nucleic Acids Res.* **25,** 4876–4882.

Thompson, J. D., Higgins, D. G., and Gibson, T. J. (1994). CLUSTALW: Improving the sensitivity of progressive multiple sequence alignment through sequence weighting, position-specific gap penalties and weight matrix choice. *Nucleic Acids Res.* **22,** 4673–4680.

Wang, R. F., Cao, W. W., and Cerniglia, C. E. (1996). PCR detection and quantification of predominant anaerobic bacteria in human and animal fecal samples. *Appl. Environ. Microbiol.* **62,** 1242–1247.

Welsh, D. T. (2000). Nitrogen fixation in seagrass meadows: Regulation, plant-bacteria interactions and significance to primary productivity. *Ecol. Lett.* **3,** 58–71.

Whitehead, T. R., and Rabinowitz, J. C. (1988). Nucleotide-sequence of the *Clostridium acidiurici* (*Clostridium acidi-urici*) gene for 10-formyltetrahydrofolate synthetase shows extensive amino-acid homology with the trifunctional enzyme C1-tetrahydrofolate synthase from *Saccharomyces cerevisiae*. *J. Bacteriol.* **170,** 3255–3261.

Wood, H. G., and Ljungdahl, L. G. (1991). Autotrophic character of the acetogenic bacteria. *In* "Variations in Autotrophic Life" (L. L. Barton and J. Shively, eds.), pp. 201–250. Academic Press, San Diego.

Further Readings

Kostka, J. R., Gribsholt, B., Petrie, E., Dalton, D., Skelton, H., and Kristensen, E. (2002). The rates and pathways of carbon oxidation in bioturbated saltmarsh sediments. *Limnol. Oceanogr.* **47,** 230–240.

Schink, B. (1994). Diversity, ecology, and isolation of acetogenic bacteria. *In* "Acetogenesis" (H. L. Drake, ed.), pp. 197–235. Chapman and Hall, New York.

[29] Functional Marker Genes for Identification of Sulfate-Reducing Prokaryotes

By Michael Wagner, Alexander Loy, Michael Klein, Natuschka Lee, Niels B. Ramsing, David A. Stahl, and Michael W. Friedrich

Abstract

Sulfate-reducing prokaryotes (SRPs) exploit sulfate as an electron acceptor for anaerobic respiration and exclusively catalyze this essential step of the world's sulfur cycle. Because SRPs are found in many prokaryotic

METHODS IN ENZYMOLOGY, VOL. 397   0076-6879/05 $35.00
Copyright 2005, Elsevier Inc. All rights reserved.   DOI: 10.1016/S0076-6879(05)97029-8

phyla and are often closely related to non-SRPs, 16S rRNA gene-based analyses are inadequate to identify novel lineages of this guild in a cultivation-independent manner. This problem can be solved by comparative sequence analysis of environmentally retrieved gene fragments of the dissimilatory (bi)sulfite (*dsrAB*) and adenosine-5′-phosphosulfate reductases (*apsA*), which encode key enzymes of the SRP energy metabolism. This chapter provides detailed protocols for the application of these functional marker molecules for SRP diversity surveys in the environment. Data from the analysis of *dsrAB* sequence diversity in water samples from the Mariager Fjord in northeast Denmark are presented to illustrate the different steps of the protocols. Furthermore, this chapter describes a novel gel retardation-based technique, suitable for fingerprinting of the approximately 1.9-kb-large *dsrAB* polymerase chain reaction amplification products, which efficiently increases the chance of retrieving rare and novel *dsrAB* sequence types from environmental samples.

Introduction

Dissimilatory reduction of sulfate (also known as anaerobic sulfate respiration) is an essential step in the global sulfur cycle and is exclusively mediated by a phylogenetically and physiologically diverse group of microorganisms, the sulfate-reducing prokaryotes (SRPs). This metabolic trait is found in five bacterial and two archaeal phyla (Fig. 1A) and is thus widely distributed in the prokaryotic tree of life. In many of these phyla, SRPs are closely related to microorganisms that are not capable of anaerobically respiring sulfate. The polyphyletic nature and the high biodiversity of SRPs (more than 130 species have been described to date) render their identification in environmental samples difficult. For example, no 16S rRNA gene-targeting probe or primer is available that allows detecting all recognized members of this guild by hybridization or polymerase chain reaction (PCR)-based techniques. This goal can only be achieved by applying highly multiplexed methods, such as the recently developed 16S rRNA gene-targeting microarray for SRP detection in the environment (Loy *et al.*, 2002). However, because the cultivation-independent discovery of novel SRPs in the environment cannot be achieved using 16S rRNA-based methods, the naturally occurring biodiversity of this guild remained largely hidden until recently.

All SRPs share the unique metabolic pathway of dissimilatory sulfate reduction, which involves the activation of sulfate to adenosine-5′-phosphosulfate (APS) by ATP sulfurylase (EC 2.7.7.4), followed by reduction of APS to (bi)sulfite by APS reductase (EC 1.8.99.2), and subsequently (bi)

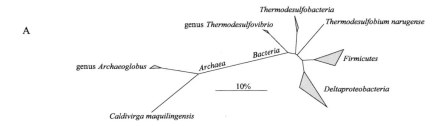

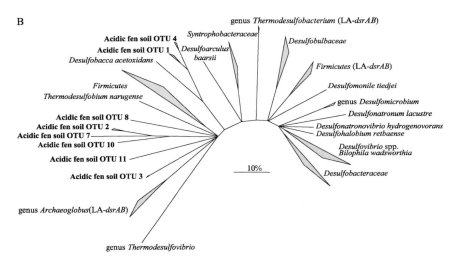

FIG. 1. (A) Schematic 16S rRNA-based phylogenetic tree showing the affiliation of all phyla currently known to harbor SRPs. It should be noted that *Thermocladium modestius,* an extremely thermohilic crenarchaeote not depicted in the tree, also has been reported to show scanty growth with sulfate as an electron acceptor (Itoh *et al.*, 1998). (B) DsrAB protein consensus tree (based on distance matrix analysis) showing the affiliation of yet uncultured, deep-branching lineages of putative sulfate- or sulfite-reducing prokaryotes from low-sulfate, acidic fen soils (boldface type) (Loy *et al.*, 2004). Numbering of operational taxonomic units (OTUs) is according to Loy *et al.* (2004). Bar indicates 10% estimated sequence divergence. Polytomic nodes connect branches for which a relative order could not be determined unambiguously by applying distance matrix, maximum parsimony, and maximum likelihood treeing methods. LA-*dsrAB*, cultured prokaryotes carrying a laterally acquired *dsrAB*.

sulfite reduction to sulfide by dissimilatory (bi)sulfite reductase (EC 1.8.99.3). The latter two key enzymes of the pathway have been shown to be evolutionarily conserved among SRPs (Hipp *et al.*, 1997; Karkhoff-Schweizer *et al.*, 1995; Wagner *et al.*, 1998) and therefore their genes were studied intensively to infer their evolutionary history and to develop selective PCR-based identification systems for SRPs in the environment. These PCR assays are suitable for amplifying either a fragment of *apsA* [the gene encoding the α subunit of the APS reductase (Deplancke *et al.*, 2000; Friedrich, 2002) or large fragments of *dsrAB* (the genes that encode the α and β subunits of the dissimilatory (bi)sulfite reductase) (Klein *et al.*, 2001; Loy *et al.*, 2004; Wagner *et al.*, 1998; Zverlov *et al.*, 2004)] from SRP pure cultures or from DNA extracted from complex microbial communities. Phylogenetic comparison of ApsA, DsrAB, and 16S rRNA gene trees of a wide diversity of cultivated SRPs has revealed that *apsA* and *dsrAB* were generally transmitted vertically during evolution. In consideration of their tremendous phylogenetic breadth, evidence for vertical descent is strongly suggestive of an early origin of dissimilatory sulfate reduction (Friedrich, 2002; Stahl *et al.*, 2002; Wagner *et al.*, 1998). In the meantime, our hypothesis that these genes encode primitive enzymes has received considerable support from sulfur isotope data from a ∼3.47-Ga-old North Pole barite deposit. An isotopic fraction was used to trace the origin of dissimilatory sulfate reduction back to the early Archaean, thus providing the earliest record of a specific microbial metabolism (Shen and Buick, 2004; Shen *et al.*, 2001).

However, the genes encoding ApsA and DsrAB were also affected by several lateral gene transfer events so that not all phylogenetic groupings of SRPs as seen in the 16S rRNA tree are also found in the ApsA or DsrAB trees (Friedrich, 2002; Klein *et al.*, 2001; Stahl *et al.*, 2002; Zverlov *et al.*, 2004) (Figs. 1B and 2). Today, comprehensive ApsA and DsrAB databases, covering most of the genus-level diversity of SRPs, are available and provide a high-resolution framework for the assignment of environmentally retrieved gene sequences of these enzymes (Castro *et al.*, 2002; Chang *et al.*, 2001; Deplancke *et al.*, 2000; Dubilier *et al.*, 2001; Joulian *et al.*, 2001; Loy *et al.*, 2002; Minz *et al.*, 1999; Nakagawa *et al.*, 2002; Perez-Jimenez *et al.*, 2001). With one exception (Deplancke *et al.*, 2000), these studies employed *dsrAB* as marker genes. Many novel SRPs (or eventually sulfite reducers because their *dsrAB* genes are also amplified by the *dsrAB* primers), which are only distantly related to cultured members of this guild, were discovered by this approach (e.g., Fig. 1B) (Baker *et al.*, 2003; Dhillon *et al.*, 2003; Fishbain *et al.*, 2003; Loy *et al.*, 2004). These findings demonstrated that the natural diversity of sulfate (and sulfite)-reducing prokaryotes is far greater than captured by traditional cultivation approaches.

The following detailed protocols are used for applying the *apsA* and *dsrAB* approaches for SRP diversity analysis in environmental samples. To illustrate the different steps, we describe an analysis of *dsrAB* sequence

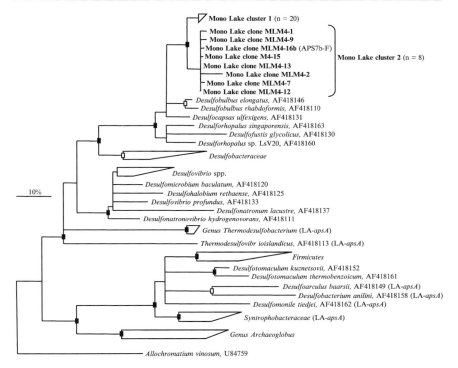

FIG. 2. Phylogenetic consensus tree (based on neighbor-joining analysis) of ApsA amino acid sequences (246 positions) deduced from *apsA* sequences showing the affiliation of clones from an 18-m water depth of Mono Lake (California) (indicated in bold). Water samples were taken in July 2000. Fourteen clones each were retrieved from two libraries, one generated from PCR products amplified with the primer set APS7-F/APS8-R and the other with APS7b-F/APS8-R (e.g., clone MLM4-16b). Apparently, cluster Mono Lake cluster 2 contains a larger number of clones amplified with primer set APS7-F/APS8-R ($n = 7$) than with APS7b-F/APS8-R ($n = 1$); however, only a limited number of clones were analyzed. The tree topology was validated by distance matrix, maximum parsimony, and maximum likelihood treeing methods. Bar indicates 10% estimated sequence divergence. Polytomic nodes connect branches for which a relative order could not be determined unambiguously by applying distance matrix, maximum parsimony, and maximum likelihood treeing methods. Parsimony bootstrap values for branches are indicated by solid ($> 90\%$) or open (75 to 90%) squares. Branches without squares had bootstrap values of less than 75%. LA-*apsA*, cultured prokaryotes carrying a laterally acquired *apsA*.

diversity in water samples from the Mariager Fjord in northeast Denmark. Because of the protected location of this 30-km-long fjord, vertical mixing in the water column is low, thus showing a very stable stratification of oxic and anoxic water layers with a clear chemocline at a 14.5-water depth at the date of sampling (September 13th, 1996). While the total number of bacteria in this fjord is largely constant over depth, numbers of SRPs in the water column increase significantly from above to below the chemocline (Ramsing *et al.*, 1993; Teske *et al.*, 1996). Water samples for our study were taken from sampling site Dybet M3 (56°39'66"N, 09°58'56"E). Details on the sampling procedure were published previously (Ramsing *et al.*, 1993; Teske *et al.*, 1996).

It should be noted that this chapter does not cover the use of [NiFe] hydrogenase gene fragments as markers for SRPs (Voordouw *et al.*, 1990; Wawer *et al.*, 1997; Wawer and Muyzer, 1995) because not all SRPs possess a hydrogen metabolism and the developed PCR primers for detection of this gene target almost exclusively members of the *Desulfovibrionaceae*.

Extraction of DNA from Environmental Samples

The *dsrAB*/*apsA* approach is based on PCR amplification of the respective gene fragments from microbial community DNA. Because some known SRPs are members of the *Archaea* or the bacterial phylum *Firmicutes* and may possess rigid cell walls, the application of DNA extraction procedures, which include effective cell lysis steps, is recommended. This is particularly important if these approaches are applied to hunt for novel SRPs because, obviously, nothing is known about their cell wall structures. Optimally, an appropriate set of different DNA extraction protocols should be used in parallel to minimize biases introduced by the individual protocols (Frostegard *et al.*, 1999; Juretschko *et al.*, 2002; LaMontagne *et al.*, 2002; Martin-Laurent *et al.*, 2001). Furthermore, it is important to take measures to ensure that fragmentation of the extracted DNA is not overly extensive because the use of highly fragmented template DNA can increase the risk of chimera formation during subsequent PCR (Liesack *et al.*, 1991) and also hamper successful amplification of the 1.9-kb-large *dsrAB* fragment. In addition, depending on the sample analyzed, it may be necessary to apply protocols that minimize contamination of the extracted DNA with PCR-inhibiting compounds such as humic acids (Griffiths *et al.*, 2000). DNA extracts retrieved from the same sample but by different methods can be pooled prior to PCR. The template quality of purified DNA can initially be tested by amplification using universal 16S rRNA-targeted primers for PCR.

For details on the extraction and purification of DNA from environmental samples, the reader is referred to Purdy (2005).

PCR Amplification of *dsrAB* and *apsA*

PCR Protocol

Polymerase chain reaction amplification of an approximately 1.9-kb *dsrAB* fragment from environmental DNA is performed using the degenerate primers DSR1Fmix (equimolar mixture of all DSR1F forward primer variants) and DSR4Rmix (equimolar mixture of all DSR4R reverse primer variants) (Table I). This set of primers was designed to perfectly match all cultivated SRPs for which *dsr* operon sequences are available at GenBank and has been used successfully to amplify *dsrAB* from almost 100 pure cultures representing all known SRP lineages (Zverlov *et al.*, 2004). For amplification of a partial *apsA* fragment of approximately 0.9-kb length, the primers APS7-F and its derivatives and APS8-R (Table II) are used for PCR. Analogous to the *dsrAB*-targeted primers, several *apsA*-targeted primer combinations are required to cover the sequence variation at the forward primer-binding site. Note also that certain SRP species (e.g., *Desulfotomaculum halophilum, Desulfosporosinus orientis, Thermodesulfovibrio yellowstonii, Desulfotalea psychrophila, Desulfobacca acetoxidans*) are not detectable using the primer pair APS7uni-F/APS8-R (Friedrich, 2002).

Both positive controls (purified DNA from suitable SRP reference organisms) and negative controls (no DNA) are included in all PCR amplification experiments. Reaction mixtures containing environmental DNA ($\sim$20 ng nucleic acids), 0.5 or 2 μM of each primer variant (2 μM should be used for degenerate primers), deoxynucleotide triphosphates, thermostable DNA polymerase, and an appropriate PCR buffer are prepared in a total volume of 50 μl. We have observed that amplification of *dsrAB* and *apsA* from complex DNA mixtures can be difficult for some environmental samples. In these cases, PCR additives (e.g., 20 mM tetramethylammonium chloride, betaine, or FailSafe PCR premix, Epicentre Technologies, Madison, WI) can additionally be included in the amplification mixture to enhance the specificity and yield of the PCR reaction (Henke *et al.*, 1997; Kovárová and Dráber, 2000). All reactions are prepared at 4° in 0.2-ml reaction tubes to avoid unspecific priming. Amplification is started by placing the reaction tubes into the preheated (94°) block of a thermocycler. A hot-start PCR can be performed to minimize unspecific product formation. Thermal cycling is carried out by an initial denaturation step at 94° for 3 min, followed by 30 to 35 cycles of denaturation at 94° for 40 s, annealing at 48° (*dsrAB*) or 45° (*apsA*) for 40 s, and elongation at 72° for 90 s (*dsrAB*) or 60 s (*apsA*). Please note that the time intervals for the individual steps might vary depending on the PCR

TABLE I

DISSIMILATORY (BI)SULFITE REDUCTASE GENE (*dsrAB*)-TARGETED PRIMERS

Primer	Sequence (5'-3')[a]					Specificity[b]	Reference
DSR1F	AC**S**	CAC	TGG	AAG	CAC	G	*Archaeoglobus fulgidus, A. profundus, Desulfovibrio vulgaris, Bilophila wadsworthia*	Wagner *et al.* (1998)
DSR1Fa	ACC	CA**Y**	TGG	AAA	CAC	G	*Desulfotomaculum thermocisternum, Desulfobulbus rhabdoformis, Desulfobacter vibrioformis, Desulfovibrio desulfuricans*	Loy *et al.* (2004)
DSR1Fb	GGC	CAC	TGG	AAG	CAC	G	*Thermodesulforhabdus norvegica*	Loy *et al.* (2004)
DSR1Fc	ACC	CAT	TGG	AAA	CAT	G	*Desulfobacula toluolica*	Zverlov *et al.* (2005)
DSR1Fd	ACT	CAC	TGG	AAG	CAC	G	*Desulfotalea psychrophila*	Zverlov *et al.* (2005)
DSR4R	GTG	TAG	CAG	TTA	CCG	CA	*A.fulgidus, D. vulgaris, D.rhabdoformis, B. wadsworthia*	Wagner *et al.* (1998)
DSR4Ra	GTG	TAA	CAG	TTT	CCA	CA	*A. profundus*	Loy *et al.* (2004)
DSR4Rb	GTG	TAA	CAG	TTA	CCG	CA	*D. vibrioformis*	Loy *et al.* (2004)
DSR4Rc	GTG	TAG	CAG	TT**K**	CCG	CA	*T. norvegica, D. thermocisternum, D. desulfuricans*	Loy *et al.* (2004)
DSR4Rd	GTG	TAG	CAG	TTA	CCA	CA	*D. toluolica*	Zverlov *et al.* (2005)
DSR4Re	GTG	TAA	CAG	TTA	CCA	CA	*D. psychrophila*	Zverlov *et al.* (2005)

[a] Degenerated positions are shown in bold.
[b] Sulfate- or sulfite-reducing prokaryotes having a fully complementary target site to the respective primers and for which *dsr* operon sequences are available in GenBank.

TABLE II

ADENOSINE-5'-PHOSPHOSULFATE REDUCTASE GENE (*apsA*)-TARGETED PRIMERS

Primer	Sequence (5'-3')[a]							Primer binding site[b]	Reference
APS7-F	GGG	YCT	**KTC**	CGC	YAT	CAA	YAC	205–236	Friedrich *et al.* (2002)
APS7a-F	GGG	YCT	**SAG**	CGC	YAT	CAA	Y	205–234	Friedrich *et al.* (2002)[c]
APS7b-F	GG	YCT	**STC**	CGC	YAT	CAA	Y	206–234	Friedrich *et al.* (2002)[c]
APS8-R	GCA	CAT	GTC	GAG	GAA	GTC	TTC	1139–1159	Friedrich *et al.* (2002)

[a] Degenerated positions are shown in bold.
[b] Positions of the *Desulfovibrio vulgaris* *apsA* open reading frame.
[c] Primer is a modification of primer APS7-F.

machine used. After a final elongation step at 72° for 10 min, cycling is completed and reactions are kept at 4° until further analysis. It is suggested to run several PCR reactions in parallel to minimize tube-to-tube variation (stochastic PCR biases) and to pool the reaction products prior to further analysis. Please note that the low annealing temperatures recommended are important to maximize coverage of the naturally occurring diversity of *dsrAB* and *apsA* genes that might possess mismatches in the respective primer target sites. However, this strategy has the unavoidable disadvantage that unspecific amplification is more likely to occur (see later). If unspecific priming becomes dominant, one can increase the annealing temperature or perform a touch-down PCR but both measures increase the risk of overlooking some SRPs.

It is important to keep in mind that standard PCR amplification techniques do introduce biases and thus are not suited to measure the abundance of the detected genes in the environment (Lueders and Friedrich, 2003; Suzuki *et al.*, 1998; Suzuki and Giovannoni, 1996; von Wintzingerode *et al.*, 1997). Several parameters have been recognized that minimize these biases (Polz and Cavanaugh, 1998) and these might be used to adapt the *dsrAB* and *apsA* PCR amplification protocols presented in this contribution.

The presence and sizes of the amplification products are determined by low melting point agarose (1.5%) gel electrophoresis (NuSieve 3:1; FMC BioProducts, Biozym Diagnostics GmbH, Oldendorf, Germany). SYBR Green I [10 μl SYBR Green I stain, FMC BioProducts, in 100 ml Tris-acetate-EDTA buffer (40 mM Tris, 10 mM sodium acetate, 1 mM EDTA, pH 8.0)]-stained bands (staining time approximately 45 min) are recorded digitally with a gel-imaging system. Due to the low primer annealing temperatures and the degeneracy of the primers, unspecific PCR products of larger or smaller size may occur (Fig. 3). Therefore it is recommended to extract PCR products of the expected size (approximately 1900 and 900 bp for *dsrAB* and *apsA*, respectively) prior to gel retardation or cloning. Extraction is achieved by excision of the respective band from the agarose gel with a glass capillary and subsequent melting of the excised agarose in 80 μl double-distilled water for 10 min at 80°. Dependent on the yield of the PCR reaction, it might be necessary to reamplify the extracted PCR product (using the PCR protocols described earlier) prior to gel retardation.

Application Example: PCR Amplification of dsrAB *from Mariager Fjord*

The *dsrAB* fragment was amplified from different water depths of the Mariager Fjord (12, 13, 14, 15, 16, 18, 21, 23, and 25 m) (Fig. 3). In addition to shorter PCR products, the expected 1.9-kb large fragment was retrieved from most of the samples, while no *dsrAB* could be amplified from DNA extracted from oxic water above the chemocline (<14 m).

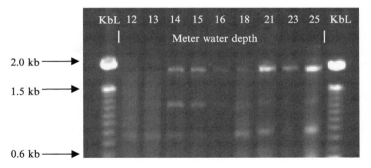

FIG. 3. PCR amplification of *dsrAB* from different water depths of the Mariager Fjord.

Fingerprinting Analysis of *dsrAB* Gene Fragments by Gel Retardation

Fingerprinting analyses such as temperature or denaturing gradient gel electrophoresis (DGGE), terminal restriction fragment length polymorphism (T-RFLP), or single strand conformation polymorphism are used commonly in microbial ecology research to reveal the sequence heterogeneity of complex PCR product mixtures of the same length. The common feature of these techniques is the visualization of a banding pattern of nucleic acid fragments after gel electrophoresis, allowing for direct comparison of multiple samples in a single and relatively rapid experiment. While the 16S rRNA gene and the gene encoding the large subunit of [NiFe] hydrogenase have already been exploited for either DGGE (Kleikemper *et al.*, 2002; Teske *et al.*, 1996; Wawer *et al.*, 1997)- or T-RFLP (Scheid and Stubner, 2001)-based diversity surveys of selected SRP groups in environmental samples, a dedicated protocol for fingerprint analysis of *dsrAB* or *apsA* PCR products has not yet been established.

A simple, inexpensive, but rarely applied fingerprinting method is gel retardation (Schmid *et al.*, 2000; Wawer *et al.*, 1995). Separation of PCR products differing in G + C content in an agarose gel is based on the selective binding of a DNA ligand to A + T- or G + C (depending on the DNA ligand used)-rich sequence motifs. Comparable to DGGE, separated fragments can be extracted from the gel, reamplified, and sequenced either directly or after cloning. Unlike DGGE, gel retardation efficiently separates DNA fragments much larger than 500 bp and is thus suitable for fingerprinting of the approximately large 1.9-kb *dsrAB* PCR amplification products.

Performance of a dsrAB *Gel Retardation Experiment*

Gel retardation of amplified *dsrAB* fragments is performed in a cooled gel electrophoresis unit (Hoefer HE33 submarine gel electrophoresis unit, Amersham Biosciences, Freiburg, Germany) using a modification of the original protocol (Wawer *et al.*, 1995). A 2% low melting point agarose gel consisting of 35 ml 0.5× TBE (44.5 mM Tris-HCl, 44.5 mM boric acid, 1 mM EDTA, pH 8.3) and 0.7 g agarose (NuSieve 3:1; FMC Bioproducts, Biozym Diagnostics GmbH, Oldendorf, Germany) is used. Agarose is dissolved in the TBE buffer by heating. After cooling the agarose mixture to 70°, one unit of the DNA ligand Q-Yellow (1 unit μl^{-1}, Q-Bioanalytik GmbH, Bremerhaven, Germany; formerly known as HA-Yellow) is added per milliliter agarose gel. The bisbenzimide dye Q-Yellow binds preferentially to A-T base pairs and retards (because the dye is coupled to polyethyleneglycol 6000) AT-rich PCR amplicons compared to DNA sequences with low AT content. Due to the photosensitivity of this DNA ligand, all experimental steps should be performed under light protection. Hence, the gel is solidified in the dark after pouring. The electrophoresis is performed by applying a voltage of 10 V cm (agarose gel)$^{-1}$ for 2.5 to 3 h min with 0.5× TBE as running buffer in the cooled gel electrophoresis unit. After the run, the gel is stained with the nucleic acid stain SYBR Green I (5 μl in 50 ml 0.5× TBE) for 1 h in the dark. The DNA bands on the gel are visualized by UV illumination (364 nm) and are cut out with a glass capillary. For subsequent analyses, the gel fragments are dissolved in 80 μl sterile double-distilled water by heating to 80° for 10 min. Subsequently, the extracted DNA fragments are reamplified using the PCR protocol described earlier.

Application Example: Gel Retardation of dsrAB *PCR Products from Mariager Fjord*

dsrAB-PCR products from 14-, 15-, and 16-m water depths of the expected length were extracted from the agarose gel, reamplified, and separated by gel retardation. All three samples showed exactly the same banding pattern, consisting of seven distinct bands (Fig. 4A), indicating a stable community structure of SRP at these depths.

Cloning and Sequencing of PCR Products Amplified from Environmental DNA Extracts

PCR products are ligated as recommended by the manufacturer (Invitrogen Corp.) either into the cloning vector pCR2.1 of the TOPO TA cloning kit (*apsA* gene PCR products) or into the cloning vector

A

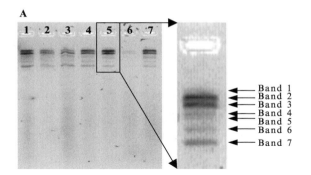

Band 1
Band 2
Band 3
Band 4
Band 5
Band 6
Band 7

B

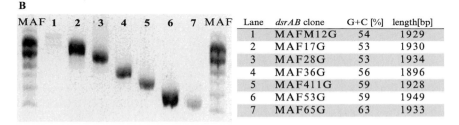

Lane	*dsrAB* clone	G+C [%]	length[bp]
1	MAFM12G	54	1929
2	MAF17G	53	1930
3	MAF28G	53	1934
4	MAF36G	56	1896
5	MAF411G	59	1928
6	MAF53G	59	1949
7	MAF65G	63	1933

FIG. 4. (A) Gel retardation of *dsrAB*-PCR products derived from 14 (triplicates in lanes 1–3)-, 15 (triplicates in lanes 4–6)-, and 16 (lane 7)-m water depths of the Mariager Fjord. (B) Gel retardation of *dsrAB*-PCR products from a 15-m water depth of the Mariager Fjord (MAF) and from Mariager Fjord *dsrAB* clones retrieved from gel retardation bands 1 to 7 (lanes 1 to 7).(Right) Mariager Fjord *dsrAB* clones analyzed in this experiment, their G + C contents, and the length of the sequences.

pCR-XL-TOPO of the TOPO XL cloning kit (*dsrAB* PCR products). To avoid redundant sequencing it is convenient to prescreen *dsrAB* or *apsA* gene clone libraries for similar clones, e.g., by restriction digest with endonucleases such as *Msp*I (Chang *et al.*, 2001; Nakagawa *et al.*, 2002), *Hae*II (Perez-Jimenez *et al.*, 2001), *Hha*I (Castro *et al.*, 2002), or *Mbo*I (Cottrell and Cary, 1999). At least two clones of each restriction pattern group should be sequenced. Nucleotide sequences are determined by modification of the dideoxynucleotide method (Sanger *et al.*, 1977) as described (Purkhold *et al.*, 2000).

After grouping of the clones by sequence similarity or restriction pattern, simple statistical evaluation of the clone library can be performed, e.g., via rarefaction analysis (Heck *et al.*, 1975; Loy *et al.*, 2002; Tipper, 1979), the construction of accumulation or rank-abundance curves (Hughes *et al.*, 2001), or by calculation of the homologues coverage (Giovannoni *et al.*, 1995; Juretschko *et al.*, 2002; Singleton *et al.*, 2001). The results of such analyses can be used to indicate whether a sufficiently exhaustive number of clones have been analyzed, i.e., providing reasonably good coverage of the gene diversity present in a clone library.

Comparative Sequence Analysis

Aligning and phylogenetic analyses of *dsrAB* (DsrAB) and *apsA* (ApsA) sequences are performed by using the respective tools implemented in the ARB program package (Ludwig *et al.*, 2004). The different steps included in phylogeny inference are described later using *dsrAB* as the example. For determining *apsA* phylogeny, an analogous modus operandi is recommended (Friedrich, 2002).

New *dsrAB* sequences are added to an ARB database that contains all aligned *dsrAB* sequences of recognized (Friedrich, 2002; Klein *et al.*, 2001; Wagner *et al.*, 1998; Zverlov *et al.*, 2005) and uncultured SRPs available in GenBank (Benson *et al.*, 2002). As a starting point, a *dsrAB* database containing 97 aligned sequences from cultured sulfate/sulfite-reducing microorganisms is publicly available for download from http://www.micro-bial-ecology.net/download.asp. Deduced amino acid sequences of newly determined *dsrAB* sequences are aligned manually using the editor GDE 2.2 (Smith *et al.*, 1994) (please note that the GDE editor is not implemented in all ARB versions). In the next step, the respective nucleic acid sequences are aligned according to the alignment of amino acids. For phylogeny inference of DsrAB amino acid sequences, insertions and deletions are removed from the data set by using a suitable alignment mask (indel filter) leaving a total of 543 amino acid positions (α subunit, 327; β subunit, 216) for comparative analyses. Distance matrix (DM) (using FITCH with global rearrangements and randomized input order of species) and maximum parsimony (MP) trees are calculated with the PHYLogeny Inference Package (PHYLIP) (Felsenstein, 1995). In addition, the programs MOLPHY (Adachi and Hasegawa, 1996) and TREE-PUZZLE (Strimmer and von Haeseler, 1996) are used to infer maximum-likelihood (ML) trees with JTT-f as the amino acid replacement model. Parsimony bootstrap analysis for protein trees are performed with PHYLIP. One hundred to 1000 bootstrap resamplings are analyzed for each calculation. The results of the phylogenetic analysis should be depicted as a consensus tree, which is drawn according to established protocols (Ludwig *et al.*, 1998). In addition, analyses should be performed to check for chimeric sequences. For this purpose, alignment masks should be applied that select only the 327 amino acid positions of the α subunit and the 216 amino acid positions of the β subunit, respectively, for phylogenetic analysis.

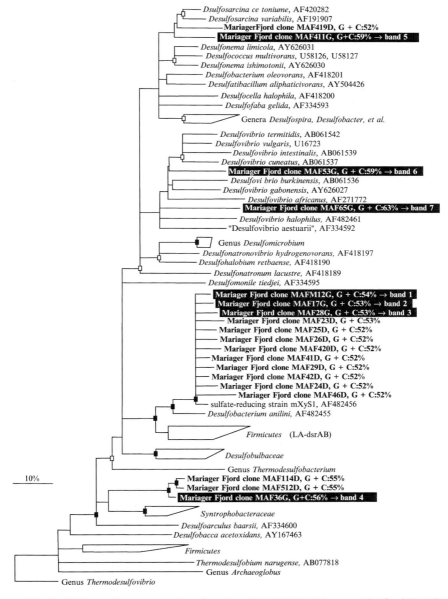

FIG. 5. Phylogenetic consensus tree (based on the FITCH distance method) of DsrAB amino acid sequences deduced from *dsrAB* sequences greater than 1750 bases showing the affiliation of *dsrAB* clones from a 15-m water depth of the Mariager Fjord (indicated in boldface type). Clones that were retrieved using gel retardation are boxed in black. The

Substantial incongruencies of the affiliation of newly determined environmental DsrAB sequences between α- and β-subunit trees are indicative for a chimeric origin of the analyzed sequence.

Some previous *dsrAB*-based environmental studies inferred SRP phylogeny only from partial sequence data (part of either the *dsrA* or the *dsrB* gene). We would like to encourage all researchers in this field to only deposit complete *dsrAB* sequences because they allow for more substantive phylogenetic inferences and comparisons of results between studies. As it stands now, partial *dsrAB* sequences occurring in the databases do not always overlap. If published partial sequences need to be included in a phylogenetic DsrAB tree, they should be added individually to a suitable full-length sequence tree without changing the overall tree topology, e.g., by using the ARB treeing tool PARSIMONY_INTERAKTIV (Ludwig *et al.*, 2004).

During surveys of environmental *dsrAB*-based SRP diversity, several novel phylogenetic lineages have been discovered (e.g., Fig. 1B). These novel gene sequences might indeed represent novel SRPs, but could theoretically represent pseudogenes that differ in sequence from recognized *dsrAB* sequences due to a higher mutation frequency caused by lack of selective pressure. Such a scenario is particularly likely if the retrieved *dsrAB* sequences contain unexpected stop codons or show mutations in the cys motif consensus sequences Cys-X_5-Cys and Cys-X_3-Cys, which are present in both subunits and are essential for binding the $[Fe_4S_4]$-siroheme cofactor of (bi)sulfite reductase (Crane *et al.*, 1995). However, functional (bi)sulfite reductases (Hipp *et al.*, 1997; Klein *et al.*, 2001) contain a truncated Cys-X_5-Cys motif (without the first Cys) in the β subunit. Furthermore, the nonsynonymous/synonymous substitution rate ratios (ω) of branches leading to the new *dsrAB* sequence clusters can be determined and used to identify putative pseudogenes (Loy *et al.*, 2004). Low ω values (<1) suggest a strong purifying (negative) selection indicative of expressed, functionally active proteins, whereas a ω value of greater than 1 indicates diversifying selection. An ω value of exactly 1 indicates that amino acid change is neutral and would be expected for a pseudogene (Yang, 1997, 1998; Yang *et al.*, 2000).

G + C content was calculated for each *dsrAB* clone sequence. Bar indicates 10% estimated sequence divergence. Polytomic nodes connect branches for which a relative order could not be determined unambiguously by applying distance matrix, maximum parsimony, and maximum likelihood treeing methods. Parsimony bootstrap values for branches are indicated by solid ($>90\%$) or open (75 to 90%) squares. Branches without squares had bootstrap values of less than 75%. The newly determined *dsrAB* sequences from Mariager Fjord have been deposited at GenBank under the accession numbers AY865325–AY865333.

Application Example: Phylogenetic Analysis of dsrAB *Sequences from the Mariager Fjord*

After direct cloning of the *dsrAB* PCR product from a 15-m water depth in *Escherichia coli*, 12 randomly selected *dsrAB* clones were sequenced and analyzed phylogenetically. Nine clones formed a monophyletic branch with *Desulfobacterium anilini* and the SRP strain mXyS1, while two clustered with members of the *Syntrophobacteraceae* and one with the genus *Desulfosarcina* (Fig. 5). Additionally, each of the seven bands that were separated by subjecting the *dsrAB* PCR product from a 15-m depth to gel retardation was extracted from the gel, reamplified, cloned, and sequenced. The seven *dsrAB* sequences, representing the seven bands, either differed in G + C content or varied slightly in length (Fig. 4B). Phylogenetic analysis of the deduced DsrAB amino acid sequences assigned five of these sequences to the three clusters identified earlier (Fig. 5). The remaining two sequences formed independent lineages affiliated with members of the family *Desulfovibrionaceae* and were not identified by direct cloning (without subsequent gel retardation) of the *dsrAB* PCR product. Although two clone pairs, MAF17G (band 2)/ MAF28G (band 3) and MAF411G (band 5)/MAF53G (band 6), with an identical G + C content were detected, the individual sequence fragments could be distinguished clearly by gel retardation. The different migration of *dsrAB* PCR products in the retardation gel is thus not solely caused by differences in G + C content, but also by slight variations in fragment length and probably also by different distribution of A + T-rich regions in the different *dsrAB* sequences. It is noteworthy that *dsrAB* G + C content does not correlate with a particular phylogenetic grouping (Fig. 5). Although no phylogenetic information can thus be inferred from the banding pattern, gel retardation of *dsrAB* PCR fragments prior to cloning provides an efficient means to increase the chance of retrieving rare sequence types (e.g., sequences representing the gel retardation bands 6 and 7 from the Mariager Fjord sample).

Acknowledgments

We are grateful for Samantha B. Joye (Athens, GA) for providing DNA extracts of Mono Lake water samples. The development of the methods described in this chapter was partially supported by a grant in the framework of the BIOLOG and BIOLOG II programs of the bmb + f (project 01 LC 0021A-TP2) to M.W., a grant from NSF Grant DEB-0213186 (systematics) to D.A.S., and a Marie Curie Intra-European Fellowship (VENTSULFUR-MICDIV) within the 6th European Community Framework Programme to A.L.

References

Adachi, J., and Hasegawa, M. (1996). MOLPHY version 2.3: Programs for molecular phylogenetics based on maximum likelihood. *Comput. Sci. Monogr.* **28**, 1–150.

Baker, B. J., Moser, D. P., MacGregor, B. J., Fishbain, S., Wagner, M., Fry, N. K., Jackson, B., Speolstra, N., Loos, S., Takai, K., Lollar, B. S., Fredrickson, J., Balkwill, D., Onstott, T. C., Wimpee, C. F., and Stahl, D. A. (2003). Related assemblages of sulphate-reducing bacteria associated with ultradeep gold mines of South Africa and deep basalt aquifers of Washington State. *Environ. Microbiol.* **5**, 267–277.

Benson, D. A., Karsch-Mizrachi, I., Lipman, D. J., Ostell, J., Rapp, B. A., and Wheeler, D. L. (2002). GenBank. *Nucleic Acids Res.* **30**, 17–20.

Castro, H., Reddy, K. R., and Ogram, A. (2002). Composition and function of sulfate-reducing prokaryotes in eutrophic and pristine areas of the Florida Everglades. *Appl. Environ. Microbiol.* **68**, 6129–6137.

Chang, Y. J., Peacock, A. D., Long, P. E., Stephen, J. R., McKinley, J. P., Macnaughton, S. J., Hussain, A. K., Saxton, A. M., and White, D. C. (2001). Diversity and characterization of sulfate-reducing bacteria in groundwater at a uranium mill tailings site. *Appl. Environ. Microbiol.* **67**, 3149–3160.

Cottrell, M. T., and Cary, S. C. (1999). Diversity of dissimilatory bisulfite reductase genes of bacteria associated with the deep-sea hydrothermal vent polychaete annelid *Alvinella pompejana. Appl. Environ. Microbiol.* **65**, 1127–1132.

Crane, B. R., Siegel, L. M., and Getzoff, E. D. (1995). Sulfite reductase structure at 1.6 A: Evolution and catalysis for reduction of inorganic anions. *Science* **270**, 59–67.

Deplancke, B., Hristova, K. R., Oakley, H. A., McCracken, V. J., Aminov, R., Mackie, R. I., and Gaskins, H. R. (2000). Molecular ecological analysis of the succession and diversity of sulfate-reducing bacteria in the mouse gastrointestinal tract. *Appl. Environ. Microbiol.* **66**, 2166–2174.

Dhillon, A., Teske, A., Dillon, J., Stahl, D. A., and Sogin, M. L. (2003). Molecular characterization of sulfate-reducing bacteria in the Guaymas Basin. *Appl. Environ. Microbiol.* **69**, 2765–2772.

Dubilier, N., Mulders, C., Ferdelman, T., de Beer, D., Pernthaler, A., Klein, M., Wagner, M., Erseus, C., Thiermann, F., Krieger, J., Giere, O., and Amann, R. (2001). Endosymbiotic sulfate-reducing and sulphide-oxidizing bacteria in an oligochaete worm. *Nature* **411**, 298–302.

Felsenstein, J. (1995). PHYLIP: Phylogeny inference package Department of Genetics, University of Washington, Seattle.

Fishbain, S., Dillon, J. G., Gough, H. L., and Stahl, D. A. (2003). Linkage of high rates of sulfate reduction in Yellowstone hot springs to unique sequence types in the dissimilatory sulfate respiration pathway. *Appl. Environ. Microbiol.* **69**, 3663–3667.

Friedrich, M. W. (2002). Phylogenetic analysis reveals multiple lateral transfers of adenosine-5′-phosphosulfate reductase genes among sulfate-reducing microorganisms. *J. Bacteriol.* **184**, 278–289.

Frostegard, A., Courtois, S., Ramisse, V., Clerc, S., Bernillon, D., Le Gall, F., Jeannin, P., Nesme, X., and Simonet, P. (1999). Quantification of bias related to the extraction of DNA directly from soils. *Appl. Environ. Microbiol.* **65**, 5409–5420.

Giovannoni, S. J., Mullins, T. D., and Field, K. G. (1995). Microbial diversity in oceanic systems: rRNA approaches to the study of unculturable microbes. *In* "Molecular Ecology of Aquatic Microbes" (J. Joint, ed.), Vol. G38, pp. 217–248. Springer Verlag, Berlin.

Griffiths, R. I., Whiteley, A. S., O' Donnell, A. G., and Bailey, M. J. (2000). Rapid method for coextraction of DNA and RNA from natural environments for analysis of ribosomal DNA- and rRNA-based microbial community composition. *Appl. Environ. Microbiol.* **66**, 5488–5491.

Heck, K. L., van Belle, G., and Simberloff, D. (1975). Explicit calculation of the rarefaction diversity measurement and the determination of sufficient sample size. *Ecology* **56**, 1459–1461.

Henke, W., Herdel, K., Jung, K., Schnorr, D., and Loening, S. A. (1997). Betaine improves the PCR amplification of GC-rich DNA sequences. *Nucleic Acids Res.* **25**, 3957–3958.

Hipp, W. M., Pott, A. S., Thum-Schmitz, N., Faath, I., Dahl, C., and Trüper, H. G. (1997). Towards the phylogeny of APS reductases and sirohaem sulfite reductases in sulfate-reducing and sulfur-oxidizing prokaryotes. *Microbiology* **143**, 2891–2902.

Hughes, J. B., Hellmann, J. J., Ricketts, T. H., and Bohannan, B. J. (2001). Counting the uncountable: Statistical approaches to estimating microbial diversity. *Appl. Environ. Microbiol.* **67**, 4399–4406.

Itoh, T., Suzuki, K., and Nakase, T. (1998). *Thermocladium modestius* gen. nov., sp. nov., a new genus of rod-shaped, extremely thermophilic crenarchaeote. *Int. J. Syst. Bacteriol.* **48**, 879–887.

Joulian, C., Ramsing, N. B., and Ingvorsen, K. (2001). Congruent phylogenies of most common small-subunit rRNA and dissimilatory sulfite reductase gene sequences retrieved from estuarine sediments. *Appl. Environ. Microbiol.* **67**, 3314–3318.

Juretschko, S., Loy, A., Lehner, A., and Wagner, M. (2002). The microbial community composition of a nitrifying-denitrifying activated sludge from an industrial sewage treatment plant analyzed by the full-cycle rRNA approach. *Syst. Appl. Microbiol.* **25**, 84–99.

Karkhoff-Schweizer, R. R., Huber, D. P., and Voordouw, G. (1995). Conservation of the genes for dissimilatory sulfite reductase from *Desulfovibrio vulgaris* and *Archaeoglobus fulgidus* allows their detection by PCR. *Appl. Environ. Microbiol.* **61**, 290–296.

Kleikemper, J., Schroth, M. H., Sigler, W. V., Schmucki, M., Bernasconi, S. M., and Zeyer, J. (2002). Activity and diversity of sulfate-reducing bacteria in a petroleum hydrocarbon-contaminated aquifer. *Appl. Environ. Microbiol.* **68**, 1516–1523.

Klein, M., Friedrich, M., Roger, A. J., Hugenholtz, P., Fishbain, S., Abicht, H., Blackall, L. L., Stahl, D. A., and Wagner, M. (2001). Multiple lateral transfers of dissimilatory sulfite reductase genes between major lineages of sulfate-reducing prokaryotes. *J. Bacteriol.* **183**, 6028–6035.

Kovárová, M., and Dráber, P. (2000). New specificity and yield enhancer of polymerase chain reactions. *Nucleic Acids Res.* **28**, E70.

LaMontagne, M. G., Michel, F. C., Jr., Holden, P. A., and Reddy, C. A. (2002). Evaluation of extraction and purification methods for obtaining PCR-amplifiable DNA from compost for microbial community analysis. *J. Microbiol. Methods* **49**, 255–264.

Liesack, W., Weyland, H., and Stackebrandt, E. (1991). Potential risks of gene amplification by PCR as determined by 16S rDNA analysis of a mixed-culture of strict barophilic bacteria. *Microb. Ecol.* **21**, 191–198.

Loy, A., Daims, H., and Wagner, M. (2002). Activated sludge: Molecular techniques for determining community composition. *In* "The Encyclopedia of Environmental Microbiology" (G. Bitton, ed.), pp. 26–43. Wiley, New York.

Loy, A., Küsel, K., Lehner, A., Drake, H. L., and Wagner, M. (2004). Microarray and functional gene analyses of sulfate-reducing prokaryotes in low sulfate, acidic fens reveal co-occurence of recognized genera and novel lineages. *Appl. Environ. Microbiol.* **70**, 6998–7009.

Loy, A., Lehner, A., Lee, N., Adamczyk, J., Meier, H., Ernst, J., Schleifer, K.-H., and Wagner, M. (2002). Oligonucleotide microarray for 16S rRNA gene-based detection of all recognized lineages of sulfate-reducing prokaryotes in the environment. *Appl. Environ. Microbiol.* **68,** 5064–5081.

Ludwig, W., Strunk, O., Klugbauer, S., Klugbauer, N., Weizenegger, M., Neumaier, J., Bachleitner, M., and Schleifer, K. H. (1998). Bacterial phylogeny based on comparative sequence analysis. *Electrophoresis* **19,** 554–568.

Ludwig, W., Strunk, O., Westram, R., Richter, L., Meier, H., Yadhukumar, Buchner, A., Lai, T., Steppi, S., Jobb, G., Forster, W., Brettske, I., Gerber, S., Ginhart, A. W., Gross, O., Grumann, S., Hermann, S., Jost, R., Konig, A., Liss, T., Lussmann, R., May, M., Nonhoff, B., Reichel, B., Strehlow, R., Stamatakis, A., Stuckmann, N., Vilbig, A., Lenke, M., Ludwig, T., Bode, A., and Schleifer, K. H. (2004). ARB: A software environment for sequence data. *Nucleic Acids Res.* **32,** 1363–1371.

Lueders, T., and Friedrich, M. W. (2003). Evaluation of PCR amplification bias by terminal restriction fragment length polymorphism analysis of small-subunit rRNA and mcrA genes by using defined template mixtures of methanogenic pure cultures and soil DNA extracts. *Appl. Environ. Microbiol.* **69,** 320–326.

Martin-Laurent, F., Philippot, L., Hallet, S., Chaussod, R., Germon, J. C., Soulas, G., and Catroux, G. (2001). DNA extraction from soils: Old bias for new microbial diversity analysis methods. *Appl. Environ. Microbiol.* **67,** 2354–2359.

Minz, D., Flax, J. L., Green, S. J., Muyzer, G., Cohen, Y., Wagner, M., Rittmann, B. E., and Stahl, D. A. (1999). Diversity of sulfate-reducing bacteria in oxic and anoxic regions of a microbial mat characterized by comparative analysis of dissimilatory sulfite reductase genes. *Appl. Environ. Microbiol.* **65,** 4666–4671.

Nakagawa, T., Hanada, S., Maruyama, A., Marumo, K., Urabe, T., and Fukui, M. (2002). Distribution and diversity of thermophilic sulfate-reducing bacteria within a Cu-Pb-Zn mine (Toyoha, Japan). *FEMS Microbiol. Ecol.* **41,** 199–209.

Perez-Jimenez, J. R., Young, L. Y., and Kerkhof, L. J. (2001). Molecular characterization of sulfate-reducing bacteria in anaerobic hydrocarbon-degrading consortia and pure cultures using the dissimilatory sulfite reductase (*dsrAB*) genes. *FEMS Microbiol. Ecol.* **35,** 145–150.

Polz, M. F., and Cavanaugh, C. M. (1998). Bias in template-to-product ratios in multitemplate PCR. *Appl. Environ. Microbiol.* **64,** 3724–3730.

Purdy, K. (2005). Nucleic acid recovery from complex environmental samples. *Methods Enzymol.* **397**(16), 2005 (this volume).

Purkhold, U., Pommering-Röser, A., Juretschko, S., Schmid, M. C., Koops, H.-P., and Wagner, M. (2000). Phylogeny of all recognized species of ammonia oxidizers based on comparative 16S rRNA and *amoA* sequence analysis: Implications for molecular diversity surveys. *Appl. Environ. Microbiol.* **66,** 5368–5382.

Ramsing, N. B., Kühl, M., and Jørgensen, B. B. (1993). Distribution of sulfate-reducing bacteria, O_2, and H_2S in photosynthetic biofilms determined by oligonucleotide probes and microelectrodes. *Appl. Environ. Microbiol.* **59,** 3840–3849.

Sanger, F., Nicklen, S., and Coulson, A. R. (1977). DNA sequencing with chain-terminating inhibitors. *Proc. Natl. Acad. Sci. USA* **74,** 5463–5467.

Scheid, D., and Stubner, S. (2001). Structure and diversity of Gram-negative sulfate-reducing bacteria on rice roots. *FEMS Microbiol. Ecol.* **36,** 175–183.

Schmid, M., Twachtmann, U., Klein, M., Strous, M., Juretschko, S., Jetten, M., Metzger, J. W., Schleifer, K. H., and Wagner, M. (2000). Molecular evidence for genus level diversity of bacteria capable of catalyzing anaerobic ammonium oxidation. *Syst. Appl. Microbiol.* **23,** 93–106.

Shen, Y., and Buick, R. (2004). The antiquity of microbial sulfate reduction. *Earth-Sci. Rev.* **64,** 243–272.

Shen, Y., Buick, R., and Canfield, D. E. (2001). Isotopic evidence for microbial sulphate reduction in the early Archaean era. *Nature* **410**, 77–81.

Singleton, D. R., Furlong, M. A., Rathbun, S. L., and Whitman, W. B. (2001). Quantitative comparisons of 16S rRNA gene sequence libraries from environmental samples. *Appl. Environ. Microbiol.* **67**, 4374–4376.

Smith, S. W., Overbeek, R., Woese, C. R., Gilbert, W., and Gillevet, P. M. (1994). The Genetic Data Environment (GDE). An expandable graphic interface for manipulating molecular information. *CABIOS* **10**, 671–675.

Stahl, D. A., Fishbain, S., Klein, M., Baker, B. J., and Wagner, M. (2002). Origins and diversification of sulfate-respiring microorganisms. *Antonie van Leeuwenhoek* **81**, 189–195.

Strimmer, K., and von Haeseler, A. (1996). Quartet puzzling: A quartet maximum likelihood method for reconstructing tree topologies. *Mol. Biol. Evol.* **13**, 964–969.

Suzuki, M., Rappe, M. S., and Giovannoni, S. J. (1998). Kinetic bias in estimates of coastal picoplankton community structure obtained by measurements of small-subunit rRNA gene PCR amplicon length heterogeneity. *Appl. Environ. Microbiol.* **64**, 4522–4529.

Suzuki, M. T., and Giovannoni, S. J. (1996). Bias caused by template annealing in the amplification of mixtures of 16S rRNA genes by PCR. *Appl. Environ. Microbiol.* **62**, 625–630.

Teske, A., Wawer, C., Muyzer, G., and Ramsing, N. B. (1996). Distribution of sulfate-reducing bacteria in a stratified fjord (Mariager Fjord, Denmark) as evaluated by most-probable-number counts and denaturing gradient gel electrophoresis of PCR-amplified ribosomal DNA fragments. *Appl. Environ. Microbiol.* **62**, 1405–1415.

Tipper, J. C. (1979). Rarefaction and rarefiction: The use and abuse of a method in paleontology. *Paleobiology* **5**, 423–434.

von Wintzingerode, F., Göbel, U. B., and Stackebrandt, E. (1997). Determination of microbial diversity in environmental samples: Pitfalls of PCR-based rRNA analysis. *FEMS Microbiol. Rev.* **21**, 213–229.

Voordouw, G., Niviere, V., Ferris, F. G., Fedorak, P. M., and Westlake, D. W. S. (1990). Distribution of hydrogenase genes in *Desulfovibrio* spp. and their use in identification of species from the oil field environment. *Appl. Environ. Microbiol.* **56**, 3748–3754.

Wagner, M., Roger, A. J., Flax, J. L., Brusseau, G. A., and Stahl, D. A. (1998). Phylogeny of dissimilatory sulfite reductases supports an early origin of sulfate respiration. *J. Bacteriol.* **180**, 2975–2982.

Wawer, C., Jetten, M. S., and Muyzer, G. (1997). Genetic diversity and expression of the [NiFe] hydrogenase large-subunit gene of *Desulfovibrio* spp. in environmental samples. *Appl. Environ. Microbiol.* **63**, 4360–4369.

Wawer, C., and Muyzer, G. (1995). Genetic diversity of *Desulfovibrio* spp. in environmental samples analyzed by denaturing gradient gel electrophoresis of [NiFe] hydrogenase gene fragments. *Appl. Environ. Microbiol.* **61**, 2203–2210.

Wawer, C., Ruggeberg, H., Meyer, G., and Muyzer, G. (1995). A simple and rapid electrophoresis method to detect sequence variation in PCR-amplified DNA fragments. *Nucleic Acids Res.* **23**, 4928–4929.

Yang, Z. (1997). PAML: A program package for phylogenetic analysis by maximum likelihood. *Comput. Appl. Biosci.* **13**, 555–556.

Yang, Z. (1998). Likelihood ratio tests for detecting positive selection and application to primate lysozyme evolution. *Mol. Biol. Evol.* **15**, 568–573.

Yang, Z., Nielsen, R., Goldman, N., and Pedersen, A. M. (2000). Codon-substitution models for heterogeneous selection pressure at amino acid sites. *Genetics* **155**, 431–449.

Zverlov, V., Klein, M., Lücker, S., Friedrich, M. W., Kellermann, J., Stahl, D. A., Loy, A., and Wagner, M. (2005). Lateral gene transfer of dissimilatory (bi)sulfite reductase revisited. *J. Bacteriol.* **187**, 2203–2208.

Author Index

H

S

Subject Index

A

Accumulation curve, *see* Rarefaction
Acetogens
  acetyl-CoA pathway enzymes, 455–456
  diversity, 455
  formyltetrahydrofolate synthase sequences
    and community analysis
    phylogenetic analysis, 461–465
    polymerase chain reaction
      amplification reactions, 458
      cloning of amplicons, 459–460
      primer design and testing, 457, 459
      thermal cycling, 458–459
    prospects, 465–466
    sequence alignment, 460–461
  metabolic overview, 455
Acetyl-CoA carboxylase
  assay, 217
  3-hydroxypropionate cycle, 213
AFM, *see* Atomic force microscopy
Ammonia-oxidizing bacteria
  abundance, 396
  ammonia oxidation measurement in
    seawater and sediments
    inhibitor studies, 397–398
    nitrogen isotope tracer studies
      $^{15}NO_2$ analysis, 401–402
      principles, 398–400
      rate calculations, 402–403
      seawater sample collection and
        incubation, 400–401
    nutrient inventory approach,
      396–397
  amoA detection in environment
    DNA extraction, 405–406
    polymerase chain reaction, 406–409
    sample collection, 405
  metabolic overview, 395–396
  phylogeny, 403–404
amoA, *see* Ammonia-oxidizing bacteria
Anaerobic ammonium oxidation bacteria
  Candidatus species, 35–36, 51–52, 250

community analysis using ribosomal RNA
  gene analysis
  fluorescence *in situ* hybridization
    detection in environmental samples,
    47–49
  intergenic spacer region targeting,
    49–51
  microautoradiography combination,
    51–52
  phylogenetic analysis, 45–47
  polymerase chain reaction
    amplification reactions, 40, 45
    cloning of products, 45
    controls, 45
    primers, 39–45
  lipid analysis, 52
  metabolic activity measurement, 53–54
  Percoll density gradient centrifugation for
    separation, 38–39
  reactor systems for enrichment and
    cultivation
    CANON system for cocultivation with
      aerobic ammonia oxidizers, 38
    sequencing batch reactor, 36–38
Anaerobic hydrocarbon-degrading bacteria
  enrichment culture
    initial incubations, 23–24
    sampling, 19, 22
    sediment-free enrichments
      media, 24–25
      overview, 24
      substrates, 25–26
  hydrocarbon substrates, 17–18,
    21–22
  purification and isolation, 26–29
  purity control, 29–30
  strain features and classification, 19
Anaerobic Lap System, anoxic media
  preparation, 12–13
Anaerobic sulfate respiration, *see*
  Sulfate-reducing bacteria
Anammox organisms, *see* Anaerobic
  ammonium oxidation bacteria

I

Iron-oxidizing bacteria
anaerobic iron oxidation coupled to nitrate
reduction and *Ferroglobus placidus*
culture, 120–121
anoxygenic phototroph culture,
121–122
microaerophilic lithotroph culture
bioreactors, 118
biphasic slant culture, 117
culture *in situ*, 118
epifluorescence microscopy, 120
Gallionella, 113
gel-stabilized gradient tube preparation,
113, 115–116
gradient plate preparation, 116–117
liquid culture, 117
media recipes, 113–115
stock solutions
Fe(II) concentration
spectrophotometric assay, 119
$FeCl_2$, 119
FeS, 118–119
neutrophilic bacteria culture
challenges, 112

L

Light field mapping, *see*
Fiber-optic microsensors

M

Malonyl-CoA reductase assay, 217–218
3-hydroxypropionate cycle, 213, 215
Malyl-CoA ligase
assay, 219–220
3-hydroxypropionate cycle, 215
MAR, *see*
Microautoradiography–fluorescence
in situ hybridization
mch
function, 444
phylogenetic analysis, 449–452
polymerase chain reaction and community
analysis
clone library construction, 447
DNA isolation, 445–446
Lake Washington study, 448–449

primer design and testing, 444–446, 452
restriction fragment length
polymorphism analysis, 447–448
thermal cycling, 445
two-step amplification, 446–447
MCR, *see* Methyl-coenzyme M reductase
mcrA, *see* Methyl-coenzyme M reductase
Metabolic flux measurement *in situ*, *see*
Radiotracer microinjection
Metabolic reductively dechlorinating
bacteria, *see* Reductively dechlorinating
bacteria
Methane monooxygenase
isoforms, 414
membrane-bound enzyme gene, *see pmoA*
soluble enzyme gene, *see mmoX*
species distribution, 414
Methanogens
community analysis, *see* Methyl-coenzyme
M reductase
habitats, 428–429, 431
phylogenetic analysis
mcrA, 429–430, 438–439
ribosomal RNA genes, 429
Methanol dehydrogenase gene, *see mxaF*
Methanopterin-dependent enzyme genes,
see fae; *fhcD*; *mch*; *mtdB*
function, 444
phylogenetic analysis, 449–452
polymerase chain reaction and
community analysis
clone library construction, 447
DNA isolation, 445–446
Lake Washington study, 448–449
primer design and testing, 444–446, 452
restriction fragment length
polymorphism analysis, 447–448
thermocycling, 445
two-step amplification, 446–447
Methanotrophs
distribution, 413, 415
diversity and phylogeny, 414–415
ecological significance, 413–414
gene analysis from environmental samples
DNA extraction, 415–416
polymerase chain reaction
mmoX, 419, 421–422
mxaF, 419–420, 422
pmoA, 416, 419–421
primer sets, 417–418, 420

S

SBR, *see* Sequencing batch reactor
Scotophobic response, *see* Chemotaxis assay
Sequencing batch reactor, anaerobic
    ammonium oxidation bacteria
    enrichment and cultivation, 36–38
STARFISH, *see*
    Microautoradiography–fluorescence
    in situ hybridization
Succinyl-CoA:citramalate CoA transferase
    assay, 220
    3-hydroxypropionate cycle, 215
Succinyl-CoA:malate CoA transferase
    assay, 218–219
    3-hydroxypropionate cycle, 215
Sulfate-reducing bacteria
    diversity, 470, 472
    marker gene analysis using *dsrAB* and *apsA*
        comparative sequence analysis of *dsrAB*
            genes, 482–485
        denaturing gradient gel electrophoresis
            of *dsrAB* amplicons, 479–480
        DNA extraction from environmental
            samples, 474
        gene functions, 470–472
        polymerase chain reaction
            amplicon cloning and sequencing,
                480–482
            apsA, 475, 477–478
            dsrAB, 475–476, 478
        metabolism overview, 470
        phylogenetic analysis
            apsA, 473
            dsrAB, 482–485
            ribosomal RNA genes, 470–471

T

Temperature microoptodes, design and
    applications, 193
Terminal restriction fragment length
    polymorphism
    amplification product purification,
        313, 322

caveats, 326
comparative analysis of
    community profiles
    algorithms, 325
    aligned profiles, 325
    data downloading, 324–325
    overview, 317
    phylogenetic analysis,
        325–326
experimental design, 311
methanogenic *Archaea* community
    analysis using *mcrA*
    cloning and sequencing of gene
        fragments, 437–438
    DNA extraction, 434–435
    interpretation, 436–437,
        439–440
    overview, 431–432
    polymerase chain reaction,
        435–436
    practical example, 432–434
nucleic acid isolation, 311,
    318–319
polymerase chain reaction
    labeled primer reactions, 322
    pilot experiment and optimization,
        312–313, 321–322
    primers
        selection, 320
        validation, 320–321
    principles and advantages in community
        analysis, 309–310
    restriction endonuclease digestion, 314,
        322–324
    sequencing gels and capillaries,
        314–317, 324
Termite, *see* Radiotracer
    microinjection
T-RFLP, *see* Terminal restriction fragment
    length polymorphism

W

Widdel flask, anoxic media
    preparation, 14

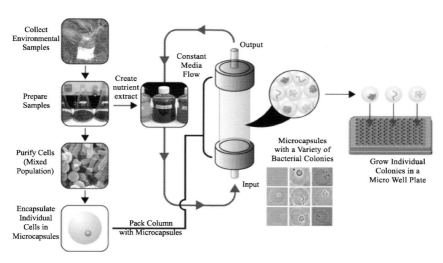

ZENGLER *ET AL.*, CHAPTER 7, FIG. 2. Flow diagram of Diversa's high-throughput cultivation approach based on the encapsulation of single cells in microcapsules for massively parallel microbial cultivation (Keller and Zengler, 2004).

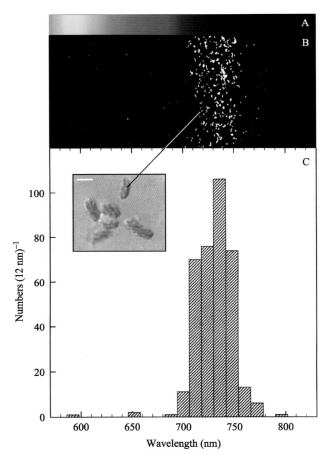

OVERMANN, CHAPTER 8, FIG. 4. Bacteriospectrogramm of the phototrophic consortium "*Chlorochromatium aggregatum.*" (A) Schematic representation of the continuous spectrum focused in the optical plane of the coverslip of the microscopic chamber. (B) Computerized dark-field photomicrograph of the chamber. Each dot represents a single consortium. (C) Integrated number of accumulated consortia per 12-nm interval. (Insert) Differential interference contrast photomicrograph of six "*C. aggregatum*" consortia. Bar: 5 μm. Modified from Fröstl and Overmann (1998).

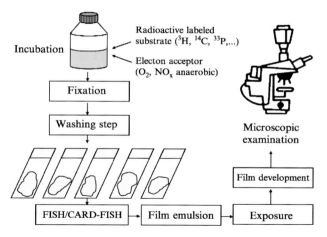

NIELSEN AND NIELSEN, CHAPTER 14, FIG. 1. Experimental overview of the procedure for conducting a microautoradiography experiment on a sample from a complex microbial community.

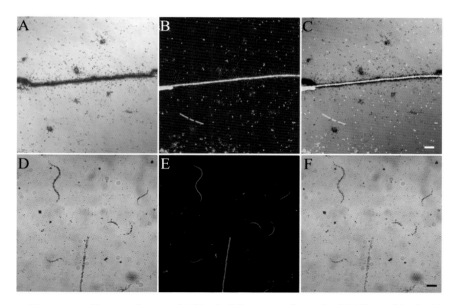

NIELSEN AND NIELSEN, CHAPTER 14, FIG. 2. Microautoradiography (MAR) combined with fluorescence *in situ* hybridization (FISH) or catalyzed reporter deposition FISH (CARD-FISH). Images A–C and D–F show the same microscopic view of an activated sludge sample (Grindsted wastewater treatment plant, Grindsted, Denmark) and a river water sample (North Pine Dam, in southeast Queensland, Australia). (A) MAR image after incubation of the sample with [^{14}C]propionate under aerobic conditions. Note the presence of a MAR-positive filament and a few single cells (seen as areas covered by a thick layer of silver grains). (B) The same microscopic field as A, but recorded for fluorescence after hybridization with a Cy3-labeled probe specific for *Meganema perioderoedes* (probe Gr1-Cy3) and a FLUOS-labeled bacterial probe (probe EUB338-FLUOS). (D) Water sample incubated with [^{3}H] thymidine under aerobic conditions, and (E) hybridization of CARD-FISH after hybridization with a Cy3-labeled probe specific for *Actinomycetes* (probe HG1–654-Cy3). (C and F) Overlay of A-B and D-E, respectively. All images were recorded by laser-scanning microscopy (LSM510 Meta, Carl Zeiss, Jena, Germany). Bar: 10 μm.

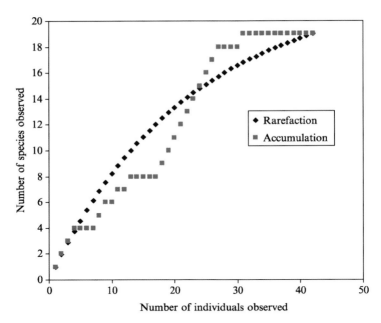

HUGHES AND HELLMANN, CHAPTER 17, FIG. 1. An example of hypothetical individual-based accumulation and rarefaction curves. The accumulation curve is one possible order of observing the 42 clones. The rarefaction curve was created with the EstimateS program (Colwell, 2004).

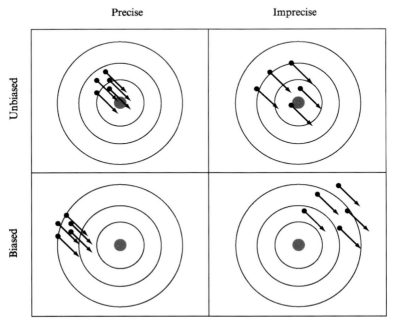

HUGHES AND HELLMANN, CHAPTER 17, FIG. 3. An illustration of precision and bias. The bulls-eye represents a true value that is trying to be determined. Arrows are individual estimates of the true value. An ideal estimator statistic is precise and unbiased (i.e., accurate), as in the upper left-hand corner.

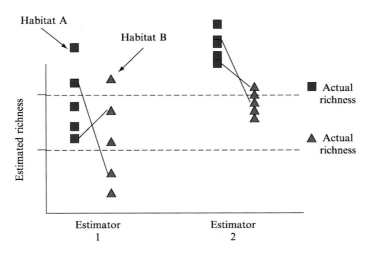

Habitat A

Habitat B

Estimated richness

■ Actual richness

▲ Actual richness

Estimator 1

Estimator 2

HUGHES AND HELLMANN, CHAPTER 17, FIG. 4. An illustration of the importance of the precision of richness estimators. Estimator 1 is unbiased, but imprecise. On average, estimator 1 correctly estimates the actual richness of habitat A and B, but the variance of the different estimates is large. Thus, if one compares one estimate of richness from one sample of each habitat, it is easy to incorrectly rank the relative richness of the two habitats. Two example comparisons are shown with solid lines; one pair of estimates correctly orders the relative richness of the habitats, the other pair orders it incorrectly. In contrast, estimator 2 is positively biased, but precise. On average, the estimator overestimates the actual richness; however, because the variance in the estimates is small, any pairwise comparison of estimates correctly ranks the richness of the habitats.

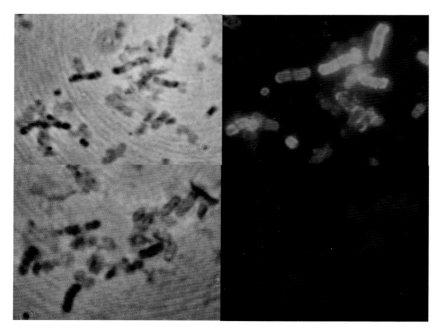

ZWIRGLMAIER *ET AL.*, CHAPTER 20, FIG. 1. RING-FISH hybridization of *Synechococcus* WH7803 (top) and *Synechococcus* PCC7942 (bottom) using a RNA polymerase gene-specific polynucleotide transcript probe. The probe targets a 690-bp fragment of the RNA polymerase gene rpoC1 of *Synechococcus* WH7803. Probe-conferred halos (fluorescein) appear only in the case of *Synechococcus* WH7803 target cells. (Left) Phase contrast and (right) epifluoresence image. From K. Zwirglmaier and D. Scanlan, unpublished results.

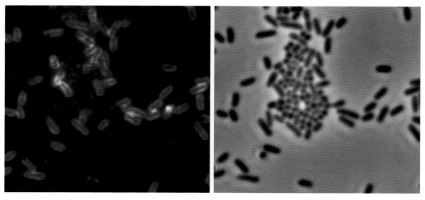

ZWIRGLMAIER *ET AL.*, CHAPTER 20, FIG. 2. RING-FISH hybridization of a cell mixture of *Escherichia coli* and *Neisseria canis* using a glyceraldehyde-3-phosphate dehydrogenase gene-specific polynucleotide transcript probe (Table I; Zwirglmaier *et al.*, 2004). The probe was constructed as specific for the respective gene of *E. coli*. RING-FISH (left) and phase-contrast (right) micrographs of the same microscopic field.

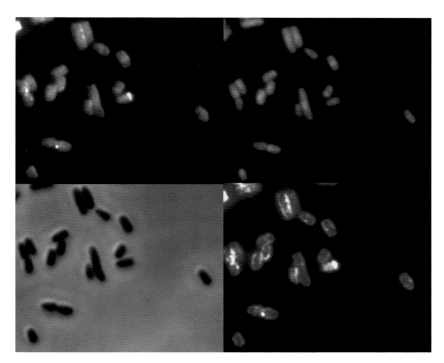

ZWIRGLMAIER *ET AL.*, CHAPTER 20, FIG. 3. Combination of RING-FISH and conventional rRNA FISH of *Escherichia coli* cells carrying plasmid pCCR2.1. A Cy3-labeled polynucleotide transcript probe specific for plasmid-encoded β-lactamase (Table I; Zwirglmaier *et al.*, 2004) was used in combination with a fluorescein-labeled oligonucleotide probe targeting 16S rRNA positions 444–462. Micrographs of the same microscopic field are shown: phase contrast (lower left), polynucleotide probe signal (lower right), oligonucleotide probe signal (upper right), and overlay of both signals (upper left).

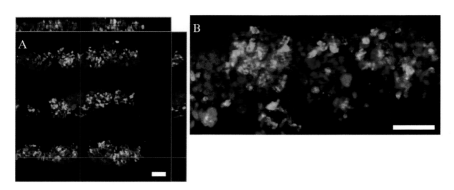

Pernthaler and Pernthaler, Chapter 21, Fig. 3. Confocal images of gill cross section of a juvenile *B. puteoserpentis*. (A) Optical sectioning; red line indicates z section in x direction (upper frame), green line shows z section in y direction (right frame). (B) Detailed image of bacteriocytes. Composite of three images: Red: Methanotrophic symbionts hybridized with 16S rRNA-targeted, cy5- labeled probe Baz_meth_845I. Blue: Thiotrophic symbionts hybridized with 16S rRNA-targeted, cy3-labeled probe Baz_thio_193. Green: *PmoA* mRNA antisense probe, visualized with $Alexa_{488}$-labeled tyramide. Bars: 10 μm.